1986

Complimentary Copy

Patricia Ledly

W9-BVK-302

FUNDAMENTALS OF PLANT SYSTEMATICS

Albert E. Radford

University of North Carolina at Chapel Hill

with contributions by

Gloria Caddell, James Doyle, Deborah Qualls, and Laurie S. Radford
University of North Carolina at Chapel Hill

T. M. Barkley
Kansas State University

Duane Isely
Iowa State University
and
Michael G. Simpson
Albright College

Systematics, as a natural science, is a continuing quest for truth and understanding as well as the never-ending pursuit of reality, organized and simplified.

James Bruce, a British diplomat, once stated that ''to most people nothing is more troublesome than the effort of thinking.'' This text is dedicated to those students willing to make that troublesome effort.

Sponsoring Editor: Claudia M. Wilson
Project Editor: Steven Pisano
Cover Design and Illustration: Alma Orenstein
Text Art: Fineline, Inc.
Production: Debra Forrest
Compositor: York Graphic Services, Inc.
Printer and Binder: R. R. Donnelley & Sons Company

FUNDAMENTALS OF PLANT SYSTEMATICS

Library of Congress Cataloging-in-Publication Data

Radford, Albert E.
  Fundamentals of plant systematics.

  Includes bibliographies and index.
  1. Botany—Classification. I. Barkley, T. M.
(Theodore Mitchell), 1934–   . II. Title.
QK95.R29 1986   581'.012   85-21903
ISBN 0-06-045305-2

86 87 88 89 9 8 7 6 5 4 3 2 1

Contents in Brief

Contents in Detail

SYSTEMATICS
PROSPECTIVE and SUMMARY

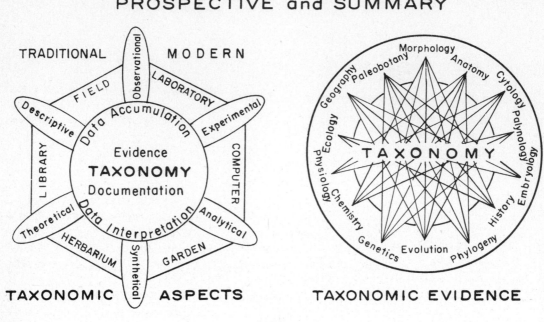

TAXONOMIC ASPECTS

TAXONOMIC EVIDENCE

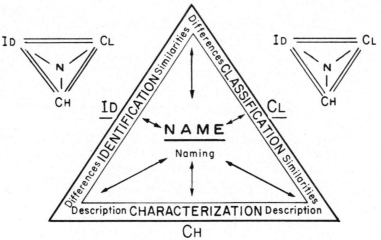

TAXONOMIC FUNDAMENTALS

BASIC PROCEDURES and SYSTEMS

All interrelated and related to the Name

A systematist is one who uses many approaches to fit together appropriately, into an orderly storage and retrieval system with a name reference base, the information about organisms gathered by morphologists, anatomists, cytologists, palynologists, embryologists, geneticists, physiologists, biochemists, ecologists, geographers, paleobotanists and systematists. A systematist is a collector, analyzer and synthesizer of information from all fields of evidence for the characterization, identification and classification of organisms.

Preface

Much of this text has been adapted, adopted, and updated from material in *Vascular Plant Systematics* (VPS). I greatly appreciate the contributions of the original authors whose information and interpretations have been used in this book. A diligent effort has been made to produce a straightforward manuscript with precise and accurate statements in the context of the treatments presented herein. Where adaptation from VPS may have altered the emphasis or meaning intended by the original author, I apologize for any disservice done. I am responsible for all treatments in the present book except the following: Chapter 2 (T. M. Barkley), Chapter 4 (Laurie S. Radford), Chapter 10 and Appendix B (Michael G. Simpson), Appendix A (James Doyle), "Plant Taxonomy: Sequential Saviors" (Duane Isely), Summary for a Revisionary Study (Deborah Qualls), and Summary for a Paleoethnobotanical Study (Gloria Caddell).

I am exceedingly grateful to my wife, Laurie S. Radford, for constructive editing of the entire text, and to Jimmy R. Massey for his critical review of many versions of the manuscript. Both have made invaluable contributions to the overall quality of the book. Reviewers made suggestions, many of which have been incorporated into the text and a number of which, although negative, stimulated this author to improve the manuscript. I am particularly indebted to my students down through the years who have added much to my understanding of the principles related to the study of many systematic subjects. The professionalism of Linda Naylor and Lula Stevenson in the typing and word processing of the manuscript, as well as their patience with the seemingly unending corrections and changes, are sincerely appreciated.

The author is obligated to the George R. Cooley Fund and the Department of Biology of the University of North Carolina at Chapel Hill for absorbing much of the preparatory costs for the manuscript—typing, word processing, photocopying, and indexing. I am particularly grateful to George R. Cooley for his personal continuing support of our taxonomic publications.

All royalties from this text will be deposited in the Friends of the UNC Herbarium Fund in the Arts and Sciences Foundation, Inc., of the University of North Carolina at Chapel Hill for the maintenance and development of plant systematics.

ALBERT E. RADFORD

Prologue: Organization and Information

Systematists, as students, thoughtfully consider, examine, interpret, and evaluate concepts, processes, and principles as well as problems in systematics. Systematists, as researchers, are students who make diligent, dedicated, organized, and meticulous inquiry or investigation into systematic subjects to discover or revise facts, theories, processes, applications, and so forth. Systematists, as scholars, are researchers who develop a profound knowledge of particular aspects of systematics.

This book is a treatment of the traditional fundamentals of plant systematics for students of introductory plant taxonomy at the university advanced undergraduate and beginning graduate levels. This text includes revisions, adaptations, and reduced versions of the materials in the reference text and source book, *Vascular Plant Systematics,* by Albert E. Radford, William C. Dickison, Jimmy R. Massey, and C. Ritchie Bell, that was published in 1974 by Harper & Row. The principles, questions, and exercises in this book represent an extensive revision of the text pertaining to those subjects in *Vascular Plant Systematics*. It is presumed that the users of this book will have as part of their background the equivalent of both an introductory and one organismic course in biology or botany. This book has been designed primarily as a text for a one-semester course with a lecture-discussion/recitation approach to the study of systematics.*

An explanation of the approach, organization, and contents of this text will aid students in understanding the book. A knowledge of the goals and objectives for general systematics, local flora, and this publication as well as a hierarchical résumé of the discipline should give students a broad view of systematics as a subject and some perspective of taxonomy. A familiarity with the underlying assumptions and themes considered relevant to the study of systematics will help students realize their academic potential.

*A note on the use of a few key terms in systematics is in order. In conversation and writing, professional systematists use taxonomy-systematics, flora manual, and revision-monograph interchangeably. The terms in each couplet are not sharply distinguished from each other, but the second in each case is usually considered the more inclusive. For writing convenience, systematics and taxonomy are used interchangeably in this text. (See the overview of systematics in F. A Hierarchical Résumé for Systematists in the Prologue and definitions of the other terms in Chapter 1.)

A. APPROACH TO THE STUDY OF PLANT SYSTEMATICS

The introduction to plant taxonomy presented in this book is a concept/process/principle, question/exercise approach to an understanding of systematics. A careful consideration of the questions and exercises related to each subject in this text should make the student more proficient taxonomically in botany as well as in other endeavors. The sets of questions and exercises can be of specific value in developing skills in:

Organization of information

Comprehension of basic concepts, principles, and relationships

Expansion of insight into the taxonomic significance of the evidence

Utilization of methods, techniques, and analytical procedures

Determination of precautions to be taken and problems to be found in the accumulation, use, analysis, presentation, documentation, and evaluation of data

Acquisition of basic knowledge of literature

Application of concepts, principles, terms, and methods

Evaluation of conclusions, determinations, and relationships

Summarization of information, methods, results, and significance

In this text, principles are presented in both prose and synoptic forms. The questions and exercises are directly related to the treatment of each subject. Basic references are provided for each topic, and teachers are expected to amplify the material as necessary. Each major concept or topic is treated from a what-how-why approach. The "what" includes the basic definition(s) and/or description(s) of the subject along with the basic assumptions or premises. The "how" gives the fundamental operations in the development and use of the system that are related to the concept or subject. The guiding principles are directives for the use or construction of the system or subject. The "why" is the purpose for the development or use of the system or topic; and it also represents a summary or conclusion as to the function of the system. The what-how-why approach provides a broad foundation for understanding taxonomic concepts, processes, and principles.

The conditions of time (when), place (where), and manner (how) provide circumstances that are basic to effective communication about any subject; to the determination of significance of classwork, research, or scholarship; and to an understanding of concepts-processes-principles. The when-where-how conditions also help to establish the context for particular events and situations. Conditions indicate the type of definition required for a concept, the degree of success expected in an operation, and the kind of principle being considered. Even though the thrust of this text is what-how (operation)-why, students must be aware of the when-where-how (manner) conditions in all taxonomic activities. The consistent application of the what-how-why question approach to study coupled with the when-where-how question approach to study is a basic key to learning efficiency and effectiveness.

The question method is one of many ways to develop concepts, skills, and values in any subject. In general, the question approach to any study is an attempt to stimulate and satisfy intellectual curiosity by the procurement of knowledge. Questions usually make authors and scholars more keenly aware of their ignorance in matters pertaining to any field of inquiry. Posing the right question for the solution of a problem is a fundamental skill that every student should develop. Questions have to be stated clearly before answers to them can be found. "Answers without questions are not answers" is a truism known for many centuries but one that is violated too frequently in taxonomic studies today. Habitually asking what-how-why under when-where-how conditions, and conscientiously seeking the answers, should help clarify the problems involved in any study.

The exercises in the text provide a means of learning by doing. Practical applications of the principles increase the understanding of concepts and relationships and lead to the discovery of new ones. Evaluations of the conclusions and results of skillfully designed problems produce a greater awareness of the significance of any subject.

B. ORGANIZATION OF THE TEXT

Each chapter in the text is organized according to the general plan presented below.

Synoptic Statement

Activities related to subject of study

Introduction

A general overview of what the subject of study encompasses

The reasons for the treatment of the subject of study

General Information

Explanations of basic concepts, principles, and processes involved in the subject of study

Additional information that aids in the understanding of the subject of study

Summary of Constituent Elements

Fundamental Concepts and Processes

Definition(s) of subject of study—what

Purposes of or reasons for study of subject—why

Operations or processes involved in use of system developed for subject of study and occasionally processes for development or construction of system for subject of study—how

Basic Premise(s) and Fundamental Principle(s)

> Premise(s) and inclusive principle(s) basic to subject of study

Guiding Principle(s)

> Guidelines for use of system related to subject of study and occasionally guidelines for construction of system for subject of study

Basic Assumption(s)

> Premises, principles, or statements usually understood or taken for granted that form the foundation for subject of study

References

> Suggested reading and documentation pertinent to explanation and understanding of subject of study

Questions

> Queries relevant or applicable to subject of study

Exercises

> Training schemes for application of concepts and principles related to subject of study

C. INFORMATIONAL CONTENT OF THE BOOK

This Prologue embraces the approach, goals, organization, content, and underlying assumptions and themes used in the book for understanding the fundamentals of systematics. Chapters 1 and 2 include a prospectus for the text and a perspective for the study of plant systematics. The prospectus is an overview of the subject and a paradigm or model for studying the fundamental components of systematics. The perspective gives a purview of the historical development of taxonomy and an indication of the present state of many aspects of the discipline. Chapters 3 to 7 are a treatment of the traditional components of taxonomy: nomenclature, description, classification, and identification. Chapters 8 to 10 supply an introduction to the dynamics of systematics. These dynamics include the uses of evidence and the functions of variation, speciation, and phylogeny in the establishment of character correlations for the circumscription, delimitation, and classification of taxa, and the determination of phenetic and phylogenetic relationships. Chapters 11 to 13 furnish the roles of taxonomic institutions in systematic service, training, and research. The Epilogue contains a summary of the conclusions, summaries, and principles presented in the text, and also serves as a conspectus for the general study of systematics as presented in this book. The Appendixes provide the basic methods for the handling of data, information, and collections associated with systematic activities as well as resource information for laboratory and field study in vascular plant systematics.

D. A SUMMARY OF CONCEPTS, PROCESSES, AND PRINCIPLES TREATED IN THIS TEXT (THE FUNDAMENTALS)

I. Major and Integral Concepts

Major Concepts	Integral Concepts
Hierarchical Level I	
Systematics	Taxonomy, evolution
Hierarchical Level II	
Taxonomy	Classification, identification, description, nomenclature
Evolution	Variation, speciation, phylogeny
Hierarchical Level III	
Classification	Hierarchy, taxon, position, rank, system
Identification	Taxon, character, character/taxon, system
Description	Evidence, character set, character, character state, circumscription, system
Nomenclature	Name, taxon, circumscription, position, rank, system
Variation	Variation type, variation hierarchy, variation sources, variation analysis
Speciation	Species types, populations, breeding systems, isolating mechanisms, species diversity, natural selection
Phylogeny	Morphic characters, cladograms, homology, homoplasy, polarity, parsimony, association, correlation

II. Major Processes and Associated Procedures

Major Processes	Associated Procedures
Hierarchical Level I	
Classification	Determination of position and rank for new and old taxa that have been remodeled, united, divided, transferred, or altered
Identification	Delimitation and determination of new and old taxa that have been remodeled, united, divided, transferred, or altered
Description	Circumscription of new and old taxa that have been remodeled, united, divided, transferred, or altered

Nomenclature	Naming of new and old taxa that have been remodeled, united, divided, transferred, or altered
Variation	Ascertainment of types, trends, and patterns of variation within organisms, populations, and groups
Speciation	Correlation of characters for establishment of new species and reinterpretation of the old; gene recombination, mutation, and natural selection
Phylogeny	Interpretation of the evolutionary history of selected taxa; cladistic analysis

Hierarchical Level II

All processes and procedures at Level I	Selection and delimitation of kinds of characters
	Determination and delimitation of the character states of selected characters
	Establishment of character correlations
	Determination of the number of characters to be used in correlations
	Decision upon admissibility of characters to be used in correlations
	Adjustment of the weighting of characters to be used in correlations
	Analysis of characters

III. Principles

General Types of Principles	**Definitions**
Basic premise(s)	Inclusive fundamental principle that is the foundation for the study of a taxonomic subject or concept; the established explanation for phenomena associated with evolutionary theory
Fundamental principle(s)	Correlative statements that complement the theoretical basis or premise for the study of a subject or concept
Guiding principle(s)	Guidelines for integral processes and procedures in the development of a system or product; or guidelines for the use of a system or product
Basic assumption(s)	Suppositions (premises, principles, or statements usually understood or taken for granted) that form a pertinent background for the study of a subject or concept

Major Subject Principles	**Subjects with Principles Summarized**
Systematic	Systematics, taxonomy, evolution

Taxonomic	Classification, identification, description, nomenclature, botanical names
Evolutionary	Variation, speciation, phylogeny
Evidentiary	Evidence, taxonomic literature, herbarium collections, living collections, history
Research	Revision, flora, problem (paleoethnobotany)

E. OBJECTIVES AND GOALS OF PLANT SYSTEMATICS

The major objectives for plant systematics as a science, the goals for the use of this book in introductory and advanced courses, and the aims for many flora and laboratory sections of beginning courses in taxonomy are summarized below for a broader understanding of the scope of systematic endeavors.

I. Major Objectives of Plant Systematics

1. To provide a scheme of classification that attempts to express phenetic, natural, or phylogenetic relationships.
2. To provide a convenient method for identifying, naming, and describing plant taxa.
3. To provide an inventory of plant taxa and information for local, regional, and continental floras.
4. To provide an understanding of evolutionary processes and relationships.
5. To provide an integrating and unifying role or focal point in the training of students in botany, particularly in regard to the relationships among the many biological fields, the types of evidence, and the diversity of organisms.

II. Goals for the Fundamentals of Plant Systematics

1. To acquire the basic knowledge, skills, and values usually associated with the study of plant systematics.
2. To understand the fundamental concepts, processes, and principles involved in the study of systematics.
3. To perceive the significance of taxonomic relationships in plant systematics.
4. To develop an appreciation of the philosophy, applicability, and aesthetics of the study of plant systematics.

III. Aims for Local Flora Studies

1. To learn how to use a manual.
2. To learn how to use keys in identification.
3. To learn how to recognize families, genera, and species.
4. To learn how to name the taxa of plants.
5. To learn how to describe plants.
6. To learn how to collect and prepare specimens.
7. To learn how to relate species diversity to habitat diversity in a region.
8. To become acquainted with fundamental taxonomic concepts and principles.
9. To become acquainted with at least one system of plant classification.
10. To become acquainted with the historical development of plant taxonomy.

F. A HIERARCHICAL RÉSUMÉ FOR SYSTEMATISTS

A natural scientist is a systematist who investigates cause-effect phenomena for an understanding of the relationships between phenetic, genetic, and ecologic diversity.

A systematist is a taxonomist who studies species origin, differentiation, genealogy, population dynamics, and community relationships for a comprehension of organismic diversity.

A taxonomist is a classifier who delimits, circumscribes, and names taxa for storage, retrieval, and use of information on taxon diversity.

A classifier is a student who establishes position and rank for delimited, circumscribed, and named taxa for perspective on biotic diversity.

A student of biotic diversity is a classifier, taxonomist, systematist, and natural scientist—a holist—whose role in society is the understanding of the diversity of organisms, populations, and taxa as well as the relationships between phenetic, genetic, and ecologic diversity for the benefit of mankind.

Systematics, as a natural science, is a continuing quest for truth and understanding as well as the never-ending pursuit of reality, both organized and simplified.

G. GENERAL COURSE ASSUMPTIONS AND EVALUATIONS

Every program or course is based on certain assumptions. Objectives for each course are related to those assumptions. Students should understand what those aims are and how and why each is to be attained. Since most students will eventually evaluate courses, they should know initially what the written and unwritten objectives of those courses are in order to be able to make sound evaluations.

For example, the emphasis in a course in systematics may be on the ability to identify; or a knowledge of the analytical procedures used; or the determination of variation in characters and populations; or training in descriptions of taxa; or an understanding of concepts and principles; or the development of expertise in the study of taxonomic literature; or the solution of problems in nomenclature; or the determination of relationships; or the application of the theory of classification; or on any number of combinations of the preceding. Students should be aware of the degree of proficiency and depth of understanding expected for the different skills and values presented in their study of systematics.

H. AUTHOR ASSUMPTIONS ON VALUES

The author firmly believes that an understanding of taxonomic concepts, principles, and processes will enable individuals to become better students and researchers through the development of their natural abilities in problem definition/analysis/solution and relationship perception. I am equally convinced that the application of these taxonomic principles will increase comparability and consistency in writing and speaking which will in turn result in improved scholarship.

Students in a well-taught course do more than accumulate facts and increase their knowledge about a subject. They acquire skills and values pertinent to their development. By associating concepts and principles presented in such a course with everyday experiences, they discover and may ascertain the significance of new relationships. Growing knowledge is placed in more meaningful perspective as a result of their reflections and meditations on past and present experiences. The more an individual thinks, with the aid of maturing skills and values, the more real education progresses as a challenging, ever-expanding process. Each student should develop a natural philosophy that provides a set of principles for guidance in practical affairs.

By consciously trying to accomplish well-defined goals that are directly related to concepts, skills, and values within the course of study, students are likely to be more effective in their personal and professional development and ultimately of greater benefit to society.

I. THEMES

Both teacher and student should be conscious of four recurring themes in the *Fundamentals of Plant Systematics*. First, the role of taxonomy in effective and efficient communication is presented many times in the text. Second, the universality of taxonomic concepts and activities to the fields of learning is indicated many places in the book. Third, the significance of taxonomic principles and theory is suggested or expressed in several parts of the study. And fourth, the practical and societal relevance of systematics is stressed in the question/exercise sections of this learning experience. Two general themes pervade this text: (1) the constituent element approach to the study of plant systematics is basic to sound research and scholarship; and (2) the fundamentals of plant systematics as herein applied to plants are equally germane to the study of a wide variety of subjects. These themes provide a basic perspective for interdisciplinary study.

J. TEXTUAL INFORMATION FOR INSTRUCTORS

This book is designed primarily for students at the advanced undergraduate level. Desirable background requirements are an introductory and one organismic course in biology and/or botany.

The book can be used most effectively in a one-semester course with a lecture/discussion/recitation approach. The material is sufficient for a two-semester course with supplementary exercises and questions. Resource material is included in the Appendixes for use in a laboratory-and-field approach to taxonomic study.

The text is easily adaptable and useful for beginning college students in introductory courses in taxonomy with careful selection of the topics, questions, and exercises in the book.

Teachers are expected to amplify concepts, processes, and principles with their own examples and to add, delete, or modify suggested questions and exercises to satisfy the standards and background of their classes.

Aids for understanding the book and for the realization of academic potential are included for student perspective on the study of systematics (see the preceding pages).

The constituent element system is introduced as a basic method for the study of the fundamentals of systematics. (See Prologue B. Organization of Text; Chapter 1, A. II. and III.; Epilogue, B. Constituent Element System).

The foundation for the study of plant systematics is presented in the basic premise, the fundamental principles, guidelines, and basic assumptions. (see Prologue II, B. Organization of the Text, E. Objectives and Goals of Plant Systematics, and F. A Hierarchical Résumé for Systematists; Chapter 1, A. and Summaries for Taxonomy and Evolution; and Epilogue, A. General Systematic Principles and B. Constituent Element System). In general, the Prologue provides the blueprint and specifications, Chapter 1 the tools and materials, for laying a solid foundation for understanding the *Fundamentals of Plant Systematics*.

K. GENERAL COMMENTS FOR STUDENTS AND INSTRUCTORS

Chapter 1 should be considered, in part, as a minicourse for meaningful systematic communication between author, instructor, and student. Many of the key words are in italics in A. I., II., and III.

A logical introduction to the study of systematics is a holistic overview of the conceptually independent but closely interrelated components of the subject (classification-identification-description-nomenclature and variation-speciation-phylogeny), as in A. II. and III. of Chapter 1. The holistic overview provides a sound basis for studying Chapters 2 to 13 in any order desired. Each chapter is a self-explanatory unit with sufficient material for a working perception of the subject.

A knowledge of the concepts, processes, and principles associated with *taxonomy* and *evolution,* as itemized in D. A Summary of Concepts, Processes, and Principles Treated in this Text (The Fundamentals) of the Prologue, is considered basic to an understanding of systematics. These fundamentals are applicable to all levels of systematic training, undergraduate through graduate.

part one

TAXONOMIC CONCEPTS, PROCESSES, PRINCIPLES

Systematics is the study of phenetic, genetic, and phylogenetic relationships among taxa.

SYSTEMATICS AS A CONCEPT

Essential Qualities

Fundamentally, systematics is the study of the nature, causes, patterns, and trends in variation among taxa.

Structurally, systematics is the study of the fundamental components of taxonomy and evolution. (Classification-identification-description-nomenclature and variation-speciation-phylogeny)

Functionally, systematics is the study of characters from many fields of evidence for determining relationships among taxa.

Developmentally, systematics is the study of variation among taxa for the establishment of character correlations and relationships basic to the construction of taxonomic and evolutionary systems.

Theoretically, systematics is founded on the proposition that in the tremendous diversity in the biological world conceptually discontinuous units exist that can be identified, classified, described, named, and logically related on the basis of evolution.

Philosophically, systematics is a perpetually self-correcting, ever-evolving endeavor that accepts no axiom as absolute or immutable and is a continuing quest for understanding taxonomic and evolutionary concepts, processes, principles, and relationships.

A Construct of Essential Qualities for Conceptual Study of a Subject (A Constituent Element Summary)

Essential Qualities	Element Inclusions Summarized
Logical	
Fundamental	General definition of subject (what)
Structural	Basic components of subject (what)
Functional	Operations and procedures involved in subject as a process (how)
Developmental	Purpose for development of product(s) related to subject as a system (why)
Theoretical	Basic premise(s) or foundation for study of subject
Philosophical	Principles related to subject of study as a system (Fundamental, Guiding, Assumptive)

The concept of a subject as an idea is formed by mentally combining all its characteristics, particularly its essential qualities. An effective procedure for developing a conceptual understanding of a systematic subject, a construct of its essential qualities, is to produce a constituent element summary for its logical, theoretical, and philosophical attributes.

chapter *1*

A Foundation for the Study of Plant Systematics*

Systematists as students, researchers, and scholars classify, identify, describe, name, observe, analyze, synthesize, theorize, evaluate, develop perspective, determine relationships, generate informational systems, act as conservators, improve communication, and ascertain the types, patterns, trends, and causal mechanisms for variation within populations, species, and higher taxa.

Generally, the study of a subject is founded upon two premises: that in the variation within that subject conceptually discrete units occur that can be recognized, classified, circumscribed, and named; and that logical relationships exist among those units. Basic taxonomic concepts, principles, and processes are directly related to these premises. Man is innately a taxonomist who either consciously or unconsciously classifies, identifies, describes, and names, and is intuitively as well as cognitively aware of relationships.

Each of us is a taxonomist from the cradle to the grave. Each of us classifies, identifies, names, and describes in making decisions about the food we eat, the beverages we drink, the clothes we wear, the homes we purchase, the books we read, the games we play, the love we express, the politics we exercise, the religion we profess, and the work we do. We talk to children, adults, and specialists about sickness, sex, and sin at different hierarchical levels; we rank basketball teams; we grade apples; we judge quality of secondhand automobiles; we sort clothes into piles—we classify. We smell an odor, garlic; we taste a white substance, sugar; we touch an object, skin; we listen to music, Mozart; we see an animal, a giraffe; we key a plant, white clover—we identify and we name. To

---

*The approach, goals, organization, and underlying assumptions and themes used in this text for the study of the fundamentals of systematics are presented in the Prologue.

us the ball is round, the grass is green, the road is long—we describe. We make cause-effect observations that involve the determination of taxonomic relationships. We find the principles of taxonomy equally pertinent to our studies of animals, books, foods, heavenly bodies, transportation, or plants.

Plant taxonomy is a most relevant field of inquiry for modern man. Plants are our fundamental sources of food and energy, of shelter and clothing, of drugs and beverages, of the oxygen we breathe; and they form the aesthetic base of our environment. The characterizing, naming, classifying, and identifying of all plants, whether actually or potentially useful to man, are the professional duties of the taxonomist or systematist.

The ecologist studying pollutant decomposers must identify the organisms involved; the geneticist splicing DNA molecules or introducing new germ plasm into a crop for increased production or disease resistance has to know the characteristics of the source plant; and the chemist analyzing plant material for a cancer-inhibiting drug should obtain the name of the organism from which the material was procured. All these scientists will be interested in the classification of related organisms as they search for more effective agents in their work. Related decomposers may prove to be more efficient removers of pollution; closely related species may be a source of better germ plasm for increased production of genes more resistant to disease; and genotypes from other identified and named organisms may provide cancer-inhibiting drugs that will be more effective, with fewer side effects.

The scientific name is the symbol of communication to which most information is attached. Taxonomists have to develop a descriptive language for precise and useful characterization of organisms; they have to devise simple schemes for identification; they have to work out a system of classification to show relationships; and they must have a method of naming all organisms. These systematic activities are basic to the development of a more productive life for man.

Many intellectual approaches are used in the study of taxonomy: the descriptive and observational, the analytical and experimental, the synthetical and theoretical. Data in systematics are accumulated from research in the field, the laboratory, the garden, the herbarium, and the library. These data can be analyzed with the aid of the computer and are documented in the form of dried specimens in the herbarium, as living specimens in the botanical garden, and as information deposited in the library. As a data-gathering science, taxonomy is a pedestal on which biology is built; and as a recognizer of relationships, it is a unifying and integrating discipline that represents a peak of biological endeavor.

Systematics is a dynamic science. The duties of the taxonomist are never-ending. The flow of data about plant materials and plants necessitates reordering descriptive information, revising schemes of identification, reevaluating systems of classification, and perceiving new relationships for a better understanding of plants. As long as the plant world exists, there will be more to learn about plants, plant products, and plant taxa for both practical and theoretical purposes.

One of the great challenges today is how to conserve the plant life of the world. We need to learn more about these organisms that are basic to our survival and how to store and retrieve the vast amount of information about plants for man's use. One of the major problems of the taxonomist is how to incorporate this body of knowledge into the classification of plants so that this data bank or storehouse of information can be efficiently used for routine activities and for the development of new relationships of significance to man.

A. BASIC SYSTEMATIC CONCEPTS AND ACTIVITIES

The study of plant systematics is founded on the premise that in the tremendous variation in the plant world, conceptually discontinuous units exist—which are usually called species—that can be recognized, classified, circumscribed, and named; and on the premise that logical relationships based on evolution exist among these units. The basic concepts in systematics are directly related to these premises. To understand these concepts we have to follow the old adage and teaching axiom, ''I tells 'em what I'm gonna tell 'em, I tells 'em, then I tells 'em what I told 'em.'' The concepts' real meaning will come over time as they are presented in different ways in varied settings.

This introductory preview provides an orientation to the study of these concepts and associated principles and processes. Understanding will come eventually, seldom immediately, with repeated consideration of this subject matter in subsequent chapters. After careful study, each student should have a thorough knowledge of the basic aspects of systematics. Before discussing these fundamental taxonomic and evolutionary concepts, there is a need to define a few frequently used words such as term, concept, principle, process, product, system, species, and so on. Experience indicates that many individuals are unable to articulate definitions for these words and that they lack a knowledge of the many ways in which they are used. Ruell Howe stated in *The Miracle of Dialogue,* ''Language is the process of knowing and being known through the use of words, and it follows that an apt use of language is essential for effective living''—and for effective taxonomic communication.

I. General Terms Pertinent to the Study of Systematics

A *term* is a word or phrase with a limiting and definitive meaning, such as a botanical term that is an accepted name for a thing or concept in a recognized discipline, for example, serrate, saw-toothed margin; or a *term* is a word or group of words that serves as the name of something, such as a concept, for example, circumscription. In general, a *concept* is something conceived, such as an idea, observation, thought, notion, or opinion. In a broad sense, a concept is a mental formulation that includes all that is characteristically associated with or suggested by a term or event denoting the concept. The standard taxonomic concepts of classification, nomenclature, variation, and so on will be considered mental formulations of wide scope, *concept systems* that are much more than implied in idea, observation, notion, and so forth. The treatment of any concept system should include the *constituent elements:* definition, components (major parts), types (kinds), the reasons for the study and change in the system, the processes or operations involved, and the pertinent principles. This constituent element approach should lead to an understanding of the concept system and to a realization of the significance and impact of the concept to systematic study (see Section A.II, The Fundamental Components of Taxonomy, and Section A.III, The Fundamental Components of Evolution).

The *definition* of a word or term is usually a formal statement of its precise meaning or significance. A definition may have logical, intuitive, speculative, theoretical, and/or practical bases. The logical basis of a definition can be structural, functional, or developmental, for example, a fruit consists of a pericarp, placenta, and seed (structural); a fruit is a plant-reproductive body (functional); a fruit is a matured ovary (developmental). The concept of definition includes three components: the specific name of the element to be defined, the term; a circumscription that is limiting for the term; and the inclusive and/or exclusive conditions for the limitation of the circumscription of the term. Limiting the definition of a fruit, a term, to a developmental basis makes the definition very exclusive and the circumscription, a matured ovary, very limited. Functional, structural, and developmental definitions of a fruit would make the definition logically inclusive but still exclusive when other types of definitions, such as theoretical, are considered. An understanding of the inclusive/exclusive conditions on which a definition is based is fundamental to effective communication about any word, term, or subject.

A *reason* for any action, decision, or study is usually a statement of justification, declaration of explanation, or the basis, motive, or cause. Reasons for study include the same bases as definition: logical, theoretical, and so on. Definitions and reasons are given for or by fields of study or disciplines such as: ''theoretically, the taxonomic reasons for the study are. . . ''; ''practically, the botanical definition of a species is. . . .'' Real understanding of a concept includes logical, speculative, theoretical, and practical bases for definitions and reasons as well as familiarity with the processes and principles involved (see definitions of a species below and Table 1.1).

In general, a *system* is an integrated assemblage or combination of diverse units, parts, or components that form a coherent whole. The basic types of taxonomic systems are those for classification, identification, nomenclature, and description. Many applied or practical systems exist in systematics, such as a student study system for taxonomic components that would be an ordered and comprehensive assemblage of definitions, reasons, processes, and principles for each type. The systems approach to the study of a subject leads to a holistic understanding of it.

A variety of activities and related terms is associated with each taxonomic component. *Determination* is the act of ascertaining or establishing the position, rank, characteristic, name, or relationship of a taxon, object, or entity. *Process* refers to a series of interrelated actions, changes, or functions that bring about an end result or product; a system of operations that result in the production of something. *Product* is the general term for that which is produced in any fashion, for example, the process of classification for organisms results in named and circumscribed taxa (products) at determined positions and ranks. *Procedure* or method is a particular course of action or an accepted or established way of doing things in an overall process. Processes are integral parts of developmental systems with the final stage a result or product.

A *principle* is used in many senses: as a fundamental truth, as the ultimate basis or cause, as a source or origin, as that from which anything proceeds, as a fundamental assumption, as a generalization grounded on cases, as a comprehensive law or doctrine from which others are derived, as a guideline for governing an activity or operation, and so forth. The principles applicable to understanding the concepts used in this text are: *fundamental principles,* principles related to the ultimate basis or cause from which anything proceeds; *basic assumptions,* principles dealing with suppositions for a study generally accepted as true; and *guiding principles* or *guidelines,* accepted or professed rules of action for a process or operation or adopted rules or standards for the control, regulation, or application of an operation or action. Fundamental principles are the results of many hypothetical and theoretical considerations as well as many evaluations of the entire ''concept-process-principle'' approach; basic assumptions are usually based on long-term observations; and guidelines or guiding principles evolve with continued use and analy-

Table 1.1   A LIMITED ANALYSIS OF THE SPECIES CONCEPT*

Definition of species	Field of study	Basis of definition
1. Species, as a distinct kind or class of individuals with some characteristics in common	Logic	Intuitive
2. Species, as populations with common characteristics, the evolutionary base of variation	Evolution	Theoretical
3. Species, as natural populations of living, reproducing, genetically related individuals isolated from other populations by barriers to gene exchange	Biology	Logical (Functional)
4. Species, as the basic category to which most taxonomic information is attached	Taxonomy	Intuitive
5. Species, as the smallest populations structurally distinct and distinguishable from all others	Morphology	Logical (Structural)
	Systematics	Practical

*Based on species as a concept in the paragraph below. Note—the concept of a definition includes a circumscription that limits under inclusive and/or exclusive conditions. No. 3, the biological definition, could be considered exclusive; Nos. 1 to 5 and others would be necessary for an inclusive definition and an understanding of species as a concept.

ses-syntheses related to the operation or process. Expert knowledge of any subject or concept comes with an understanding of its principles.

In this text, the *basic premise* is the inclusive fundamental principle or proposition that provides the foundation for the study of a major taxonomic subject. *Theory* is defined as a more-or-less verified or established explanation accounting for known facts or phenomena, for example, evolutionary. A *hypothesis* is the working premise for any systematic problem or study.

A single statement can represent more than one type of principle, depending on the context. For example, "one correct name for a taxon with a particular circumscription, position, and rank" is a basic assumption for users of manuals and keys; it is a guiding principle for taxonomists who name new species or consider the nomenclature of the old; and it is a fundamental principle or basic premise for developers of a code of nomenclature for plants at the rank of family or below. Point of view makes a difference in the meanings and types of principles.

The theory of biological evolution is essentially that the organisms living today have descended by gradual changes from ancient ancestors quite unlike themselves and that natural selection determined the course of the changes. This theory is a unifying, inclusive, fundamental principle for the interpretation of the relationships of extant organisms and the development of systems of phylogenetic classification. The concept of biological evolution is such an integral part of most systematists' thinking that it can be considered a basic assumption for much of their work. Conditions make a difference in understanding principles.

The *species* as an evolutionary unit and basic concept in classification is difficult to define because of the many ways in which the term is used. Basically, species means a distinct kind or class of individuals having some characteristics or qualities in common. Species as populations with common characteristics are considered the evolutionary base of variations in the biological world. Species as a group of evolutionarily related populations are treated as the fundamental units in the racial or phylogenetic history of organisms. Biologically, species are natural populations of living, reproducing, genetically related individuals isolated from other populations by barriers to gene exchange. Taxonomically, the species is regarded as the basic category to which most taxonomic information is attached. Practically, the species of higher plants are recognized as the smallest populations structurally distinct and distinguishable from all others. This limited treatment, in a general context, is a step toward an understanding of the species concept. Context does make a difference in definition. (The treatments of species in nomenclatural and speciational contexts are presented in Chapters 3 and 9, respectively.)

The learning situation in systematics and other endeavors can be improved by reducing or eliminating many problems in communication as well as misunderstandings between individuals by knowing

1. The bases of definitions;
2. The nature of the reasons for any action;
3. The contexts in which principles or concepts are used; and
4. The kinds of components, processes, and products involved in any system.

''The process of knowing and being known through the use of words'' is made easier by the application of these four simple principles for effective communication.

II. The Fundamental Components of Taxonomy

Taxonomy is often defined as a science dealing with the study of classification, including its bases, principles, rules, and procedures. The general purpose of all taxonomy is to arrange elements, components, objects, or taxa in a way that gives the greatest possible command of knowledge, makes the most effective use of information, and leads most directly to the acquisition of pertinent data, information, and knowledge. The fundamental components of taxonomy are classification, identification, description, and nomenclature with the basic activities of determining position and rank, diagnostically delimiting, circumscribing, and naming, respectively. These conceptually independent components are actually closely interrelated. These fundamentals of taxonomy will be treated as concept systems in the following paragraphs according to the format: (1) definition, (2) reason for study, (3) components, (4) types, (5) processes, (6) reasons for changes, and (7) appropriate accessory terms. The principles for each component are presented in subsequent chapters. The principles pertinent to taxonomy in general are included in the summary of this chapter.

a. Classification

Fundamentally, *classification* is the arrangement of groups of plants with particular circumscriptions by rank and position according to artificial criteria, phenetic similarities, or phylogenetic relationships (Figure 1.1). Generally, classification provides a system of named and circumscribed reference bases (taxa) for informational storage, retrieval, and use. Logically, classification produces perspective, as well as position and rank, for the hierarchical elements of any component or subject. Practically, it indirectly makes communication more effective through greater comparability and consistency of treatment of objects, entities, and taxa.

MERCHANDISE MART
(A Classification)

Rank order	Hierarchy	Positions			
I.	Store	1. Grocery	2. Hardware	3. Clothing	4. Furniture
II.	Department	1. Produce	2. Meat	3. Bakery	4. Canned Goods
III.	Section	1. Fruits	2. Leafy vegetables	3. Stem vegetables	4. Root vegetables
IV.	Aisle	1. Apples	2. Oranges	3. Pears	4. Melons
V.	Shelf	1. Delicious	2. York	3. Golden	4. Limbertwig
VI.	Compartment	1. Extra fancy	2. Fancy	3. Commercial	4. Cut-rate

Figure 1.1 Ranks and selected positions in a merchandise mart.

A *classification system* resulting from the classification process is composed of named, circumscribed taxa at determined positions and ranks. The usual types of systems of classification for plants are: *artificial,* which are utilitarian systems based on habit, color, form, or characters of a similar nature; *phenetic,* which are based on overall similarities; and *phylogenetic,* which are presumably based on common evolutionary descent.

The processes of classification include the determination of position and rank for new taxa and the consideration of change of position and/or rank for old taxa that have been remodeled, divided, united, transferred, or altered in rank. *Remodeling* of taxa usually involves addition, deletion, or transfer of characteristics used in their circumscription; *dividing* a taxon is its separation into two or more recognizable groups; *uniting* taxa is the joining of two or more taxa into one; *transferring* is the changing of position of the taxon, for example, from one family to another (at the same level); *altering* is the raising or lowering of the rank of a taxon, for example, variety to a species or species to subspecies (at different levels). Classifications of taxa are changed for the proper reasons of either a more profound knowledge of the facts resulting from adequate taxonomic study or nonconformance to the principles and rules on which the classification system is based.

Six named ranks (I to VI, Store to Compartment) are listed in Figure 1.1. Four named positions at each rank are given—space prohibits including all. Four stores (rank I) in the merchandise mart are listed: (1) Grocery, (2) Hardware, (3) Clothing, (4) Furniture. Four departments (rank II) in the grocery store are listed: (1) Produce, (2) Meat, (3) Bakery, (4) Canned goods. Hardware, clothing, and furniture departments are not listed. Four sections (rank III) in the produce department are itemized: (1) Fruits, (2) Leafy vegetables, (3) Stem vegetables, (4) Root vegetables. The sections for the other departments (meat, bakery, canned goods) are not listed. The ranks are inclusive, for example, commercial (VI), Delicious (V), apples (IV) in the fruit section (III) of the produce department (II) in the grocery store (I) of the merchandise mart. In this classification system, the taxonomic groups (taxa) are located vertically by rank and horizontally by position, for example, oranges are in the fruit section between the pears and apples. In general a rank is located above, below, or between two other ranks, a position of a taxon is located in relation to the next higher rank (species A in Genus A) and next to or between two other positions of the same rank (*Quercus alba* is next to *Quercus bicolor*—vertical and horizontal coordinates). Any classification system shows the relationship of the included taxa by rank and position, for example, in the merchandise mart, the leafy vegetables are between the stem vegetables and fruits in the produce section. Presumably the closer the relationship, the closer the taxa or items are to one another. Circumscription for a taxon at a given position and rank is a characterization delimiting that taxon from all other taxa (position) of the same rank, for example, York (apples) would be delimited from Delicious, Golden, and Limbertwig.

Hierarchy within a system of classification for organisms refers to an ascending series of levels or ranks with the one above being inclusive of all below. *Rank* is a level within a hierarchy or the location of a category in the taxonomic hierarchy, as genus between family and species within the plant classification system. *Position* of a taxon is its place as an element of a taxon of the next higher rank, for example, the position of *Hydrolea quadrivalvis* is as a species in the genus *Hydrolea*. When the species is transferred to *Nama* as *Nama quadrivalve,* its position would be as a species in *Nama*. The principal ranks for classification of plants in descending sequence are: division, class, order, family, genus, and species. Examples of taxa of flowering plants for each rank in descending sequence are, respectively: Magnoliophyta, Magnoliopsida, Ranunculales, Menispermaceae, *Menispermum,* and *Menispermum canadense.*

b. Identification

Functionally, *identification* is the assignment of additional unidentified objects to a correct rank and position once a classification has been established. Practically, plant identification is the determination of a name for a specimen in relation to a previously established identification system. These are user definitions based on previously established systems of identification-classification. The reasons for the development of an *identification system* are to provide a means of easy, accurate, positive identification for each taxon and to supply a means of ready determination of the name of a taxon for communication or information-retrieval purposes.

Identification also includes the development of systems that are composed of diagnostically delimited and named taxa, each with a particular position and rank, and an identification device such as a key. The most widely used systems of identification are those with *dichotomous keys,* which are devices consisting of a series of contrasting statements requiring the identifier to make comparisons and choices based on statements in the key relating to the material to be identified (Figure 1.2). The newer *polyclave devices* use computer or mechanically based systems that allow the user to select the characteristics for identifying each specimen by choosing from some character set and repeating an elimination process until an identification is made.

Identifiers as developers of systems of identification have to diagnostically distinguish new taxa and determine the diagnostic characteristics for old taxa that have been remodeled, united, divided, transferred, or altered in rank. Identifiers as users determine the group to which a specimen belongs according to a previously developed identification system. Identifications of taxa are changed for the proper reasons of either a more profound knowledge of the facts resulting from adequate taxonomic study or because the system of identification does not result in consistent positive identification of the taxon.

6. PARNASSIA L.
Grass-of-Parnassus

Glabrous, rhizomatous, perennial herbs. Leaves primarily basal, ovate, cordate, or reniform, rounded to acute, entire, base cordate, reniform, or widely cuneate, long-petiolate; cauline leaf solitary, similar to basal leaves, smaller and usually cordate-clasping. Flowers solitary on elongate peduncles, perfect, regular; sepals 5, green tinged with white; petals 5, white usually tinged with green, veins conspicuous; stamens 5; staminodia 5, deeply 3-parted, apical glands subglobose, ovoid, or narrowly lanceolate; carpels 4, stigmas 4, sessile, ovary superior, 1-locular, many-ovulate, placentation parietal. Mature capsule not seen.

Petals clawed; staminodia equaling or shorter than stamens;
    leaves cordate to reniform .... 1. *P. asarifolia*
Petals not clawed; staminodia longer than stamens;
    leaves slightly cordate to round.
Staminodia apices lanceolate; ovary whitish; main veins of
    petal usually 11 or more .... 2. *P. caroliniana*
Staminodia apices obovate to subglobose; ovary greenish; main
    veins of petal 7 to 11 .... 3. *P. grandifolia*

1. P. asarifolia Vent. Plant 1–5 dm tall. Basal leaves usually reniform, 2.5–6 cm long, 2–8 cm wide. Sepals 4–5 mm long; petals clawed, 10–16 mm long, main veins 11–15; stamens exceeding staminodia; staminodial glands ovoid; ovary broadly ovoid, ca. 5 mm long at anthesis. Aug.–Oct. Bogs and seepage slopes, rare; mts. [Va., Ga., Ala., Tenn., Ky., W.Va.].
2. P. caroliniana Michaux. Plant 2–5 dm tall. Basal leaves usually ovate, 2–6 cm long and wide. Sepals 3–4.5 mm long; petals not clawed, 15–20 mm long, main veins 9–19; stamens shorter than staminodia; staminodial glands elongate, apiculate; ovary broadly ellipsoid, ca. 9 mm long at anthesis. Sept.–Nov. Savannahs, rare; Bladen, Columbus, Lee, Onslow Cos., N.C., Georgetown Co., S.C. [Fla., Miss.].
3. P. grandifolia DC. Plant 1–5 dm tall. Basal leaves ovate to reniform, 3–8 cm long, 2.5–7 cm wide. Sepals 4–6 mm long; petals not clawed, 12–25 mm long, main veins 7–9; stamens shorter than staminodia; staminodial glands subglobose to slightly cylindric; ovary subglobose, 5–6 mm long Sept.–Oct. Seepage areas, very rare; Ashe, Clay, Haywood Cos., N.C., Greenville Co., S.C. [Va., Fla., Tenn., W.Va.].

Figure 1.2 A dichotomous key to the species of *Parnassia* and genus and species descriptions. In the descriptions note the hierarchy for plant structures: plant, plant organ, organ part; common characters: number of plant parts, color, size; and flower part format: sepals, petals, stamens, carpels. (*Source:* Radford, A. E., H. E. Ahles, and C. R. Bell, 1968, *Manual of the Vascular Flora of the Carolinas,* p. 523, the University of North Carolina Press, Chapel Hill. Used with permission.)

c. Description

In general, a *description* is a statement of attributes for a taxon, individual organism, object, or entity. A *botanical description* is the recording of many types of characters with appropriate character states for plant parts, plants, and plant taxa. *Characterization* is the orderly recording of distinctive characters with appropriate character states for plant parts, plants, and plant taxa. *Circumscription* is the orderly recording of limiting and diagnostic characters with appropriate character states for plant taxa. The purpose of descriptive study is to provide a vocabulary of descriptors for intelligent communication about objects, entities, and taxa; and to supply the descriptors and descriptive system for a named and/or identified reference base to each object, entity, or taxon.

Character state	Character	Character set	Type of evidence
Lanceolate	Shape	General shape	Morphology
Serrate	Margin	General shape	Morphology
Obtuse	Apex	General shape	Morphology
Whorled	Arrangement	Disposition	Morphology
Terminal	Position	Disposition	Morphology
Descending	Orientation	Disposition	Morphology

Figure 1.3 Example of a hierarchy for descriptive terminology.

The *descriptive system,* the product of a descriptive effort, is composed of named concepts called terms that, preferably, have a particular circumscription, position, and rank. The *typical descriptive system* is an alphabetized glossary with named terms, each with a particular circumscription. A *desirable descriptive system* is one with terms, each with one correct name and a particular circumscription classified as to position and rank—a system not yet produced.

Description as a process involves the selection of characters and the determination of character states for the circumscription of new taxa and the making of changes in circumscriptions of old taxa that have been remodeled, divided, united, transferred, or altered in rank. Descriptions of taxa are changed for the proper reasons of either a more profound knowledge of the taxa resulting from adequate taxonomic study or nonconformance of the descriptors to accepted practice or sound etymological study.

A scientific description should have a hierarchical system for descriptors. The basic unit in description is the descriptor or characteristic that is the character state. *Character states* are distinct phases of a character that can be measured, counted, or otherwise assessed. A taxonomic *character* consists of two or more phases or states. Characters may be grouped as *character sets,* which may be fundamental components of types of evidence.

The hierarchy for scientific description in ascending sequence according to this scheme includes the ranks character state, character, character set, and evidence. For example, ''ovate'' is a *character state* of the *character,* shape; ''crenate'' is a *character state* of the *character,* margin; ''acute'' is a *character state* of the *character,* apex. Shape, margin, and apex are characters in the *character set,* general shape. ''Opposite'' is a *character state* for the *character,* arrangement; ''basal'' is a *character state* for the *character,* position; ''ascending'' is a *character state* for the *character,* orientation. Arrangement, position, and orientation are characters in the character set, disposition. These two character sets and others would be components of a type of *evidence,* morphology. Morphological, anatomical, and palynological evidence and so on would form a comprehensive descriptive terminology for plants and their parts (Figure 1.3).

d. Nomenclature

Nomenclature is a precise and simple system used by botanists in all countries that deals on the one hand with terms denoting the ranks of taxonomic groups or units, and on the other hand with the scientific names that are applied to the individual taxonomic groups of plants. The reason for naming any object, entity, or taxon is to provide for each a symbol of communication and reference base for information storage, retrieval, documentation, and use. Theoretically, the study of nomenclature for a group of organisms provides one correct name for a taxon with a particular circumscription, position, and rank that is universal, stable, and unambiguous. The *primary nomenclatural system* for plants is that described in the *International Code of Botanical Nomenclature* (Voss, E. G., et al., 1983), a system composed basically of principles, rules, recommendations, and conserved and rejected family and generic names.

Nomenclaturists name new taxa and determine the correct name for old taxa that have been remodeled, united, divided, transferred, or altered in rank. Names of taxa are changed for the proper reasons of either a more profound knowledge of the facts resulting from adequate taxonomic study or the necessity of giving up a nomenclature that is contrary to the rules (modified from the Preamble of the International Code of Botanical Nomenclature, 1983).

The *principal ranks* of taxa in the system of nomenclature in an ascending series are: species, genus, family, order, class, division. A *taxon* (plural: taxa) refers to a taxonomic group of any rank, for example, *Acer* is a taxon at the rank of genus; Orchidaceae, at the rank of family; Bryophyta, at the rank of division. *Name* in general is a word or combination of words by which a taxon is known. A *correct name* of a taxon with a particular circumscription, position, and rank, is the legitimate name that must be adopted for it under the rules. The *scientific name* of a species is a binary combination consisting of the name of the genus followed by a single specific epithet, for example, *Celtis occidentalis. Epithet* is a word in a name of more than one word, subsequent to the generic name, for example, *Celtis occidentalis* var. *georgiana; occidentalis* is the specific epithet and *georgiana* is the varietal epithet. The ''one *correct name* for a taxon with a particular circumscription, position, and rank'' is a fundamental as well as a guiding principle for all of taxonomy that is so basic that it should be considered a basic assumption for all taxonomic work.

III. The Fundamental Components of Evolution

Basically, *biological evolution* is usually considered a series of processes involving descent of organisms with modification marked by successive adaptations to environmental conditions, governed by competition and natural selection acting on variation. The general purpose of evolutionary study is to determine the origin of new taxa and to interpret the historical development of the old. The fundamental components of evolution are variation, speciation, and phylogeny with the basic activity of analysis of character variation for (1) the generation of character correlations for organisms, populations, and species; (2) the determination of the origin of new taxa; and (3) the interpretation of the historical development (origin, maintenance, trends, pathways) of the old. These conceptually independent components are actually closely interrelated. The fundamental components of evolution are treated as concept systems in the following paragraphs according to the format used for the components of taxonomy. The principles for each component are presented in subsequent chapters. Principles pertinent to evolution in general are included in a summary at the end of this chapter.

a. Variation

Biologically, *variation* within a group is the divergence from the standard, mean, or typical structural, functional, and developmental characters, particularly those not attributable to differences in age, life history, sex, or environment. Systematists ascertain the types, patterns, and trends in variation of selected characters within populations, species, and higher groups to delimit, circumscribe, and classify taxa as well as to determine their phenetic and phylogenetic relationships.

Character variation analyses include the selection of objects (individuals, populations, groups) and characters, and the determination of character states pertinent to the objects for character correlations that are relevant to the purpose of the study. The choice of objects, characters, parameters of characters, and character states should always be applicable to the objectives of the systematic study for which the character variation analysis is being done.

The *character variation system* includes the objects, characters, character states, and the character state/object data. Taxonomically, this system is normally expected to generate a descriptive product, including delimitations, characterizations, and a classification of taxa along with some indication of their relationships. Systematically, the system is expected to generate a taxonomic system as well as make some contribution to an explanation of the origin, evolutionary development, and maintenance of the taxa (speciation and phylogeny). The system changes as the objects, characters, and character states vary.

Evidentiary characters used in variation studies may be classified as physical (shape, color), chemical (process, composition), and biological (structure, function). In character variation studies, *evidence* is information used for systematic purposes. *Phenotypic variation* is the morphological, functional, or developmental variation resulting from the action of different environments on one or more genotypes within a population(s). *Genotypic variation* (genetic) is the differences in genotypes within a population(s) as a result of mutation and recombination.

b. Speciation

Speciation is the formation of new species. Biologically, speciation is the development of populations of freely interbreeding organisms that are adapted to their environment and reproductively isolated from other such populations. Genetically, speciation is the formation of gene pools in populations in which the genes reproduce (asexually) during DNA replication and in which phenotypes (organisms) are generated that can reproduce sexually to produce new, reproductively isolated combinations of genes. Evolutionarily, speciation is the formation of biological species in the perspective of evolutionary time and process. A *species* as a biologic, genetic, and evolutionary product of speciation can be considered a basic, discrete biological unit with a distinctive set of correlated characters that is fixed for a moment in evolutionary time.

Speciation involves the development of new (mutation) and different gene combinations (genetic recombination) in variants in separate populations that become isolated and adapted to their environment through natural selection.

Some of the reasons biologists study speciation in groups of organisms are (1) to discover the causal mechanisms for the production of variation and variants within selected populations, (2) to determine isolating mechanisms controlling interbreeding between selected populations, and (3) to ascertain the nature of the adaptation of selected populations to their environmental stress and natural competition.

Speciation can be regarded as a system that produces basic taxa. This system includes populations and gene pools as well as components such as variation-producing, isolating, regulating, stabilizing, and selecting mechanisms. The system and resulting products change as the mechanisms vary. Some of the concepts related to these basic inclusions and mechanisms require further treatment here.

Population generally refers to all the individuals of one species occupying a given area, which is usually isolated from other similar groups. *Gene pool* includes the total genetic material of an interbreeding population at a given time.

Mutation and genetic recombination are genetic sources of variation in natural populations. A *mutation* is a sudden heritable change in genetic materials that can be due to an alteration of a single gene by duplication, replacement, or deletion of a number of DNA base pairs. *Genetic recombination* is any process that gives rise to a new combination of hereditary determinants, such as a reassortment of parental genes during meiosis through crossing-over.

Isolation refers to the restriction or limitation of gene flow between discrete populations caused by barriers to interbreeding, such as sexual structural incompatibility, geographic distance, and different times of maturation for the reproductive components. The development of isolating mechanisms leads to the fixation of variants. A *variant* is any organism or group showing marked divergence or deviation from some standard in structure, function, or development. *Stabilizing selection* is selection for the mean or intermediate phenotype with the consequent elimination of peripheral variants, maintaining an existing state of adaptation to a stable environment.

Fundamental processes in speciation and evolution are adaptation, natural selection, and competition. *Adaptation* is the process of adjustment of a population to environmental stress or change. It is also a series of processes associated with the evolutionary modification of organisms to environmental stress resulting in improved survival and reproductive efficiency. *Natural selection* refers to a series of processes during the course of evolution that preserve variants best adapted to their environment in the face of natural competition. *Natural competition* occurs most frequently when there is a simultaneous demand by two or more organisms or species for an essential common resource that is actually or potentially in limited supply.

c. Phylogeny

Phylogeny is the study of the genealogy and the evolutionary history of groups of organisms. *Genealogy* is the study of ancestral relationships and lineages. Phylogenetic studies result in genealogical branching patterns indicating the presumed ancestor/descendent relationships. The realistic goal for phylogenetic effort is the development of the most probable pathways for the evolution of selected taxa based on an analysis of a finite number of useful characters. One product of most phylogenetic studies is a *cladogram* (see Chapter 10), which is a hypothetical, diagrammatic representation of the evolutionary branches of a lineage showing divergence from ancestral stock. Usually, the cladogram is a branching diagram representing the relationships between characters and character states from which phylogenetic inferences are made.

For systematists, the primary reason for studying the phylogeny of groups of organisms is the ultimate production of a classification that reflects the evolutionary history of the included taxa. *Phylogenetic classifications* are those interpreted in terms of evolutionary theory. A *phylogenetic system of classification* is composed of named and circumscribed taxa at determined positions and ranks. The circumscriptions of the included taxa at their respective positions and ranks presumably indicate *phylogenetic relationships,* which are supposedly character expressions of affinities based on recency of common ancestry.

Most of the presently used systems of plant classification have been established on phylogenetic principles but are largely founded on *phenetic relationships* based on structural similarities.

Basically, a phylogenetic study of groups of organisms includes the selection of taxa and characters pertinent thereto and the determination of the generalized or primitive and the derived or advanced state of each character. Then, the hypothetical ancestor of the taxa studied is determined. Derived characters are then usually grouped in a variety of ways to produce an evolutionary tree or branching pattern showing the probable pathways of development for each of the taxa from the hypothetical ancestor. Evolutionary trees or cladograms may vary with the change in characters used, with different interpretations for generalized and derived features, with different character correlations at the nodes or branch bases, and with the analytical methods used.

The concepts of morphocline, plesiomorphy, and apomorphy are fundamental to all phylogenetic study. A *morphocline* is a sequence of character state modifications representing heritable changes from a preexisting structure or feature. A morphocline is usually based on some obvious intergradation of character states. For example, in the character "petal fusion," the gradation and the character states may be designated "free"–"partly fused"–"fused." The ancestral, primitive, or generalized condition of "free" is known as a *plesiomorphic* condition or character state; the derived or advanced conditions of "partly fused" and "fused" are described as *apomorphic*.

IV. Taxonomic Activities

Taxonomic activities are treated here in synoptic form as a general introductory summary of what systematists actually do. This synopsis also provides a convenient reference to the core of taxonomic training from a functional standpoint.

1. Taxonomists as *classifiers* determine position and rank for new taxa and establish the correct position and rank for old taxa that have been remodeled, divided, united, transferred, or changed in rank; they also determine the group to which a specimen belongs according to a system of classification.
2. Taxonomists as *identifiers* diagnostically distinguish new taxa and determine the diagnostic characteristics for old taxa that have been remodeled, divided, united, transferred, or changed in rank according to a system of identification; they also determine the group to which a specimen belongs according to a system of identification.
3. Taxonomists as *describers* select characters and determine character states for the circumscription of new taxa and change the circumscription of old taxa that have been remodeled, divided, united, transferred, or changed in rank according to a system of description; they also determine the circumscription for a specimen according to a system of description or classification.

4. Taxonomists as *nomenclaturists* name new taxa and determine the correct name for old taxa that have been remodeled, divided, united, transferred, or changed in rank according to the International Botanical Code; they also determine the correct name for a specimen according to a system of identification or classification.

5. Taxonomists as *coiners* of botanical names select the proper Latin or Greek roots, prefixes, suffixes, and terminations to create scientific names that are connotative, mnemonic, reasonably short, and easy to pronounce. Taxonomists as *users* of scientific names study Latin and Greek roots, prefixes, suffixes, and terminations to make scientific names connotative, mnemonic, reasonably succinct, and acceptably pronounceable.

6. Taxonomists as *observers* study the characteristics of organisms, populations and taxa for the discovery of evidence of spatial, temporal, abiotic, and biotic relationships for a better understanding of the taxonomy of the groups; they also study site-species relationships for a better understanding of the systematics and species biology of individuals and populations.

7. Taxonomists as *analyzers* of character variation select the object (individual, population, taxon) and character(s) to be analyzed; choose the parameters of the character(s) to be analyzed; determine the feasibility of the character variation analysis; analyze; summarize the results of the analysis; and evaluate the results in relation to the original purpose or hypothesis.

8. Taxonomists as *synthesizers* gather information about organisms from many fields of evidence—morphology, anatomy, embryology, palynology, cytology, genetics, physiology, biochemistry, ecology, geography, and paleobotany—for the development of user systems of classification, identification, and characterization.

9. Taxonomists as *theorizers* study the concept–principle–process approach—develop new concepts and reinterpret and amplify the old; devise new principles, reexamine and revise the philosophical bases for the old; initiate new procedures and improve the old—to achieve clarity in concepts, stimulate an understanding of principles, and produce a rationale for and comprehension of process methodology.

10. Taxonomists as *evaluators* of research and concept-principle-process studies evaluate results of research in relation to the original purpose or hypothesis and the concept-principle-process studies in relation to efficacy, efficiency, accuracy, and applicability; ascertain the significance of the research or study relative to the field of endeavor; make decisions on the relevance of the research or study to scientific and societal needs; and determine the proficiency of the individual in research, concept-principle-process study, and the inherent skills and values in each type of effort.

11. Taxonomists as *historians* develop perspectives for a better understanding of taxonomy by studying the oldest to the most recent pertinent documents, selecting criteria for the determination of the significance of each taxonomic subject studied, establishing a value system for the criteria, and applying the value system to the criteria selected for significance values. Perspective provides enlightenment to the systematics of the future.

12. Taxonomists as *determiners* of relationships select organisms and characters to be studied; determine the character states; calculate the resemblances between organisms on the basis of characteristics; recognize the taxa on the basis of the resemblances (character correlations); and make generalizations about the taxa

based on the resemblances, for example, phenetic, genetic, or phylogenetic relationships.

13. Taxonomists as *generators* of systems of information for taxonomic purposes provide effective methods for collection, analysis, and presentation of data; supply a total strategy for storage, retrieval, and exchange of information; contribute a design for documentation of all elements, data, and information within the system; and develop a system that produces a basis for taxonomic decisions, stimulates new questions, and yields a variety of products.

14. Taxonomists as *conservators* of our natural heritage provide an informational reference base as well as documentation for threatened and endangered species, significant populations and communities, and many unique elements in our states and regions for the preservation of genetic and ecologic diversity.

15. Taxonomists as *communicators* make communication more effective and efficient by classifying and identifying objects, groups of objects, entities, or taxa that are circumscribed and named or coded. These taxonomic processes add comparability and consistency to the written and spoken treatment of any subject.

16. Taxonomists as *systematists* ascertain the types, patterns, and trends in variation within populations, species, and higher taxa; classify, identify, describe, and name; and observe, analyze, synthesize, and evaluate to determine relationships based on similarity and differentiating characteristics from all fields of evidence.

B. SYSTEMATICS AS A FIELD OF ENDEAVOR

Systematics is societally relevant. Taxonomic training will prepare students to do impact studies and inventories of natural areas; to be nature counselors, consultants on the most effective use of local plant resources, participants in environmental programs, teachers of biology, research associates and technicians, and better students in any field of endeavor. Professional systematists will be trained to make their research germane to the botanical problems of our society. Most of the training and research in systematics today is done in our academic and public institutions such as universities, herbaria, museums, and botanical gardens. These institutions are the conservators of our cultural and natural heritage and the training centers for our future development.

Most information of societal relevance derived from taxonomic research is published in floras, manuals, revisions, monographs, notes or short articles, or reviews. A *flora* in its simplest sense is a list of plants found in an area. Most floras do have additional basic information on habitat, distribution, and flowering-fruiting data. A *manual* is a flora with basic information plus keys, descriptions, glossary, history of botanical exploration of the area, and documentation for the treatment of each of the included taxa. Many manuals have all or perhaps only a few carefully selected species illustrated; a few have distribution maps or detailed distributional data for each species. A manual is usually considered a sound beginning for the study of the taxa of a region, not the end. *Revisions* are treatments for selected taxa throughout their range or in a major portion thereof. The treatments usually include studies of nomenclature and classification along with descriptions based on several types of evidence. Ecological and geographical evidence are usually treated extensively. The major contributions in most revisions are improvements in the classification of taxa, the updating of nomenclature, and the presentation of the systems of identification or key for the taxa studied. *Monographs* are detailed taxonomic studies of selected taxa using many types of evidence. Very few taxonomic studies of today are comprehensive enough to be considered monographic.

Much useful and basic taxonomic information on nomenclature, distribution, numerical taxonomy, and classification is found in *notes* or short *journal articles* in systematic, botanical, and regional scientific periodicals. *Reviews* are basically of two types: synopses of recently published books; and summaries of the facts, information, and knowledge related to the rapidly advancing fields of taxonomic research, for example, chemotaxonomy, numerical taxonomy, and phylogenetic classification.

For purposes of taxonomic training, a few textbooks have been produced for undergraduate and beginning graduate education. Many how-to-identify paperbacks have been published recently for public education in short courses in botanical gardens and museums as well as for liberal arts courses in colleges and universities.

Many states and nations are still in the *pioneer phase* of development of systematics, which includes the discovery, classification, description, naming, and identification of plants. Others are in the flora-manual-revision-monograph preparation *(the consolidation)* aspects of their development. The professional systematists in some nations are analyzing breeding systems, studying patterns of variation, determining the evolutionary potential, and doing pertinent work in the chemical, numerical, cytological, anatomical, embryological, and palynological aspects of systematics for selected groups of organisms. This is the *experimental* or *biosystematic stage* of development. A very few individuals in the world are analyzing and synthesizing information and data from many fields of evidence for the production of systems of classification based on evolutionary and phylogenetic relationships. This is the truly *encyclopedic* or *holotaxonomic level* of effort in systematics. Numerical taxonomy is now producing a new approach to the old problems of classification by stimulating a reexamination of the philosophical bases of taxonomy. These new studies are affecting all phases of systematics, from the pioneer to the holotaxonomic.

The fields or subdisciplines of systematics recognized today include the traditional, biosystematic, chemosystematic, ecosystematic, phylosystematic, and taximetric or numerical taxonomic. The *traditional* aspects of systematics cover the pioneer and consolidation phases of development. *Biosystematics* is the field of study dealing with variation and evolution, primarily experimental and analytical, and mostly treating the species and infraspecific taxa. *Chemosystematics* is the application of chemical evidence to the solution of taxonomic problems and determination of systematic relationships. Chemical evidence is being used rather extensively, thus the recognition of one field based on one type of evidence. Many other fields of evidence such as cytology (cytotaxonomy), genetics, embryology, palynology, ecology, and geography are used in the solution of taxonomic problems and determination of systematic relationships. *Ecosystematics* is the systematics of ecological and genetic diversity that is comprised of the classification, identification, nomenclature, and description of components and elements of natural diversity; the study of the relationships between biotic and abiotic diversity; and the determination of the significance of those relationships. *Phylosystematics* (phylogenetics) is the field of study dealing with phylogeny and classification, primarily theoretical and synthetic, treating mostly the genus, family, order, and class. *Taximetrics* or numerical taxonomy, according to Sneath and Sokal (1973) is "the grouping by numerical methods of taxonomic units into taxa on the basis of their character states."

C. A SUMMARY STATEMENT FOR PLANT SYSTEMATICS

Traditionally, systematists have studied organisms and populations from structural, functional, developmental, environmental, geographic, and evolutionary standpoints for purposes of classification. Fundamentally, taxonomists have found classification basic to understanding organismic diversity. Historically, systematists have been the specialists in organismic diversity who have appreciated the genetic resources occurring in different species populations. Now and in the future, research taxonomists with the expertise and knowledge of genetic diversity associated with organisms will provide information needed for the development of new biotechnologies related to gene splicing, cloning, and so on. Systematists will continue to supply genetic stock for the introduction of new crops and for the improvement of the old.

Many new species await discovery, particularly in the tropics. Floras of many regions remain to be done. Revisions of taxonomic groups and the determination of species-habitat relationships represent major work areas in systematics for years to come. The characterization of new and old taxa using new evidence and the development of more effective methods of distinguishing and classifying taxa are the ongoing duties of the taxonomists. Phylogenetics is producing much theoretical and methodological discussion on the origin, evolution, and classification of many groups. The study of variation and speciation in natural populations will continue to be the stimulus and challenge for many systematists.

The written record indicates man has been classifying organisms for more than 2,000 years. Yet taxonomy continues to be a dynamic science, with the incorporation of new ideas, evidence, methods, and tools developed in systematics and other fields of science, as well as mathematics. Statistics and the computer have changed taxonomic procedures tremendously. Scanning and standard transmitting electron microscopes have made many micro- and ultrastructural characters available for classification. Spectrochemistry and chromatography have meant much to the chemical characterization of taxa and to the determination of intra- and intertaxon relationships. New methods of studying the nucleic acids will furnish characters of basic significance to understanding relationships among groups of organisms.

The systematics of the future will require research specialists with broad training in the fields of science and statistics. Technically trained support personnel will be necessary for the advancement of the science. Students, young and old, with an appreciation for the aesthetics, applicability, and philosophy of systematics will provide a citizen foundation for the development of systematics and many other fields of learning.

SUGGESTED READING

Davis, P. H., and V. H. Heywood. 1965. Principles of Angiosperm Taxonomy. D. Van Nostrand Company, Inc., New York. Chapter 1, pp. 1–12.

Jones, S. B., and A. E. Luchsinger. 1979. Plant Systematics. McGraw-Hill Book Company, New York. Chapter 1, pp. 1–6.

Radford, A. E. et al. 1974. Vascular Plant Systematics. Harper & Row, New York. Chapter 1, pp. 1–12.

Sneath, P. H. A., and R. R. Sokal. 1973. Numerical Taxonomy. W. H. Freeman, San Francisco. Chapter 1, pp. 1–15.

Stace, C. A. 1980. Plant Taxonomy and Biosystematics. University Park Press, Baltimore. Chapter 1, pp. 1–51.

Voss, E. G., et al. 1983. International Code of Botanical Nomenclature. Bohn, Scheltema, and Holkema, Utrecht, The Netherlands.

SUMMARY FOR TAXONOMY

Definition of Taxonomy   Taxonomy is the study of nomenclature, description, classification, identification, and relationships.

Purpose of Taxonomic Study   To develop systems of classification in which elements, components, objects, or named taxa are arranged in a way that gives the greatest possible command of knowledge, makes the most efficient and effective use of information, and leads most directly to the acquisition of more data, information, and knowledge.

Operations in Taxonomic Study   To circumscribe accurately; to determine the position and rank of taxa; to delimit diagnostically; to name; to determine relationships; to develop systems of characterization, classification, identification, and nomenclature (see taxonomic activities).

Basic Premise for the Study of Plant Taxonomy   That in the tremendous variation in the plant world, conceptually discontinuous groups with character correlations exist that can be recognized, classified, circumscribed, named, and logically related.

Fundamental Principles of Taxonomy

1. Variation in plants makes possible the establishment of taxonomic systems.
2. The fundamental components in taxonomic systems—classification, identification, description, nomenclature—even though conceptually distinct, are related and interrelated.

Guiding Principles or Guidelines for the Development of Taxonomic Systems for Plants

1. Recognition and selection of groups of plants.
2. Establishment of a hierarchy for the groups of plants.
3. Determination of the position of each group of plants.
4. Diagnostic delimitation of each group of plants with a determined position and rank.
5. Circumscription (written) of each group of plants with a determined position and rank or give reference(s) to previous circumscription.
6. Naming of each group of plants with a determined position and rank; or give reference(s) to previous nomenclature.

Basic Assumptions for Study of Taxonomy

1. Discontinuities in variation and character correlations occur that make circumscriptions of taxa at determined positions and ranks possible.
2. A taxon can be diagnostically delimited from all other taxa.
3. Position and rank can be determined for diagnostically delimited taxa.
4. Names for all taxa can be provided in a system for naming and names.
5. Various relationships, for example, phenetic, genetic, phylogenetic, exist that make the development of taxonomic systems possible.
6. Many characters from all types of evidence are used in description, identification, and classification of taxa.

SUMMARY FOR EVOLUTION

Definitions for Evolutionary Study   *Evolution* is usually considered a series of processes involving descent of populations with modification marked by successive adaptations to environmental conditions, governed by competition and natural selection acting on variation. Biologically, *variation* within a group of organisms is the divergence of structural, functional, and developmental characters from the typical, mean, or standard. *Speciation* is the development of populations of freely interbreeding organisms adapted to their environment that are reproductively isolated from other such populations. *Phylogeny* is the study of the genealogy and evolutionary history of groups of organisms.

Purpose of Evolutionary Study   To ascertain the types, patterns, and trends in variation of selected characters within populations for the determination of the origin of new species and the interpretation of the historical development of the old through evolutionary time.

Operations in Evolutionary Study   To ascertain the types, patterns, and trends in variation of selected characters within populations, species, and higher taxa for evolutionary purposes; to discover the causal mechanisms for variation in populations, to determine the isolating mechanisms controlling interbreeding between selected populations, and to ascertain the nature of adaptation of selected populations to their environmental stress and competition; and to produce evolutionary branching patterns showing the probable pathways of historical development of selected taxa.

Basic Premise (Theory) of Evolutionary Study   That groups of organisms have descended through time with modification marked by successive adaptations to environmental conditions, governed by competition and natural selection acting on variation.

Fundamental Principles for Evolutionary Study

1. Existing species are products of evolutionary forces that have descended from preexisting species.
2. Variation in organisms makes evolution possible.
3. Sources of variation in natural populations are genetic and environmental.

4. Limitations of variation in natural populations are intrinsic and due to such causes as self-fertilization, asexual reproduction, genetic fixation, and stabilizing selection.

Guidelines for Evolutionary Study

1. Evolution may tend toward elaboration and diversity or toward reduction and simplicity of organisms and organismal parts.
2. "A character may evolve at very different rates in different taxa, and within one taxon various characters may evolve at very different rates, and in different directions" (Stace, 1980).
3. Evolutionary trends, progressive or retrogressive, are generally consistent but occasionally reversible with changes in the environment.
4. All parts of organisms at all stages of their development may produce evidence of evolutionary significance.

Basic Assumptions for Evolutionary Study   (Based on Charles Darwin, *On the Origin of Species,* 1859)

1. Many more individuals are born in each generation than will survive and reproduce.
2. Variation occurs among individuals; they are not identical in all of their characteristics.
3. Some individuals with certain characteristics have a better chance of surviving than individuals with other characteristics.
4. Some of the characteristics of individuals resulting in differential reproduction are heritable.
5. Extremely long spans of time are available for slow, gradual change in individuals.

GENERAL TAXONOMIC QUESTIONS

1. What are the fundamental premises for the study of systematics? For any subject?
2. What are the reasons for the study of taxonomy? What are the definitions of the fundamental components of systematics?
3. What do taxonomists accomplish as describers, classifiers, namers, and identifiers? Analyzers, synthesizers, theorizers, observers?
4. What are the basic assumption(s) related to each of the taxonomic operations or functions?
5. What are the advantages and disadvantages in communication of having only one correct name for any object with a given set of properties or specifications?
6. What are the component parts of any system of classification? Why can the Sears catalog be considered a system of classification? Identification? Description? Nomenclature?
7. What are the phases of development in plant systematics? Meanings of phases?
8. How is the premise on which the study of plant systematics is founded equally pertinent to all other studies? To advertising? To the study of chemistry?
9. What are the basic reasons for making a change in the classification of a taxon? Chemical compounds? Dogs? Shoes? Breakfast cereals?
10. What are the types of systems in classification? Identification? Description? Nomenclature?

What is the basis of classification for the types?

11. How are ''remodeling,'' ''uniting,'' ''dividing,'' ''change of position,'' and ''alteration of rank'' basic to an understanding of taxonomy?

12. What is a hierarchy for classification? Identification? Description? Nomenclature? Why does one establish a hierarchy for any subject? Position?

13. What are the reasons for the study of classification? Identification? Description? Nomenclature? Analysis of character variation? Determination of relationships? Taxonomic theory? Taxonomic processes? Documentation? Informational systems? Historical perspective? Evaluation of significance?

14. Why is a television commercial a system of identification? How are commercials misleading as to comparability and consistency in classification?

15. What is meant by each of the following: term, concept, principle, premise, and species? Indicate some of the problems in gaining an understanding or determining the meaning of each.

16. Under what conditions or context is a basic premise a fundamental principle? A fundamental principle, a basic assumption? A basic assumption, a guideline?

17. Under what conditions or context is classification a term? A specific name? A concept? A concept system? A process? A set of principles?

18. Is a species treatment in a manual a definition? If so, select one species, give the name of the term, the circumscription for the definition, and the exclusive/inclusive conditions limiting the circumscription.

EXERCISES IN TAXONOMY

1. Classification

A. A book is organized into ten chapters, three sections per chapter, and from five to eight paragraphs per section, with each chapter and section appropriately titled.

1. What is the name of each rank in the book? What is the hierarchy for classification?

2. How many positions are in the highest rank?

3. What did the author do taxonomically when he transposed modifiers, changed phrases to clauses, and split verbs within a sentence? Within a paragraph?

4. What did the author do taxonomically when he combined two short paragraphs into one long one; when he put Section B of Chapter 1 into Chapter 3 as Section C; when he decided Sections A and B of Chapter 4 should be put together as Section A; when he made Chapter 11 into Section C of Chapter 10; when he made Section C of Chapter 6 into Chapter 7?

5. How is a book a system of classification? In classification, is it possible to do more than remodel, unite, transfer (change position), and change rank? What more can be done?

6. As the author reorganizes, does he have to consider title changes or modification of his nomenclature? Changes in his descriptive system? Changes in his system of identification?

B. In the chart on the next page, determine the character and character set for each of the listed groups of character states. (This exercise introduces the student to the problems of comparability and the need for documentation.)

1. What are the ranks of this descriptive hierarchy?

2. List the characters in the character set, measurement; the character states for the character, volume; character states for the character, orientation.

3. Why should the completed chart be considered a system of classification for description? Give the position names for the rank of character for the character set, measurement.

4. Why should the completed chart be considered a system of nomenclature for description?

5. Why is classification basic to comparability and consistency of description? To effective communication?

3rd order rank 3 character state	2nd order rank 2 character	1st order rank 1 character set
a. Cubic meter, one gallon, one liter		
b. Square meter, acre, hectare		
c. Ounce, gram, dram, karat		
d. Revolutions per minute, horsepower, British thermal unit		
e. Volume, weight, area, capacity		
f. Length, width, height		
g. Cycly, merosity, unit (dozen, six-pack)		
h. Basal, terminal, median		
i. Opposite, alternate, whorled		
j. Horizontal, ascending, descending		
k. Arrangement, position, orientation		
l. Ovate, linear, elliptic		
m. Crenate, serrate, entire		
n. Acute, obtuse, mucronate		
o. Outline, margin, apex		
p. Shape, size, disposition, color		
q. Blue, yellow, green		
r. Light green, dark green, grayish green		
s. Banded, marbled, striped		
t. Chroma, value, hue		

2. Manual

Select a plant manual pertinent to your region and indicate:

1. Name of the principal author or investigator; the editor; the publisher.
2. The author and title of the earliest manual for the area; author responsible for the treatment of the grasses in the manual, for the composites.
3. The number of contributors to the manual; the number of states covered by the manual; the number of counties per state; the names of the physiographic provinces by state.
4. The number of taxa treated in the manual; the number of families.
5. The definition of annulus; runcinate; pome; catkin.
6. The names of ranks with descriptions; with authorities; with common names.
7. The name of a taxon without a distribution map; one with an illustration.
8. Full name of the author indicated by "L"; indicated by Const. or Constance.
9. The number of genera in the Ranunculaceae; number of species in the genus, *Carex*.
10. The scientific name for dandelion; common name for *Lycopodium;* common name for *Pinus virginiana*.
11. The corolla color in *Linaria vulgaris;* fruit shape in *Capsella bursa-pastoris;* stem height in *Rumex crispus*.
12. A common weed by name; indicate its distribution by province in your state, the blooming date for the weed, and the fruiting date for the weed.
13. A common tree by name; indicate its chromosome number and its habitat.
14. A spring flowering herb by name; indicate its family name, generic name, and specific name.

15. A shrub by name; indicate its leaf arrangement, leaf shape, and leaf duration.
16. A vine by name; indicate its leaf margin, leaf venation, and leaf base.
17. A lily by name; indicate its number of stamens, color of sepals, and size of leaves.
18. The descriptive sequence used for each leaf part; each flower part.
19. The types of evidence used or major components found in the description of the species; in the description of the genus.
20. The ranks with keys for identification; the types of keys used; the system of classification followed in the manual.

chapter *2*

History of Plant Taxonomy*

Taxonomists as historians develop perspective for a better understanding of taxonomy by studying the oldest to the most recent pertinent documents, by selecting criteria for the determination of the significance of each taxonomic subject studied, by establishing a value system for the criteria, and by applying the value system to the criteria selected for significance values.

The history of science holds a fascination for professionals and the general public alike, and no work of science is regarded as complete without at least some historical consideration. As a result, scientists have left a huge and complicated bibliographic morass for the historians. Data are to be had in the most diverse sorts of publications and, needless to say, the indexing of it varies from merely inadequate to nonexistent. Nevertheless, there are several studies in the history of science that are both detailed and quite readable, and it is from these and other secondary sources that much of the factual information in this chapter has been derived. The important secondary sources are itemized at the end of this chapter.

A. THE DEVELOPMENT OF CLASSIFICATION

I. Preliterate Mankind

Little can be said with accuracy about the botanical knowledge of our preliterate ancestors, however much can be inferred. Even the most primitive of ancient mankind, that is, those who made their living by gathering food from the landscape, were of necessity

*Contributed by T. M. Barkley, Kansas State University, Manhattan, Kansas.

practical plant taxonomists. Through experience, these peoples learned which plants were edible and which were not. And, as has been said, the people who failed to learn the elements of botany also failed to become our ancestors.

It is noteworthy that recent studies of extant primitive societies have shown that primitive peoples often possess the linguistic mechanisms for accurate distinctions among the kinds of plants with which the people are familiar. Furthermore, the distinctions recognized by these people are often as sophisticated as, and sometimes more than, the equivalent distinctions made by professional taxonomists.

II. Ancient Literate Civilization

The age of ancient history, when people could read and write but before the advent of printing, saw the rise of the intellectual and social bases of what is now our civilization. It is difficult for us to put our minds into the first century A.D. to investigate the outlook on life at that time, but it is relevant to the history of plant taxonomy. The advent of writing was certainly one of the great breakthroughs of ancient peoples, for it allowed both instant and accurate recall of data to any person skilled in writing. The literate man and his illiterate colleagues regarded the skill as a means of recording what was being said and of recalling it accurately at a later date, possibly by another person. The written word simply was fossilized speech, that is, a preservation of what people were talking about. To be certain, polemics and pronouncements soon were committed to script, but it appears that in ancient times, writing was done largely for the record, while publication was an oral affair. This is a noteworthy point. When we consider ancient botany, we should read it as material for the record and not as offerings of new data or new interpretations, published with the intent of adding broader vistas to the affairs of mankind. The ancient authors are best regarded as recorders of what was held as general knowledge by the sophisticated people of the times, with little in the way of individual interpretation. This general knowledge most assuredly was derived from the experiences of preliterate mankind.

a. Theophrastos (ca. 370–285 b.c.)

Theophrastos was the greatest botanical writer of classical antiquity, and he is the intellectual grandfather of modern botany (see Greene, E. L., 1909). As a young man he went to Athens, where he studied with Plato and later with Aristotle, and after their deaths he became the chief of the Lyceum in Athens and of its gardens and library.

Theophrastos is credited with more than 200 works by the students of Greek scholarship, but most of these survive only as fragments or as quotations in the works of other writers. Two of his botanical works have survived and it is on them that a large measure of his reputation rests. They are *Enquiry into Plants* and *The Causes of Plants;* both are available in English translation. Among the things noted by Theophrastos were: (1) the distinction between external structures (as organs) and internal structures (as tissues); (2) the distinction of different kinds of tissues; (3) the classification of plants into four great groups: trees, shrubs, subshrubs, and herbs; (4) the distinction of flowering plants (phanerogams) and nonflowering plants (cryptogams); (5) the recognition of various kinds of sexual and asexual reproduction; (6) a basic understanding of gross anatomy, that is, that the calyx and corolla are modified leaves, etc.; and (7) the recognition of the fruit in its modern, technical sense.

Theophrastos was aware of some of the biological realities of plants. He was impressed by the fact that many plants, particularly cultivated plants, do not breed true from seeds. He was convinced that species were unstable and were readily changeable. He even offered a speculative suggestion on how to change few-seeded wheat into the more valuable many-seeded wheat. Altogether, Theophrastos wrote about nearly 500 kinds of plants, and he called them by the names then in common usage. The generic names *Crataegus, Daucus, Asparagus,* and *Narcissus* are among those used today in the same sense that Theophrastos used them.

It is further noteworthy that while associated with Aristotle at the Lyceum, Theophrastos had occasion to become acquainted with Alexander, the son of Philip of Macedon. Alexander, who is remembered as Alexander the Great, never forgot Theophrastos, and during the conquests either he or his lieutenants arranged for materials to be sent back to Athens, thereby allowing Theophrastos to write about such plants as cotton, cinnamon, pepper, and bananas.

The student of plant taxonomy will be well rewarded to read some of the writings of Theophrastos. They give an insight into what was being done at the Lyceum in Athens (equivalent to the university) during one of the truly golden ages of learning, and they provide an idea as to what was common knowledge. The modern reader is inevitably struck by the breadth of botanical knowledge that was current in antiquity.

b. Caius Plinius Secundus (A.D. 23–79)

Pliny the Elder, as he has come to be known, deserves a place in the history of plant taxonomy not because of any elemental contribution on his part, but because he authored one of the most influential compendiums of information ever written. Pliny was, by any standard, a gentleman and a scholar. He served in the Roman army, was a lawyer, a student of history and a friend of the Emperor Vespasian. He died during the eruption of Vesuvius in A.D. 79; his death was attributed by some to being overcome by the heat and sulfurous smoke, while other historians note that he died of heart failure. Pliny wrote voluminously, but his chief work was his *Natural History,* a multivolume magnum opus, of which 37 volumes survive. In it, Pliny tried to record everything that was known about the world, and about a quarter of the work dealt with biological subjects. The botany was mostly medical or agricultural in nature. Some of his writing was in error, and we may in all charity say that Pliny was gullible and inclined to believe (or at least record) fanciful tales from travelers. However, for more than a thousand years Pliny's *Natural History* was held in near-reverential awe by the peoples of western Europe. It was no accident that this book was among the first to be printed by movable type in the late fifteenth century. A bit of trivia is that Pliny was apparently the first to use the word stamen in its modern sense.

c. Pedanios Dioscorides (A.D. First Century)

Presumably, Dioscorides was a contemporary of Pliny, but it is not known exactly when he lived (see Gunther, 1959). He was of Greek parentage, a native of the Roman province, Cilicia, and a physician in the Roman army. During the course of his professional labors, he compiled a memorable work, *Materia Medica,* which presented an account of various plants and natural materials useful in medicine. The significance of his work is attested by ancient writers, but Dioscorides's fame and influence persists to the present, and the accolades still passed in his direction are justly deserved.

Materia Medica of Dioscorides was derived in part from earlier writers, notably one remembered as Krateuas, but also in large part from firsthand observation as he practiced medicine. It discussed with clarity and a degree of precision about 600 species of plants. The extant versions are divided into chapters in a rather helter-skelter way, but there appears to be at least a superficial degree of natural relationships, for example, the mints (Lamiaceae) and the umbels (Apiaceae) are more or less grouped together. Furthermore, the work was written in a straightforward style, making it possible for one not versed in the intricacies of literary Greek to use the book. The plant names are given in Greek, and in some manuscripts they have been added in Latin, Turkish, Arabic, and Hebrew.

It does us well to try to call to mind a world in which few people could read. The illiterate masses were by no means stupid and oftentimes not ignorant; they simply did not know the art of reading. Into this milieu, the *Materia Medica* came as a most useful document. It was readily understandable, and it dealt with a subject of perennial concern, namely, human medicine. As a result, the possession of a copy of *Materia Medica* was a virtual guarantee of success and fortune for a clever literate person. The owner could engage in the practice of pharmacy or medicine because he possessed a highly esteemed source book on the subject. The reason that this is important to our story is that shortly after A.D. 500, the Byzantine Emperor Flavius Anicius Olybrius had a beautiful, illustrated copy of Dioscorides's book prepared as a gift for his daughter, the Princess Juliana Anicia. (This would be regarded as one of the nicest presents that a man could give his daughter.) During the ensuing years the copy prepared for Juliana remained in Constantinople (Istanbul), but the political and military skullduggery between the Austrians and the Ottoman Turks eventually led to the transfer of the manuscript to Vienna, where it remains to this day. The manuscript is known to antiquarians as the Codex Juliana or as Codex Vindobonensis, and it has been studied by scholars in perhaps as great detail as any book on natural history ever written. It is a great link between us and the intellectual heritage of our ancestors. There are other ancient manuscripts of Dioscorides, but the Codex Juliana is the most detailed and the oldest, and it is available in a splendid facsimile edition, as well as in translation. To a person who has learned the rudiments of Greek, it is a pleasure to look through the pages of a facsimile of the Codex Juliana and to read the plant names: *Anemone, Dipsacon, Aloe, Aristolochia,* and *Phaseolus*. Most of the plants in Dioscorides can be recognized with certainty, but a few entities cannot be identified.

d.   Other Ancient Authors

It is true that modern science had its roots in the ancient cultures of the Mediterranean basin, and plant taxonomy is no exception. However, other ancient peoples dealt with plants in their literatures, and while of secondary importance to us, their works are nonetheless interesting. Virtually every literate society has produced works on medicine and agriculture, subjects close to botany.

The literature of China is poorly known in the West, but reputedly it is rich in botanical works, especially those with a medical relevance. One of the earliest of these is attributed to the legendary Emperor Shen-nong, who is reputed to have lived ca. 3600 B.C.

Another botanical book is called *Cheng lei pen ts'ao,* and it was printed from movable type in A.D. 1108.

Ancient India produced botanical works of a medical sort, beginning before the Christian era of the West. A notable book, titled *Vrikshauyrveda* and attributed to a writer named Parasara, is in reality a general botany textbook, with discussions of morphology, classification, and plant distribution. The presentations of plant morphology are seemingly detailed, suggesting to the modern reader that the author had some kind of hand lens or microscope.

Meso-American botany is represented in the Badianus manuscript, which is a medical-botany book written in 1552 by two Aztec Indians in Mexico. The text is in Latin, and there are many stylized drawings. The treatment of the subject is quite simple in a botanical sense, but it is a fascinating historical document.

III. Medieval Botany

The thousand or so years from the decline of Rome to the Renaissance is a period when little academic accomplishment was made in Europe; however, there were a few notables in botany. Wahlafrid Strabo, a ninth-century monk, composed a poem called *De Cultura Hortorum;* it is a short work noting the herbs in the monastery garden, and recording the medicinal uses for them. The quality of the botany is not in itself special, but the poem is a pleasant reminder of a man recording something that he felt worth noting, and at a time when it was not easy to do so.

The best remembered of the naturalists of the Middle Ages is Albertus Magnus (ca. A.D. 1200–1280) (see Core, 1955). He was called "Doctor Universalis" by his contemporaries, and has been termed the "Aristotle of the Middle Ages" by some historians. St. Albert (he has been elevated to the sainthood by the Roman Catholic Church) wrote voluminously on many subjects, and his botany was but a minor contribution. He did not alter the course of botanical science, and his actual influence is not readily apparent. However, in an era of general ignorance, he was at least as sophisticated as Theophrastos. St. Albert employed a classification scheme that recognized monocots and dicots, and generally separated the vascular from the nonvascular plants.

During the somnolence of the West, the Muslim world was enjoying a cultural zenith. Muhammad and his followers had converted the populations from India to Morocco and Spain to the faith of Islam, and groups of brilliant scholars developed in many cultural centers throughout the realm. The Muslims were practical people, and they treasured their high degree of literacy. They were impressed by Greek learning, but were not quite prepared to pursue the same paths that the Greeks had trod. The Muslim scholars translated and generally curated the speculative and theoretical works of the ancient Greeks, and in large measure prevented the loss of these works during the medieval period in Europe. The Muslims made some truly great contributions to natural science, particularly in medicine and agriculture, but they did rather little in botany. We should tip the hat in passing to Ibn-Sina, or Avicenna as he is known in the West, for he authored the *Canon of Medicine,* which, like the *Materia Medica* of Dioscorides, was and remains one of the great scientific classics of all time. One Muslim scholar who dealt at length with botany was Ibn-al-'Awwam, a twelfth-century native of Spain, who was concerned with agriculture and discussed some 600 kinds of plants. He accurately interpreted sexuality in plants, and he had a grasp of the role of insects in the pollination of figs.

IV. The Renaissance and Its Progeny

The aforementioned ancient and medieval writers have two things in common: each writer felt that what he was writing was in the realm of common knowledge or received wisdom, and he was writing to transmit information. With the Renaissance, the cultural and intellectual fabric of the West was altered. People sought originality in the arts, politics, the sciences, and in all manners of personal expression. It is a matter of fact, although the causal interrelationships are not clear, that two momentous technical innovations occurred at the time the Rennaissance was having its greatest impact. These were the invention of printing and the development of the science of navigation. The impact was far-reaching in the history of plant taxonomy.

Printing made books cheap and literacy common. A literate person could consult the newly printed works of the ancients without requiring the services of a literate scholar, and anyone who so wished could compose a treatise and have a reasonable hope of seeing it printed and distributed. The result was that medically oriented books on plants became quite popular during the early days of printing, and the problems associated with the ancient texts prompted many people to write and publish their own botanical-medical books, or ''herbals,'' as such books came to be called. The popularity of these herbals enabled their writers to propose putative relationships among the various kinds of plants and to experiment with classification and nomenclatural schemes.

Navigation made it possible for sailors to start a long voyage with the reasonable expectation that they would return from it. This expectation, naturally, led to the successful exploitation of the New World by the nations of western Europe. And, in turn, this led to a great increase in the number of plants known to Europeans and the practical need to expand the classification systems employed in the herbals.

a. The Herbalists

Shortly after the advent of printing, a series of herbals were published under the name *Gart der Gesundheit,* or *Hortus Sanitatis* (see Arber, 1938). These books were often compilations from local sources and the writings of the ancients, usually with crude illustrations, and they were normally published without attribution of authorship. Apparently, they were popular with the small print shops of the day, for many editions published in numerous cities have survived. In general, the quality of the botany in the *Gart der Gesundheit* was low, and this was largely the result of two problems. One was a continuing veneration for the ancients and the reluctance of people to challenge that which was written. The other problem was that the ancient authors were residents of the Mediterranean region, where the plants were different from those at hand in northern and central Europe. The *Gart der Gesundheit* would certainly have been confusing to the poor medical practitioners of the day who tried to use these books to identify plants of medical utility in the garden or the field. Clearly something better was needed.

The sixteenth century saw the development of an intellectual ferment in Europe that some historians call the Age of Adventure. This was the time of the great herbalists, who were the first modern men to perceive clearly that the botanical writings of the ancients were inadequate and who did something about it. Among the best known and most influential of these herbalists were three Germans: Otto Brunfels (1464–1534), Jerome Bock (1489–1554), who Latinized his name as Hieronymus Tragus, and Leonhart Fuchs (1501–1566). All three were men of letters and scholarship, and Brunfels and Fuchs were practicing physicians. The three have been called the "German Fathers of Botany."

The herbal of Otto Brunfels was called *Herbarum Vivae Eicones,* and it contained many excellent illustrations. Jerome Bock's great herbal was his *Neu Kreuterbuch,* which contained no illustrations but had a wealth of technically detailed, accurate descriptions, based on firsthand observation. Fuchs' herbal, *De Historia Stirpium,* had both good illustrations and descriptions.

What separates these three writers from their predecessors is that they were admittedly creating new insights and schemes for viewing the plant kingdom. They admired the erudition of the ancients, but were not appealing to the texts of antiquity for authority, which is not so simple as it sounds, for well into the seventeenth century, medical botanists were still claiming that their works were based ultimately on Dioscorides.

The herbalists of the sixteenth century were motivated largely by practical considerations, that is, medicinal and agricultural uses of the plants. Little emphasis was placed on the system of classification. The number of plants that these herbalists dealt with was not particularly great (fewer than a thousand) so it was entirely reasonable for a person interested in botany to learn all of the plants by sight.

b.   The Seventeenth Century

By the 1600s, the great numbers of plants arriving in Europe from the New World, Asia, and Africa could not be ignored by the herbalists, and it was becoming obvious that some formalized scheme was required to keep them all straight in the minds of the botanists. The Italian scientist, Andrea Caesalpino (1519–1603), struggled with the notion of a rational classification scheme based on the intrinsic nature of the plants, rather than on purely utilitarian concepts. He was an Aristotelian in his approach, and he assumed that certain features of plants were, by their very nature, more meaningful than other features. Such an approach is termed a priori reasoning, and while it could lead to untenable conclusions, it did provide for the development of hypotheses that could be tested, and therefore Caesalpino was being scientific in his approach.

Another Aristotelian legacy that entered the history of systematics at about this time is the typological concept. In basic, and not quite precise, terms it may be regarded as the assumption that every entity is represented by an ultimate or ideal form, and that there is a type embodiment in the mind of mankind that reflects this ultimate ideal. In effect, this mind set predisposed taxonomists of the period to accept a certain fixity of species, and to regard species or kinds of things as discrete, nonvarying entities. Subsequent experience has shown this approach to be false; however, Aristotelian typology has had an influence in taxonomy that is with us to the present day.

Gaspar Bauhin (1560–1624) (see Core, 1955), a Swiss botanist, compiled a monumental opus in his *Pinax Theatri Botanici,* which was a register of the different kinds of plants known to science up to that time. More important, it contained an account of what names the various authors had used for each plant. In other words, Bauhin accounted for the rapidly developing synonomy in systematic botany.

Gaspar Bauhin's works are interesting for another reason, as he appears to have distinguished clearly between the concept of genus and that of species. To a more sophisticated age, that distinction seems obvious, but it was of signal importance, for Bauhin at least tacitly recognized the distinction of different levels (sets) of hierarchies in taxonomic grouping.

A further point is that Bauhin made some use of binomial nomenclature in the *Pinax*. It is not used consistently, but with enough regularity so we know that he was experimenting with it as a nomenclatural scheme.

Another notable contributor to the growth of systematics was the Englishman, John Ray (1627–1705). He published numerous works, but the most significant for us are his *Methodus plantarum Nova* (1682) and the *Historia Plantarum* (3 volumes, 1686–1704). In these works Ray produced a classification scheme in which plants that looked alike were grouped together, and he recognized the fact that individual characters in themselves may or may not be worthwhile taxonomic features. He was ahead of his time, but was groping at what has come to be called the natural system of taxonomy.

In 1700, Joseph Pitton de Tournefort (1656–1708) published his *Institutiones Rei Herbariae,* which has left a mark on taxonomy. The *Institutiones* was a less sophisticated work than Ray's *Historia,* but it was aimed at the very practical matter of putting a name on a plant-in-hand. It arranged nearly 9,000 kinds of plants in some 700 genera, which in turn were grouped into classes. The nomenclatural scheme was imperfect and inconsistent, but it was an important organizational step in reducing the botanical chaos of the era.

By the end of the seventeenth century, systematic botany had arrived at the point where a great organizational overhaul was needed. Through the seventeenth century, plants usually were named by short Latin phrases or sometimes by a single word, but there was little or no internal scheme for using the names to imply relationships among different kinds of plants. Understanding of relationships depended largely on experience, and it was nearly impossible for one not versed in the intricacies of botany to identify a plant-in-hand. A degree of clarity was to come with the advent of a referable system of nomenclature in the eighteenth century.

c. The Linnaean Period

In the history of botany, the eighteenth century clearly belongs to Carl Linnaeus (1707–1778), the scientist who effectively created the modern system of nomenclature (see Stafleu, 1971). The influence of Linnaeus is so great that he is commemorated in scientific societies (The Linnaean Society of London), botanical journals (Linnaea), and formal taxonomy (a linnaeon is a broadly based concept of species). Linnaeus was the son of a Swedish country parson; he was educated at the universities of Lund and Uppsala, and eventually obtained the M.D. degree of Harderwijk in the Netherlands. After a brief turn at practicing medicine, he returned to Uppsala, where he spent the rest of his life as a professor of natural history. He was a prodigious and dedicated worker, and his life and accomplishments have been the subject of many biographical and bibliographical studies.

Linnaeus's work is the culmination of that of the botanists who were striving to create a workable system for the classification of plants. Linnaeus divided the plant kingdom on the basis of traditional a priori reasoning; he assumed that reproductive features were intrinsically more important than other characters for taxonomic purposes.

In the case of flowering plants, the prime divisions were based on the number of stamens that a flower possesses. This division produced incongruities that were apparent to everyone, including Linnaeus (e.g., *Cactus* is classified adjacent to *Prunus* because of a similar stamen number and placement). But the scheme had one crowning achievement: it was usable by anyone who had the bare rudiments of botanical training, and it was completely referable. Any user of the scheme would repeatedly arrive at the same name for the same plant, without the necessity of asking a knowledgeable authority. In short, Linnaeus provided a solution to the age-old problem in systematics, that is, how to have a precise, referable, and expandable system of classification.

A key ingredient in the utility of Linnaeus's system is the consistent use of binomial nomenclature, wherein each entity receives two names, a genus name and a species name, and the correlated requirement that, when a person names a plant, he states both of these names. The Linnaean nomenclatural scheme was adopted almost immediately by the world's taxonomists, and it has remained essentially intact to the present.

Linnaeus's systematic contributions did not arrive all at once. They were proposed in several publications over the years, but the place where he assembled his findings, so far as plant taxonomy is concerned, was his *Species Plantarum,* a two-volume catalog of the plant kingdom, published in May 1753. This book marks the first consistent use of binomial nomenclature, and it has subsequently been adopted by botanists as the starting point for modern botanical nomenclature. (Names published before May 1, 1753, have no significance in nomenclatural priority.)

Linnaeus and his contemporaries were concerned with the mechanics of classification and were not prime contributors to the understanding of the biological facts of life. They operated under an assumption that species were more-or-less fixed entities of finite number. They clearly grasped the fact that some plants are more closely related to each other than to other plants, but this seems not to have been too weighty on their minds. However, one cannot feel superior to the Linnaeans, for they brought a large degree of order and stability out of nomenclatural confusion. For this they deserve our recollection and respect.

d. The Natural System

By the late 1700s a sufficient body of information had accumulated to cause botanists to consider the ultimate purposes of their taxonomy (see Garnsey & Balfour, 1906). A readily referable cataloguing scheme had been provided by Linnaeus, but it was admittedly no more than a cataloguing scheme. There was scant information content to the system, and plants that apparently were related were not necessarily classified together. France at this time was undergoing an intellectual ferment with sweeping consequences in natural science, philosophy, and eventually in revolutionary politics, and in the midst of this upheaval a new philosophical basis for taxonomy took root.

Michel Adanson (1727–1806) was unimpressed by the a priori arbitrary choice of characters used for making taxonomic decisions, and he created taxonomic schemes for several groups of organisms (both plants and animals) based on the equal use of as many measureable features as possible. His taxonomic schemes have been forgotten, but he is remembered for clearly working with the assumption that no single character is intrinsically more important than any other character. The important word is intrinsically, for subsequent experience has shown that indeed some characters are more useful than others, but such conclusions are to be based on empirical data and not on a priori reasoning. Historians of science note that Adanson foreshadowed what is now called phenetic taxonomy.

The de Jussieu family of France had four members who made noteworthy contributions to botany: Antoine, Bernard, Joseph, and Antoine-Laurent. In one way or another, all were connected with the Jardin des Plantes in Paris, but it is Antoine-Laurent de Jussieu (1748–1836) who attracts our special attention. Sometime before the French Revolution, it was decided to arrange the plants growing in the garden according to the scheme of Linnaeus. The plan was never completed because continual adjustment was made so that plants apparently related to each other would be planted adjacent to each other. The eventual product was a gardenful of plants arranged so that those that seemed to be alike were growing together. In 1789, the year of the revolution, Antoine-Laurent de Jussieu published his *Genera Plantarum Secundum Ordines Naturales Disposita,* which presented a classification scheme derived in some measure from experience gained in arranging the garden. This is the first major work to be intentionally natural in its approach, that is, plants that look alike, based on a suite of characters, are grouped together.

There was no hard-and-fast scientific principle or philosophical dictum supporting the idea of a natural system—that would come a century later with the elaboration of the theory of organic evolution. The natural system rested largely on common sense, and we note, as a matter of historical record, that the botanists of the time readily accepted the notion of a natural system and regarded it as a vast improvement over the fundamentally artificial system of Linnaeus.

A contemporary of Antoine-Laurent de Jussieu was J. B. P. de Lamarck (1744–1829), who recognized the philosophical utility of a theory of organic evolution, but who so badly muddled the explanation of it that the theory was regarded as utter nonsense by most members of the scientific community until the middle of the next century.

Another great botanical family, like the de Jussieus, was the French-Swiss family named de Candolle. The father, Augustin Pyramus de Candolle (1778–1841), published a splendid little book in 1813 called *Theorie elementaire,* in which the basic principles of botany are presented. It was a most influential general botany textbook. A. P. de Candolle offered a new classification scheme, but philosophically he was following A. L. de Jussieu with a natural system aimed at putting like plants together. The truly monumental effort of A. P. de Candolle was the *Prodromus Systematis Naturalis Regni Vegetabilis,* which was an attempt to write an account of all of the higher plants in the world. A. P. de Candolle wrote it from 1816 until 1841; his son Alphonse (1806–1893) worked on the project until 1873, and after that time the project was scaled down to a series of monographs, published by Alphonse and eventually by his son Anne Casimir de Candolle (1836–1918). The *Prodromus* was never completed, but for some groups of plants it remains the only broad treatment available.

By middle years of the nineteenth century, the idea of a natural system had become thoroughly entrenched in the minds of botanists, and the last and best elaborated of the preevolutionary natural systems was conceived at this time by George Bentham (1800–1884) and Sir Joseph Dalton Hooker (1817–1911), the director of Kew Gardens (see Green, J. R., 1909). Their work, titled *Genera Plantarum,* was published at intervals between 1862 and 1883, and it is a still useful compendium of generic descriptions arranged in a natural scheme ultimately derived from those of A. P. de Candolle and A. L. de Jussieu. The generic descriptions were models of completeness and precision, and, for the most part, were based on observations of materials in the great herbaria of Britain, with a minimum of data taken from the literature.

V. The Theory of Evolution

Charles Darwin published his monumental book, *On the Origin of Species,* in 1859, and it made the concept of organic evolution intellectually respectable. The development of the concept of evolution is interesting as a convoluted tale of people groping at a persistent idea, but that is tangential to our story here. What is important is that the Darwin-Wallace theory of evolution, as modernized by the incorporation of genetic theory, has basically stood the test of time and has become virtual scientific dogma. Certainly the details are subject to revision, but the conceptual schemes of modern systematics rest firmly on the base of evolutionary theory. Biologists by and large accept the ideas that the nature of life has changed continually over the geologic ages, and that life on earth consists of populations of organisms that are in various states of flux (see Mayr, 1982).

a. The Engler-Prantl Approach

A group of German botanists were among the first to struggle with the concept of evolution regarding the development of a classification scheme, although they did not intentionally begin their efforts with evolutionary matters in mind (see Eiseley, 1958). S. Endlicher (1805–1849) and later A. W. Eichler (1839–1887) proposed taxonomic schemes that were later amplified and revised by A. Engler (1844–1930) and K. Prantl (1849–1893). The latter two produced a major reference work in their *Die Natürlichen Pflanzenfamilien*, published in many volumes between 1887 and 1915. The original classification scheme was natural in that it grouped together those plants that the authors believed to be related, and by the time the scheme was published completely in the *Pflanzenfamilien* (1915), it had acquired a definite progressive character. It started with the simplest plants, and worked through the plant kingdom to those that were the most complex structurally. With regard to the dicotyledons, it started with the Salicaceae (the willow family), with *Salix,* the willows, and *Populus,* the cottonwoods, because they possess flowers of the simplest structure, and it ends with the Asteraceae (or Compositae, the sunflower family) because of their great floral complexity. Everyone agreed that it was a fine natural system, and it was documented beautifully in a multivolume description of the plant kingdom. It, therefore, became the standard classification scheme for most botanical institutions and publications, and until the 1980s, most floras (i.e., books treating the flora of a particular region) had their families arranged in the sequence employed by Engler and Prantl in the *Pflanzenfamilien*.

As the theory of evolution became more pervasive in plant taxonomy, an effort was made to show that the progressive sequence of the Engler and Prantl scheme reflected the evolutionary development of the plant kingdom. It was thought by some that simplicity of structure could be equated with primitiveness; however, the studies of paleobotanists and comparative morphologists showed this notion to be untenable.

b. Bessey and His Followers

The ideas of C. E. Bessey (1845–1915), a professor at the University of Nebraska, form the groundwork for the elaborate evolutionary classifications of contemporary systematics. Bessey's goal was to arrange the groups of flowering plants in a classification scheme that reflected their evolutionary relationships. To do this, he derived a set of dicta or propositions based on empirical evidence as to what features might be considered primitive and what might be advanced. Primitive, in this case, refers to being present in the most ancient plants; advanced refers to a feature more recently derived. Primitive-versus-advanced is not the same as simple-versus-complex, for a structural feature may become advanced either by reduction toward simplicity or by elaboration toward complexity. Bessey carefully noted the reasoning behind each of his dicta, and then proceeded to generate a classification system based on the principles of evolution. The notable part is that C. E. Bessey and a German contemporary, Hans Hallier (1868–1938), produced a classification scheme for the express purpose of reflecting evolution; such schemes are said to be phylogenetic.

Since Bessey's time, taxonomic studies have produced numerous sophistications and improvements, especially in our understanding of the nature of primitive land plants and the origin of flowering plants. Likewise, detailed studies have greatly amplified our understanding of relationships at the species level in many groups. However, most botanists agree with Bessey that a principal goal of a classification scheme is to reflect evolution and thereby give a sound theoretical basis to what is meant by a natural system.

Several notable systems have been produced by contemporary botanists, including A. Takhtajan (1910–   ) of Leningrad, A. Cronquist (1919–*1991*) of New York, and R. Thorne (1920–   ) of Claremont, California. Cronquist's system is the best developed, and it has been selected for use in the family circumscriptions and family sequence in several recent floristic projects in the United States and elsewhere.

An important point should be noted regarding the use of plant names: the actual name applied to a plant remains the same, regardless of what classification scheme is employed. For example, the plant known as *Helianthus annuus* (the common weedy sunflower) is called by that name in any classification system, and this fact is the result of the Linnaean scheme of nomenclature. In modern natural and phylogenetic systems, the identification of a plant carries with it a knowledge of its natural affinities and evolutionary relationships. Therefore, the modern systems are richer in information content than are the simple, artificial systems of the past.

B. THE KNOWLEDGE OF THE WORLD'S FLORA

I. Introduction

Ancient mankind was obviously familiar with the local flora, but it was widespread commercial enterprise, fostered by the development of navigation, that brought European society into contact with large numbers of previously unfamiliar plants. As a simple exercise in appreciating the impact of the early navigators, one should try to imagine what was on the dining tables of Europe prior to the exploration of the Americas. Just for starters, there would have been no corn, potatoes, or tomatoes. The intriguing story, however, is not restricted to the botanical information gained as incidental knowledge to business ventures, but it includes the story of pure exploration, where people traveled far and wide, just to find out what was there. The following discussion is necessarily abbreviated and restricted to temperate North America, but the reader should realize that other parts of the world also have had their share of colorful explorers.

A roll call of botanists who made significant contributions to the science in North America is long, and here we can discuss only a few of them.

a. The Exploratory Period

John Clayton (1685–1773), an English lawyer who lived in Virginia from 1705 until his death, was the first person to do any great amount of botanical work in what is now the United States (see Core, 1955). As a botanist by avocation, he collected plants, wrote descriptions, and corresponded about botany with J. Gronovius (1690–1762), a professor at Leyden in the Netherlands. Clayton's materials were the basis for the book titled *Flora Virginica,* published by Gronovius in 1739. Linnaeus later used the *Flora Virginica* in the preparation of his *Species Plantarum* (1753).

Two contemporaries of John Clayton were Mark Catesby (1680–1749) and Peter Kalm (1715–1779). Catesby spent some ten years in Virginia and the Carolinas, and he eventually published a sumptuous *Natural History of Carolina, Florida and the Bahama Islands* (1731–1743). Kalm was a student of Linnaeus, and he traveled and collected in eastern North America between 1748 and 1751.

John Bartram (1701–1777) and his son, William Bartram (1739–1823), were among the earliest natives of North America to be regarded as intellectual equals by the naturalists of Europe. The Bartrams were farmers near Philadelphia, but were scholars by taste and avocation. John Bartram was a friend of the notables of the region, including Benjamin Franklin, and he was a charter member in the American Philosophical Society. The Bartrams traveled and collected widely in eastern North America, and William published his *Travels Through North and South Carolina, Georgia, Florida,* in 1791. In an era when European society was fortified by the philosophy of J. J. Rousseau, the sense of the rustic and the back-to-nature aspect of William Bartram's book made it very popular among the intellectual classes. The Bartrams had an active correspondence with Europeans plus a large trade in seeds and other plant materials, and for years they served as the effective intermediaries between European and North American students of natural history.

Another father-son team came from the Michaux family. André Michaux (1746–1802), the father, was sent to the New World by the government of France in 1785, and eventually he came to reside in South Carolina. He published his *Flora Boreali-Americana* in 1803. Francois-André Michaux (1770–1855), the son, wrote a three-volume work on the trees of North America that was published in French in 1810 to 1813 and in English as *Sylva of North America* in 1818 to 1819.

Frederick Pursh (1774–1829) came to the United States in 1799 from his native Germany, and for several years he was employed in New York City as the curator-gardener of the Elgin Botanic Garden, a private garden belonging to Dr. David Hosack, a prominent physician, botanist, and teacher. After some years of exploration and the acquisition of specimens, Pursh returned to Europe, where in England in 1814 he published his now classic *Flora Americae Septentrionale.* This work contained the descriptions of the plants collected on the Lewis and Clark expedition, as well as other new materials from the western part of the continent.

By the early years of the nineteenth century, the procedures of botanical nomenclature had been generally agreed on in custom, if not yet in actual codified rules. (Formal rules of nomenclature date from the second half of the nineteenth century.) Botanists were assured by scientific custom that the discoverer of a new species would receive credit for his discovery by having his own name appended to the scientific name if he named it himself, or oftentimes having a new species named after him if somebody else actually described the new entity. It seems evident, at least from our vantage point, that some large portion of the botanical activity of the middle-nineteenth century was inspired by the desire to discover new species. Two men's studies exemplify this difficult work: Constantine Rafinesque-Schmaltz (1773–1840) and Thomas Nuttall (1786–1859).

Rafinesque (the Schmaltz is dropped in routine reference to the man) was of mixed European ancestry, but was born in Constantinople. After some years of wandering and study, he was appointed to a professorship at Transylvania University in Lexington, Kentucky, where his unconventional habits and rather egotistical personality made him something of a local character and eventually saddled him with the probably well-deserved reputation as an eccentric. He published voluminously and in a wide assortment of periodicals and books; but specimens documenting his work have not always been available, making an objective evaluation of his contributions difficult. For many years he was remembered largely as a tragicomic figure in the history of American botany, but the late E. D. Merrill pursued the works of Rafinesque and provided a good basis for a degree of belated respect for this indefatigable but peculiar student of natural history.

Quite a different sort was Thomas Nuttall. He was an Englishman who spent much time in North America, collecting in the Mississippi-Missouri River drainage systems and studying matters of natural history at Harvard and later at the Academy of Natural Sciences in Philadelphia. With admirable precision, he described many new species and systematized his understanding of them. His collections were properly prepared and are available in several herbaria, both in North America and in Europe. Unlike Rafinesque, there is no question about the utility of the work of Nuttall. The most significant single work by Thomas Nuttall was his *Genera of North American Plants* (1818), in which the genera are described, species listed, and new species are treated.

b.   The Consolidation Period

The first third of the nineteenth century saw the rise of scientific establishments in the United States, and with them a separation of botanists into two groups: those who were chiefly field workers and collectors, and those who worked in laboratories in the scientific establishments. A most productive relationship began at this time with the cooperative efforts of John Torrey (1796–1875) and Asa Gray (1810–1888), who exemplified the period's scholar-botanists (see Dupree, 1959). Both men were trained as physicians, and both became academic scientists, Dr. Torrey at Columbia University and other institutions in the New York area, and Dr. Gray at Harvard. The events relevant to their cooperation are beyond the scope of this chapter, but a general outline follows: Torrey and Gray decided to collaborate on a flora for the whole of North America. At about the same time, there were numerous collectors exploring and preparing specimens in the vast areas beyond the Appalachian uplift, and many (or indeed, most) of the specimens came to John Torrey or Asa Gray. The result was *A Flora of North America,* by Torrey and Gray, published in several sections between 1838 and 1843. The work established its authors as the chief authorities on the flora of the continent, and this, in turn, attracted more specimens to their respective herbaria.

In the middle decades of the nineteenth century, numerous expeditions were mounted with the support of the federal government and charged with surveying the Mexican boundary, searching for a possible route for a railroad to the Pacific coast, or with other equally broad-based schemes. Plant collectors were a part of these expeditions, and the specimens mostly went to Torrey or Gray for description and study, or in later years just to Asa Gray. Special groups were sent to experts for consideration, for example, the Cactaceae went to Dr. G. Engelmann (1809–1884) of St. Louis and the ferns to D. C. Eaton (1834–1895) of Yale. The reports of these expeditions still make fascinating reading, for in addition to the botany, they contain narratives of the travel, descriptions of the geography, and a glimpse of what western North America was like prior to the advent of rapid transportation and permanent settlement.

Late in his career, Asa Gray began the preparation of his detailed *Synoptical Flora of North America,* a project unfinished at the time of his death. The sections that were completed remain useful to botanists engaged in floristic or monographic studies. Another project of Asa Gray has carried his name into nearly every botanical laboratory in the east half of the continent, and that was the publication (in 1848) of his *Manual of Botany*. The *Manual* has been revised by subsequent botanists at the Gray Herbarium of Harvard University; the most recent is the eighth edition by M. L. Fernald, published in 1950. Throughout the botanical community it is known simply as Gray's manual, and in its present form it consists of keys, descriptions, and statements of range for the vascular plants of the northeastern quarter of the United States and adjacent southern Canada.

The preeminence of Asa Gray in the history of botany in North America is beyond dispute. He is the exemplary embodiment of the right man in the right place at the right time, for he was both academically prepared and philosophically suited to take complete advantage of the opportunities that came along.

A lesser-known but hard-working contemporary of Asa Gray was Alphonso Wood (1810–1881), a man of honest scholarship and experience who published an immensely popular *Classbook of Botany*. It became the elementary textbook for a generation of young American botanists.

c. The Splitters of the West

No survey of the history of systematic botany in North America should overlook Edward Lee Greene (1843–1915) and Per Axel Rydberg (1860–1931) (see McKelvey, 1955). These men are interesting because they were able and prolific field workers, writers of floras, and describers of great numbers of new species, and because they culminate a philosophical approach to the study of the flora of the continent.

E. L. Greene came into botany while a clergyman in the mining districts of New Mexico. He collected plants there, and sent them to Asa Gray at Harvard for identification, but because Professor Gray was occupied with many projects, he could not spend a great amount of time on Greene's plants. E. L. Greene was prompted to take up botany, and soon became versed in the principles and techniques of writing and publishing results of his own studies. He eventually became a professional botanist, serving at the University of California in Berkeley, the federal government in Washington, and the Catholic University of America in Washington. He edited and published two journals, *Pittonia* and *Leaflets of Botanical Observation and Criticism,* both of which were devoted almost exclusively to his own writings.

P. A. Rydberg came to the United States from Sweden to work as a mining engineer and eventually he found his way into botany through training at the University of Nebraska and Columbia University. This training led to a position at the New York Botanical Garden, where his chief concern was the flora of the Rocky Mountain uplift and the adjacent prairies and plains. He produced numerous publications on the flora of this vast region.

Greene and Rydberg each described great numbers of species that later botanists have regarded simply as synonyms for previously published species. The approach that they both used in their science was to study variation by observation in the field and herbarium, and if a plant seemed to be different, it was described as new. They were the quintessential splitters. From our vantage in the late-twentieth century, we may view these men with a degree of admiration and charity, for they were struggling with a large and complex flora at a level of detail quite impossible for their predecessors to attain. They did provide referable handles and usable floras for the plants of a huge region.

The flora of Canada was collected and studied extensively by a contemporary of Greene and Rydberg named John Macoun (1832–1920). He published his *Catalogue of Canadian Plants* in three parts between 1883 and 1886. His reputation rests on his voluminous collections and his critical compilations of floristic lists (i.e., his *Catalogue*); he did not describe large numbers of new species.

Needless to note, this brief survey of the development of our knowledge of the flora of North America has omitted many people and many contributions. Some of the excitement that the nineteenth century collectors had may still be generated in the American tropics, where plant exploration is far from complete.

C. BOTANICAL GARDENS AND HERBARIA

Among the earliest botanical gardens were the "simples" gardens of the medieval monasteries, where herbs and drug plants were grown, largely for medicinal purposes. The modern botanical institution was a product of the late Renaissance, when some of these simples gardens were used in the training of physicians and thus acquired an identity with a university, complete with a professor as a director. The oldest of these academic gardens that is still extant is at Padua, Italy, with a history dating from 1533. Nearly as old is the garden at Pisa, dating from 1543. These and other university gardens eventually justified their existence by disseminating knowledge about plants to the public, and, therefore, they are the precursors of the modern botanical gardens and herbaria.

By the middle 1700s, public gardens were popular, and nearly every capital city and every university in Europe had some sort of botanical facility. As Asia, Africa, and the New World were explored, these institutions grew rapidly, and acquired large herbaria (collections of dried plant materials systematically filed for easy reference). Naturally, the herbaria in countries with large colonial empires grew the fastest and came to possess the broadest collections. The result has been the development of several truly great botanical institutions, with large libraries, extensive herbaria, and collections of living plants.

The Royal Botanic Gardens at Kew, England, is usually regarded as the foremost botanical institution in the world. Its origin was in the 1600s but it became a modern research facility in 1841, with the appointment of William J. Hooker as the director. Since then it has actively sought specimens from all over the world, and now has a herbarium with more than 5 million specimens, and an extensive botanical library. The British Museum of Natural History likewise has excellent facilities for botanical research, and the fact that the two institutions are both in the area of greater London makes London something of a world center for plant taxonomists. Other great botanical institutions in Europe are the Paris Natural History Museum, the V. L. Komarov Institute in Leningrad, and the Botanical Garden and Museum in Berlin-Dahlem, which was damaged during World War II with the loss of specimens.

In North America, the most important early botanical institution was the Academy of Natural Sciences of Philadelphia (see Ewan, 1969). It was here that plant specimens usually were sent in the early years of the republic, and it was here that visiting European naturalists called when they arrived in the New World. Thomas Nuttall was an associate of the Academy, and his *Genera of North American Plants* was published in Philadelphia. The academy is active to the present day in matters of natural history, but it is no longer preeminent in botany. The herbarium contains more than 2 million specimens.

There are five other major botanical institutions in the United States.

1. The Gray Herbarium and Arnold Arboretum of Harvard University are two corporate entities that for practical purposes have combined their resources and function as a unified research facility. The impetus for founding the botanical tradition at Harvard came from Asa Gray in the middle 1800s, and the tradition of excellence has never waned. The combined herbaria contain well over 3 million specimens of vascular plants.
2. The New York Botanical Garden in New York City was founded in 1891, largely through the efforts of Dr. N. L. Britton of Columbia University. Among its holdings are the herbarium of John Torrey, as well as old collections of various departments of Columbia University. The herbarium contains more than 4 million specimens and is especially rich in materials from western North America and the American tropics.
3. The Botany Department of the Natural History Museum of the Smithsonian Institution, Washington, D.C., is the national herbarium for the United States and is supported by the federal government. The herbarium was begun in the middle 1800s, but its greatest growth and impetus for research came after the U.S. Department of Agriculture herbarium was transferred to the Smithsonian Institution in the 1890s. The herbarium now totals more than 4 million specimens.
4. The Missouri Botanical Garden, St. Louis, Missouri, was founded in 1859 by Henry Shaw, a well-to-do merchant of that city. It has been associated with Washington University in St. Louis, and for many years it has been highly regarded for training graduate students and for the breadth of its approach to research. The herbarium has nearly 3 million specimens.
5. The Field Museum of Natural History, Chicago, has some 2.5 million specimens and is particularly rich in materials from Central America.

Most major American universities maintain herbaria and the associated library and laboratory resources for research in systematic botany. Some of these university facilities are quite extensive, for example, both the University of California at Berkeley and the University of Michigan have herbaria with more than a million specimens, and several other university facilities are not far behind.

It should be obvious that the mere size of a herbarium is not in itself an accurate measure of the importance of the collection. Provenance of the specimens, quality of data, ease of accessibility, and other considerations are all of importance. However, in general, extensive botanical research involving floristic or monographic work requires the resources of a major herbarium. Therefore, taxonomists often borrow specimens from distant herbaria, and they regularly visit major research institutions to consult the herbaria and libraries. Thus, a tradition of cooperation has developed in the science, and every effort is made to make the resources for research available to all who are botanists.

D. CONTEMPORARY SYSTEMATIC BOTANY

By the early years of the present century, the regional floras of northern Europe and northeastern North America were sufficiently well known so that it was possible, at least in theory, to place a referable name on any specimen-in-hand, and thus some taxonomists, especially in northern Europe, turned their attentions toward intensive and experimental studies of the genetics, variation, and ecology of individual species.

The concept of species had been acknowledged as imprecise since the time of John Ray in the seventeenth century; however, a working definition, based on herbarium experience and the belief that a species was a group of plants that looked alike was at least tacitly held by botanists, and it was sufficiently useful to allow a reasonably accurate account of the flora to be compiled. However, several areas of research produced results suggesting that a more precise concept of species might be possible. The development of comparative cytology, led by O. Winge in Denmark, showed that chromosome numbers and morphology offered accessible and rather stable data for systematics. G. Turesson in Sweden began experimental studies to document variation within species, and his research started the field called genecology. Comparative cytology, genetics, and genecology were eventually melded into the field of experimental systematics, or biosystematics, wherein the goal is to explain rather precisely the boundaries of the species and the biological mechanisms responsible for the life style of the group under study. Thus, biosystematics, or experimental taxonomy, has been a prime source for data in interpreting speciation and evolution.

The history of zoology is, of course, a bit different from that of botany, in part because animal biology (or at least the biology of terrestrial vertebrates) is not so complicated as that of plants. Few terrestrial vertebrates have polyploidy, apomixis, or vegetative reproduction as normal parts of their life cycle. The zoologists found that a good concept of species could be based on the ability of organisms to interbreed and therefore form a pool of organisms of common descent. This notion, called the biological species concept, is admirable because it allows a certain precision in its definition, but it is obviously not entirely applicable to plants.

The desire for a more precise definition of botanical species has given rise to a wealth of literature, beginning in the 1930s. Among the seminal papers is one by W. H. Camp and C. Gilly (1943), wherein a dozen different kinds of plant species are described, based on the behavior of the plants in nature. No one has taken their kinds of species seriously for purposes of naming plants, but botanists have noted well that plant species are not all alike in how they function.

A most influential program of research was carried out by the Carnegie Institution of Washington at their plant research facility on the campus of Stanford University in California, by J. Clausen, D. D. Keck and W. Hiesey. Beginning in 1940, and for some years thereafter, they published a series of reports on their studies on the nature of plant species. Their research involved an elaborate program of transplants in a common research garden and at numerous stations along a transect from the California coast eastward to the Tioga Pass in the Sierras. The studies involved several species and yielded clear results showing that widespread species have regional variants (ecotypes) that are biologically different from the variants of other localities. Discussions summarizing the nature of plant species have been published by Stebbins (1950) and Grant (1971) among others, and have been presented in nearly every textbook on systematics.

With the passage of the pioneer period of exploration and description of the flora of North America, taxonomy subsequently developed two areas of research: floristic studies and monographic studies.

Floristic studies are concerned with all of the plants of a particular region, which may be a political entity such as a state, or a large, natural region such as the Great Plains. The goal of floristics is to account for all of the plants of the region with keys, descriptions, statements of range, habitats and phenology, and, in the more recent floristic studies, to offer analytical explanations of the origin and geohistorical development of the flora. The results are published as a flora, if sumptuous and in detail, or as a manual if abbreviated and portable. The distinction between a flora and a manual is not sharp, and the two terms are used almost interchangeably. While there are current regional floras for parts of the United States, there is no floristic treatment for the whole country, although several attempts have been made to produce a flora of the United States and for North America as a whole. The members of the American Society of Plant Taxonomists, in conjunction with their Canadian colleagues, in 1983, voted to encourage the preparation of a continental flora for North America north of Mexico, and as this is written, a project called Flora North America is being organized for that purpose.

Monographic studies are concerned with a natural group of plants, regardless of geographical occurrence, and often treat a genus, a section of a genus, or a complex of allied species. The goal of monographic studies it to provide keys, descriptions, correct nomenclatural alignments, as well as insights into the biology of the group and its evolution. Detailed monographs are sometimes years in preparation, and incorporate data from diverse sources. There is often an account of geohistorical events and their effects on the plants, and modern monographs often include studies of pollination behavior, seed dispersal, and physiological adaptations. Monographic studies that are less intensive are usually termed revisions, but as with floras and manuals, the distinction between a monograph and a revision is not sharp. Botanists generally agree that the experience gained in preparing a monograph is useful in training new taxonomists, and monographic studies are often done as doctoral dissertations. Monographs and revisions are a prime source of information for botanists engaged in floristic studies.

The tropical and subtropical regions of the earth have exceedingly rich floras, and botanical science there is still in the phase of exploration and description (sometimes termed alpha taxonomy) and it is likely to remain so for the near future. The botanical knowledge of these floras is still too limited to allow the compilation of sophisticated floristic works, but data are accumulating rapidly enough so that useful local floras are possible, enabling botanists to prepare basic revisions of complicated groups. The rapid destruction of the great tropical floras seems near imminent, so there is an urgency for tropical botany that transcends mere botanical curiosity.

SUGGESTED READING

Arber, A. 1938. Herbals, Their Origins and Evolution. The University Press, Cambridge, England. [The standard reference on herbals].

Camp, W. H. and C. Gilly. 1943. The structure and origin of species. Brittonia 4: 323–385.

Core, E. L. 1955. Plant Taxonomy. Prentice-Hall, Englewood Cliffs, N.J. [Has a quite readable account of the history of plant taxonomy].

Dupree, A. H. 1959. Asa Gray, 1810–1888. Harvard University Press, Cambridge, Massachusetts. [The standard biography of Gray, with good coverage of the development of botany in the United States].

Eiseley, L. 1958. Darwin's Century. Doubleday & Co., New York. Anchor Books (paperback), 1961. [An interesting account of the rise and impact of Darwinism].

Ewan, J. A. 1969. A Short History of Botany in the United States. Hafner Publishing Company, New York and London.

Garnsey, H. E. F. (translator) and I. B. Balfour (reviser). 1906. Sachs', 1530–1860. Clarendon Press, Oxford. [One of the basic references on the subject].

Grant, V. 1971. Plant Speciation. Columbia University Press, New York.

Green, J. R. 1909. A History of Botany, 1860–1900. Clarendon Press, Oxford. [A continuation of the work of J. von Sachs].

Greene, E. L. 1909. Landmarks of Botanical History. Smithsonian Miscellaneous Collections, Vol. 54., Washington. [A detailed discussion of ancient botany, particularly good on Theophrastos].

Gunther, R. L. 1959. The Greek Herbal of Dioscorides. Hafner Publishing Company, New York.

Mayr, E. 1982. The Growth of Biological Thought. Belknap Press of Harvard University Press, Cambridge, Massachusetts. [Includes a fine discussion of the development of the concept of species, as seen by an influential zoologist].

McKelvey, S. D. 1955. Botanical Exploration in the Trans-Mississippi West, 1850–1950. Arnold Arboretum of Harvard University. Jamaica Plains, Massachusetts.

Sachs, J. von. See Garnsey and Balfour and Green, J. R.

Sirks, M. J. and C. Zirkle. 1964. The Evolution of Biology. Ronald Press, New York. [One of the most readable books on the subject; excellent list of references].

Stafleu, F. A. 1971. Linnaeus and the Linnaeans. International Association for Plant Taxonomy, Utrecht. (Regnum vegetabile, Vol. 79).

Stebbins, G. L. 1950. Variation and Evolution in Plants. Columbia University Press, New York. [A classic and influential book].

Steere, W. C. (ed.). 1958. Fifty Years of Botany. Golden Jubilee Volume of the Botanical Society of America. McGraw-Hill Book Company, New York.

SUMMARY FOR A HISTORICAL STUDY OF TAXONOMY

Definition of History (Taxonomy) *History* is a branch of knowledge dealing with the transformation of systematic ideas, concepts, processes, principles, institutions, and activities through time.

Purposes for a Historical Study of a Taxonomic Subject To review the narrative, records, and events pertinent to the problem for perspective on the subject; to provide an intelligible documented background for the solution of the problem or proof of the hypothesis; or to produce a documented systematic treatment of the subject that includes the accomplishments, trends, and significant features related to the subject.

Operations in a Historical Study of a Taxonomic Subject To delimit and delineate the problem sharply; to review available relevant literature for perspective on the subject; to select material prudently from primary and secondary sources; to examine the material critically; to evaluate the material objectively on the basis of clearly established criteria; and to develop a carefully documented systematic narrative, record, or account pertinent to the objectives of the taxonomic study.

Basic Premise for the Historical Study of a Taxonomic Subject That in the very diverse treatments of systematic concepts, processes, principles, and systems conceptually discrete units usually called events can be recognized, classified, circumscribed, named, and logically evaluated for the interpretation of the past and present state of systematics or any component theoreof, and for the provision of a perspective on the future accomplishments and trends in the subject.

Fundamental Principles for a Historical Study of Taxonomic Subjects

1. Each document pertaining to the subject has intrinsic informational value.
2. Variation in the quality and quantity of each document pertaining to the subject necessitates the establishment of criteria for the determination of the value of each document.
3. New evidence from overlooked sources and new documents pertaining to the subject, the establishment of different criteria for the determination of values, and new methods for historical study result in changes of interpretation of significant values, trends, and perspectives of the subject of study.

Guiding Principles for the Historical Study of a Taxonomic Subject

1. Procure documents relevant to the study.
2. Document materials used in the study.
3. Give the bases for the selection and establishment of the criteria and the value system used in the study.
4. Indicate and document the relationships between the historical background and procedures used in the solution of the taxonomic problem or proof of the hypothesis for the study.

5. Document the reasons given for differences in significant values, trends, and perspectives presented in the study.

Basic Assumptions for the Historical Study of a Taxonomic Subject

1. Historical documents relevant to the study are available, useful, and usable.
2. That the study will have significance values, trends, and/or perspectives interpretable in relation to past, present, and future events related to the subject.

QUESTIONS ON THE HISTORY OF TAXONOMY

1. How can history be defined from a functional standpoint? A theoretical standpoint?
2. How does a study of the history of plant taxonomy give a student perspective in the subject?
3. Why should the history of plant taxonomy be studied?
4. What are the advantages and disadvantages of the study of history of taxonomy by periods? By nations and regions? By comparative chronologies? By major contributions? By taxonomic subdisciplines (nomenclature, identification, classification, characterization), methodologies, facilities (herbaria, gardens, libraries), field exploration?
5. How is the history of plant taxonomy related to, but distinctive from, the history of science? To the scientific disciplines on which it is dependent?
6. What are the relationships (ideas, philosophy, theories) of the history of plant taxonomy to ancient history? Greek-Roman history? Medieval history? The Renaissance? Modern history?
7. What are the criteria for the division of the history of plant taxonomy into periods?
8. Characterize the natural system period and Linnaean period. What were the major questions and ideas of each period? The major contributions? The major contributors? What made a contribution major?
9. What was the significance of the theory of evolution to the development of plant taxonomy?
10. What major changes occurred in plant taxonomy during the first fifty years of the twentieth century?
11. What are the basic problems of historical scholarship?
12. What are the material sources for the history of plant taxonomy? Where are most of these historical resources located today? Give specific examples.
13. What precautions have to be taken in the selection of materials for historical study?
14. Who are the major botanical historians today? Of the past?
15. Historically, why is the evolution of ideas, concepts, and principles of more significance than who did what and when?
16. Why is the historical introduction included in most scholarly or research publications?

EXERCISES IN THE HISTORY OF TAXONOMY

1. Trace the historical development of one of the following by periods, major contribution, and major event.* Document each contribution of entry with reference and date and indicate whether source is primary or secondary:

---

*Devise a comparison chart to show the major developments by subject, for example, *nomenclature*-binomial nomenclature, generic concept, family concept, international code, and principles of nomenclature, with adequate documentation within a time frame by contributor.

      a. Nomenclature   *g.* Terminology
      b. Description   *h.* Methodology
      c. Identification   *i.* Botanical libraries
      d. Classification   *j.* Botanic gardens
      e. Concepts   *k.* Herbaria
      f. Principles   *l.* Botanical data banks

2. Select one major figure in taxonomic history for study and try to explain how the individual received the impetus or stimulus for his major contribution, for example, C. E. Bessey and his system of classification.

3. Trace the history of floristic exploration of your area. Which individuals made major contributions to the understanding of the flora of your region? List the major contributions to the knowledge of the local flora and indicate why each contribution was considered major.

4. Describe the historical development of botany in your region. Indicate how the development of the knowledge of taxonomy parallels and differs from the development of botany in general in your area.

5. Determine the outstanding contributions to the development of systematics made by the botanists of your area. List the contributions and indicate why each is considered exemplary.

chapter *3*

Plant Nomenclature

Taxonomists as nomenclaturists name new taxa and determine the correct name for old taxa that have been remodeled, divided, united, transferred, or changed in rank according to the International Botanical Code; they also determine the correct name for a specimen according to an identification or classification system.

''How strange and chaotic life would become if it were possible to abandon the use of names for the identification of everything we see, make, or handle. The acquisition and dissemination of knowledge would become impossible, and the business of the world could not go on'' (Macself in Johnson, 1971).

Man has always been a nomenclaturist; he has given names to plants, animals, and objects, and has placed them in categories with or without employing special terminology and systems. ''Centuries ago each plant was known by a long, descriptive sentence, which was unwieldy, to say the least. Then Gaspar Bauhin (1560–1624) devised a plan of adopting two names only for each plant. But it was not until the great Swedish naturalist, Linnaeus (1707–1778), undertook the task of methodically naming and classifying the whole living world 'from buffaloes to buttercups' that the dual name system became permanently established. Linnaeus brought order out of chaos and indexed the vegetable world on a basis so sound and universally acceptable to the peoples of all nations that most of his names are in use to this day'' (Johnson, 1971).

Since the publication of *Species Plantarum* by Linnaeus in 1753, the forming of names in Latin for international use has been a fundamental task for botanists or taxonomists. The scientific or specific name of an organism is a binary combination of (1) the genus, or generic name, and (2) the specific epithet. *Quercus alba* L. is the scientific or specific name for white oak; *Picea rubens* Sarg. is the scientific or specific name for red spruce. *Quercus* and *Picea* are the respective generic names and *alba* and *rubens* are the specific epithets. The "L." after *Quercus alba* and "Sarg." after *Picea rubens* are abbreviations for Linnaeus and Sargent, the authors of the two species, respectively. No species name is accurate or complete unless followed by the full or abbreviated name(s) of the author(s). It is necessary to cite the name of the author so that the date of the first valid publication of the name of the taxon can be verified.

Scientific names are italicized in print or underlined when typed or handwritten or printed. The initial letter of the generic name is always capitalized, the remainder small. All letters in the specific epithet are lowercase except in a few special cases where initial capitals are permitted. However, many botanists today begin *all* specific epithets with lowercase letters.

The name of a genus is a noun and singular in number or a word treated as such. The specific epithet is usually an adjective, for example, in *Quercus alba* L., alba means white; but the epithet may be a noun in apposition—for example, in *Pyrus malus* L., malus is the name of the genus for apple—or a noun in the genitive singular commemorating or honoring a person—for example, in *Panicum ashei* Pearson, W. W. Ashe is commemorated by having a species named for him by Pearson.

Generic names and specific epithets may be taken from any source whatsoever but are always treated as Latin. Botanists are enjoined to refrain from forming names that are long and difficult to pronounce in Latin or to adapt to Latin, and to avoid making hybrid names and epithets by combining words from different languages. Latin terminations should be used as far as possible.

Species are grouped into an ascending hierarchy of taxa: the genus, family, order, class, and division with subgroups of each. Taxonomic groups of any rank are known as taxa (singular, taxon). As an example of ascending ranking, *Quercus alba* L., *Quercus laevis* Walter, *Quercus falcata* Michaux, and *Quercus bicolor* Willd. are four species in the genus, *Quercus. Quercus* L., *Fagus* L. and *Castanea* Miller are three of the genera in the family, *Fagaceae; Fagaceae* and *Betulaceae* are two families in the order, *Fagales; Fagales, Urticales,* and *Piperales* are three of the many orders in the class *Magnoliopsida,* which is in the division *Magnoliophyta* (see Table 3.2).

Why use this relatively complicated system of names? Supposedly Latin names are so long and unpronounceable. Why not use common names? Benson (1962) states very succinctly why vernacular or common names cannot replace scientific names.

"**1.** Names in common language are ordinarily applicable in only a single language; they are not universal.
2. In most parts of the world relatively few species have common or vernacular names in any language.
3. Common names are applied indiscriminately to genera, species or varieties.
4. Often two or more unrelated plants are known by the same name, and frequently even in one language a single species may have two to several common names applied either in the same or different localities."

In the northeastern United States, nine species of the genus, *Lilium,* in the family Liliaceae occur that are all called lily. In some parts of the northeast some of those species are known as orange lily, yellow lily, red lily, and so on. Two or more species are called orange lily and some natives know the orange lily as the wood lily. In the lily family, several species are known as lily that occur in other genera, for example, fawn-lily or trout-lily in the genus, *Erythronium* and day-lily in the genus, *Hemerocallis*. There are several lilies in the iris family, for example, blackberry lily in the genus, *Belamcanda* and celestial lily in the genus, *Nemastylis*. In other regions of the United States there are other species in the genus, *Lilium,* known as lily; species in other genera of the Liliaceae are also called lily; and species in genera in families other than Liliaceae and Iridaceae are known as lily. Add names for lilies in other languages and one can readily see the need for standardized scientific names without the confusion generated by common names.

Modern botanists in all countries use an *International Code of Botanical Nomenclature* (Voss et al., 1983, hereafter referred to as ICBN), which is a precise and simple system "dealing on the one hand with (1) terms which denote the ranks of taxonomic groups or units, and on the other hand with (2) the scientific names which are applied to the individual taxonomic groups of plants." (See the preamble to the ICBN in Table 3.1

Table 3.1   INTERNATIONAL CODE OF BOTANICAL NOMENCLATURE (1983)*

Preamble

1. Botany requires a precise and simple system of nomenclature used by botanists in all countries, dealing on the one hand with the terms which denote the ranks of taxonomic groups or units, and on the other hand with the scientific names which are applied to the individual taxonomic groups of plants. The purpose of giving a name to a taxonomic group is not to indicate its characters or history, but to supply a means of referring to it and to indicate its taxonomic rank. This Code aims at the provision of a stable method of naming taxonomic groups, avoiding and rejecting the use of names which may cause error or ambiguity or throw science into confusion. Next in importance is the avoidance of the useless creation of names. Other considerations, such as absolute grammatical correctness, regularity or euphony of names, more or less prevailing custom, regard for persons, etc., notwithstanding their undeniable importance, are relatively accessory.
2. The *Principles* form the basis of the system of botanical nomenclature.
3. The detailed provisions are divided into *Rules,* set out in Articles, and *Recommendations*. Examples are added to the rules and recommendations to illustrate them.
4. The object of the *Rules* is to put the nomenclature of the past into order and to provide for that of the future; names contrary to a rule cannot be maintained.
5. The *Recommendations* deal with subsidiary points, their object being to bring about greater uniformity and clearness, especially in future nomenclature; names contrary to a recommendation cannot, on that account, be rejected, but they are not examples to be followed.
6. The provisions regulating the modification of this Code form its last division.
7. The Rules and Recommendations apply to all organisms treated as plants (including fungi but excluding bacteria), whether fossil or non-fossil.† Nomenclature of bacteria is governed by the International Code of Nomenclature of Bacteria. Species provisions are needed for certain groups of plants: The International Code of Nomenclature of Cultivated Plants (1980) was adopted by the International Commission for the Nomenclature of Cultivated Plants; provisions for the names of hybrids appear in Appendix 1.
8. The only proper reasons for changing a name are either a more profound knowledge of the facts resulting from adequate taxonomic study or the necessity of giving up a nomenclature that is contrary to the rules.
9. In the absence of a relevant rule or where the consequences of rules are doubtful, established custom is followed.
10. This edition of the Code supersedes all previous editions.

---

*Used with permission.
†In this Code the term "fossil" is applied to a taxon when its name is based on a fossil type and the term "non-fossil" is applied to a taxon when its name is based on a non-fossil type (see Art. 13.3).

for its organization.) An understanding of the concept of rank and the application of scientific names to individual taxonomic groups are both basic to comprehension of the nomenclature.

A. RANKS OF TAXA

The accepted system of nomenclature provides a hierarchical arrangement of the ranks of taxa. Every individual plant is treated as belonging to a number of taxa of consecutively subordinate ranks; division, class, order, family, genus, species, each with subcategories— or taxa of consecutively ascending ranks. For example, an individual plant of swamp rose would belong to the consecutively higher ranks of the species, *Rosa palustris;* the genus, *Rosa;* the family, Rosaceae; the order, Rosales; the class, Magnoliopsida; and the division Magnoliophyta according to recent systems of classification. The ending on the name of the taxon indicates its rank, for example, the Ros-*aceae* indicates the rank of family (Table 3.2).

The ranks of taxa represent levels of relationships required for purposes of classification in treating different groups of plants. Conceptually, the ranks of taxa cannot be defined precisely, just arranged hierarchically. A group of plants, however, can be circumscribed and delimited as named ranks, for example, the roses in the genus *Rosa*. A genus such as *Pinus* or *Quercus* or the family Aceraceae can be sharply defined while other groups may not be circumscribed easily, such as many of the genera in the mustard family or many of the species in the genus, *Aster*. No set number of diagnostic characteristics, no special types of characters, and no fixed number of entities are involved in the designation of any group of plants as a certain rank or category. Since the basic rationale for most taxonomic systems is the concept of evolution, systematists have anticipated a lack of sharp definition of some groups of plants as taxa. Some organisms are evolving rapidly with intermediates of all sorts, while others represent dead ends with the intermediates long since extinct; the remainder are at different stages and conditions of evolutionary development.

The treatment of natural groups as ranks at the generic, familial, and higher categories is very subjective, with many of the recognized higher taxa the result of custom and tradition. Natural groups, for example, pines (*Pinus*), oaks (*Quercus*), and clovers (*Trifolium*), have a certain taxonomic aspect recognizable by the layman and professional alike. Kinds of plants, for example, the pines, oaks, and clovers, represent genera or the rank of genus. Historically and traditionally, the mints, legumes, composites, umbels, mustards, and others have been identified at the family level. Orders, classes, and divisions have not been recognized by the layman as such but he has observed that the fungi, algae, mosses, liverworts, ferns, conifers, and flowering plants were part of a larger group or higher rank of taxa. Professional botanists have recognized the last-named groups as members of divisions or classes or subordinate ranks thereof.

Species, the basal rank, has been characterized or defined by many authors since the time of Linnaeus as a name in a book, a judgment, a group of individuals that in sum total of their attributes resemble each other to a degree usually regarded as specific, the smallest group to which distinctive and invariable characters can be assigned, a discrete and immutable entity of divine origin, or what a competent authority thinks it is. Some of the

Table 3.2 RANKS AND ENDINGS

Rank	Ending	Example
Division*	phyta; mycota (Fungi)	Pterophyta; Eumycota
Subdivision*	phytina; mycotina (Fungi)	Pterophytina; Eumycotina
Class*	opsida (Cormophyta); phyceae (Algae); mycetes (Fungi)	Pteropsida; Cyanophyceae; Basidiomycetes
Subclass*	opsidae (Cormophyta); phycidae (Algae); mycetidae (Fungi)	Pteropsidae; Cyanophycidae; Basidiomycetidae
Order	ales	Rosales
Suborder	ineae	Rosineae
Family	aceae	Rosaceae
Subfamily	oideae	Rosoideae
Tribe	eae	Roseae
Subtribe	inae	Rosinae
Genus†	us, a, um, es, on, etc.	*Rosa; Ipomoea, Geum*
Subgenus		
Section		
Subsection		
Series		
Subseries		
Species‡		
Subspecies		
Varietas		
Subvarietas		
Forma		
Subforma		

*Recommended endings.

†*Note on subdivision(s) of a Genus:* The name of a subdivision of a genus is a combination of a generic name and a subdivisional epithet connected by a term (subgenus, section, series, etc.) denoting its rank. The epithet is either of the same form as a generic name, or a plural adjective agreeing in gender with the generic name and written with a capital initial letter. Examples: *Ricinocarpos* sect. *Anomodiscus, Euphorbia* sect. *Tithymalus* subsect. *Tenellae* us, a, um, es, on, etc.

‡*Note on infraspecific taxa:* The name of an infraspecific taxon is a combination of the name of a species and an infraspecific epithet connected by a term denoting its rank. Infraspecific epithets are formed as those of species and, when adjectival in form and not used as substantives, they agree grammatically with the generic name. Examples: *Saxifraga aizoon* var. *aizoon* subvar. *brevifolia* forma *multicaulis* subforma *surculosa* Engler & Irmscher or *Saxifraga aizoon* subforma *surculosa* Engler & Irmscher.

characteristics are applicable to an inclusive definition of species but all of them combined do not give an adequate introduction to the concept of species.

Cronquist (1968) has stated:

Aside from infraspecific taxa, the only taxonomic category with an inherent rank is the species. An exact definition of the species is impossible, and the more precise one attempts to be, the larger the number of species which do not fit the definition. Still, the basic concept is simple enough. A species is the smallest population which is permanently (in terms of human time) distinct and distinguishable from all others. It is the smallest unit which simply cannot be ignored in the scheme of classification. It is the primary taxonomic unit, and it may also be thought of as the basic evolutionary unit. In sexual populations, gene exchange between different species is restricted or even impossible. Interspecific hybrids are not always wholly sterile, but they are not so fertile and so competitively adapted as to swamp out the parents. Although there are some differences in interpretation, a reasonable degree of reproductive isolation from other species under natural conditions is an essential specific quality. Without such isolation, the population would lose its identity through interbreeding.

Taxonomists should always remember that the basic purposes of classification are (1) to order organisms into named taxa on the basis of their relationships; (2) to provide an orderly arrangement which expresses those relationships in a practical or natural way; and (3) to produce a system for efficient and effective storage, retrieval, and use for the taxa included in the classification. And above all, taxonomists should be aware that changes in ranks or in the classification of taxa frequently necessitate changes in names of taxa that lead to confusion in their use as communication symbols. Sound taxonomic procedure indicates that changing the rank or position of natural, recognizable groups should be avoided if no significant change in the concept of their relationships to each other and to other groups results from systematic study.

B. NAMES AND NAMING

The primary duties of plant nomenclaturists are to name new taxa and to determine the correct name of old taxa that have been remodeled, divided, united, transferred, or changed in rank. Naming new taxa and changing the names of old taxa is highly technical work usually done by experts in nomenclature or with at least the advice of specialists in the subject. The following treatment is included to introduce the beginning student to the basic elements and rules of the nomenclatural process as included in the ICBN (1983).

Scientific names are Latin monomials, binary combinations, trinary combinations, and so on. Names of the genera and higher ranks are monomials, for example, *Rosa* L. and Rosaceae Juss. The names of species are binary combinations (binomials), for example, *Lilium grayi* Watson; the names of subspecies, trinary combinations (trinomials), for example, *Hibiscus moscheutos* ssp. *palustris* (L.) Clausen; the names of varieties, quadrinary combinations (quadrinomials), for example, *Lilium catesbaei* Walter ssp. *catesbaei,* var. *longii* Fernald. In most manuals, the names of varieties appear to be trinary combinations, for example, *Lilium catesbaei* Walter var. *longii* Fernald, because the variety is a variation of the type subspecies *L. catesbaei* ssp. *catesbaei* that is automatically described by the author of the species and usually not indicated in the manual unless one or more variations of subspecies other than the typical are treated.

The fundamental principle for nomenclature is that "Each taxonomic group with particular circumscription, position, and rank can bear only one correct name, the earliest that is in accordance with the Rules, except in specific cases" (ICBN, 1983). A *correct name* or *epithet* for a taxon with a particular circumscription, position, and rank is a legitimate name or epithet that is a validly published name that must be adopted for it under the rules. A *legitimate name* is one in accordance with the rules. A *validly published name* or epithet is in accordance with Articles 32-45 of the Code in which the basic provisions are (1) effective publication, (2) publication in the form specified for the name of each category of taxa, (3) publication with a description or diagnosis, or a reference to a previously published description or diagnosis, of the taxon to which the name applies, (4) accompaniment by a Latin description or diagnosis or by a reference to a previously and effectively published Latin description or diagnosis of the taxon, and (5) an indication of the nomenclatural type. An *effectively published name* is one published in printed matter generally available to botanists. An *admissible name* is one having a form accord-

Table 3.3   PRINCIPLES FROM INTERNATIONAL CODE OF BOTANICAL NOMENCLATURE (1983)*

"I. Botanical nomenclature is independent of zoological nomenclature. The Code applies equally to names of taxonomic groups treated as plants whether or not these groups were originally so treated.†

II. The application of names of taxonomic groups is determined by means of nomenclatural types.

III. The nomenclature of a taxonomic group is based upon priority of publication.

IV. Each taxonomic group with a particular circumscription, position, and rank can bear only one correct name, the earliest that is in accordance with the Rules, except in specific cases.

V. Scientific names of taxonomic groups are treated as Latin regardless of their derivation.

VI. The Rules of nomenclature are retroactive unless expressly limited."

*Used with permission.
†For the purposes of this Code, "plants" does not include the bacteria.

ing to the rules so that it can enter into botanical nomenclature and be validly published (form would include proper rank ending, grammatical agreement of specific epithet with generic name, proper number, and so on). For example, see names with proper form in preceding paragraphs. The name of a taxon has to be validly published to have status under the Code.

The principles are the guidelines for the legitimate naming of any taxon (Table 3.3, I–VI). A legitimate name has to be validly published. A validly published name for a taxon of the rank of family or below must have a nomenclatural type (II) indicated that is usually a specimen of a taxon at the specific rank to which the name of the taxon is permanently attached. With a few exceptions, the correct name for any taxon from family to genus inclusively is the earliest legitimate name with the same rank (IV). With a few exceptions, the correct name for any taxon below the rank of genus is the earliest legitimate epithet in the same rank along with the correct name of the genus or species (IV). Valid publication of names for different groups (taxa) must have established beginning dates (III), for example, vascular plants, May 1, 1753 (Linnaeus, Species Plantarum, ed. 1); Musci (the Sphagnaceae excepted), January 1, 1801 (Hedwig, Species Muscorum). The principle of priority (III) is not mandatory for names of taxa above the rank of family. The scientific names of taxa are treated as Latin, regardless of their derivation (V). New rules for the ICBN can be made by future International Botanical Congresses and are retroactive unless expressly limited by the congress (VI). The nomenclature of plants is independent of that for animals (I).

Taxonomic study frequently results in changes in nomenclature of old taxa that have to be made according to the rules in the latest Code.* Some of the basic provisions included in the rules for each of the major procedures in classification are described below for fundamental information on retention, choice, and rejection of names of taxa that have been a subject of systematic study. The latest edition of the Code has to be used for detailed provisions pertinent to any taxonomic research.

When taxonomic study indicates that diagnostic characters or the circumscription of a taxon should be changed (remodeled), no change in its name is warranted. The simple

*The following paragraphs, including examples, are adapted directly from ICBN (1983).

addition or deletion of characteristics used in a circumscription of a taxon does not require a change in the name of the taxon.

When taxonomic study indicates that a genus should be divided into two or more genera, the generic name must be retained for the genus that includes the species designated as the type. For example in the genus *Aesculus* L. are *Aesculus* sect. *Aesculus,* sect. *Pavia* (P. Mill.) Persoon, sect. *Macrothyrsus* (Spach) C. Koch, and sect. *Calothyrsus* (Spach) C. Koch, the last three of which were regarded as distinct genera by the authors cited in the parentheses; in the event that these four sections are treated as genera, the name *Aesculus* must be kept for the first of the taxa, which includes the species *Aesculus hippocastanum* L., as this species is the type of the genus founded by Linnaeus. When a species is divided into two or more species, the specific epithet must be retained for the species that includes a specimen, description, or figure originally designated as the type.

When taxonomic study indicates that two or more taxa of the same rank should be united, the oldest legitimate name or epithet should be retained. If the two genera *Dentaria* L. (Sp. Pl. 653. 1753; Gen. Pl. ed. 5.295. 1754) and *Cardamine* L. (Sp. Pl. 654. 1753; Gen. Pl. ed. 5.295. 1754) are united, the resulting genus must be called *Cardamine* because the name was chosen by Crantz (Class. Crucif. 126. 1769), who was the first to unite the two genera.

When taxonomic study indicates that a subdivision of genus should be transferred to another genus or placed under another generic name for the same genus without change of rank, its epithet must be retained if legitimate. For example, *Saponaria* sect. *Vaccaria* DC. when transferred to *Gypsophila* becomes *Gypsophila* sect. *Vaccaria* (DC.) Godr. When taxonomic study indicates that a species should be transferred to another genus, or placed under another generic name for the same genus without a change in rank, the specific epithet, if legitimate, must be retained. For example, *Spergula stricta* Sw. (1799) when transferred to the genus *Minuartia* must be called *Minuartia stricta* (Sw.) Hiern (1899).

When taxonomic study indicates that the rank of a genus or infrageneric taxon should be changed, the correct name or epithet is the earliest legitimate one available in the new rank. For example, *Magnolia virginiana* var. *foetida* L. (Sp. Pl. 536. 1753) when raised to specific rank is called *Magnolia grandiflora* L. (Syst. Nat. ed. 10. 1082. 1759), the earliest available name at the species level, not *M. foetida* (L.) Sargent (Gard. & For. 2:615. 1889). In no case does a name or an epithet have priority outside its own name.

In some cases, taxonomic study indicates that names and epithets should be rejected because they are nomenclaturally superfluous—for example, the generic name *Cainito* Adans. (Fam. 2:166. 1793) is illegitimate because it was a superfluous name for *Chrysophyllum* L. (Sp. Pl. 192. 1753)—or because they are later homonyms—for example, the name *Tapeinanthus* Boiss. ex Benth. (1848), given to a genus of the Labiatae, is a later homonym *Tapeinanthum* Herb. (1837), a name previously and validly published for a genus of Amaryllidaceae.

Many exceptions to the rules treated above are included in the Code, so accuracy in nomenclature requires the use of the latest edition.

C. NOMENCLATURE IN FLORAS AND MANUALS

Most students and amateur as well as professional botanists initially become acquainted with scientific names in floras or manuals. Some of the names included in these books will be in boldface type, some in italics; some with simple author citations, some highly complicated with parentheses, ex, in, &, and so on; and some at the beginning of the taxon description, and some at or near the end. The correct name in boldface type at the beginning of the taxon treatment and the synonym(s) in italics near the end of the taxon description represent the taxonomic judgment of the author(s) of the nomenclature of the taxon in the flora or manual. The synonymy and author citations provide a ready reference to the treatment of that taxon in other manuals or floras covering the same region and to earlier editions of these books, as well as clues to the taxonomic history of the taxon treated.

Author citations for the correct names and synonyms provide a foundation for understanding the taxonomy of the taxa included in a flora or manual; therefore, seven types of author citations with examples and some basic definitions related to synonymy are described. The common kinds of hybrids found in most floras are defined. Nomenclatural types included in revisions and some manuals are treated here.

I. Author Citations

The proper citation for author of validly published new names and taxonomic changes in old names are indicated in the following paragraphs.*

Original Author   For the name of a taxon to be accurate and complete and the data readily verifiable, it is necessary to cite the author(s) who first validly published the name concerned. Examples: *Rosaceae* Juss., *Rosa* L., *Rosa gallica* L., *Rosa gallica* var. *eriostyla* R. Keller, *Rosa gallica* L. var. *gallica*.

Joint Author   When a name has been published jointly by two authors, the names of both should be cited, linked by means of the word *et* or by an ampersand (&). When a name has been published jointly by more than two authors, the citation should be restricted to that of the first one followed by et al. Examples: *Didymopanax gleasonii* Britton et Wilson (or Britton & Wilson); *Streptomyces albo-niger* Hesseltine, J. M. Porter, Deduck, Hauck, Bohonos, & J. H. Williams (Mycologia 46: 19. 1954) should be cited as *S. albo-niger* Hesseltine et al.

Name Proposal   When a name has been proposed but not validly published by one author and is subsequently validly published and ascribed to him by another author, the name of the former author followed by the connection word ex may be inserted before the name of the publishing author. Examples: *Havetia flexilis* Spruce ex Planch. et Triana or *Havetia flexilis* Planch. et Triana.; *Gossypium tomentosum* Nutt. ex Seem or *Gossypium tomentosum* Seem.; *Lithocarpus polystachya* (Wall. ex A. DC.) Rehder or *Lithocarpus polystachy* (A. DC.) Rehder (ex means validly published by).

Publication   When a name with a description or diagnosis (or reference to a description or diagnosis) supplied by one author is published in a work by another author, the word *in* should be used to connect the names of the two authors. In such cases, the name of the author who supplied the description or diagnosis is the more important and should be

*Unless otherwise indicated, examples in I, II, and III are adapted directly from the ICBN (1983).

retained when it is desirable to abbreviate such a citation. Examples: *Viburnum ternatum* Rehder *in* Sargent, Trees and Shrubs 2: 37. 1907, or *Viburnum ternatum* Rehder; *Teucrium charidemii* Sandwith *in* Lacaita, Cavanillesia 3: 38. 1930, or *Teucrium charidemii* Sandwith.

Description Alteration   When the alteration of the diagnostic characters or of the circumscription of a taxon without the exclusion of the type has been considerable, the nature of the change may be indicated by adding such words, abbreviated where desirable, as *emendavit* (emend. means corrected), followed by the name of the author responsible for the change), *mutatis characteribus* (mut. char. means change of character), *pro parte* (p.p. means in part), *excluso genere* or *exclusis generibus* (excl. gen. means excluding genus or genera), *exclusa specie* or *exclusis speciebus* (excl. sp. means excluding species), *exclusa varietate* or *exclusis varietatibus* (excl. var. means excluding variety), *sensu amplo* (s. ampl. means in a broad sense), *sensu stricto* (s. str. means in a strict sense). Examples: *Phyllanthus* L. emend. Mull. Arg.; *Globularia cordifolia* L. excl. var. (emend. Lam.).

Rank Alteration   When a genus or a taxon of lower rank is altered in rank but retains its name or epithet, the author who first published it as a legitimate name or epithet (the author of the basionym; the term basionym is defined in the next section) must be cited in parentheses, followed by the name of the author who made the alteration (the author of the combination). Examples: *Medicago polymorpha* var. *orbicularis* L. when raised to the rank of species, by Allioni, becomes *Medicago orbicularis* (L.) All.

Taxon Transfer   When a taxon of lower rank than genus is transferred to another taxon, with or without alteration of rank but retaining its name or epithet, the author who first published this as a legitimate name or epithet (the author of the basionym) must be cited in parentheses, followed by the name of the author who made the alteration (the author of the combination). Examples: *Cheiranthus tristis* L. transferred to the genus *Matthiola* by Robert Brown becomes *Matthiola tristis* (L.) R. Br.

II.   Basic Definitions Related to Synonyms

A *synonym* is a rejected name due to misapplication or difference in taxonomic judgment. A *nomenclatural synonym* is a different name based on the same nomenclatural type with the second name a synonym of a previous name. A *taxonomic synonym* is a different name based on a different type but taxonomic judgment indicates identity equal to a previously described taxon. Examples*: *Paspalum laeve* Michaux includes *P. longipilum* Nash, *P. circulare* Nash; *P. laeve* var. *circulare* (Nash) Fernald. The author's taxonomic judgment indicates that *P. longipilum* and *P. circulare* published in Small's manual are taxonomic synonyms of *P. laeve,* since the names are based on three different types; and that *P. laeve* var. *circulare* in Gray's manual is a nomenclatural synonym of *P. circulare* Nash in Small's manual, since it is based on the same type.

*Examples from Radford et al., 1968, *A Manual of the Vascular Flora of the Carolinas,* University of North Carolina Press, Chapel Hill.

A *basionym* is a specific or infraspecific epithet that has priority and is retained when transferred to a new or different taxon. Example*: *Desmodium ochroleucum* M. A. Curtis includes *Meibomia ochroleuca* (M. A. Curtis) Kuntze. In this case *Meibomia ochroleuca* is a nomenclatural synonym based on the same type but the basionym *ochroleuca* was correctly retained by Kuntze in his treatment of the taxon.

A *homonym* is a case in which two or more identical names are based on different types, only one of which can be legitimate. Example: *Spergula stricta* SW. (1799) transferred to *Arenaria* cannot become *Arenaria stricta* because of a different species, *Arenaria stricta* Michx. (1803). If transferred, *S.* "stricta" would be a homonym of *A.* "stricta" Michx. (1803).

A *tautonym* is an illegitimate binomial in which the generic name and specific epithet is the same. Example*: *Armoracia rusticana* (Lam.) Gaertn., Mey. & Scherb. includes *Armoracia armoracia* (L.) Britton as a synonym. *A. armoracia* is a homonym that is an illegitimate binomial rejected under the rules and is a nomenclatural synonym based on the same type as *A. rusticana*.

An *autonym* is a legitimate, automatically created tautonym for infrageneric or infraspecific taxa. Examples: *Hypericum* subgenus *Hypericum* section *Hypericum; Hypericum perforatum* L. ssp. *perforatum* var. *perforatum*.

III. Hybrids

A *hybrid* is the offspring of two plants or animals of different breeds, races, forms, varieties, subspecies, species, or genera. An intergeneric hybrid is a hybrid between species of two or more genera, for example, × *Agropogon* (meaning *Agrostis* × *Polypogon*). An interspecific hybrid is a hybrid between two species of the same genus, for example, *Salix* × *capreola* (meaning *Salix aurita* × *S. caprea*). The symbol × in a name indicates hybridity.

IV. Types

The types defined below provide an introduction to the type concept based on principle II of the ICBN. *Holotype* is the one specimen or other element used by the author or designated by him as the nomenclatural type. As long as a holotype is extant, it automatically fixes the application of the name concerned. *Isotype* is a duplicate (part of a single gathering made by a collector at one time) of the holotype; it is always a specimen. *Lectotype* is a specimen or other element selected from the original material to serve as a nomenclatural type when no holotype was designated at the time of publication or as long as it is missing. *Neotype* is a specimen or other element selected to serve as nomenclatural type as long as all of the material on which the name of the taxon was based is missing. *Nomenclatural type* is that element with which the name is permanently associated. It is not necessarily the most typical or representative element of a taxon. *Syntype* is any one of two or more specimens cited by the author when no holotype was designated, or any one

*Examples from Radford et al., 1968, *A Manual of the Vascular Flora of the Carolinas,* University of North Carolina Press, Chapel Hill.

of two or more specimens simultaneously designated as types. *Topotype* is a specimen of a named taxon collected, usually later, from the original type locality, or from the area from which the species was described.

These comments on nomenclature in floras and manuals only provide a foundation for understanding the kinds of author citations as relating to changes in the taxonomy of old taxa and the naming of the new.

SUGGESTED READING

Benson, L. 1962. Plant Taxonomy, Methods and Principles. The Ronald Press Company, New York. Chapter 11, pp. 341–345; Chapter 12, pp. 346–367.

Cronquist, A. 1968. Evolution and Classification of Flowering Plants. Houghton Mifflin Company, Boston.

Davis, P. H., and V. H. Heywood. 1965. Principles of Angiosperm Taxonomy. D. Van Nostrand Company, Inc., New York. Chapter 8, pp. 259–292.

Hess, W. J. 1975. "A tree by any other name . . ." Morton Arboretum Quarterly 11:9–13, 16.

Isely, D. 1972. The Disappearance. Taxon 21:3–12.

Jeffrey, C. 1968. An Introduction to Plant Taxonomy. J. & A. Churchill Ltd., London. Chapter 4, pp. 34–61; Chapter 5, pp. 62–93.

———— 1973. Biological Nomenclature. Edward Arnold Ltd. and The Systematics Association, London.

Johnson, A. T. 1971. Plant Names Simplified, second edition. W. H. & L. Collingridge Ltd., London.

Lawrence, G. H. M. 1951. Taxonomy of Vascular Plants. Macmillan, New York. Chapter 9, pp. 192–222.

McVaugh, R., R. Ross, and F. A. Stafleu. 1968. An Annotated Glossary of Botanical Nomenclature. Regnum Vegetabile Vol. 56. Bohn, Scheltema & Holkema, Utrecht.

Radford, A. E. et al. 1974. Vascular Plant Systematics. Harper & Row, New York. Chapter 3, pp. 35–56.

———— Voss, E. G. et al (eds.). 1983. International Code of Botanical Nomenclature (ICBN). Adopted by the Thirteenth International Botanical Congress, Sydney, August, 1981. Regnum Vegetabile Vol. 97. Bohn, Scheltema & Holkema, Utrecht.

SUMMARY FOR NOMENCLATURE

Definitions for Nomenclature   *Nomenclature* is a precise and simple system used by botanists in all countries that deals on the one hand with terms denoting the ranks of taxonomic groups or units, and on the other hand with the scientific names that are applied to the individual taxonomic groups of plants (modification from the preamble of the ICBN). The *scientific name* of a taxon is a word or combination of words by which it is known. *Taxon* refers to a taxonomic group of any rank.

Purpose of Nomenclatural Study   To provide one correct name for a taxon with a particular circumscription, position, and rank that is universal, stable, and unambiguous; to provide names for groups of plants (taxa) that are the communication symbols and reference bases for information storage, retrieval, and use; and to provide names for groups of plants (taxa) that are rank indicative.

Operations in Nomenclatural Study To name new taxa and to determine correct names for old taxa that have been remodeled, divided, united, transferred, or changed in rank according to the ICBN.

Basic Premise for Nomenclatural Study Each taxon with a particular circumscription, position, and rank can bear only one correct name.

Fundamental Principles of Nomenclature

1. Natural plant taxa that have been delimited, circumscribed, and classified are named according to a system of nomenclature.
2. Nomenclaturists change the name of a taxon for the proper reasons of a more profound knowledge of the facts resulting from adequate taxonomic study or because a nomenclature is contrary to the rules of the ICBN.

Guiding Principles for the Nomenclature of New and Old Taxa

1. Correct Name.* Each taxonomic group with a particular circumscription, position, and rank can bear only *one correct* name, the earliest that is in accordance with the Rules, except in specified cases (principle IV).
2. Latin. Scientific names of taxonomic groups are treated as *Latin* regardless of their derivation (principle V).
3. Priority. The nomenclature of a taxonomic group is based upon *priority* of publication (principle III).
4. Type. The application of names of taxonomic groups is determined by means of *nomenclatural types* (principle II).
5. Retroactive. The rules of nomenclature are *retroactive* unless expressly limited (principle VI).
6. Independent. Botanical nomenclature is *independent* of zoological nomenclature. The Code applies equally to names of taxonomic groups treated as plants whether or not these groups were originally assigned to the plant kingdom (principle I).
7. Publication. Names have to be effectively and validly *published* with proper author citations.

Basic Assumptions for the Study of Nomenclature

1. Discontinuities in variation and character correlations occur that allow delimitations and circumscriptions of plant taxa at determined positions and ranks.
2. A system of nomenclature exists for the naming of new taxa and for the determination of the correct names for old taxa that have been remodeled, united, divided, transferred, or changed in rank.

QUESTIONS ON NOMENCLATURE

1. What is nomenclature? What are the basic objectives of nomenclature?
2. What is binomial nomenclature and why is it considered fundamental to botany?

---

*All principles except number 7 are from the ICBN.

3. What are the differences between nomenclature and naming?
4. Which scientific names are monomials? Binomials? Trinomials?
5. What are the differences between a scientific name and a specific or infraspecific epithet?
6. What are the purposes for giving a name to any taxonomic group?
7. Why cannot vernacular or common names replace scientific names? Give four basic reasons.
8. What are the proper reasons for changing the name of a plant?
9. How do most nomenclatural problems arise?
10. What are the aims of the International Code for Botanical Nomenclature (ICBN)?
11. What are the objects of the principles, rules, and recommendations in ICBN?
12. What are the characteristics of a good name? A good specific epithet?
13. What are the nomenclatural reasons for rejecting a name? List five reasons.
14. What are the basic components of a sound definition of species? Explain.
15. What is meant by author citation? Give examples of four common types of author citations from your local or regional manual.
16. What are the basic differences between name and nym?
17. What is the ascending hierarchy of plant taxa? Give examples of each rank.
18. What is meant by the type concept? Retroactive rules of nomenclature?
19. What are the basic provisions of a validly published name?
20. Which ranks can be conserved according to ICBN? Where are they found in ICBN?
21. What are the problems in defining rank?
22. What is meant by particular circumscription, position, and rank in principle IV of the ICBN?
23. What is the fundamental principle of all nomenclature? What is the theory of nomenclature?
24. Which of the principles in ICBN have to do with the process of naming? Explain.
25. What are the problems common to the nomenclature of paper clips, organic compounds, dogs, definitions, characteristics, and plants?
26. What are the advantages and disadvantages of a stable or fixed nomenclature?
27. What are the nomenclatural reasons for name changes in plants? Are these reasons equally valid for any objects or premises?
28. What are the taxonomic reasons for name changes in plants? Are these reasons equally valid for name changes for any objects?
29. What are the nomenclatural consequences of systematic changes?
30. Are the principles of plant nomenclature applicable to the nomenclature of descriptive terms? Any structures or objects? How?

EXERCISES IN NOMENCLATURE

1. Select and name one species from each of the following families and indicate the order, subclass, class, subdivision, and division for each:
   a. Selaginellaceae
   b. Orchidaceae
   c. Pinaceae
   d. Salicaceae
   e. Boraginaceae
2. Select and name four species from different families and indicate why you consider each a good scientific name.
3. Give an example of four types of author citations. Explain each.
4. Select two species with synonyms, name the species, and give possible explanations as to why the synonyms are synonyms—based on author citations and generic names.
5. Select two examples of each of the following types of names and give the rank of each: monomial, binomial, trinomial.

6. Select and name one family, then indicate the nomenclatural components found in the treatment.
7. Name two examples of each of the following: generic synonym, species synonym, varietal synonym, subspecific synonym.
8. Give two examples of genera with common names that are different from the specific common names in the genus.
9. Select a taxon in which the common name is applied to more than one subordinate taxon; select and name a taxon that has more than one common name.
10. Name a taxon and give an example of a taxonomic synonym and an example of a nomenclatural synonym. Cite an example of each type of synonym.
11. Select a taxon that is an example of a taxon transfer as indicated in synonymy of the taxon; select another taxon that is an example of a change in rank. How do you know that your examples represent a taxon transfer and a change in rank?
12. Indicate the basionym in the taxon transfer and change in rank cited in the previous exercise. Why are the epithet(s) or name(s) considered the basionyms?
13. Define or indicate what is meant by each of the following:

    a. Legitimate name
    b. Correct name
    c. Illegitimate name
    d. Circumscription
    e. Position
    f. Rank
    g. Taxon
    h. Effective publication
    i. Valid publication
    j. Typification
    k. Priority
    l. Limitation of priority
    m. Authors' name citation

    n. Literature citation
    o. Retained name or epithet
    p. Remodeled name or epithet
    q. Divided name or epithet
    r. Taxon transfer
    s. Choice of names
    t. Rank alteration
    u. Rejection of names or epithets
    v. Gender of generic names
    w. Names of hybrids
    x. Nomenclature of taxa
    y. Conservation of taxa

chapter 4

Botanical Names*

Taxonomists as users of scientific names study Latin and Greek roots, prefixes, suffixes, and terminations that make scientific names connotative, mnemonic, reasonably succinct, and acceptably pronounceable. Taxonomists as coiners of botanical names select the proper Latin or Greek roots, prefixes, suffixes, and terminations to create scientific names that are connotative, mnemonic, reasonably short, and easy to pronounce.

Botanical names are the scientific symbols by which taxonomic groups of plants at any rank are known. These scientific names are either Latin words or words that have been latinized from some other language, most often Greek. They are the symbols of communication by which botanists around the world can talk about any given plant in any language, but all, supposedly, using the one correct Latin name for that plant. A fundamental principle of modern nomenclature is that there can be only one correct scientific name for each kind of plant.

Such was not always the case. For centuries, people used long sentences to indicate the plants they were talking about, or gave them names that often varied from locality to locality or from region to region. Many times the same name was applied to more than one kind of plant, and often the same plant was known by several different names. A state of confusion frequently resulted. Then, about the middle of the eighteenth century, the great Swedish naturalist Linnaeus developed Gaspar Bauhin's "two names only for each plant" idea into the binomial system, a system so fundamental that it has survived virtually unchanged to the present. Nomenclature was at last soundly based, with definitive, reliable, and reasonably succinct symbols in *one* language—Latin—that would remain constant for the use of botanists of all nationalities.

It is obvious that all students of taxonomy need to learn the scientific names of

*Contributed by Laurie Stewart Radford, University of North Carolina at Chapel Hill.

plants, without the knowledge of which they cannot identify, classify; or describe plants, or determine correct names, much less coin new ones. Without this knowledge, they cannot determine which species are rare, endangered, or threatened and in need of management or protection. As was indicated in Chapter 1, plants are the aesthetic and economic base of our very existence and survival—they are the fundamental sources of food, clothing, shelter, drugs for healing, and water and oxygen for life itself. Taxonomists as workers in this "most relevant field of inquiry for modern man," as researchers, classifiers, evaluators, and so on, must have symbols of communication that all can understand and use. These symbols of communication are the scientific names of the plants with which they work, names that must become part of their everyday vocabulary.

Many students are dismayed when confronted with the necessity of learning scientific names because they know nothing of Latin or Greek; but if they will simply learn the meanings of a number of scientific terms and their roots, they will soon find that they can remember with comparative ease, and with much satisfaction, the names as well as some of the characteristics of a great many plants. They will also realize that they do not need to be scholars of Latin or Greek, although an elementary course in either language could be a decided advantage. It is for just such students that this chapter is presented, to help them understand how scientific names are formed, to provide lists of words and terms that are commonly used in coining plant names, and to give instruction in the correct pronunciation of scientific names. Diligent study of these lists of Latin and latinized words, along with their English meanings, will very soon result in the ability to recognize and remember the scientific names of an ever-increasing number of plants and to understand the meanings of numerous botanical terms as well.

The principles governing the formation and use of botanical names apply to taxa at all levels of the hierarchy; however, in this chapter, the discussion of scientific names will be limited to specific epithets. The rule that scientific names are generally underlined or italicized will be set aside in the present treatment, where for clarity or emphasis it seems important to underline only endings or other parts of scientific names.

A. FORMATION OF SPECIFIC EPITHETS

I. General Information

Each species of plant has only one correct scientific name, one peculiar to that species alone. This name is a *binary combination* (a binomial) and consists of a generic name and a specific epithet. For example, *Salix nigra* is the binomial for black willow, *Salix* being the generic name, and *nigra* the specific epithet. Thus, a specific epithet can be defined as the second part of a binomial. The student should understand, however, that it is never correct to use the specific epithet alone to designate a particular species; it must always be combined with a generic name to form the binary combination for that species. One cannot say that *nigra* is the scientific name for black willow or for any other species; standing alone, *nigra* is simply the Latin word for black. When combined with *Salix* it is the name for black willow (*Salix nigra*), or with *Juglans* the name for black walnut (*Juglans nigra*), or with *Fraxinus* the name for black ash (*Fraxinus nigra*).

Specific epithets are formed from nouns, adjectives, participles, gerunds, and so on, and by combining such words with a variety of prefixes and suffixes. It should be understood that unlike English nouns, every Latin or latinized noun has gender, either masculine (m.), feminine (f.), or neuter (n.). In most cases, each gender is indicated by a different ending; for example, most nouns ending in *-us* are masculine, nouns ending in *-a* are nearly always feminine, while those ending in *-um* are neuter. An adjective or other modifier must agree in gender with the noun it modifies. While a noun will have only one nominative ending depending on its gender, a modifier will often have three different endings, depending on the gender of the word it is modifying. Thus, the Latin adjective for hairy appears with three different endings as it modifies the nouns *Lathyrus, Lactuca,* and *Vaccinium,* respectively masculine, feminine, and neuter: Lathyr*us* hirsut*us*, Lactuca hirsut*a*, and Vaccini*um* hirsut*um*. It should be noted that occasionally the specific epithet is a noun in apposition, in which case it carries its own gender, regardless of that of the generic name it follows. In the binomial *Cypripedium calceolus,* a species of lady slipper, the neuter name *Cypripedium* is followed by the masculine noun *calceolus* (a small shoe) in apposition, which retains its own gender. The specific epithet of some trees is feminine and does not agree with a masculine genus, for example, Fraxin*us* nigr*a* (black ash).

Most specific epithets are adjectives, or words used in an adjectival sense and will have masculine, feminine, or neuter endings. The most common Latin endings are:

m.	f.	n.	Examples
-us	-a	-um	albus, alba, album (white)
-er	-ra	-rum	niger, nigra, nigrum (black)
-is	-is	-e	brevis, brevis, breve (short)
-r	-ris	-re	acer, acris, acre (pungent)

Modifiers ending in *-ans,* (eleg*ans*, elegant), *-ens* (rep*ens*, creeping), *-or* (bicol*or*, with two colors), and *-x* (simple*x*, simple) use the one ending for all three genders. Example: Ranuncul*us* rep*ens*, Ludwigi*a* rep*ens*, and Trifoli*um* rep*ens*.

II. Commemorative Epithets

The names of persons are sometimes used as specific epithets, generally to honor or commemorate the man or woman who first discovered a particular species. In such cases, the specific epithet usually has a genitive rather than a nominative ending. The treatment given here for commemorative names is based on Article 73 of the 1983 edition of the *International Code of Botanical Nomenclature* (ICBN), Recommendation 73 C.

A specific epithet taken from the name of a man should be formed as follows:

1. If the name ends in any vowel except *a,* (i.e., *e, i, o, u,* and final *y*), the letter *i* is added to the end of the name, for example, Lilium gray*i* for Asa Gray or Aster blake*i* for Joseph Blake.
2. If the name ends in *a,* the letter *e* is added: balansa*e* for Mr. Balansa.
3. If the name ends in a consonant, *ii* is added: Chelone lyon*ii* for John Lyon; Rubus grimes*ii* for Jerome Grimes. Exception: if a name ends in *er,* only one *i* is used, for example, Setaria faber*i* for Ernst Faber, Solidago cutler*i* for Manasseh Cutler.

4. If a name is used as an adjective, it must agree in case and gender with the genus it modifies: Rub*us* cardian*us* for Fred Wallace Card; Chenopodi*um* boscian*um* for Louis Augustin Bosc; Ruelli*a* purshian*a* for Frederick Traugott Pursh.

If a woman's name is used in the substantive form as an epithet, the ending will be the feminine genitive singular for that word: Crataegus cole*ae* for Emma Jane Cole; C. beckwith*ae* for Florence E. Beckwith; Cornus price*ae* for Sarah Frances Price. If the name is used as an adjective, the same rules apply as for a man's name: Apios purshian*a* for Miss Price.

III. Descriptive Epithets

Most specific epithets indicate something characteristic about a species, such as color of flowers or fruit (Aronia *melanocarpa,* black-fruited), habit of growth (Celastrus *scandens,* climbing), shape of leaf (Desmodium *rotundifolium,* round-leaved), discoverer of the species (Lilium *grayi,* for Asa Gray), or place of discovery (Solidago *roanensis,* of Roan Mt., N.C.). Names of plant parts are used in combination with prefixes and suffixes to form a number of specific epithets: *biflorus,* two-flowered, from the prefix *bi* (two) and the plant part *flos* (flower); *petiolaris,* having a petiole, from the plant part *petiolus* (petiole) and the suffix *-aris* (indicating possession). It is possible to determine the meanings of a great many epithets simply by referring to the lists of plant parts, prefixes, and suffixes given in the sections that follow.

IV. Names of Plant Parts Used as Epithets

The most obvious parts of a plant (here arranged alphabetically for convenience), are the bud, flower, fruit, leaf, root, seed, seedling, and stem. Each term is followed by its Latin equivalent, and by the Greek where known, indicated by (L.) and (Gr.) respectively. The compound words are usually separated into their basic parts with English meanings to give the student a better understanding of the word: for example, *carpopodium: carpo-* (fruit) + *podion* (little foot).

1. Bud—*gemma* (L.); *blastos* (Gr.)
  primordium—*primordium: primus* (first) + *ordiri* (to begin)
  promeristem—*promeristema: pro-* (before) + *meristos* (divided)
  scale—*squama* (L.); *lepis* (Gr.)
2. Flower—*flos* (L.); *anthos* (Gr.)
  androecium—*androecium: andro-* (male) + *oikos* (house)
    stamen—*stamen* (L.); *stemon* (Gr.)
    staminodium—*staminodium*
  androgynophore—*androgynophorum: andro-* (male) + *gyno-* (female) + *phoras* (bearing)
    stipe—*stipes* (stem)
    column—*columna* (pillar)
  androphore—*androphorum: andro-* (male) + *phoras* (bearing)
    receptacle—*receptaculum* (vessel for holding something)
  calyx—*calyx*
    sepal—*sepalum* (L.); *sepalon* (Gr.)

carpel—*carpellum*
    carpopodium—*carpopodium: carpo-* (fruit) + *podion* (little foot)
    funiculus—*funiculus* (cord, rope)
    locule—*loculus* (chamber, room)
    ovary—*ovarium*
    ovule—*ovulum*
    placenta—*placenta* (a flat cake)
    podogyne—*podogynium: podo-* (foot) + *gyne* (female)
    stigma—*stigma* (mark or spot)
    stipe—*stipes* (stem)
    style—*stylus* (L.); *stylos* (Gr.)
clinanthium—*clinanthium*
corolla—*corolla* (a little crown)
    petal—*petalum* (L.); *petalon* (Gr.)
disc—*discus* (circular plate)
gynophore—*gynophorum: gyno-* (female) + *phoras* (bearing)
hypanthium—*hypanthium*
    base—*fundus*   neck—*collum*
    limb—*limbus*   tube—*tubus*
ovary—*ovarium*
ovule—*ovulum*
pedicel—*pedicellus*
perianth—*perianthium*

banner—*vexillum*	labellum—*labellum*	*lepis* (Gr.)
base—*fundus*	ligula—*ligula*	sepal—*sepalum* (L.);
beard—*barba*	limb—*limbus*	*sepalon* (Gr.)
bristle—*seta*	lip—*labium*	spur—*calcar*
callosity—*callositas*	lobe—*lobus*	standard—*vexillum*
carina—*carina*	nectary—*nectarium*	tepal—*tepalum*
claw—*unguis*	palate—*palatum*	throat—*faux*
corona—*corona*	petal—*petalum* (L.);	tube—*tubus*
fringe—*fimbria*	*petalon* (Gr.)	vexillum—*vexillum*
hood—*cucullus*	pouch—*saccus*	wing—*ala*
horn—*cornu*	sac—*saccus*	
keel—*carina*	scale—*squama* (L.);	

petal—*petalum* (L.); *petalon* (Gr.)
pistil—*pistillum*
polyphore—*polyphore: poly-* (many) + *phoras* (bearing)
receptacle—*receptaculum* (vessel for holding something)
sepal—*sepalum* (L); *sepalon* (Gr.)
stamen—*stamen* (L.); *stemon* (Gr.)
    anther—*anthera*
    filament—*filamentus* (thread)
stigma—*stigma* (mark or spot)
style—*stylus* (L.); *stylos* (Gr.)
tepal—*tepalum*
whorl—*verticillus*

3. Fruit—*fructus* (L.); *carpos* (Gr.)
    carpophore—*carpophorum: carpo-* (fruit) + *phoras* (bearing)
    dissepiment—*disseptimentum* (a partition)
    ectocarp—*ectocarpium: ecto-* (outside) + *carpos* (fruit)

endocarp—*endocarpium: endo-* (within) + *carpos* fruit)
funiculus—*funiculus* (a rope or cord)
mericarp—*mericarpium: meri-* (part of) + *carpos* (fruit)
mesocarp—*mesocarpium: meso-* (middle) + *carpos* (fruit)
pericarp—*pericarpium: peri-* (around) + *carpos* (fruit)
placenta—*placenta* (a flat cake)
seed—*semen* (L.); *sperma* (Gr.)
septum—*septum* (L.); *phragma* (Gr.): (wall or partition)

4. Leaf—*folium* (L.); *phyllon* (Gr.)
blade—*lamina*
leaflet—*foliolum*
midrib—*costa*
petiole—*petiolus*
petiolule—*petiolulus*
pulvinus—*pulvinus* (cushion)
rachilla—*rhachilla* (little axis)
rachis—*rhachis* (axis)
stipel—*stipella* (little stalk)
stipule—*stipula* (stalk)

5. Root—*radix* (L.); *rhiza* (Gr.)

6. Seed—*semen* (L.); *sperma* (Gr.)
aril—*arillus* (fleshy growth)
chalaza—*chalaza* (a lump)
embryo—*embryo*
   coleoptile—*coleos:* (sheath) + *ptilon* (feather)
   cotyledon—*cotyledon* (cup leaf)
   epicotyl—*epicotylus: epi-* (on) + *cotylus* (cup)
   hypocotyl—*hypocotylus: hypo-* (beneath) + *cotylus* (cup)
   plumule—*plumula* (little feather)
   radicle—*radicula* (little root)
endosperm—*endospermium: endo-* (within) + *sperma* (seed)
hilum—*hilum* (scar, trifle)
raphe—*raphe* (seam)
seed coat—*testa*

7. Seedling—*plantula* (L.)
cataphyll—*cataphyllum: cata-* (below) + *phyllon* (leaf)
collet—*collum* (neck)
eophyll—*eophyllum: eos-* (early, primitive) + *phyllon* (leaf)
metaphyll—*metaphyllum: meta-* (changed) + *phyllon* (leaf)

8. Stem—*caulis* (L.); *caulos* (Gr.)
bark—*cortex*   pith—*medulla*
branch—*ramus*   wood—*lignum*

V. Prefixes

A prefix is an element of one or more letters placed at the beginning of a word (usually a noun or an adjective) to modify its meaning or application. Many prefixes are adverbs or prepositions. In botanical Latin, adjectives with modified endings often serve as prefixes. In such cases, the stem of the adjective and usually a connecting vowel (*i* or *o*) form the prefix. To illustrate: the adjective *altus* (tall) is reduced to its stem *alt* and the vowel i added, the result being *alti,* which now can serve as a prefix meaning having the quality of being tall or of tallness. When used before the term for stem (caulis), we get the word *alticaulis,* meaning tall-stemmed. When standing alone, a prefix is usually followed by a hyphen, indicating attachment to the beginning of another word or term.

a. Number

uni- (L.): one; *uniflorus,* one-flowered

mono- (Gr.): one; *monanthos,* one-flowered

bi- (L.): two; *bifoliatus,* two-leaved

di- (Gr.): two; *diphyllus,* two-leaved

tri- (L.): three; *triangularis,* having three angles

tri- (Gr.): three; *triacanthos,* having three thorns

quadri- (L.): four; *quadrifolius,* four-leaved

tetra- (Gr.): four; *tetraphyllus,* four-leaved

quinque- (L.): five; *quinquefolius,* five-leaved

penta- (Gr.): five; *pentaphyllus,* five-leaved

sex- (L.): six; *sexangularis,* six-angled

hex- (Gr.): six; *hexagonus,* six-sided

septem- (L.) seven; *septemlobus,* seven-lobed

hepta- (Gr.): seven; *heptapetalus,* seven-petaled

octo- (L.): eight; *octoflorus,* eight-flowered

octo- (Gr.): eight; *octandrus,* with eight stamens

novem- (L.): nine; *novemnervis,* with nine nerves

ennea- (Gr.): nine; *enneaphyllus,* with nine leaves

decem- (L.): ten; *decemlobus,* with ten lobes

deca- (Gr.): ten; *decapetalus,* with ten petals

amphi- (Gr.): double, of two kinds

centri- (L.): one hundred

dicha-, dicho- (Gr.): in two

diplo- (Gr.): double

haplo- (Gr.): single

multi- (L.): many

myrio- (Gr.): countless

oligo- (Gr.): few

pan-, panto- (Gr.): all

pauci- (L.): few

pleio- (Gr.): few

pluri- (L.): several

poly- (Gr.): many, numerous

sesqui- (L.): one and a half

terni- (L.): in three or in threes

b. Position

a-, ab- (L.): away from

ad- (L.): toward, against

ambi- (L.): around, round about

amphi- (Gr.): around, on both sides of

ante- (L.): before, in front of

anti- (L. and Gr.): against

apo- (Gr.): away from, down

co-, com-, con- (L.): together

de- (L.): downward, away from

di-, dis- (L.): apart from, asunder

dia- (L. and Gr.): through

ecto- (Gr.): out of, from

endo-, ento- (Gr.): inside, within

epi- (Gr.): on, on top of

ex- (L.): from, out of

extra- (L.): outside

hyper-, hypero- (Gr.): beyond, above

infra- (L.): below

inter- (L.): between, among

intra- (L.): within

intro- (L.): inside

meta- (L. and Gr.): next to, after

ob- (L.): against

para- (Gr.): near, beside

per- (L.): through

peri- (L. and Gr.): around

prae- (L.): before, in front of

sub- (L.): below, under

super-, supra- (L.): above, over

syn-, sys- (Gr.): together, joined

trans- (L.): across, beyond

c. Shape

aniso- (Gr.): uneven, unequal

astro- (Gr.): stellate, starlike

cerato- (Gr.): hornlike

cteno- (Gr.): pertaining to a comb

cyath- (L. and Gr.): cuplike

cyclo- (Gr.): circular

fili- (L.): threadlike

fimbri- (L.): fimbriate

glosso- (Gr.): tonguelike

goni- (Gr.): angled, angular

hetero- (Gr.): various, different

holo- (Gr.): entire

lanci- (L.): lance-shaped

nephro- (Gr.): kidney-shaped

ophio- (Gr.): snakelike

ortho- (Gr.): straight, erect

ovi- (L.): egg-shaped

rhyncho- (Gr.): having a snout or beak

rhytido- (Gr.): wrinkled, rumpled

scyphi-, scypho- (L. and Gr.): cuplike

spheno- (Gr.): wedge-shaped

tylo- (Gr.): with knots or projections

uro- (Gr.): tailed

d. Size

angusti- (L.): narrow	macro- (Gr.): large, giant
brachy- (Gr.): short	mega-, megalo- (Gr.): very large, great
brevi- (L.): short	micro- (Gr.): small, little
crassi- (L.): thick, short	nano- (Gr.): dwarf
grandi- (L.): large	pachy- (Gr.): thick
iso- (Gr.): equal	parvi- (L.): small
lati- (L.): wide, broad	platy- (Gr.): broad
lepto- (Gr.): slender	steno- (Gr.): narrow
longi- (L.): long	tenui- (L.): slender, thin

e. Miscellaneous

a-, an- (Gr.): without, not	flavi- (L.): yellowish
actino- (Gr.): rayed, starlike	fusci- (L.): dark or dark brown
andro- (Gr.): male	galacto- (Gr.): milky
anemo- (Gr.): pertaining to wind	gamo- (Gr.): fused, united
archae-, arche- (Gr.): old, primitive	geo- (Gr.): pertaining to earth
argyro- (Gr.): silvery	gymno- (Gr.): naked, bare
atri-, atro- (L.): black	gyno- (Gr.): female
botry- (Gr.): bunch	hirti- (L.): hairy with long hairs
callo- (Gr.): beautiful	homo- (Gr.): like, same
canio- (Gr.): pertaining to dogs	laevi- (L.): smooth
cardia- (Gr.): pertaining to a heart	lani- (L.): woolly
carpo- (Gr.): pertaining to fruit	lasio- (Gr.): woolly
caryo- (Gr.): nutlike	laxi- (L.): loose, unstrung
chlamydo- (Gr.): wearing a cloak	leio- (Gr.): smooth
chloro- (Gr.): green	lepido- (Gr.): scaly
chryso- (Gr.): golden	leuco- (Gr.): white
coelo- (Gr.): pertaining to a hollow	lino- (Gr.): made of flax
crypto- (Gr.): hidden	melano- (Gr.): black, very dark
cyano- (Gr.): dark blue	ne- (Gr.): not, free from
dasy- (Gr.): shaggy, hairy	neo- (Gr.): new
e-, ef-, ex- (L.): without, lacking	non- (L.): not
erio- (Gr.): woolly	nudi- (L.): naked
erythro- (Gr.): reddish	ochro- (Gr.): yellowish
eu- (Gr.): good, well	odonto- (Gr.): tooth-shaped

oxy- (Gr.): sharp

paleo- (Gr.): old

phaeo- (Gr.): dark

phanero- (Gr.): easily seen, visible

photo- (Gr.): light

phyllo- (Gr.): pertaining to a leaf

phyto- (Gr.): pertaining to a plant

picro- (Gr.): bitter

podo- (Gr.): of a foot

porphyro- (Gr.): purple

prae- (L.): before, very

pro- (L.): for, instead of

pseudo- (Gr.): false

ptero- (Gr.): winged

ptycho- (Gr.): pertaining to grooves or folds

pyro- (Gr.): fiery

pyrrho- (Gr.): fire red, ruby red

rami- (L.): pertaining to branches

re- (L.): back

rhizo- (Gr.): pertaining to roots

rhodo- (Gr.): rose-colored

sapro- (Gr.): rotten, decayed

sarco- (Gr.): fleshy

schisto- (Gr.): split, cleft

schizo- (Gr.): deeply divided

sclero- (Gr.): hard

semper- (L.): always

sessile- (L.): sessile

stachyo- (Gr.): spiked

stato- (Gr.): fixed, standing

sticto- (Gr.): spotted

sur- (L.): somewhat, above

tephro- (Gr.): ash gray

trachy- (Gr.): rough

tricho- (Gr.): hairy

viridi- (L.): green

viti- (L.): pertaining to a vine

xero- (Gr.): dry

xylo- (Gr.): woody

zantho-, xantho- (Gr.): yellow

zygo- (Gr.): joined, married

VI. Suffixes

A suffix is an element of one or more letters placed at the end of a word to modify its meaning or application. The grammatical nature, gender, and meaning of the resulting compound is determined by that particular suffix. To illustrate: by joining the noun *herba* (herb, plant) and the suffix *-arium* (place where something is done or kept), we have the compound word *herbarium,* a noun of neuter gender meaning a place in which dried plants are kept. Compounds thus formed by adding suffixes to a number of word stems can result in nouns or adjectives, but most often they are the latter as they apply to specific epithets. For this reason, the majority of suffixes listed here are adjectival in nature. The Latin or Greek origin of each suffix is indicated by the symbols (L.) and (Gr.), respectively, followed by the character or quality each denotes, plus an example to illustrate. Masculine, feminine, and neuter endings are given for each suffix, where applicable. When standing alone, a suffix is usually preceded by a hyphen, indicating attachment at the end of another word or term.

-aceus, -a, -um (L.): likeness, resemblance; *crustaceus,* crustlike

-aeus, -a, -um (Gr.): belonging to; *aetnaeus,* pertaining to Mt. Etna

-alis, -is, -e (L.): possession, or pertaining to; *digitalis,* pertaining to a finger

-anus, -a, -um (L.): belonging to, position; *virginianus,* of Virginia

-aris, -is, -e (L.): relating to, possession; *petiolaris,* having a petiole

-arium (L.): place where something is done or kept; *herbarium,* collection of dried plants

-arius, -a, -um (L.): possession or connection; *plumarius,* pertaining to plumes

-ascens (L.): process of becoming, incomplete; *violascens,* becoming violet

-aticus, -a, -um (L.): place of growth; *aquaticus,* growing in water

-atilis, -is, -e (L.): place of growth; *fluviatilis,* growing in streams

-atus, -a, -um (L.): likeness or possession; *rostratus,* having a beak

-bilis, -is, -e (L.): ability or capacity; *sensibilis,* capable of irritability or sensitivity

-bundus, -a, -um (L.): fullness, abundance; *floribundus,* full of flowers

-ellus, -a, -um (L.): diminutive; *echinellus,* minutely spiny

-ensis, -is, -e (L.): origin, country or place of growth; *alabamensis,* from Alabama

-escens, -is, -e (L.): process of becoming; *flavescens,* becoming yellow, yellowish

-estris, -is, -e (L.): place of growth; *campestris,* growing in fields

-eus, -a, -um (L.): resemblance in quality or color; *roseus,* rose-colored

-eus, -a, -um (Gr.): possessed by or belonging to; *giganteus,* belonging to giants, thus gigantic

-icans (L.): almost identical resemblance; *candicans,* whitish

-icola (L.): a dweller; *saxicola,* a dweller among rocks, growing among rocks

-icus, -a, -um (Gr.): belonging to; *virginicus,* belonging to Virginia

-ilis, -is, -e (L.): capacity or ability, property; *flexilis,* capable of being bent, flexible

-ineus, -a, -um (L. and Gr.): color or material; *stramineus,* straw-colored

-inus, -a, -um (L.): possession or resemblance; *velutinus,* like velvet

-oideus, -a, -um (L.); -oides, -odes (Gr.): like, resembling; *helianthoides,* resembling the genus *Helianthus*

-osus, -a, -um (L.): abundance, fullness; *foliosus,* full of leaves

-utus, -a, -um (L.): possessing; *cornutus,* having horns, horned

B.   MEANINGS OF SPECIFIC EPITHETS

I.   Adjectival Epithets

As previously stated, most specific epithets are adjectival in nature and thus convey something characteristic about a particular species. The listings given here are grouped by characteristics such as color, size, and place of growth, and with very few exceptions, they are adjectival in nature. Masculine gender only is given here.

a. Epithets Relating to Color

aeneus: brassy green, bronze

albescens: becoming white, turning white

albicans: whitish

albidus: whitish

albolutescens: whitish-yellow

albus: white

argenteus: silvery

argyreus: silvery

atropurpureus: dark purple

atrorubens: dark red

atroviolaceus: dark violet

atrovirens: dark green

aurantiacus: yellowish-orange

aureus: golden yellow

azureus: sky blue

brunnescens: becoming deep brown

caerulescens: becoming deep blue

caeruleus: deep blue, sky blue

caesius: lavender, bluish-gray

calcareus: chalk white

candicans: becoming pure white

candidus: shining white

canescens: becoming grayish or hoary

cardinalis: cardinal red

carneus: flesh-colored

cinereus: ash gray

cinnamomeus: light reddish-brown

coccineus: scarlet; deep red

croceus: saffron yellow

cruentus: blood red, blood-stained

cyaneus: dark, deep blue

dealbatus: covered with white powder, whitened

eburneus: ivory white

exalbidus: whitish

ferrugineus: rusty, light reddish-brown

flavescens: becoming yellow

flavidulus: slightly yellow

flavidus: somewhat yellow

flavovirens: yellowish-green

flavus: pale yellow

fulgidus: shining, bright-colored

fulvus: tawny, dull yellow

glaucescens: becoming sea green, grayish-blue

glaucus: gray-green, with a bloom

griseus: pearl gray

helvolus: pale brownish-yellow

hepaticus: liver-colored, dark reddish-brown

hyacinthinus: purplish-blue

incanus: hoary, whitish-gray

incarnatus: flesh-colored

lividus: lead-colored

luridus: dull yellow, brownish-yellow

luteolus: pale yellow

lutescens: becoming pale yellow

luteus: deep yellow

niger: black, glossy black

niveus: snow white, purest white

ochroleucus: yellowish-white

porphyreus: purple

prasinus: grass green, bright green

puniceus: crimson, purple-red

purpurascens: becoming purple, purplish

purpureus: purple

roseus: rose-colored, rosy

rubellus: reddish

rubens: reddish

rubiginosus: brown-red

rufidulus: somewhat red

rufus: reddish

stramineus: straw-colored

violaceus: violet

violascens: becoming violet-colored

virens: green

virescens: becoming green

viridescens: becoming green

viridis: green

vitellinus: egg-yolk yellow

b. Epithets Relating to Direction

aquilonius: northern

australis: southern, from the southern hemisphere

austrinus: southern

borealis: northern

meridionalis: southern

occidentalis: western

orientalis: eastern

septentrionalis: northern

c. Epithets Relating to Geography

acadiensis: of Nova Scotia, Canada

aegyptiacus: of Egypt

africanus: of Africa

alabamensis: of Alabama

aleppicus: of Aleppo, northern Syria

allegheniensis: of the Alleghenies

alpinus, alpestris: of the Alps, of high mountains

altamaha: of the Altamaha River, Georgia

americanus: of America

amurensis: of the Amur River

anglicus: of England

arabicus: of Arabia

argentinus: of Argentina

asiaticus: Asian

atlanticus: of the Atlantic

australiensis: of Australia

austriacus: of Austria

austro-carolinianus: of South Carolina

babylonicus: of Babylon

barbadensis: of Barbados

bavaricus: of Bavaria, West Germany

bermudensis: of Bermuda

bonariensis: of Buenos Aires

brasiliensis: of Brazil

burmanicus: of Burma

californicus: of California

canariensis: of the Canary Islands

canadensis: of Canada

cantabrigiensis: of Cambridge, England

capensis: of the Cape of Good Hope, of a cape

caribaeus: from the Lesser Antilles, West Indies

carolinae-septentrionalis: of North Carolina

carolinianus, carolinensis, carolinus: of the Carolinas (U.S.)

carthagensis: of Carthage, northern Africa

catawbiensis: from Catawba River, North Carolina

cherokeensis: of the Cherokee Country, or Cherokee County, North Carolina

chilensis: of Chile

chinensis, sinensis, cathayanus: of China

cisatlanticus: on this side of the Atlantic

columbianus: of the District of Columbia

corsicus: of Corsica, France

cubensis: of Cuba

curassavicus: of Curaçao, Caribbean Sea

domingensis: of Santo Domingo

europaeus: of Europe

florentinus: of Florence, Italy

floridanus: of Florida

gallicus: of France

germanicus: of Germany

gileadensis: of Gilead

graecus: of Greece

groenlandicus: of Greenland

guadalupensis: of Guadeloupe, West Indies

halepensis: of Aleppo

helveticus: of Switzerland

hispanicus: of Spain

idaeus: of Mt. Ida

illinoensis: of Illinois

indicus: of India or the Indies

islandicus: of Iceland, of an island

italicus: of Italy

jamaicensis: of Jamaica

lancastriensis: of Lancaster, Pennsylvania

ludovicianus: of Louisiana or former Louisiana Territory

marianus, marilandicus: of Maryland

martinicensis: of Martinique, West Indies

mexicanus: of Mexico

mississippiensis: of Mississippi

monspeliensis: of Montpellier, southern France

neapolitanus: of Naples, Italy

neerlandicus: of the Netherlands

neogaeus: of the New World

norvegicus: of Norway

novae-angliae: of New England

noveboracensis: of New York

ogelthorpensis: of Ogelthorpe, Georgia

ohiensis: of Ohio

pennsylvanicus: of Pennsylvania

philadelphicus: of Philadelphia

polonicus: of Poland

provincialis: of Provence, southern France

roanensis: of Roan Mt., North Carolina

sibiricus: of Siberia (U.S.S.R.)

sinensis: of China

syriacus: of Syria

tennesseensis: of Tennessee

texanus, texensis: of Texas

thapsus: of Thapsus, Sicily

virginianus, virginiensis, virginicus: of Virginia

zetlandicus: of the Shetland Isles, Great Britain

zeylanicus, ceylanicus, taprobanicus: of Ceylon

d. Epithets Relating to Habit

arborescens: arborescent

ascendens: ascending

caespitosus: cespitose

decumbens: decumbent

dichotomus: dichotomous

erectus: erect

expansus: clambering

fastigiatus: fastigiate

fruticosus: fruticose

furcatus: forked

geniculatus: geniculate

patens: spreading

procumbens: procumbent

prostratus: prostrate

ramosus: branched

repens: creeping, prostrate and rooting

reptans: creeping, prostrate and rooting

scandens: climbing

soboliferus: soboliferous

stoloniferus: stoloniferous

volubilis: twining

e. Epithets Relating to Habitat

agrestis: pertaining to fields or cultivated land

alpinus: of the Alps, growing in an alpine zone

alsodes: of woods

amphibius: living in water and on land

aquaticus: growing in water

arenarius: growing on sand

arenicola: dweller in sandy places

arvensis: pertaining to fields, especially plowed fields

austromontanus: of the southern mountains

campestris: of fields

collicola: dweller in the hills

collinus: dwelling or growing in the hills

cumulicola: dweller on a heap or mound

demersus: growing under water

elodes: of marshes

epihydrus: on the water

fluviatilis: of rivers

fontinalis: of springs

hypogeus: underground

inundatus: growing in places apt to be flooded

jugosus: mountainous

lacustris: of lakes or ponds

lithophilus: rock-loving, growing on rocks

littoralis: of the seashore

lucorum: sacred thickets

maritimus: of or belonging to the sea

montanus: of the mountains

monticola: dweller in the mountains

nemorosus: of woodlands and groves

orae: coastal, of the coast

paludosus: growing in boggy places

palustris: of marshes or swamps

porophilus: lover of soft rock

pratensis: growing in meadows

riparius: pertaining to river banks

rivularis: growing by streams

ruderalis: growing among rubbish

rupestris: growing among rocks

sabulosus: growing in shady places

saltuense: of or pertaining to forests or woodland pastures

sativus: cultivated

saxatilis: growing among rocks

saxicola: dweller among rocks

segetalis: growing in grain fields

silicola: dweller on siliceous or flinty soils

sylvaticus, silvaticus: of the woods

sylvestris, silvestris: growing in woods

terrestris: growing in dry grounds

umbrosus: growing in the shade

vinealis: growing in vineyards

f.   Epithets Relating to Seasons

aestivalis: pertaining to summer

aestivus: pertaining to summer

autumnalis: pertaining to fall or autumn

hiemalis: belonging to winter

solstitialis: of summer

vernalis: pertaining to spring

vernus: pertaining to spring

g. Epithets Relating to Size

altissimus: very tall, high

altus: tall, high

angustatus: narrow, slender

depauperatus: reduced

dilatatus: broadened, widened

elatior: taller

elatius: tall

exaltatus: very tall

exiguus: little

giganteus: very large, gigantic

grandis: large

humilis: dwarf

intermedius: halfway, intermediate

magnus: large

major: larger, greater

minor: smaller, less

minus: small, minute

minutus: very small, minute

nanus: dwarf

parvulus: very small

pauxillus: small

perpusillus: very small

praealtus: very tall or high

procerus: very tall

pumilus: dwarf

pusillus: very small, insignificant

reductus: reduced

robustus: stout

II. Nominative Epithets: Old Generic, Common, and Aboriginal Names

alsine: Greek meaning of a luxuriant plant, chickweed

ammi: Greek for an unbelliferous plant

amomum: an aromatic shrub from which the Romans made fragrant balsam

anagallidea: resembling the genus *Anagallis*, pimpernel

ananassa: from *Ananassa*, the pineapple

aparine: old generic name for a plant, cleavers

armoracia: ancient name for horseradish

atamasco: aboriginal name

azedarach: from the Persian meaning free or noble tree

batatos: native name for sweet potato

benzoin: old name for some member of the *Lauraceae* (Fernald, 1950)

bursa-pastoris: shepherd's pouch, from *Bursa,* an old generic name (Fernald, 1950)

calamintha: old generic name meaning beautiful mint

carota: old generic name for carrot

catalpa: aboriginal name

cepa: Latin word for onion

chamaedrys: old generic name meaning ground oak

convolvulus: name for any twining plant, literally, ''a caterpillar that wraps itself up in a leaf''

cotula: from the genus *Cotula,* a composite, Greek for a small cup

cracca: Latin word for pulse or wild vetch

cucubalus: old generic name (Fernald, 1950)

cucullaria: old generic name meaning hoodlike

cyparissias: name used by Pliny for a plant related to spurge

dracontium: Greek for dragon wort

elatine: classical name for a low, creeping plant (Fernald, 1950)

githago: old generic name

halicacabum: from old generic name *Halicacabus,* another name for bladder wort

haspan: native name for *Cyperus haspan* in Ceylon (Fernald, 1950)

helioscopia: an ancient name meaning turning toward the sun (Fernald, 1950)

hydropiper: Greek for water pepper

intybus: Latin word for endive

jalapa: old generic name meaning from Jalapa, Mexico (Fernald, 1950)

julibrissin: modification of a Persian name (Fernald, 1950)

kali: old generic name from the Persian meaning a large carpel (Fernald, 1950)

labrusca: Latin word for the wild vine, presumably a grape

lappa: old generic name meaning a bur

leucanthemum: old generic name meaning white flowered

lupulus: early generic name (Fernald, 1950)

lychnitis: Greek for a plant from which wicks were made

mays: aboriginal name

mitreola: old generic name referring to miterlike fruits (Fernald, 1950)

mollugo: old generic name for whorl-leaved plants

napus: Latin word for a kind of turnip

negundo: aboriginal name

psyllium: old generic name, Greek for fleabane is *psyllion*

quamoclit: native Mexican name for morning glory

rhoeas: old Greek name meaning wild poppy

scnega: old generic name given to plant the Seneca Indians used (Fernald, 1950)

serpentaria: Latin name for snakeweed

tetrahit: old generic name meaning four-parted (Fernald, 1950)

III. Miscellaneous Epithets

acicularis: needlelike

acris: sharp-pointed

affinis: related

alatus: winged

amabilis: lovely

amoenus: lovely, pleasing

anceps: two-headed or edged

annuus: annual

anomalus: abnormal, unlike its kind

apodus: without a foot, sessile

aristatus: awned

aromaticus: aromatic or spicy

axillaris: growing in an axil

baccatus: berrylike

barbatus: bearded

biennis: biennial

bracteosus: with conspicuous bracts

brevibarbis: with a short beard

brevicaudatus: short taillike appendage

brevipilis: short-haired

brevistylus: with short style

bulbiferus: bulb-bearing

caducus: falling off or dropping early

caespitosus: growing in tufts or patches

calcaratus: bearing spurs

campanulatus: bell-shaped

capillaceus: slender, hairlike

capitatus: growing in heads

caudatus: tailed

cernuus: nodding or drooping

ciliatus: fringed with hairs

clandestinus: hidden or concealed

clavellatus: shaped like a small cup

comosus: bearded, tufted

concolor: uniform in color

cornutus: horned

costatus: ribbed

cuneatus: wedged

dasystachys: woolly-spiked

debilis: weak

deciduus: falling off, shedding

decurrens: running down

densus: thick, dense

dentatus: toothed

desiccatus: dried up

didymus: twins, in pairs

difformis: irregular, unevenly formed

dioicus: dioecious, unisexual

discolor: not uniform in color

dumosus: shrubby, bushy

echinatus: spiny

elegans: elegant

epetiolatus: without a petiole

ericoides: heathlike

farinosus: mealy

fasciculatus: growing in bunches, fascicled

fertilis: capable of producing fruit

filiculmis: with a threadlike stem

fistulosus: hollow, cylindrical, tubular

flabellatus: like a small fan

flexuosus: bent alternately in opposite directions

floridus: flowering

fluitans: floating

foetidus: ill-smelling, strong-smelling

frondosus: full of leaves

fruticosus: shrubby, bushy

fungosus: spongy

furcatus: forked

fuscatus: dusky

generalis: pertaining to all

geniculatus: abruptly bent, like a knee joint

glaber: smooth, without hairs

gracilis: slender

gregarius: herded or flocked together

heterophyllus: with leaves of more than one kind

hybridus: designating a hybrid

hystrix: bristly (literally a porcupine)

imbricatus: overlapping, like shingles

inflatus: bladderlike, swollen

inodorus: without scent or odor

invisus: hateful, detested

laevigatus: smooth, as if polished

lanosus: woolly

laxus: loose, lax

linearis: narrow, linear

lyratus: lyre-shaped

maculatus: spotted or blotched

mollis: soft, pubescent

muricatus: rough with short, hard points

natans: floating on the surface of the water

nitens: shining, polished

nodosus: knotted, knobby

normalis: at right angles, normal

nudatus: exposed, laid bare

odoratus: fragrant

patens: spreading

pendulus: hanging down, nodding

perennis: perennial

petiolatus: having a petiole, petioled

plicatulus: folded into small longitudinal pleats

praecox: developing very early, precocious

prostratus: lying flat, thrown to the ground

pubens: downy

punctatus: marked with dots, depressions, or glands

radiatus: spreading from common center

ramosus: branched

recurvus: curved backward

regularis: typical, according to rules

reticulatus: netted

rigidus: stiff

rostratus: beaked

rosulatus: in the form of a rosette

rugosus: wrinkled

scaber: rough, harsh

scaposus: having a scape

sebiferus: wax- or tallow-bearing

sericeus: silky

serotinus: late to leaf, flower, or appear

setosus: bristly

speciosus: good-looking, beautiful

spectabilis: showy, worth seeing

squarrosus: rough, with outward projecting tips

stans: erect

sterilis: sterile, barren

stipitatus: with a little stalk or stipe

sulcatus: furrowed

tenellus: slender, tender, soft

teres: clyindric, circular in cross section

tinctorius: used for dyeing

tortus: twisted

trivialis: common, ordinary, trivial

truncatus: shortened, cut off at end

tuberosus: tuberous

uncinatus: hooked

velutinus: velvety

ventricosus: swollen, especially on one side

versicolor: variously colored

vulgaris: common

C. PRONUNCIATION OF BOTANICAL NAMES

Since botanical names are Latin or latinized words, it would seem that they should be pronounced as Latin is pronounced.* But what sort of Latin? Reformed academic? Roman Latin? Church Latin? It has been said that Latin is the universal language of botanists and other scientists; this may be true of written Latin, but wherever spoken, it has been modified somewhat by language usage peculiar to each country in which it is spoken, so that the same word may sound differently when uttered by a German, a Frenchman, an Italian, an Englishman, or an American.

Several methods of pronouncing Latin are in use today. One, the so-called English method, has rules for pronouncing Latin and latinized words that are generally analogous to those for the pronunciation of English words. The Roman method attempts to follow as closely as possible the pronunciation used by the Romans themselves from about 50 B.C. to A.D. 50, as far as can be determined. The continental method was developed during the Middle Ages from the modern languages of the time, and widely used by the Roman Catholic Church. The vowels had about the same values as those of the Roman method, but the consonants were pronounced as those of the country in which the method was used.

Botanists, at least English-speaking botanists who have expressed themselves on the matter of pronunciation, seem to agree generally that for our purposes it is probably best to use the English sounds for vowels and consonants, while following the rules of classical Latin for accenting. Until someone can devise a simpler method to which most botanists will ascribe, the beginning student can manage very well by checking with a manual like Gray's seventh or eighth edition, in which a grave accent (`) denotes a long vowel and an acute accent (´) a short vowel. If the student is curious to know why the vowel is long or short, or why the accent falls where it does, he or she will need to understand a few conditions and rules. The following may be of help to one so inclined.

I. General Information

a. Syllables

Before a word can be accented, it must be divided into syllables. Every Latin word has as many syllables as it has separate vowels or diphthongs.

1. When a single consonant comes between two vowels, the consonant is taken with the vowel that follows it: acer, a-*c*er.
2. When two consonants come between two vowels, the first goes with the first vowel, and the second with the second vowel: a*l*bidus, a*l*-*b*i-dus. Exception: if the first consonant is *b*, *c*, *d*, *g*, *k*, *p*, or *t*, and the second is either *l* or *r*, both consonants go together with the second vowel; for example, gla*b*ra, gla-*b*ra. If the two consonants are *ch*, *ph*, or *th*, each pair is counted as one letter and goes with the second vowel: microce*ph*ala, mi-cro-ce-*ph*a-la.
3. When there are more than two consonants, all but the first go with the second vowel: a*bsc*onditus, a*b*-*sc*on-di-tus.
4. *X* is always taken with the vowel preceding it: e*x*pansus, e*x*-pan-sus; e*x*albidus, e*x*-al-bi-dus.

*See an unabridged dictionary for pronunciation of family and generic names.

b.   **Vowels**

1. Final vowels have the long sound (alsin*e*, al-si-n*ee*), except final *a*, which has the sound of an unstressed "ah" (vern*a*, ver-n*ah*).
2. Final *es* sounds like the English word "ease": alsod*es*, al-so-d*eez*.
3. *Y* is always a vowel with the quality of *i*: diphyllus, di-fil-us.
4. Two vowels together that do not form a diphthong are always sounded separately. The first of the two has the short sound: filifol*ia*, fi-li-fo-li-*ah*. It should be noted that this rule does not hold for words transcribed from the Greek, for example, the *e* in *Achillēa* is long because it is a contraction of the Greek diphthong *ei*. A diphthong is treated as a long vowel wherever it occurs, even if transcribed by a single letter.

c.   **Diphthongs**

1. *Ae* and *oe* have the sound of long *e* in "me": l*ae*vis, l*ee*-vis; rh*oe*as, r*ee*-as.
2. *Au* sounds like *au* in "caudal" or *aw* in "awful": c*au*datus, caw-da-tus.
3. *Ei* usually becomes *i*, and is like the English long *i*, in "kite."
4. *Eu* sounds like *u* in "neuter", or "neurology": *eu*rycarpus, *u*-ri-car-pus. Diphthongs are always classed as long vowels.
5. It should be noted that some botanists treat *oi* as a diphthong, pronouncing it as *oi* in "oil", not as two separate vowels according to Latin rules. Thus they say Helian-*thoi*-des, rather than Helian-*tho-i*-des.

d.   **Consonants**

1. *C* and *g* have soft sounds of *s* and *j* respectively when followed by *e, i, y, ae,* or *oe*. Examples: *C*edrus, *see*-drus; *c*yaneus, *si*-a-n*c*-us; *g*eneralis, *je*-ne-ra-lis; *g*ynandra, *ji*-nan-dra. Otherwise, *c* has the hard sound of *k* and *g* the hard sound of *g* in "go." Examples: *c*andidus, *k*an-di-dus; *g*labrus, *g*la-brus.
2. When a word is begun by one of the following pairs of consonants, the first letter is silent: *cn, ct, gn, mn, pn, ps, pt, tm*. Examples: *c*nicus, *n*i-kus; *c*tenium, *t*e-ni-um; *p*syllium, *s*il-i-um.
3. *Ch, ph,* and *th* are counted as one letter each. *Ch* has the sound of *k*, but is silent before *th* at the beginning of a word; *ph* is sounded as *f; th* as th in "thing." Examples: *ch*loran*th*us, *k*lo-ran-*th*us; *ph*ellos, *f*el-los.
4. *Cc* followed by *i* or *y* sounds like *k-si*: co*cc*inea, ko*k-si*-ne-a.
5. *G* before a "soft" g (gg), takes the sound of one soft *g*.
6. Initial *x* has the sound of *z*: *x*anthium, *z*an-thi-um; otherwise *x* sounds like *ks:* Zantho*x*yllum, zan-tho*ks*-il-lum.
7. *Ci, si,* and *ti,* when following an accented syllable and followed by another vowel, often have the sound of *shi* or *zhi:* Sene*ci*o, se-ne-*shi*-o; Artemi*si*a, ar-te-mi-*zhi*-a.

II.   Accenting According to the Rules of Classical Latin

a.   **General**

1. The last syllable is never accented. The next to the last syllable is called the penult: de-*cum*-bens; the third from the last syllable is called the antepenult: *de*-cum-bens.

2. In a word of two syllables, the accent always falls on the first syllable: *à*-cer.
3. In a word of more than two syllables, the accent falls on the next to the last (penult) if the penult is long. It is long if it ends in a long vowel, a diphthong, or a consonant: als*i*ne, al-*sì*-ne; am*oe*nus, a-*mòe*-nus; dec*um*bens, de-*cúm*-bens. Note: When the penult ends in a consonant, the *vowel* is short, although the syllable is long.
4. If the penult is not long (therefore short), the accent falls on the antepenult, the third syllable from the end: dra*con*tium, dra-*cón*-ti-um (the penult, *ti,* is short because it ends in a vowel with a short sound).
5. The accent can never be farther from the end than the antepenult. Thus the penult is the crucial factor in accenting. If it ends in a consonant, it has the accent; if it ends in a diphthong it has the accent because a diphthong is always counted as a long vowel; if it ends in a vowel, it has the accent if the vowel is long. Thus one must determine whether that vowel is long or short; this requires a Latin dictionary. Most specific epithets are Latin or Greek adjectives, some are nouns in the genitive case, and others are old generic and aboriginal names. There are no rules given for the last two categories, but there are some hints that can be of help in pronouncing the adjectives and genitives.

b. Long Penult

As a rule, the penult is long in the following adjectival endings and thus takes the long accent:

-àlis	-àtus	-ìnus*	-oìdes	-ùnes	-ùsus
-ànus	-ènus	-ìtus	-òsus	-ùnus	-ùtus
-àris	-ètus	-ìvus	-òvus	-ùrus	

Examples: car-di-*nà*-lis, bar-*bà*-tus, fa-ri-*nò*-sus.

c. Short Penult

In the following adjectival endings, the vowel in the penult is usually short, thus removing the accent to the antepenult:

-ăcus	-ĕus	-ĭdus	-ĭor
-ĕger	-ĭchus	-ĭlis	-ĭus
-ĕris	-ĭcus	-ĭlus	-ŭus
-ĕrus	-ĭdis	-ĭmus	-y̆us

Examples: se-*bí*-fe-rus, he-*pá*-ti-cus, *cán*-di-dus, am-*bí*-gu-us.

Compound words ending in the following syllables also have a short vowel in the penult (thus the accent is removed to the antepenult):

*Exceptions: short in can-*ná*-bi-nus, gos-*sý*-pi-nus, se-*ró*-ti-nus; usually long in cy-er-*rì*-nus and sa-li-*cì*-nus.

-clădus	-gĕra	-pŏdus
-cŏla	-gĕrum	-phy̆tum
-cŏlor	-gy̆nus	-stŏmus
-fĕra	-lĕpis	-tŏmus
-fĕrum	-ŏlens	-vĭrens
-fĭdus	-pĭlis	
-fīlus	-phĭlus	

Examples: ce-*rí*-fe-ra, cu-mu-*lí*-co-la, di-*chó*-to-mus.

Note: If the connecting vowels, short *i* in Latin and short *o* in Greek, appear in the penult, they do not have the accent because they make the penult short: spin*i*fer, spì-n*i*-fer. When the connecting vowels appear in the antepenult in words with short penult, the antepenult receives the accent and it is short: spin*i*fera, spi-n*í*-fe-ra. Of course, in an adjectival ending in which the penult has consonant length (ends in a consonant), the penult has the accent and the vowel is short:

-éllus	-éstus
-énsis	-fórmis
-éssus	-íllus
-éster	-úster

Examples: hal-e-*pén* sis, mul-ti-*fór*-mis.

The same rule applies to compound words ending in -cárpus, -róstris, and -phýllus: chlorocárpus, heterophýllus.

d. Commemorative Names

A few comments should be made about the pronunciation of latinized commemorative names. It is here that the rules are most apt to be disregarded or ignored, chiefly because the person being honored may not be able to recognize his or her name if the rules are strictly adhered to.

An example may suffice to illustrate the dilemma. The family name, James, pronounced as one syllable in English, with a long *a*, Jàmz, becomes in Latinized form a four-syllable word with a long accent on the antepenult: Ja-mè-si-i. In cases of this sort, many botanists simply pronounce the family name as it sounds and add the proper Latin ending. It now comes out as Jàmes-i-i, which is recognizable, but very un-Latin, because in Latin words the accent cannot be farther back than the antepenult, nor can there be more than one vowel to a syllable. Still many botanists prefer taking the exception to the rules if the name cannot be recognized in its Latinized form.

III. A Concluding Statement

As stated by the authors of the seventh edition of Gray's Manual, ''Botanical names . . . are not always capable of easy or consistent pronunciation. From long-established custom they are usually pronounced in English-speaking countries according to the pronunciation of Latin after the English method. . . . The subject is one into which considerations of taste, convenience, and custom enter to such an extent that it is most difficult to lay down definite principles free from pedantry.''

SUGGESTED READING

Fernald, M. L. 1950. Gray's Manual of Botany, eighth edition. American Book Company, New York.

Jackson, B. D. 1928. A Glossary of Botanical Terms. J. B. Lippincott Company, Philadelphia.

Radford, A. E. et al. 1974. Vascular Plant Systematics. Harper & Row, New York. Chapter 3, pp. 35–78.

Stearn, W. T. 1966. Botanical Latin. Thomas Nelson & Sons, Ltd., London.

SUMMARY FOR BOTANICAL NAMES

Definition of Botanical Names *Scientific names* of plants are either Latin words or words that have been latinized from some other language, most often Greek.

Purpose of the Study of Botanical Names To learn how to coin botanical names from the proper Latin or Greek roots, prefixes, suffixes, and terminations as scientific names that are connotative, mnemonic, reasonably short, and easy to pronounce; to understand previously coined botanical names by consulting alphabetized classifications of adjectival characteristics and by learning the meaning of common prefixes, suffixes, and the names of plant parts.

Operations in the Study of Botanical Names To form botanical scientific names from the proper Latin or Greek roots, prefixes, suffixes, and terminations according to the rules of Latin grammar and the recommendations of the *International Code for Botanical Nomenclature;* to study Latin roots, prefixes, suffixes, and terminations for the meanings of botanical scientific names.

Basic Premise for the Study of Botanical Names Each taxon with a particular circumscription, position, and rank can bear only one botanical name that is the communication symbol and reference base for information storage, retrieval, and use.

Fundamental Principles for the Study of Botanical Names

1. Latin is the basis for scientific botanical names; words from other languages, most often Greek, and proper names are latinized for use in coining scientific names.
2. A botanical name used as a scientific name can come from any source, for example, descriptive, commemorative, or based on old or common names.

Guiding Principles for the Coining or Use of Botanical Names

1. Scientific botanical names should be short, easy to pronounce, distinctive in meaning, easy to remember, and rank indicative.

2. Scientific botanical names should be coined according to the recommendations in the latest ICBN and the rules for Latin grammar.
3. Most specific epithets are adjectives or words used in an adjectival sense, that must have masculine, feminine, or neuter endings.
4. Commemorative epithets usually have genitive rather than nominative endings.
5. Names of plant parts are frequently used in combination with appropriate prefixes or suffixes to form meaningful specific epithets or names for other taxa.
6. Latin adjectives with modified endings often serve as prefixes in the formation of meaningful specific epithets or names for other taxa.
7. Compound names formed by the adding of suffixes can result in nouns or adjectives that are used as meaningful specific epithets or names for other taxa.

Basic Assumptions in the Study of Botanical Names

1. A botanical name becomes the scientific name for a taxon with a particular circumscription, position, and rank.
2. The scientific name for each taxon with a particular circumscription, position, and rank conforms to the rules in the latest ICBN.

QUESTIONS ON BOTANICAL NAMES

1. What is a botanical name? Scientific name? Specific epithet? Botanical term?
2. What are two reasons for the study of botanical names?
3. How are specific and infraspecific names formed? Generic names? From which parts of speech?
4. What is the significance of gender in the formation of descriptive epithets? Commemorative epithets? Other specific epithets?
5. What precautions should be taken in the formation of specific epithets? Infraspecific epithets? Generic names? What should be considered in the selection of the specific epithet for any plant?
6. How are specific epithets usually descriptive? What is the logical choice of character and characteristic used in an epithet?
7. What are the reasons for learning the Latin equivalents, and Greek where known, for plant parts?
8. What is meant by prefix and suffix? Classify prefixes and suffixes as to type and give the basis of your classification. Characterize each type of prefix and suffix.
9. What is meant by nominative epithet? What is the basis for inclusion of miscellaneous epithets in this chapter?
10. What are the basic problems in pronouncing botanical names? Why was Latin selected for international nomenclature?
11. What are the guiding principles for the study of epithets?
12. What are the basic assumptions for the study of plant epithets?

EXERCISES IN BOTANICAL NAMES

1. After studying prefixes, suffixes, and names of plant parts (A. Formation of Specific Epithets, IV Names of Plant Parts used as Epithets, V Prefixes, VI Suffixes), select ten combined epithets (prefix or suffix and plant part) from a manual or flora pertinent to your region and give the meaning of each.
2. After studying types of epithets (B. Meanings of Specific Epithets), select ten species from a manual or flora pertinent to your region and indicate the type of epithet, for example, color,

habit, habitat, season, and the gender of each (A. Formation of Specific Epithets, I. General Information).

3. After studying the pronunciation of botanical names (C. Pronunciation of Botanical Names), select ten species from a manual or flora pertinent to your region, syllabify, accent, and pronounce the generic name as well as the specific epithet. Give the reasons for your accenting.

4. Fill in the first chart below by coining epithets from a prefix type and plant part and the second chart by combining an adjectival epithet with a plant part and the gender indicated. The first entry on each chart is an example of each.

(a) Epithet	Meaning of epithet	Prefix type	Plant part type	Gender	Syllabify and accent
1. diphyllus	two-leaved	Number	Leaf, Gr.	M	di-phýl-lus
2.		Size	Stem, L.	F	
3.		Position	Stamen, Gr.	N	
4.		Shape	Petal, L.	M	
5.		Number	Root, Gr.	F	
6.		Size	Fruit, Gr.	N	
7.		Position	Stamen, L.	M	
8.		Shape	Leaf, Gr.	F	

(b) Epithet	Meaning of epithet	Prefix combination	Plant part type	Gender	Syllabify and accent
1. leucosperma	white-seeded	Color	Seed, Gr.	F	leu-co-spér-ma
2.		Apex	Petal, L.	M	
3.		Vesture	Stem, L.	F	
4.		Sex	Flower, L.	N	
5.		Shape	Carpel, L.	M	

5. Using the chart below as a guide, with size prefixes and leaf as a part, combine size prefixes with flower, using the Latin stem and masculine ending; with fruit, Greek and feminine; with stem, Latin and neuter; with seed, Greek and feminine.

Latin	Meaning	Greek
angustifolia	narrow-leaved	stenophylla
brevifolia	short-leaved	brachyphylla
crassifolia	thick-leaved	pachyphylla
grandifolia	large-leaved	macrophylla
latifolia	wide- or broad-leaved	platyphylla
longifolia	long-leaved	dolichophylla
parvifolia	small-leaved	microphylla
tenuifolia	slender-leaved	leptophylla

6. Syllabify, accent, pronounce, and give the reason for the accent for each of the following generic names or epithets:

Leiophyllum	ipecacuanhae	abortivus
Actaea	Chelone	Asarum
graveolens	catawbiense	Hydrangea
polypodioides	Ruellia	Cleistes
tulipifera	psoralioides	Oenothera
marilandica	Luzula	Sacciolepis

chapter 5

Plant Description

Taxonomists as describers select characters and determine character states for the circumscriptions of new taxa; change the circumscriptions for old taxa that have been remodeled, divided, united, transferred, or changed in rank; and determine the circumscription for a specimen according to a system of description or classification.

Systematists as observers study the characteristics of organisms, populations, and taxa to discover evidence of spatial, temporal, abiotic, and biotic relationships for a better understanding of the taxonomy of plant diversity. These observed characteristics have to be verbalized for communication. The characteristics or descriptors have to be available for articulation. Most of the descriptive terminology and narrative included in this chapter can be applied to all objects, entities, and organisms as well as taxa.

Plant descriptions should convey an image or impression of plant characters, attributes, and nature. Descriptions of plants, their parts, and taxa provide the basic information for all of taxonomy, that is, characterization, identification, and classification, which form the core of floras, manuals, revisions, monographs, and systems of classification. Descriptive terminology constitutes the framework for determination of relationships between taxa.

In general, a description is a statement of the characteristics of a taxon, individual organism, object, or entity. A characterization of a taxon or an organism is an orderly recording of distinctive characters with appropriate character states for the plant and its parts. Circumscription is the orderly recording of limiting and diagnostic characters with pertinent character states for taxa with given positions and ranks. The purposes of the descriptive study of plants are to provide a vocabulary of descriptors for communication about taxa and organisms and their parts and to supply the descriptive reference base for named taxa.

The typical descriptive vocabulary is an alphabetical glossary of defined terms. Description as a process involves the selection of characters and the determination of character states for the circumscriptions of new taxa and the making of changes in circumscriptions of old taxa that have been remodeled, divided, united, transferred, or altered in rank. Descriptions of taxa are changed for the proper reasons of either a more profound knowledge of the taxa resulting from adequate taxonomic study or nonconformance of the descriptors to accepted practice or sound etymological study.

No absolutely comprehensive description has ever been written for any species or higher taxon. In data banking, systems are being devised for input and retrieval of all descriptive information as it is produced by various specialists. Monographic treatments are fairly comprehensive for selected types of evidence such as morphological, cytological, and ecological but definitely not for all fields of evidence (see Chapter 8). Manuals usually contain little beyond the descriptive terminology relevant to diagnostic and accurate identification. Simple floras have a more limited descriptive content, frequently with no more than geographic and ecological data for each species listed.

Two major problems in descriptive writing are (i) making the description relevant to the type of publication contemplated and (ii) selecting pertinent characters with the appropriate character states for each taxon described. Another problem is following a descriptive sequence established by the author or editor for consistent and comparative treatment. Still another problem could be the bridging of the terminology gap due to an inadequate vocabulary or poor definition of terms in the glossary. Many other problems may have to be resolved as one describes parts, plants, and taxa accurately and relevantly.

A. DESCRIPTIONS IN FLORAS AND MANUALS

Descriptions of species in floras include vegetative and reproductive features, times of flowering and fruiting, place of growth or habitat, and distribution, along with pertinent synonymy used in other floras, and references to the taxonomy of the included taxa. Basically, this format for descriptions of species has been in use since the time of the herbalists in the sixteenth century. Chromosome number (cytology) is the only new type of character found in manuals published in the last 50 years. Generic and familial descriptions usually follow an abbreviated format for the descriptions of species.

The descriptive treatments of taxa (structures described and adjectives applied) are more consistent and stylistically similar in recent floras than in the older manuals. The hierarchy generally used now for structure is plant, plant organ, and plant organ part in that order with an adjectival sequence: type, color, shape, size, number, and disposition pertinent to each structural part. The descriptive format—which is a sequence of presentation of descriptive data by structure, adjective, and punctuation—is explained in the preface of most manuals. The presentation of evidence is traditionally morphological: vegetative (plant, root, stem, leaf), reproductive (inflorescence, flower, fruit, seed), cytological (chromosome number), phenological (flowering-fruiting dates), ecological (habitat), and geographic (region, state, province, county). Synonymy and references follow the treatment of evidence. The parts of organs also follow a regular order, for example, in the flower the order is calyx, corolla, androecium, and gynoecium; and in the androecium, for instance, it is stamen, anther, and filament. The adjectival sequence for the calyx, for example, might be number, fusion, orientation, color, shape, size, and texture of sepals, or possibly the symmetry of the entire calyx and then the adjectives for each sepal. All taxa of coordinate rank should have the same sequence for consistent and comparable treatment.

B. CHARACTERS AND CHARACTERIZATION

The concept of character is fundamental to the science of taxonomy. Characters provide the basic information for classification and the diagnostic features used in identification; characters are essential for the determination of relationships; and combinations of characters associated with different organisms form the bases for the naming of taxa. Characterization of plants and their component parts is the initial and primary process in systematics. The assignment of attributes to plants and their parts (characterization) precedes, and is a requirement for, the establishment of discontinuities in form and function between organisms. Discontinuities have to be present for the delimitation and definition of taxa and are prerequisites to the development of identification and classification schemes. Only well-delimited and well-defined taxa should be named.

In general, a character is any expression of form, structure, or function that the taxonomist uses for a particular purpose such as comparison or interpretation. Practically, a character may be defined as any feature whose expression can be measured, counted, or otherwise differentiated. Fundamentally, a taxonomic character is one with two or more states that cannot be further subdivided logically in the material being studied. Character states are descriptors or characteristics.

The fundamental unit of description is the *character state,* which is a basic component of *character.* Characters are grouped as *character sets,* which together form a type of *evidence.* For example, ''lanceolate'' is a *character state* of the *character* shape; ''serrate'' is a *character state* of the *character* margin; ''obtuse'' is a *character state* of the *character* apex. Shape, margin, and apex are characters in the *character set* general shape. ''Whorled'' is a *character state* for the *character* arrangement; ''terminal'' is a *character state* for the *character* position; ''descending'' is a *character state* for the *character* orientation. Arrangement, position, orientation are characters in the *character set,* disposition. These two characters sets and others would be components of a type of *evidence,* morphology (see Figure 1.3). Morphological, anatomical, palynological, and other types of evidence (see Chapter 8) would form a comprehensive descriptive terminology for plants and their parts.

Botanical description includes several basic types of characters: those pertinent to the structural, functional, and developmental features of plants as organisms; those generally applicable to any object or organism and its parts (adjectival); and those used to describe groups of characters in a particular context. Some examples of structural characters pertinent to plant description are a leaflet arrangement type (a character) with several states—trifoliolate, biternate (character states); mature embryo sac type (a character) with several states—*Pennaea, Peperomia* (character states); pollen aperture type (a character) with several states—colporate, syncolpate (character states). An example of a developmental character is primary xylem development type with several states—exarch, endarch, mesarch; and an example of a functional character is breeding system type with several states—chasmogamy, cleistogamy (character states). General descriptive characters pertinent to plants but applicable to all objects are color, size, shape, number, and so

Table 5.1 MISCELLANEOUS CHARACTERS

1. a. Adaptive. Characters that respond to the changing environmental situation enabling the organism to survive by an alteration of structure and/or function.
   b. Nonadaptive. Characters that do not change in response to a changing environmental situation.
2. a. Analytic. Characters used in identification, characterization, and delimitation of taxa.
   b. Synthetic. Characters of constant nature and wide occurrence used in constitutive and organizational characterization of larger taxa.
3. a. Biological. Characters related to some vital function or behavior.
   b. Fortuitous. Characters not related to some vital function or behavior.
4. a. Continuous. Characters showing uniform integradation or a cline.
   b. Discontinuous. Characters not showing uniform integradation or a cline, but separated by breaks in the continuum.
5. a. Cryptic. Characters hidden, microscopic.
   b. Phaneritic. Characters obvious, macroscopic.
6. a. Diagnostic. Characters used to distinguish one taxon from another.
   b. Descriptive. Characters used to give the features or attributes of a taxon.
7. a. Good. Characters that are easily recognizable, have a narrow range or expression, and are seemingly genetically fixed.
   b. Bad. Characters that are not easily recognizable, have a wide range of expression, and are seemingly not genetically fixed.
8. a. Homologous. Characters having a common ancestral origin.
   b. Analogous. Characters similar but not having a common ancestral origin.
9. a. Logically correlated. Character or property that is a logical consequence of another.
   b. Noncorrelated. Character or property that is not the logical consequence of another.
10. a. Macro. Characters gross, macroscopic, usually external.
    b. Micro. Characters small, microscopic, usually internal.
11. a. Meaningful. Characters or attributes that are significant or valuable in characterization, identification, or classification.
    b. Meaningless. Characters or attributes that are not significant or valuable in characterization, identification, or classification.
12. a. Ontogenetic. Characters associated with the development of the individual.
    b. Phylogenetic. Characters associated with the development of the taxon over long periods of time.
13. a. Plastic. Characters seemingly variable.
    b. Fixed. Characters seemingly invariable.
14. a. Primitive. Characters possessed by present-day taxon and its ancestors.
    b. Advanced. Characters possessed by present-day taxon not possessed by its ancestors.
15. a. Qualitative. Characters relating to kinds of forms, structures, behaviors, or functions.
    b. Quantitative. Characters assessed by number and size.
16. a. Reliable. Characters that are consistent, distinct, recognizable, and usable by taxonomists.
    b. Unreliable. Characters that are inconsistent, indistinct, unrecognizable, and not usable by taxonomists.
17. a. Specific. Diagnostic characters used in delimiting a species.
    b. Generic. Diagnostic characters used in delimiting a genus.
18. a. Two-state. Characters all or none, present or absent.
    b. Multistate. Characters expressed by more than all or none or present or absent states.
19. a. Variant. Characters that do vary within population samples.
    b. Invariant. Characters that do not vary within population samples.
20. a. Weighted. Characters given a greater value due to presumed importance.
    b. Nonweighted. Characters given same value regardless of importance.

on. Special characters are used to describe groups of characters and their character states within a given context such as reliable/unreliable from an identification standpoint or adaptive/nonadaptive from an evolutionary viewpoint. The botanically applicable miscellaneous special characters are summarized as contrasting couplets in Table 5.1. General descriptive characters are treated later in this chapter and structural characters in Appendix D.

A precise set of terms has to be devised for comprehensive description. Character states have to be well defined for accuracy and ease of application. Characters have to be effectively ordered and ranked with the basis for ordering and ranking explained for a meaningful comparative classification of descriptive terms. Character sets have to be reasonably delimited but sufficiently inclusive to cover all aspects of the types of evidence used in description. Comprehensive description has to include all types of evidence for diversity of information necessary for determination of true relationships between taxa. Precision in definition of character states, logic in the hierarchical ordering and arranging of characters and character sets, and comprehensiveness of descriptive evidence are fundamental aspects of phytography.

C. GENERAL DESCRIPTIVE CHARACTERS

Phytography deals with the descriptive terminology of plants and their component parts for the purpose of providing an accurate and complete vocabulary for description, identification, and classification. Phytographic studies should furnish the student with a vocabulary for intelligent communication about plants and an understanding of the use of relative terms and help the student observe plants more critically and describe them more precisely. In this chapter, the general descriptive characters, such as color, form, and size applicable to objects of any type are classified and defined (Table 5.2). The traditional structural characters are treated under morphological evidence in Appendix D.

Table 5.2   SUMMARY OF GENERAL DESCRIPTIVE CHARACTERS TREATED IN THE TEXT

I. Coloration	VI. Disposition
Color	Arrangement
Color patterns	Position
II. Size	Orientation
Linear	Posture
Area	VII. Surface
Volume	Configuration
III. Number	Venation
Units	Epidermal excrescences
Cycly	Vesture
Merosity	VIII. Texture
IV. Fusion	General
General	IX. Symmetry
V. Shape	General
Symmetric figures	X. Temporal phenomena
Asymmetric or special	Duration
figures	Maturation
Apices and Bases	Periodicity
Margins	

In the following treatment, all terms within a subject are alphabetized except those in *shape,* where it was deemed best to group related terms. Either adjectives or noun forms, usually not both, are used throughout a classification. We know that some classifications have incomplete vocabularies, but instructions are included for the determinations of the meanings of many additional terms.

Phytography as an ordered and precise subject is still in a rather primitive state. It is generally known that different terms are often applied to the same character state, and, conversely the same term is used for many character states. It seems that there are too many terms in this chapter, but a real need exists for even more terms, with precise definition and application. Improvement is needed throughout the term-concept-classification system to advance phytography and taxonomy.

I. Coloration

Color is the quality of an object or substance with respect to light reflected by the object, usually determined visually by measurement of hue, chroma, and value. Standardized color charts should be used for accurate color description of plant parts. Three standard color charts for solid color descriptions are: (1) Ridgway, R., 1912, *Color Standards and Nomenclature,* published by the author, Washington, D. C.; (2) *Horticultural Colour Charts,* issued by the British Colour Council in collaboration with the Royal Horticultural Society, 2 volumes, London, 1938, 1941; and (3) Munsell color charts, which have very wide usage. Colors are specified in terms of hue, value, and chroma, a Munsell notation being written as hue value/chroma. The following information on color is from *Munsell Soil Color Charts,* 1975 edition, published by the Macbeth Division of the Kollmorgen Corporation, Baltimore, Maryland. Soil colors cover about one-fifth of the entire color range found in the Munsell colors. Soil color, nomenclature, and classification are applicable to all color.

The Hue notation of a color indicates its relation of Red, Yellow, Green, Blue, and Purple; the Value notation indicates its lightness; and the Chroma notation indicates its strength (or departure from a neutral of the same lightness).

The colors displayed on the individual Soil Color Charts are of constant Hue, designated by a symbol in the upper right-hand corner of the card. Vertically, the colors become successively lighter from the bottom of the card to the top by visually equal steps; their value increases. Horizontally they increase in Chroma to the right and become grayer to the left. The Value notation of each chip* is indicated by the vertical scale in the far left column of the chart. The Chromas notation is indicated by the horizontal scale across the bottom of the chart.

As arranged in the collection the charts provide three scales: (1) radial, or from one chart to the next in hue; (2) vertical in value; and (3) horizontal in chroma.

"The nomenclature for soil color consists of two complementary systems: (1) Color names, and (2) the Munsell notation of color. Neither of these alone is adequate for all purposes. The color names are employed in all descriptions for publication and for

---

*A color chip is a small rectangle with precise Munsell notation used for comparing the color of chip with that of the object in which color is to be determined. An accurate color chart with chips is too expensive for inclusion in a text of this type.

general use. The Munsell notation is used to supplement the color names wherever greater precision is needed, as a convenient abbreviation in field descriptions, for expression of the specific relations between colors, and for statistical treatment of color data. The Munsell notation is especially useful for international correlation, since no translation of color names is needed. The names for soil colors are common terms now so defined as to obtain uniformity and yet accord, as nearly as possible, with past usage by soil scientists. Bizarre names like 'rusty brown,' 'mouse gray,' 'lemon yellow,' and 'chocolate brown' should never be used."

The soil color names and their limits are given in the diagrams that appear opposite each chart.

"The Munsell notation for color consists of separate notations for hue, value, and chroma, which are combined in that order to form the color designation. The symbol for hue is the letter abbreviation of the color of the rainbow (R for red, YR for yellow-red, Y for yellow) preceded by numbers from 0 to 10. Within each letter range, the hue becomes more yellow and less red as the numbers increase. The middle of the letter range is at 5; the zero point coincides with the 10 point of the next redder hue. Thus 5YR is in the middle of the yellow-red hue, which extends from 10R (zero YR) to 10YR (zero Y).

"The notation for value consists of numbers from 0, for absolute black, to 10, for absolute white. Thus a color of value 5/ is visually midway between absolute white and absolute black. One of value 6/ is slightly less dark, 60 percent of the way from black to white, and midway between values of 5/ and 7.

"The notation for chroma consists of numbers beginning at 0 for neutral grays and increasing at equal intervals to a maximum of about 20, which is never really approached in soil. For absolute achromatic colors (pure grays, white, and black), which have zero chroma and no hue, the letter N (neutral) takes the place of a hue designation.

"In writing the Munsell notation, the order is hue, value, chroma with a space between the hue letter and the succeeding value number, and a virgule between the two numbers for value and chroma. If expression beyond the two numbers is desired, decimals are always used, never fractions. Thus the notation for a color of hue 5YR, value 5, chroma 6, is 5YR 5/6, a yellowish-red. The notation for a color midway between the 5YR 5/6 and 5YR 6/6 chips is 5YR 5.5/6; for one midway between 2.5YR 5/6 and 5YR 6/8, it is at 3.75YR 5.5/7. The notation is decimal and capable of expressing any degree of refinement desired. Since color determinations cannot be made precisely in the field—generally no closer than half the interval between colors in the chart—expression of color should ordinarily be to the nearest color chip."

a. Distribution of Colors (Color Patterns)   (Figure 5.1)

The following terms are related to distribution of color (patterns) on a plant organ or part.

Banded. Transverse stripes of one color crossing another.

Blotched. Color disposed in broad, irregular blotches.

Bordered. One color surrounded by an edging of another.

Clouded. Colors unequally blended together.

Discoidal. A single large spot of color in the center of another.

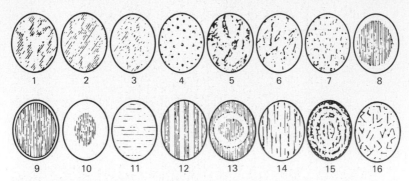

Figure 5.1 Color patterns (adapted from Lindley, 1848). 1. variegated, 2. blotched, 3. spotted, 4. dotted, 5. clouded, 6. marbled, 7. tessellated, 8. bordered, 9. edged, 10. discoidal, 11. banded, 12. striped, 13. ocellated, 14. painted, 15. zoned, 16. lettered.

Dotted. Color disposed in very small round spots.

Edged. One color surrounded by a very narrow rim of another.

Lettered. Color disposed as separate line figures similar to letters.

Marbled. A surface traversed by irregular vein of color, as a block of marble often is.

Ocellated. A broad spot of some color with another spot of a different color within it.

Painted. Color disposed in streaks of unequal intensity.

Spotted. Color disposed in small spots.

Striped. Longitudinal stripes of one color crossing another.

Tessellated. Color arranged in small squares, so as to have some resemblance to a checkered pavement.

Variegated. Color disposed in various irregular, sinuous spaces.

Zoned. The same as ocellated, but the concentric bands more numerous.

II. Size

Botanically, size usually refers to the spatial dimensions or proportions of plants and plant parts, usually expressed as number values in reference to length, width, and depth or as an expression of relative or comparative amounts. Spatial measurements are one-dimensional, linear; two-dimensional, area; and three-dimensional, volume. Length/width ratios are frequently given for planar structures such as leaves and sepals or petals. The linear characters are usually the length or width of a plant part or the height of a plant; the area characters are the length and width combined or the number of square units; and volume characters are indicated by diameter or thickness.

Many kinds of adjectives are used to describe the sizes of plants and plant parts, for example, gigantic, giant, large, medium, small, reduced, minute, dwarf, attenuate, long, tall, short, abbreviated, and so on. All of these adjectives are meaningless unless placed in a context of size classes with the dimensions of each class indicated precisely.

Tree height		
Three size classes	**Five size classes**	**Seven size classes**
Small trees <5 m	Dwarf trees <2 m	Dwarf trees <1 m
Medium trees 5–15 m	Small trees 2–5 m	Very small trees 1–2 m
Large trees >15 m	Medium trees 5–15 m	Small trees 2–5 m
	Large trees 15–50 m	Medium trees 5–15 m
	Giant trees >50 m	Large trees 15–25 m
		Very large trees 25–50 m
		Giant trees >50 m

The greater the range in sizes, the greater the number of size classes usually required. In general, actual sizes are preferred to adjectives for size classes, but since the adjectives are frequently used as a result of previous experience, including many size estimates, classes should be established with dimensional bases. In manuals or floras without size classes, the user should examine the sizes as related to the adjectives used by the author to obtain some understanding of the author's use of the different adjectives.

a. Selected Size/Plant Parts Terms and Epithets

Many terms have been coined pertaining to the size of plant parts in relation to similar plant parts of other taxa, and terms pertaining to the size of plant parts in relation to similar plant parts of the same organism. The terms have been formed by combining a size prefix with the appropriate word stem of an organ or its part.

English terms	Prefix	Plant part	Meaning	Epithet (masculine)
Anisophyllous	aniso- (Gr)	phyllon (Gr)	with leaves of more than one size or shape	anisophyllus
Brachystylous	brachy- (Gr)	stylos (Gr)	having short styles	brachystylus
Brevirostrate	brevi- (L)	rostratus (L)	having short beak(s)	brevirostratus
Heterophyllous	hetero- (Gr)	phyllon (Gr)	having leaves of different sizes or shapes	heterophyllus
Homocarpous	homo- (Gr)	carpos (Gr)	having fruits of same sizes or shapes	homocarpus
Isopetalous	iso- (Gr)	petalon (Gr)	having petals of equal size or shape	isopetalus
Macrocarpous	macro- (Gr)	carpos (Gr)	having large fruits	macrocarpus
Megacephalous	mega- (Gr)	cephalon (Gr)	having large heads	megacephalus
Microphyllous	micro- (Gr)	phyllon (Gr)	having small leaves	microphyllus
Parvifolious	parvi- (L)	folius (L)	having small leaves	parvifolius
Platyspermic	platy- (Gr)	sperma (Gr)	having wide seeds bilaterally symmetrical	platyspermus
Stenophyllous	steno- (Gr)	phyllon (Gr)	having narrow leaves	stenophyllus

III. Number

Number refers to the sum, total, count or aggregate of a collection of units, or simply to a numeral. The sum total of individual units is indicated by a number. Botanically the number of units within a structure is usually indicated, for example, five petals in a corolla or flower; ten stamens in the flower or androecium. Many terms have been formed to indicate the number of units within a structure.

a. Selected Number/Plant Part Terms and Epithets

English terms	Prefix	Plant part	Meaning	Epithet (feminine)
Monocarpic	mono- (Gr)	carpos (Gr)	having one fruit	monocarpa
Uniflorous	uni- (L)	flos (L)	having one flower	uniflora
Diphyllous	di- (Gr)	phyllon (Gr)	having two leaves	diphylla
Bicarpellate	bi- (L)	carpellum (L)	having two carpels	bicarpellata
Trispermous	tri- (Gr)	sperma (Gr)	having three seeds	trisperma
Tetrastichous	tetra- (Gr)	stichos (Gr)	having four rows (i.e., of leaves)	tetrasticha
Pentandrous	penta- (Gr)	andros (Gr)	having five stamens	pentandra
Multicaulous	multi- (L)	caulis (L)	having many stems	multicaulis
Polycephalous	poly- (Gr)	cephalon (Gr)	having many heads	polycephala

b. Cycly

Appropriate numerical terms have been formed for units of plant whorls, for example, the number of whorls of floral parts in a flower or the number of whorls of leaves on a stem. Cycly refers to the number of whorls.

Acarpous. Without carpels or carpellate whorl or gynoecium.

Apetalous. Without petals or corolla.

Aphyllous. Without leaves or whorls of leaves.

Arhizous. Without roots or whorls of roots.

Asepalous. Without sepals or calyx.

Astemonous or anandrous. Without stamens or androecium or staminate whorl.

Dicyclic. Two-whorled.

Monocyclic. One-whorled.

Oligotaxy. Reduction in number of whorls.

Pentacyclic. Five-whorled.

Pleiotaxy. Increase in number of whorls.

Polycyclic. Many-whorled.

Tetracyclic. Four-whorled.

Tricyclic. Three-whorled.

c. **Merosity**

Numerical terms are also used with reference to the number of parts within a whorl. Merosity refers to the number of parts within whorls of floral parts, leaves, or stems.

Dimerous. Whorl with two members.

Heteromerous or anisomerous. With different number of members in different whorls.

Isomerous. With same number of members in different whorls.

Monomerous. Whorl with one member.

Oligomerous. With reduction in number of members within whorl.

Pentamerous. Whorl with five members.

Pleiomerous. With increase in number of members within whorl.

Polymerous. Whorl with many members.

Pseudomonomerous. Whorl seemingly with one member which is a fusion product of two or more parts.

Tetramerous. Whorl with four members.

Trimerous. Whorl with three members.

IV. Fusion

Fusion refers to the combining or blending of a structure or structures with similar or dissimilar structures. General terms related to the states of fusion are presented below.

a. **General Fusion Terms**

Adherent. With unlike parts or organs joined, but only superficially and without actual histological continuity.

Adnate. With unlike parts or organs integrally fused to one another with histological continuity.

Coalesced. With like or unlike parts or organs incompletely joined and partially fused in a more or less irregular fashion.

Coherent. With like parts or organs joined, but only superficially and without actual histological continuity.

Connate. With like parts or organs integrally fused to one another with histological continuity.

Contiguous. Touching but not adnate, connate, adherent, or coherent.

Distinct. With like parts or organs unjoined and separate from one another.

Fasciated. Unnaturally and often monstrously connate or adnate, the coalesced parts often unnaturally proliferated in size and/or number; for example, inflorescence of *Celosia*.

Free. Unlike parts or organs not joined, separate from one another.

V. Shape

Botanically, shape refers to the form, figure, or outline of plants and plant parts. Terms are presented for shapes of two- and three-dimensional structures as well as the configurations of the apices, bases, and margins of planar structures or sections of solid figures.

a.   **Shapes—Plane and Solid**   **(Figures 5.2 and 5.3)**

1. *Symmetric Figures*. Based on terminology in Taxon (Volume 11) and Stearn (1966).

A. *Elliptic*. With widest axis at midpoint of structure and with margins symmetrically curved.

   Plane L/W
   a. Narrowly elliptic more than 6:1–3:1
   b. Elliptic 2:1–3:2
   c. Widely elliptic 6:5
   d. Circular 1:1
   e. Oblate 5:6
   f. Transversely elliptic 2:3–1:2
   g. Narrowly transversely elliptic 1:3–1:6 or more

   Solid L/D
   a. Narrowly ellipsoid more than 6:1–3:1
   b. Ellipsoid 2:1–3:2
   c. Broadly ellipsoid 6:5
   d. Spheroid 1:1
   e. Obloid 5:6
   f. Transversely ellipsoid 2:3–1:2
   g. Lenticular 1:3–1:6 or more

B. *Oblong*. With widest axis at midpoint of structure and with margins essentially parallel.

   Plane L/W
   a. Linear more than 12:1
   b. Narrowly oblong 6:1–3:1 or lobate
   c. Oblong 2:1–3:2
   d. Widely oblong 6:5
   e. Square 1:1
   f. Transversely widely oblong 5:6
   g. Transversely oblong 2:3–1:2
   h. Transversely narrowly oblong 1:3–1:6
   i. Transversely linear 1:12 or more

   Solid L/D
   a. Cylindric or terete more than 12:1
   b. Narrowly oblong 6:1–3:1
   c. Oblong 2:1–3:2
   d. Broadly oblong 6:5
   e. Cubical 1:1
   f. Transversely broadly oblong 5:6

SHAPES

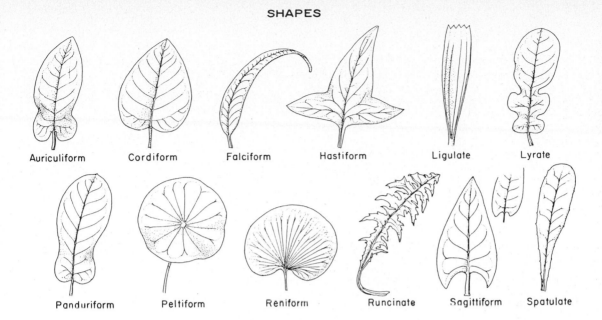

Auriculiform Cordiform Falciform Hastiform Ligulate Lyrate

Panduriform Peltiform Reniform Runcinate Sagittiform Spatulate

SYMMETRIC PLANE FIGURES
(Adapted from Taxon, 1962)

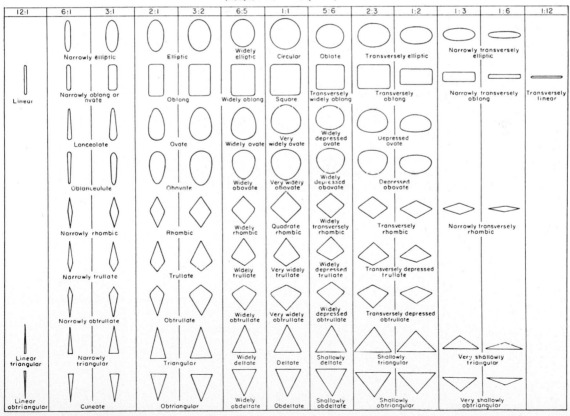

Figure 5.2 Plane shapes.

SHAPES

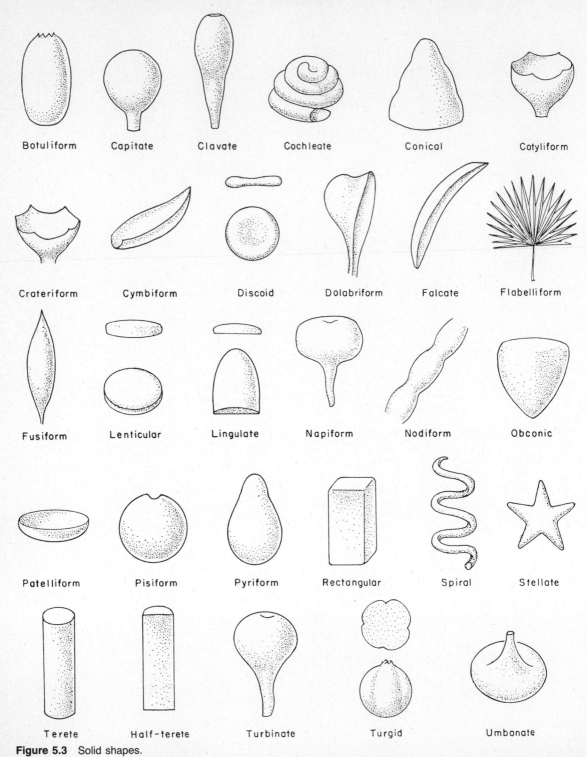

Figure 5.3 Solid shapes.

g. Transversely oblong 2:3–1:2
h. Transversely narrowly oblong 1:3–1:6
i. Transversely cylindrical or terete 1:12 or more

C. *Ovate*. With widest axis below middle and with margins symmetrically curved; egg-shaped.

Plane L/W
a. Lanceolate more than 6:1–3:1
b. Ovate 2:1–3:2
c. Widely ovate 6:5
d. Very widely ovate 1:1
e. Widely depressed ovate 5:6
f. Depressed ovate 2:3–1:2

Solid L/D
a. Lanceoloid more than 6:1–3:1
b. Ovoid 2:1–3:2
c. Broadly ovoid 6:5
d. Very broadly ovoid 1:1
e. Broadly depressed ovoid 5:6
f. Depressed ovoid 2:3–1:2

D. *Obovate*. Inversely ovate.

Plane L/W
a. Oblanceolate more than 6:1–3:1
b. Obovate 2:1 3:2
c. Widely obovate 6:5
d. Very widely obovate 1:1
e. Widely depressed obovate 5:6
f. Depressed obovate 2:3–1:2

Solid L/D
a. Oblanceoloid more than 6:1–3:1
b. Obovoid 2:1–3:2
c. Broadly obovoid 6:5
d. Very broadly obovoid 1:1
e. Broadly depressed obovoid 5:6
f. Depressed obovoid 2:3–1:2

E. *Rhombic*. With widest axis at midpoint of structure, and with straight margins; elliptic but margins straight and middle-angled.

Plane L/W
a. Narrowly rhombic more than 6:1–3:1
b. Rhombic 2:1–3:2
c. Widely rhombic 6:5
d. Quadrate rhombic 1:1
e. Transversely widely rhombic 5:6
f. Transversely rhombic 2:3–1:2
g. Narrowly transversely rhombic 1:2–1:6 or more

Solid L/D
a. Narrowly rhomboid more than 3:1
b. Rhomboid 2:1–3:2
c. Broadly rhomboid 6:5
d. Quadrate rhomboid 1:1
e. Transversely broadly rhomboid 5:6
f. Transversely rhomboid 2:3–1:2
g. Narrowly transversely rhomboid 1:3–1:6 or more

F. *Trullate*. With widest axis below middle and with straight margins; ovate but margins straight and angled below middle; trowel-shaped.

Plane L/W
a. Narrowly trullate more than 6:1–3:1
b. Trullate 2:1–3:2
c. Widely trullate 6:5
d. Very widely trullate 1:1
e. Widely depressed trullate 5:6
f. Transversely depressed trullate 2:3–1:2

Solid L/D
a. Narrowly trulloid more than 6:1–3:1
b. Trulloid 2:1–3:2
c. Broadly trulloid 6:5
d. Very broadly trulloid 1:1
e. Broadly depressed trulloid 5:6
f. Transversely depressed trulloid 2:3–1:2

G. *Obtrullate*. Inverscly trullate.

Plane L/W
a. Narrowly obtrullate more than 6:1–3:1
b. Obtrullate 3:2–2:1
c. Widely obtrullate 6:5
d. Very widely obtrullate 1:1
e. Widely depressed obtrullate 5:6
f. Transversely depressed obtrullate 2:3–1:2

Solid L/D
a. Narrowly obtrulloid more than 6:1–3:1
b. Obtrulloid 3:2–2:1
c. Broadly obtrulloid 6:5
d. Very broadly obtrulloid 1:1
e. Broadly depressed obtrulloid 5:6
f. Transversely depressed obtrulloid 2:3–1:2

H. *Triangular*. With three sides and three angles.

Plane L/W
a. Linear-triangular more than 12:1
b. Narrowly triangular 6:1–3:1
c. Triangular 2:1–3:2
d. Widely deltate 6:5
e. Deltate 1:1

    f. Shallowly deltate 5:6
    g. Shallowly triangular 2:3–1:2
    h. Very shallowly triangular 1:3–1:6 or more

Solid L/D
    a. Subulate more than 12:1
    b. Narrowly pyramidal 6:1–3:1
    c. Pyramidal 2:1–3:2
    d. Broadly deltoid 6:5
    e. Deltoid 1:1
    f. Shallowly deltoid 5:6
    g. Shallowly pyramidal 2:3–1:2
    h. Very shallowly pyramidal 1:3–1:6 or more

I. *Obtriangular*. Inversely triangular.

Plane L/W
    a. Linear-obtriangular or narrowly cuneate more than 12:1
    b. Cuneate 6:1–3:1
    c. Obtriangular 2:1–3:2 or widely cuneate
    d. Widely obdeltate 6:5
    e. Obdeltate 1:1
    f. Shallowly obdeltate 5:6
    g. Shallowly obtriangular 2:3–1:2
    h. Very shallowly obtriangular 1:3–1:6 or more

Solid L/D
    a. Linear-obpyramidal or narrowly cuneiform more than 12:1
    b. Cuneiform 6:1–3:1
    c. Obpyramidal or broadly cuneiform 2:1–3:2
    d. Broadly obdeltoid 6:5
    e. Obdeltoid 1:1
    f. Shallowly obdeltoid 5:6
    g. Shallowly obpyramidal 2:3–1:2
    h. Very shallowly obpyramidal 1:3–1:6 or more

2. *Special Plane Figures—Outline.*
    a. Acicular. Needlelike, round or grooved in cross section.
    b. Auriculiform. Usually obovate with two small, rounded, basal lobes.
    c. Cordiform. Heart-shaped.
    d. Dimidiate. Inequilateral with one-half wholly or nearly wanting.
    e. Falcate. Scimitar-shaped.
    f. Filiform. Threadlike, usually flexuous.
    g. Hastiform. Triangular with two flaring basal lobes.
    h. Lunate. Crescent-shaped, with acute ends.
    i. Lyrate. Lyre-shaped; pinnatifid with large terminal lobe and smaller lower lobes.
    j. Obcordiform. Inversely cordiform.
    k. Panduriform. Fiddle-shaped; obovate with sinus or indentation on each side near base and with two small basal lobes.
    l. Peltiform. Rounded with petiole attached to center of blade or apparently to laminar tissue.
    m. Rectangular. Box-shaped, longer than wide.

    n. Reniform. Kidney-shaped, with shallow sinus and widely rounded margins.

    o. Runcinate. Oblanceolate with lacerate or parted margins.

    p. Sagittiform. Triangular-ovate with two straight or slightly incurved basal lobes.

    q. Spathulate or Spatulate. Oblong or obovate apically with a long attenuate base.

3. *Special Solid Figures*. (See calyx, corolla, and perianth types for species three-dimensional shapes, Appendix D.)

    a. Acerose. Needle-shaped, sharp.

    b. Annular. Ringlike.

    c. Arcuate. Bent like the arc of a circle.

    d. Botuliform. Sausage-shaped.

    e. Capillate. Hair-shaped.

    f. Capitate. Headlike.

    g. Clavate. Club-shaped.

    h. Cochleate. Snail-shaped.

    i. Compressed or Complanate. Flattened.

    j. Conical. Having figure of true cone.

    k. Coroniform. Crown-shaped.

    l. Cotyliform. Cup-shaped.

    m. Crateriform. Shallow cup-shaped as the involucre of some species of *Quercus*.

    n. Cruciform or Cruciate. Cross-shaped.

    o. Cylindric. Long-tubular.

    p. Cymbiform. Boat-shaped.

    q. Discoid. Orbicular with convex faces.

    r. Dolabriform. Axe-shaped.

    s. Eccentric. One-sided; off-center.

    t. Falcate or Seculate. Sickle-shaped.

    u. Fistulose. Hollow, as a culm without pith.

    v. Flabelliform. Fan-shaped.

    w. Fusiform. Spindle-shaped, broadest in middle and tapering to each end.

    x. Half-terete. Flat on one side, terete on other, semicircular in cross section.

    y. Hippocrepiform. Horseshoe-shaped.

    z. Lenticular. Biconvex, usually elongate and flattish.

    aa. Lingulate. Tongue-shaped, plano-convex in cross section.

    bb. Meniscoidal. Thin and concave-convex.

    cc. Napiform. Turnip-shaped.

    dd. Navicular. Boat-shaped.

    ee. Nodiform or Nodulose. Knotty or knobby, as the roots of most of the Fabaceae.

    ff. Obconic. Inversely conical.

    gg. Patelliform. Knee-shaped, disk-shaped.

    hh. Pisiform. Pea-shaped.

    ii. Pyriform. Pear-shaped.

    jj. Rectangular. Boxlike, longer than wide.

    kk. Spiral. Twisted like a corkscrew.

    ll. Stellate. Star-shaped.

    mm. Strombiform. Elongate snail-shaped.

    nn. Terete. Cylindrical.

    oo. Torose. Cylindrical with contractions at intervals.

    pp. Turbinate. Top-shaped, obconic.

    qq. Turgid. Tumid or swollen.

    rr.   Umbilicate. Depressed in the center.
    ss.   Umbonate. Round with a projection in center.
    tt.   Umbraculiform. Umbrella-shaped.
    uu.   Vermiform. Worm-shaped.

b. Apices and Bases   **(Figure 5.4)**

(Pertains to leaves, petals, sepals, scales, bracts, or other flattened structures.)

1. *Apices and Bases with Sinuses.*
    a.   Retuse. Lobe rounded; sinus depth to $\frac{1}{16}$ distance to midpoint of blade; margins convex.
    b.   Emarginate. Lobe rounded; sinus depth $\frac{1}{16}$–$\frac{1}{8}$ distance to midpoint of blade; margins straight or convex.
    c.   Cordate (apex obcordate). Lobe rounded; sinus depth $\frac{1}{8}$–$\frac{1}{4}$ distance to midpoint of blade; margins convex and/or straight.
    d.   Cleft. Lobe rounded; sinus depth $\frac{1}{4}$–$\frac{1}{2}$ distance to midpoint of blade; margins convex and/or straight.
    e.   Reniform. Lobe rounded; sinus depth variable; outer margin convex to straight; inner margin convex.
    f.   Auriculate. Lobe rounded; sinus depth variable; outer margin concave, inner margin convex to straight.
    g.   Lobate. Lobe rounded; sinus depth variable; outer and inner margins concave.
    h.   Sagittate. Lobe pointed and oriented downward or inward in relation to petiole or midrib; sinus depth variable; margins variable.
    i.   Hastate. Lobe pointed and oriented outward or divergent in relation to petiole or midrib; sinus depth variable; margins variable.

2. *Apices and Bases without Sinuses.*
    a.   Truncate. Cut straight across; ending abruptly almost at right angles to midrib or midvein.
    b.   Rounded. Margins and apex forming a smooth arc.
    c.   Obtuse. Margins straight to convex, forming a terminal angle of more than 90°.
    d.   Acute (base cuneate). Margins straight to convex, forming a terminal angle of 45–90°.
    e.   Acuminate (base narrowly cuneate). Margins straight to convex, forming a terminal angle of less than 45°.
    f.   Caudate (base attenuate). Acuminate with concave margins.
    g.   Hastate. Margins variable; lobe pointed, oriented outward or divergently in relation to petiole or midrib.
    h.   Cuspidate. Acute but coriaceous and stiff.
    i.   Spinose or Pungent. Acuminate but coriaceous and stiff.

3. *Apices with Midrib, Midvein, or Vein Extension.*
    a.   Apiculate. More than 3:1 l/w, usually slightly curled and flexuous.
    b.   Aristate. More than 3:1 l/w, usually prolonged, straight and stiff.
    c.   Cirrhous. More than 10:1 l/w, coiled and flexuous.
    d.   Mucronate. Less than 3:1 l/w, straight and stiff.
    e.   Mucronulate. 1:1 l/w or broader than long; straight.
    f.   Muticous. Without a vein extension, awn, or hair.
    g.   Piliferous. More than 20:1 l/w, hairlike, flexuous.

LEAF APICES, ATTACHMENTS AND BASES

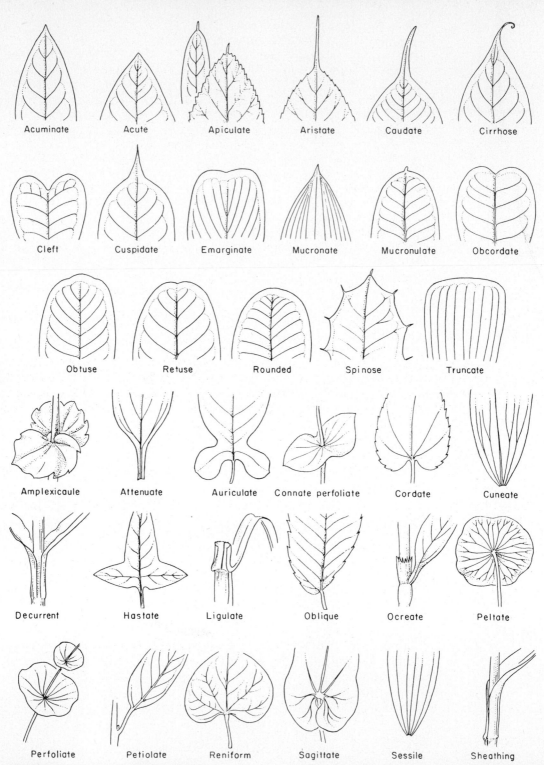

Figure 5.4 Apices, attachments, bases.

4. *Specialized Bases and Leaf Attachments*.
   a. Amplexicaul. Completely clasping the stem.
   b. Clasping. Partly surrounding the stem.
   c. Connate or Connate-perfoliate. Having bases of opposite leaves fused around the stem.
   d. Decurrent. Extending along stem downward from the leaf base.
   e. Ligulate. Having a tonguelike outgrowth at base of blade or tip of sheath.
   f. Oblique. Having an asymmetrical base.
   g. Ocreate. Having a stipular tube surrounding stem above insertion of petiole or blade.
   h. Ocreolate. Diminutive of ocreate; usually applied to bract bases.
   i. Peltate. Usually having petiole attached near the center on the underside of blade.
   j. Perfoliate. Having base completely surrounding the stem.
   k. Petiolate. With a petiole.
   l. Sessile. Without a petiole.
   m. Sheathing. Having tubular structure enclosing stem below apparent insertion of blade or petiole.
   n. Surcurrent. Extending along stem upward from leaf base.

c. **Margins**   **(Figure 5.5)**

(Pertains to leaves, petals, sepals, bracts, scales, or other flattened structures.)

Note: For precision in margin description, indicate or describe the type (see below), the symmetry of the individual tooth, the margins of the individual tooth, the apex of the individual tooth, the type of sinus (rounded or angled), the number of teeth per unit of margin measurement, the spacing (regular or irregular) of the teeth, and the nature of teeth (simple or compound in two or more size groups).

1. *Margin Types*.
   a. Aculeate. Prickly.
   b. Bicrenate or Doubly Crenate. With smaller rounded teeth on larger rounded teeth.
   c. Biserrate or Doubly Serrate. With sharply cut teeth on the margins of larger, sharply cut teeth.
   d. Ciliate. With trichomes protruding from margins.
   e. Cleft. Indentations or incisions cut $\frac{1}{4}$–$\frac{1}{2}$ distance to midrib or midvein.
   f. Crenate. Shallowly ascending round-toothed or teeth obtuse; teeth cut $\frac{1}{16}$–$\frac{1}{8}$ way to midrib or midvein.
   g. Crenulate. Diminutive of crenate, teeth cut to $\frac{1}{16}$ distance to midrib or midvein.
   h. Crispate. Curled; margins divided and twisted in more than one plane.
   i. Dentate. Margins with rounded or sharp, coarse teeth that point outward at right angles to midrib or midvein, cut $\frac{1}{16}$–$\frac{1}{8}$ distance to midrib or midvein.
   j. Denticulate. Diminutive of dentate, cut to $\frac{1}{16}$ distance to midrib or midvein.
   k. Divided. Indentations or incisions cut $\frac{3}{4}$ to almost completely to midrib or midvein.
   l. Entire. Without indentations or incisions on margins; smooth.
   m. Erose. Irregularly, shallowly toothed and/or lobed margins; appearing gnawed.
   n. Filamentose or Filiferous. With coarse marginal fibers or threads.
   o. Fimbriate. Margins fringed.
   p. Fimbriolate. Minutely fimbriate.

MARGINS

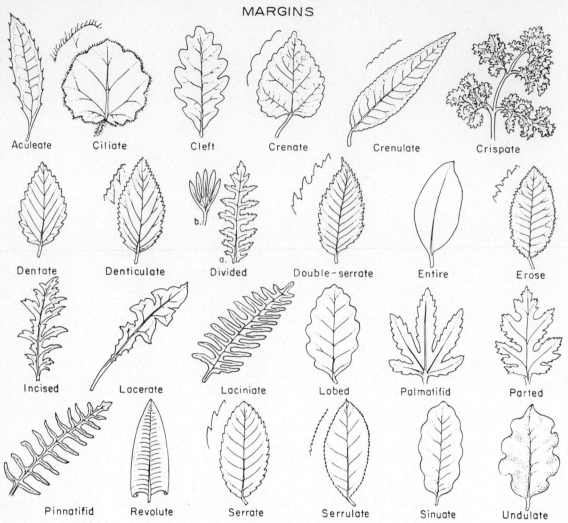

Figure 5.5 Margins.

q. Incised. Margins sharply and deeply cut, usually jaggedly.
r. Involute. Margins rolled inward.
s. Lacerate. Margins irregularly cut, appearing torn.
t. Laciniate. Margins cut into ribbonlike segments.
u. Lobed. Large, round-toothed, cut $\frac{1}{8}$–$\frac{1}{4}$ distance to midrib.
v. Palmatifid. Cut palmately.
w. Parted. Indentations or incisions cut $\frac{1}{2}$–$\frac{3}{4}$ distance to midrib.
x. Pinnatifid. Cut pinnately.
y. Repand. Sinuate with indentations less thant $\frac{1}{16}$ distance to midrib or midvein.
z. Retrorsely Crenate. Rounded teeth directed toward base.
aa. Retrorsely Serrate. Sharp or pointed teeth directed toward base.

bb. Revolute. Margins rolled under.

cc. Serrate. Saw-toothed; teeth sharp and ascending, but cut $\frac{1}{16}$–$\frac{1}{8}$ distance to midrib or midvein.

dd. Serrulate. Diminutive of serrate, but cut to $\frac{1}{16}$ distance to midrib or midvein.

ee. Sinuate. Margins shallowly and smoothly indented, wavy in a horizontal plane, without distinctive teeth or lobes, indented $\frac{1}{16}$–$\frac{1}{8}$ distance to midrib or midvein.

ff. Undulate. Margins shallowly and smoothly indented, wavy in a vertical plane.

VI. Disposition

Botanically, disposition refers to the placement of plant organs or parts in relation to similar or dissimilar structures, parts, the environment, axis, or point.

a. Arrangement (Figure 5.6)

Arrangement refers to the placement of plant organs or parts in relation to similar organs or parts or the placement of plant organs or parts with respect to one another.

Alternate. One leaf or other structure per node.

Clustered, Conglomerate, Agglomerate, Crowded, Aggregate. Parts dense, usually irregularly overlapping each other.

Decussate. Opposite leaves at right angles to preceding pair.

Distichous. Leaves two-ranked, in one plane.

Equitant. Leaves two-ranked with overlapping bases, usually sharply folded along midrib.

Fasciculate. Leaves or other structures in a cluster from a common point.

Geminate or Binate. Paired; in pairs.

Imbricate. Leaves or other structures overlapping.

Loose, Distant, or Scattered. Parts widely separated from one another, usually irregularly.

Opposite. Two leaves or other structures per node, on opposite sides of stem or central axis.

Polystichous. Leaves or other structures in many rows.

Rosulate. Leaves in a rosette.

Secund or Unilateral. Flowers or other structures on one side of axis.

Tetrastichous. Leaves or other structures in four rows.

Whorled, Radiate, or Verticillate. Three or more leaves or other structures per node.

b. Position

Position refers to placement or attachment of plant organs or parts to other dissimilar organs or major parts or the environment.

ARRANGEMENT

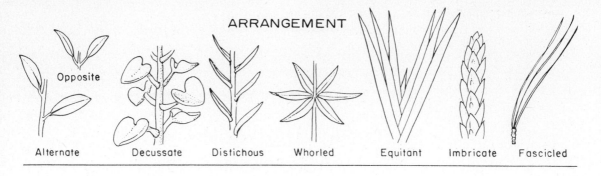

Alternate   Decussate   Distichous   Whorled   Equitant   Imbricate   Fascicled

TRANSVERSE POSTURE

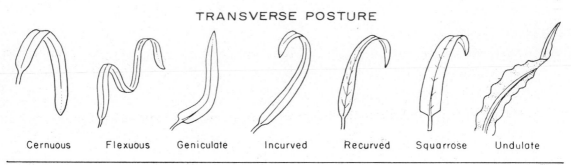

Cernuous   Flexuous   Geniculate   Incurved   Recurved   Squarrose   Undulate

LONGITUDINAL POSTURE

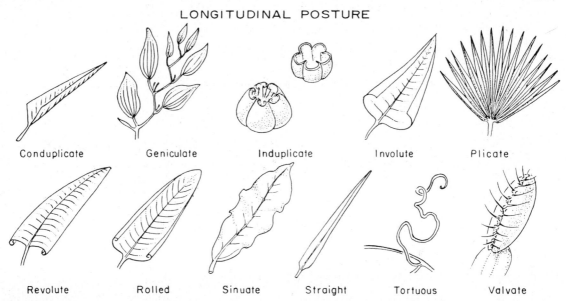

Conduplicate   Geniculate   Induplicate   Involute   Plicate

Revolute   Rolled   Sinuate   Straight   Tortuous   Valvate

Figure 5.6 Arrangement and posture.

1. General terms for position

> Definitions are based on the placement of organs or parts with respect to other dissimilar organs or parts.

>> Apical or Terminal. At the top, tip, or end of a structure.

>> Basal or Radical. At the bottom or base of a structure.

>> Continuous. Basal, lateral, and terminal.

>> Discontinuous. Basal and lateral, basal and terminal, or lateral and terminal; not continuous.

>> Lateral or Axillary. On the side of a structure or at the nodes of the axis.

2. Environmental position

> Definitions are based on the placement of organs or parts in relation to surrounding environment.

>> Aerial or Epigeous. Above the ground or water; in the air.

>> Emergent. With part(s) of plant aerial and part(s) submersed; rising out of the water above the surface.

>> Epipetric. On a rock.

>> Epiphytic. On another plant.

>> Floating. On the surface of the water.

>> Submersed. Beneath the surface of the water.

>> Subterranean or Hypogeous. Below the surface of the ground.

>> Surficial or Epigeous. On or spread over the surface of the ground.

c. Orientation

> Orientation refers to the placement of organs or parts in relation to an angle of divergence from a central axis or point.

1. Terms for upward orientation

> Definitions are based on ascending or upward angle of divergence or convergence.

>> Acroscopic. Facing apically.

>> Antrorse. Bent or directed upward.

>> Appressed or Adpressed. Pressed closely to axis upward with angle of divergence of 15° or less.

>> Ascending. Directed upward with an angle of divergence of 16–45°.

>> Assurgent. Directed upward or forward.

>> Connivent. Convergent apically without fusion.

>> Inclined. Ascending at 46–75° of divergence.

2. *Terms for downward orientation*

Definitions are based on the descending or downward angle of divergence or convergence.

Basiscopic. Facing basally.

Declinate. Directed or curved downward.

Deflexed. Bent abruptly downward.

Depressed. Pressed closely to axis downward with angle of divergence of 166–180°.

Descending. Directed downward with an angle of divergence of 136–165°.

Pendulous. Hanging loosely or freely.

Reclinate. Bent down upon the axis, no angle of divergence.

Reclined. Descending at 106–135° of divergence.

Reflexed. Bent or turned downward.

Retrorse. Bent or directed downward.

3. *Terms for twisting orientation*

Definitions are based on the twining or twisting of parts about or in relation to a central axis.

Contorted. Twisted around a central axis; twisted.

Dextrorse. Rising helically from right to left, a characteristic of twining stems.

Resupinate. Inverted or twisted 180°, as in pedicels in Orchidaceae.

Sinistrorse. Rising helically from left to right, a characteristic of twining stems.

Twining. Twisted around a central axis.

4. *Terms for horizontal or grouped orientation*

Definitions are based on the horizontal or grouped placement of parts in relation to a central axis.

Agglomerate, Conglomerate, Crowded, or Aggregate. Dense structures with varied angles of divergence.

Divergent, Patent, or Divaricate. More or less horizontally spreading with angle of divergence of 15° or less up or down from horizontal.

Horizontally. Spreading outward at 90° from vertical axis or plane.

Salient, Porrect, or Projected. Pointed outward, usually said of teeth.

d. **Posture**

(Figure 5.6)

Posture refers to the placement of a part(s) of a single structure in relation to its longitudinal or transverse axis or plane.

1. *Terms for transverse posture*

Definitions are based on the placement of the ends of a single structure in relation to its central transverse axis.

Applanate or Plane. Flat, without vertical curves or bends.

Arcuate. Curved like a crescent, can be downward or upward.

Cernuous. Drooping.

Geniculate. Abruptly bent vertically, usually near the base.

Incurved. Curved inward or upward.

Recurved. Curved outward or downward.

Squarrose. Usually sharply curved downward or outward in the apical region, as the bracts of some species of *Aster*.

2. *Terms for longitudinal posture*

Definitions are based on the placement of the sides of a single structure in relation to its central longitudinal axis.

Conduplicate. Longitudinally folded upward or downward along the central axis so that ventral and/or dorsal sides face each other.

Geniculate. Abruptly bent horizontally, usually in series.

Induplicate. Having margins bent inward and touching margin of each adjacent structure.

Involute. Margins or outer portion of sides rolled inward over upper or ventral surface.

Plicate. With a series of longitudinal folds; plaited.

Revolute. Margins or outer portion of sides rolled outward to downward over lower or dorsal surface.

Rolled. Sides enrolled, usually loosely, over upper or lower surfaces.

Sinuate. Long horizontal curves in the body of the structure parallel to the central axis.

Straight. Without a curve, bend, or angle.

Valvate. Sides enrolled, adaxially or abaxially so that margins touch.

3. *Terms for wave posture*

Definitions are based on the placement of surfaces of a single structure in relation to the hypothetical central longitudinal plane.

Flexuous. With a series of long or open vertical curves at right angles to the central axis.

Lorate. With elongate vertical waves in the margins or sides at right angles to the longitudinal axis.

Undulate. With a series of vertical curves at right angles to the central axis.

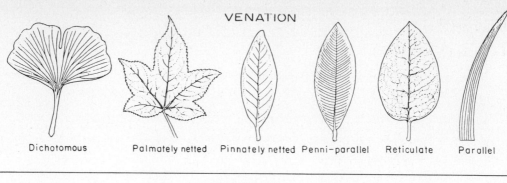

VENATION

Dichotomous   Palmately netted   Pinnately netted   Penni-parallel   Reticulate   Parallel

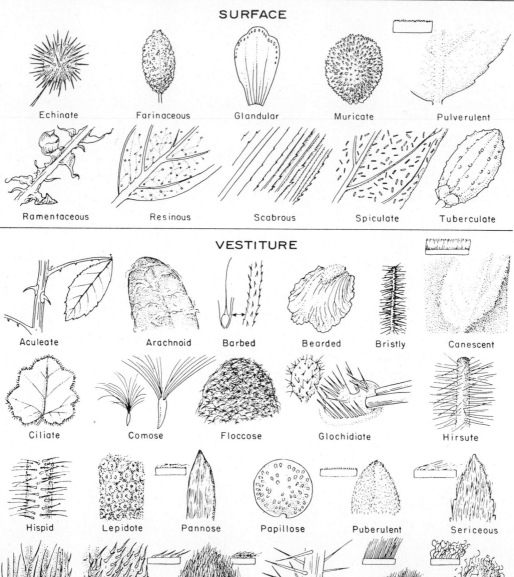

SURFACE

Echinate   Farinaceous   Glandular   Muricate   Pulverulent

Ramentaceous   Resinous   Scabrous   Spiculate   Tuberculate

VESTITURE

Aculeate   Arachnoid   Barbed   Bearded   Bristly   Canescent

Ciliate   Comose   Floccose   Glochidiate   Hirsute

Hispid   Lepidote   Pannose   Papillose   Puberulent   Sericeous

Setose   Strigose   Tomentose   Urent   Velutinous   Villous

Figure 5.7 Surface features.

e. General Structural Position

General structural position refers to regional placement of structures or their parts in relation to other structures or parts.

Abaxial. Away from the axis; the lower surface of the leaf; dorsal.

Adaxial. Next to the axis; facing the stem; ventral.

Apical. At or near the tip.

Basal. At or near the base.

Central. In the middle or middle plane of a structure.

Circumferential. At or near the circumference; surrounding a rounded structure.

Distal. Away from the point of origin or attachment.

Dorsal. Pertaining to the surface most distant from the axis; back of an outer face of organ; lower side of leaf; abaxial.

Marginal. Pertaining to the border or edge.

Medial. On or along the longitudinal axis.

Peripheral. On the outer surface or edge.

Proximal. Near the point or origin of attachment.

Subapical. Near the apex.

Subbasal. Near the base.

Ventral. Pertaining to the surface nearest the axis; inner face of the organ; the upper surface of the leaf; adaxial.

VII. Surface   (Figure 5.7)

Surface refers to patterns of configuration, vesture, and excrescences occurring on the outer face of plant organs or parts.

a. Configuration

Definitions based on configuration refer to surface patterns exclusive of venation and epidermal outgrowths.

Aciculate. Finely marked as with pinpricks, fine lines usually randomly arranged.

Alate. Winged.

Alveolate. Honeycombed.

Areolate. Divided into many angular or squarish spaces.

Bullate. Puckered or blistered.

Canaliculate. Longitudinally grooved, usually in relation to petioles or midribs.

Cancellate or Clathrate. Latticed.

Corrugate. Ridged.

Costate. Coarsely ribbed.

Fenestrate. With windowlike holes through the leaves or other structures.

Flexuous. Coarsely undulate with folds at right angles to long axis.

Foveolate. Pitted.

Plicate or Plaited. Fluted, longitudinally folded.

Punctate. Covered with minute impressions or depressions.

Pustulate. With scattered blisterlike swellings.

Reticulate. Netted.

Ribbed. With longitudinal nerves.

Ringed. With old bud-scale scar rings.

Rugose. Covered with coarse reticulate lines.

Ruminate. Coarsely wrinkled, appearing as chewed.

Scarred. With old leaf base, stipular and/or branch scar regions.

Smooth or Plane. Without configuration.

Striate. With longitudinal lines.

Sulcate. With longitudinal grooves.

Tortuous. Having the surface variously twisted.

b. Venation   **(Figure 5.7)**

Definitions based on venation refer to surface patterns of visible veins.

Dichotomous. With veins branching or forking in pairs equally.

Netted or Reticulate. With veins forming a network.

Pinnately Netted. With secondary veins arising from midrib or midvein.

Palmately or Digitately Netted. With three or more primary veins arising from a common point.

Parallel. With veins extending from base to apex, essentially parallel.

Penni-parallel. With veins extending from midrib to margins, essentially parallel.

c. Epidermal Excrescences   **(Figure 5.7)**

Definitions based on epidermal excrescences refer to types of epidermal outgrowths or secretions.

Asperous. Having a rough surface.

Bristly. Beset with bristles.

Echinate. Covered with spines.

Farinaceous. Mealy.

Glabrate or Glabrescent. Becoming glabrous.

Glabrous. Smooth; devoid of trichomes.

Glandular. Covered with minute, blackish to translucent glands.

Glaucescent. Sparingly or slightly glaucous.

Glaucous. Covered with a bloom or smooth, waxy coating.

Glutinous. Having a shiny, sticky surface.

Granular. Finely mealy, covered with small granules.

Greasy or Unctuous. Slick, oily, slippery to touch.

Mucilaginous. Gummy or gelatinous.

Muricate. Covered with short, hard protuberances.

Muriculate. Minutely muricate.

Opaque. With a dull surface.

Prickly. With prickles.

Pruinose, Frosted, or Sebiferous. With a heavy wax coat.

Pubescent. Covered with dense or scattered trichomes.

Pulverulent. Covered with fine, powdery wax granules.

Ramentaceous. Having many thin scales, as on the epidermis of some ferns.

Resinous. Having a yellowish, sticky exudate.

Roridulate or Dewy. Covered with waxy platelets, appearing dewy.

Scaberulent. Approaching scabrous.

Scabridulous. Minutely scabrous.

Scabrous. Having a harsh surface.

Shining, Nitid, or Laevigate. Lustrous, polished.

Spiculate. With crystals in or on the surface.

Squamose. Having coarse scales.

Subglabrate. Almost glabrous.

Tuberculate or Verrucose. With a warty surface.

Viscid. Sticky or glutinous.

d.   Vestiture   **(Figure 5.7)**

Definitions based on vestiture refer to the type of hairy cover on the surface of an organ or part.

Aculeate. Prickly.

Arachnoid. Cobwebby.

Barbed. With short, rigid reflexed bristles or processes.

Barbellate. Minutely barbed.

Bearded or Barbate. With long trichomes usually in a tuft, line, or zone.

Canescent or Incanous. Covered with dense, fine grayish-white trichomes.

Ciliate. With conspicuous marginal trichomes.

Ciliolate. With tiny or small marginal trichomes.

Comose. With a tuft of trichomes, usually apical.

Downy. Covered with short, weak, soft trichomes.

Floccose. Covered with dense, appressed trichomes in patches or tufts.

Glabrate. Without trichomes.

Glandular. Covered with secretory or excretory trichomes.

Glochidiate. With barbed trichomes, glochids, usually in tufts.

Hirsute. Covered with long, rather stiff trichomes.

Hirsutulous or Hirtellous. Minutely hirsute.

Hispid. Covered with very long, stiff trichomes.

Hispidulous. Approaching hispid, minutely hispid.

Inermous. Unarmed, without prickles or spines.

Lanate. Covered with long, intertwined trichomes, cottony.

Lanuginose. Cottony, similar to lanate but trichomes shorter.

Lepidote or Squamulose. Covered with minute scales.

Paleaceous. With small membranous scales, chaffy.

Pannose or Felted. With matted, feltlike layer of trichomes.

Papillose. Covered with minute tubercles.

Pilose. With soft, shaggy trichomes.

Puberulent. Minutely pubescent.

Pubescent. Usually with straight, slender trichomes.

Scurfy or Lentiginous. With exfoliating scaly incrustations.

Sericeous. With long, silky trichomes, usually appressed.

Setose or Setaceous. Having setae or bristlelike trichomes.

Strigillose. Diminutive of strigose.

Strigose. Covered with sharp, coarse, bent trichomes usually with a bulbous base.

Tomentose. Covered with dense, interwoven trichomes.

Urent or Stinging. With erect, usually long trichomes that produce irritation when touched.

Velutinous. Covered with dense, straight, long, and soft trichomes; pilelike.

Villosulous. Minutely villous.

Villous. Covered with long, soft, crooked trichomes.

VIII. Texture

Texture refers to the consistency of an organ or its parts.

Baccate. Juicy and very succulent.

Carnose or Sarcous. Fleshy.

Cartilaginous. Hard and tough but flexible.

Ceraceous. Waxy.

Chartaceous. Papery, opaque and thin.

Coriaceous. Thick and leathery.

Corneous. Horny.

Crustaceous. Hard, thin, and brittle.

Diaphanous. Translucent.

Fibrous. Having loose, woody fibers.

Flaccid. Lax and weak.

Gelatinous. Jellylike; soft and quivery.

Herbaceous. Soft and succulent.

Hyaline. Thin and translucent or transparent.

Incrassate. Thickened.

Indurate. Hardened.

Ligneous. Woody.

Membranous. Thin and semitranslucent; membranelike.

Osseous. Bony.

Pannose. With a felty texture.

Pellucid. Clear, transparent.

Scarious. Thin and dry, appearing shriveled.

Sclerous. Hard.

Spongy. Cellular; spongelike.

Suberous. Corky.

Suffrutescent. Woody basally, herbaceous apically.

Woody. Hard and lignified.

IX. Symmetry

Symmetry refers to the correspondence in arrangement, size, and form of parts on opposite sides of a point, line, or plane or the regularity of form, size, or arrangement with reference to corresponding parts.

Actinomorphic or Radial. With floral parts that radiate from the center like spokes on wheel.

Asymmetric. Without regularity in any dimension.

Dorsiventral. Planate and having distinct dorsal and ventral surfaces, the two usually different.

Equilateral. With halves or sides equal in shape and size.

Inequilateral. With leaves or sides unequal in shape and size.

Irregular. With floral parts within a whorl dissimilar in shape and/or size.

Regular. With floral parts within a whorl similar in shape and size.

Spherical. With multidimensional radial symmetry.

Zygomorphic or Bilateral. With floral parts in two symmetrical halves.

X. Temporal Phenomena

Temporal phenomena refer to duration, maturation time, and seasonality of plants and plant parts.

a. Duration

Duration refers to the length of life of a plant or its parts.

Annual. Living one year or less.

Biduous. Lasting two days.

Biennial. Living two years, usually flowering in second year.

Bimestrial. Lasting two months.

Caducous. Dropping off very early, usually applied to floral parts.

Caulocarpic. Plants having the stem living for many years, bearing flowers and fruits.

Cladoptosic. Shedding of branches, stems and leaves simultaneously, as in *Taxodium*.

Deciduous. Parts persistent for one growing season, then falling off.

Deliquescent. Softening and wasting away.

Ephemeral. Germinating, growing, flowering, and fruiting in a short period, as most desert herbs.

Evanescent. Passing away, disappearing early.

Evergreen. Persistent two or more growing seasons.

Fugacious. Ephemeral, usually applied to plant parts.

Marcescent. Usually ephemeral with persistent remains; withering persistent.

Monocarpic, Hapaxanthic. Perennial or annual, flowering and fruiting once, then dying; fruiting once.

Perennial or Polycarpic. Living more than two years; fruiting more than once.

Persistent. Remaining attached; applied to individual parts.

Pliestesial or Multiperennial. Monocarpic but living several to many years before flowering, as in *Agave*.

Rhizocarpic. Plants having the roots living for many years with the stems dying annually.

Seasonally Deciduous. Falling after one growing season.

Summer Annual. Germinating in spring or early summer and flowering and fruiting in late summer or early fall, then dying.

Winter Annual. Living less than one year but through the winter; germination usually in late fall, and usually flowering and fruiting in early spring.

b.   Maturation

Maturation refers to the time of development of organs and their parts.

Anthesis. Time of flowering; opening of flower with parts available for pollination.

Blastocarpous. Germination of seeds while within the pericarp, as in *Rhizophora*.

Coetaneous. Flowering as the leaves expand; synantherous.

Heterogamous. With maturation of stamens or anther and carpels or stigma at different times.

Homogamous. With maturation of stamens or anther and carpels or stigma at same time.

Hysteranthous. With leaves appearing after flowers.

Precocious. Developing unusually early.

Proanthesis. Flowering before normal period, as spring flowers in the fall.

Protandrous. With stamens or anthers developing before carpels or stigma.

Protantherous. With leaves appearing before flowers.

Protogynous. With carpels or stigma maturing before stamens or anthers.

Synantherous. With leaves and flowers appearing at same time.

c.   Periodicity

Periodicity refers to seasonal appearance of a plant, its organs, or its parts.

Aestival. Appearing in summer.

Aianthous or Semperflorous. With flowers appearing throughout the year.

Annotinal or Yearly. Appearing yearly.

Autumnal. Appearing in autumn.

Biferous. Appearing twice yearly.

Biflorous. Flowering in autumn as well as in spring.

Bimestrial. Occurring every two months.

Diurnal. Opening during the day.

Equinoctial. Having flowers that expand and close regularly at particular hours of the day.

Hibernal or Hiemal. Appearing in the winter.

Matutinal. With flowers opening in the morning.

Nocturnal. Opening during the night.

Seasonal. Occurring during a seasonal cycle or during each season.

Serotinal. Opening late; appearing in late summer.

Vernal. Appearing in spring.

Vespertine. With flowers opening in the evening or night; appearing or expanding in the evening.

SUGGESTED READING

Benson, L. 1962. Plant Taxonomy, Methods and Principles. The Ronald Press Company, New York. Chapter 14, pp. 391–400.

Davis, P. H. and V. H. Heywood. 1965. Principles of Angiosperm Taxonomy. D. Van Nostrand Company, Inc., New York. Chapter 4, pp. 110–141.

Radford, A. E. et al. 1974. Vascular Plant Systematics. Harper & Row, New York. Chapter 5, pp. 79–82; Chapter 6, pp. 83–166; Chapter 24, pp. 501–521.

Stearn, W. T. 1966. Botanical Latin. Thomas Nelson & Sons, Ltd., London.

Systematics Association Committee for Descriptive Terminology. 1962. II. Terminology of Simple Symmetrical Plane Shapes (Chart I), Taxon XI (5): 145–156.

SUMMARY FOR DESCRIPTION

Definitions for Description   *Description,* in general, is the orderly recording of as complete a set as possible of characters with appropriate character states for plant parts, plants, and plant taxa. *Characterization* is the orderly recording of distinctive characters with appropriate character states for plant parts, plants, and plant taxa. *Circumscription* is the orderly recording of limiting and diagnostic characteristics for plants and plant taxa. *Character state* is a characteristic that is the basic unit in description. Character states are basic components and distinct phases of a *character* (any attribute or descriptive phrase referring to form, structure, or behavior that the taxonomist separates from the whole organism for a particular purpose such as comparison or interpretation; any feature whose expression can be measured, counted, or otherwise assessed; a taxonomic character of two or more states, which within the study at hand cannot be further subdivided logically). Characters are grouped as *character sets,* which are fundamental components of types of *evidence.*

Purpose for the Study of Description   To provide a vocabulary of descriptors for precise communication about plant parts, plants, and plant taxa and to supply the descriptors and descriptive system for effective and efficient information storage, retrieval, and use for each named taxon with a particular circumscription, position, and rank.

Operations in Description   To select appropriate characters and determine the character states for circumscriptions of new taxa at given positions and ranks; to amend or change the circumscriptions of old taxa that have been remodeled, united, divided, transferred, or changed in rank; to supplement the circumscriptions or descriptions of old taxa as new evidence is procured; and to develop descriptive systems that have comparability, consistency, and comprehensiveness of treatment with etymologically sound terminology.

Basic Premise for Study of Description   Discontinuities in variation and character correlations occur that make circumscriptions of taxa at given positions and ranks possible.

Fundamental Principles of Description

1. Plant taxa that have been delimited, classified, and named can be circumscribed according to a system of description.
2. Descriptions of taxa are changed for the proper reasons of either a more profound knowledge of the taxa resulting from adequate taxonomic study, or of nonconformance of the descriptors to accepted practice or sound etymological study.

Guiding Principles for the Description of Plants

1. Comprehensive descriptions of taxa should include characters and appropriate character states from all fields of evidence.
2. Characters must be hierarchically arranged and classified for comparability and consistency in relevant and diagnostic description of taxa.
3. Character states must be precisely defined and named for accuracy and ease of application in descriptions of taxa.
4. A general descriptive format should be followed for orderly use of terminology. All taxa of coordinate rank should have the same characters described in the same sequence.
5. Basic sources of terms and definitions used in the descriptions of taxa should be documented.
6. The characters selected and used in the descriptions of taxa should be relevant and pertinent to the type of publication.
7. Descriptions should cover the known range of variation within a taxon.

Basic Assumptions for the Study of Description

1. Taxa can be circumscribed and diagnostically delimited from all other taxa.
2. A system of description exists for the describing of new taxa and the changing of the descriptions of old taxa that have been remodeled, united, divided, transferred, or changed in rank.

QUESTIONS ON DESCRIPTION AND DESCRIPTIVE FORMAT

1. What is meant by circumscription, characterization, and description? Give the diagnostic differences between the terms.
2. Why does one study description or descriptive terminology?
3. What does a describer of objects, entities, or taxa actually do?

4. What is meant by a descriptive hierarchy? How is a descriptive system a classification system? Nomenclature system? Identification system? (Use the Ridgway color chart or the Munsell color chart or any other as an example.)

5. Why use a descriptive format? What are the relationships between descriptive format and communication effectiveness?

6. What are the major guidelines for good description?

7. Under what conditions can a character become a character set or a character state become a character?

8. What constitutes an analytic character? A synthetic one? Under what conditions does an analytic character become a synthetic one; a synthetic one an analytic one?

9. What general types of characters are found in keys? Manual descriptions?

10. What precautions have to be taken in character and character state definition and ordering for use in data accumulation, use, presentation and documentation?

11. How are the principles of nomenclature pertinent to character classification?

12. How is an understanding of the principles of characterization applicable and pertinent to the writing of a theme? Stocking a grocery store? Making a budget? Executing taxonomic research?

13. How is a knowledge of character, evidence, and characterization basic to your personal scholastic development in observational acuity, descriptive proficiency, conceptual breadth, principles understanding, philosophical conspective, and communication effectiveness?

14. What are the major problems in descriptive writing?

15. What are your criticisms of the guiding principles for description?

16. What are the advantages of a standardized classification of the character sets, characters, and character states for each type of evidence? Disadvantages?

17. How does a developer of a descriptive system for any subject resolve the problems of comparability? Consistency? Compatibility with other descriptive systems?

18. How does scientific description differ from artistic description? Should there be a difference? If so, why?

19. What are the basic premises for the study of scientific description?

20. What is the relevance of the standardization of descriptive terminology to computer-based classification and identification systems?

EXERCISES IN DESCRIPTION

1. Select two species (a and b) from different families in a manual of your choice and answer the following:

   a. Types of evidence in family description for species a; species b.

   b. Hierarchical types of characters in generic description for species a; species b.

   c. Plant parts format for species a; leaf character format for species b.

   d. Petal characters/character states for species a; fruit characters/character states for species b.

   e. Cytological evidence for species a; chromosome number for species b.

   f. Habitat for species a; ecological evidence for species b.

   g. Phenological evidence for species a; fruiting date for species b.

   h. Physiographic province distribution of species a; distribution by state for species b in a region of the nation.

   i. Basic reasons for studying description or phytography of species a; basic operations for selecting the relevant and pertinent characteristics for the keys or description of species b.

   j. Documentation for characteristics used in description of species a; synonyms in another manual for your region for species b.

   k. List of guiding principles used in description of species a; list of guiding principles used in descriptive format of species b.

    l. Fundamental differences in family, generic, and species description for species a; fundamental similarities in family, generic, and species description for species b.

2. General descriptive studies.
    a. Determine the color and color patterns for the corollas of several species in the field or laboratory.
    b. Determine the size classes for the leaves of several species in the field or laboratory.
    c. Determine the cycly and merosity for the flowers of several species in the field or laboratory.
    d. Describe the fusion of floral parts of several species in the field or laboratory.
    e. Determine the shape, apex, base, and margin of several species in the field or laboratory.
    f. Determine the arrangement, position, orientation, and posture for the leaves of several species in the field or laboratory.
    g. Determine the configuration, venation, excrescences, vestiture, and texture for the leaves of several species in the field or laboratory.
    h. Determine the symmetry, duration, maturation, and periodicity of the flowers of several species in the field or laboratory.

3. Criticize the developmental sequence of a classification system for description as itemized below:
    a. Select major components of description.
    b. Determine position of each component.
    c. Establish hierarchy for components.
    d. Determine elements for each component by rank and position.
    e. Circumscribe each element with a given position and rank or give reference to previous circumscription(s).
    f. Name each circumscribed element with a given position and rank or give reference(s) to previous nomenclature.

chapter 6

Plant Classification

Taxonomists as classifiers establish and determine position and rank for new taxa; determine the correct position and rank for old taxa that have been remodeled, divided, united, transferred, or changed in rank; develop systems of classification; and determine the group to which a specimen belongs according to a system of classification.

The ability to classify is a characteristic inherent in all of us. Classification is a basic human activity. We taste food—it is good, so-so, bad, or inedible; we smell a flower—it is pleasant, offensive, or odorless; we see animals on the savannah—they are carnivores or herbivores; we classify while using different senses. We observe large plants, medium plants, small plants—we establish a size hierarchy. For many centuries we have classified some plants as trees—others as shrubs, vines, or herbs. Furthermore, we knew that some trees were evergreen, others deciduous; that some vines were woody, others herbaceous. We classify at the sensory and perceptive levels. Most of us arrange objects, organisms, and entities for a greater command of knowledge, a more efficient use of information, and an easier acquisition of pertinent data.

  In general, classification is the placement of objects and organisms into groups and categories for effective organization. Meats, vegetables, and canned goods, for instance, are placed in the food division of the store; the meat, in a section; the beef, in a compartment; the steaks, in their refrigerated unit; and so on. With this system, the butcher can add meat to his depleted supplies and the customer can find the desired kind easily. A new line of steaks can be added to the present arrangement satisfactorily for the benefit of the customer and management alike. With the types of steaks together, it is easy to develop a knowledge of the kinds, to acquire data on the price per pound for each type, and to procure the information for making a decision on how much of the desired steak to purchase in relation to pocketbook resources. Descriptions with pertinent characteristics, as in most classifications, accompanying each kind of meat would help the learning process about steaks considerably.

Technically, plant classification is the ordering of groups of plants into a hierarchy of taxa in positional compartments according to phenetic similarities, phylogenetic relationships, or artificial criteria. The purposes of plant classification are to order groups of organisms into taxa on the basis of their relationships, to provide a system that expresses those relationships in a practical or natural way, and to produce a classification for effective informational storage for taxa with particular circumscriptions, positions, and ranks. All of this, presumably, creates organizational effectiveness for the study of plants. Professional plant classifiers determine positions and ranks for new taxa with particular circumscriptions; establish the correct positions and ranks for old taxa with modified circumscriptions that have been remodeled, united, divided, transferred or changed in rank; and determine the group to which a specimen belongs according to a system of classification.

The hierarchy for plants consists of six ranks: division, class, order, family, genus, and species as the basal rank. Each rank has subcategories (see Table 3.2). Any hierarchy is established for convenience in classification as well as perspective on the groups classified. Each rank is inclusive of those below—one or more classes within a division, one or more orders within a class, on to one or more species within a genus. No set number of taxa except at least one are required for each rank and no set number of characteristics except one are necessary for distinguishing one taxon from another within a rank. Generally, in a hierarchical system the lower the rank, the greater the number of characteristics in common for the taxa at that rank; and the higher the rank, the fewer the number of taxa per rank. Species within a genus usually would have more characteristics in common than genera within a family. In the Cronquist system of classification for flowering plants (1981), the division Magnoliophyta, for example, has 2 classes, 11 subclasses, 83 orders, 383 families, and so on.

Systems of plant classification are developed for the use of layperson and scientist alike. The type of system produced—phenetic, phylogenetic, artificial—depends on the goals of the developer. All systems consist of a hierarchy of categories, each with one or more named taxa with particular circumscriptions and positions. Successful systems of classification for any subject are *comprehensive,* have internal *consistency* and *comparability,* possess the potential for *compatibility* with other systems; and are easily modified as the subject develops with additional knowledge, new techniques, and the establishment of new relationships within and among taxa or groups of any sort.

Classification of plants requires a knowledge of traditional procedures as well as the more sophisticated methods involving modern mathematics and the computer. An understanding of the systems of classification for plants in use today requires a study of the development of the systems through time.

A. CLASSIFICATION

Students are introduced to classification, hierarchy, and taxon as concepts in introductory botany or biology. Beginners soon realize that the six ranks and their subcategories in plant classification have been established in the ICBN. The term, relationship, very early becomes part of the vocabulary without a real understanding as to how it is determined. Circumscription and position are usually words foreign to budding taxonomists. The processes in classification—remodeling, uniting, dividing, transferring, and changing of rank—are seldom explained in standard texts in systematics. The determination of position and rank for a new taxon with a particular circumscription and an old taxon with a modified circumscription depends on an understanding of the basic dynamics of classification.

Many students of plant classification become familiar with characteristics of higher taxa—division, subdivision, class—while studying local flora or how-to-know courses. After limited training, students are able to recognize green, red, and brown algae; mosses and liverworts; ferns, lycopods, and horsetails; conifers and cycads; monocots and dicots; gymnosperms and angiosperms; and can classify organisms accordingly. Most botanists become acquainted with the classification of families, genera, and species while using manuals and floras to identify specimens. Most taxonomists develop an appreciation for the order and class level while arranging the families in their personal herbaria into those taxa according to a modern system of classification.

I. Traditional Classification

Historically, most new taxa are recognized, classified, described, and named in the pioneer phase of taxonomy for a region. Many discoveries of plants new to science are made during the exploration of relatively unknown areas. Ranks have to be determined for the new taxa. New families have to be placed in appropriate orders; genera in families; species in genera; and infraspecific taxa in species—the position of each new taxon has to be determined.

Continuing taxonomic work in an area produces more collections, increased knowledge of organismic diversity, discovery of new relationships among taxa, and a better understanding of the habitat for the different plants. Authors consolidate this information about plants in floras and manuals. Traditionally, writers of floras reclassify and reinterpret taxa. Their taxonomic judgment indicates that certain taxa should be combined, some divided, others transferred by position, or changed in rank. The synonymy accompanying descriptions of taxa in most manuals is ample evidence for the dynamics of classification (see synonymy in floras in Chapter 3). Manuals and floras provide a sound basis for taxonomic training for many people.

Frequently, individuals become involved in classification while identifying collections and authenticating their identifications. A specimen doesn't quite fit the description: for example, the measurements for the leaves are too small; the shape of the petals is obovate, not elliptic; stamens are fewer than the number indicated. The amateur as well as professional botanist soon has the evidence for remodeling (changing) the description of the taxon or possibly the circumscription (delimitation) for the species. Problems with the keys to the taxa of a region, discrepancies in the descriptions of some species, or dissatisfaction with the classification of the taxa can lead to a revisionary study of a selected group.

The morphological characters found in floras provide the basis for revisionary work. Characters such as chemical, cytological, anatomical, palynological, microstructural and so on are standard additions to revisionary studies. Experimental characters are

Table 6.1   NATURAL AND ARTIFICIAL CLASSIFICATIONS COMPARED

Natural classification	Artificial classification
Basis	**Basis**
Sum total of all the characters of its members.	One character or a very few characters of its members, especially chosen.
Advantages	**Disadvantages**
Groups together plants most alike in their hereditary constitutions.	May not group together plants most alike in their hereditary constitutions.
In general, groups together plants most closely related phylogenetically.	May not group together plants most closely related phylogenetically.
Contains a great deal of information about its members.	Contains only a very limited amount of information about its members.
Can easily incorporate additional information about its members.	Cannot incorporate more information about its members.
Has a high predictive value.	Has little or no predictive value.
Disadvantages	**Advantages**
Identification may be difficult.	Identification can be made easy.
The placing of poorly known plants may be uncertain or impossible.	Poorly known plants may be definitely placed.
Is liable to change with increase in our knowledge.	Is stable and is not changed by increase in our knowledge.

Source: Jeffrey, C. 1982. An Introduction to Plant Taxonomy, 2nd ed. Cambridge University Press. Used with permission.

used occasionally in this type of research. Supposedly, more characters from more fields of evidence will enable the taxonomist to produce a more definitive classification for the taxa of interest. Analysis of the characters as related to the different taxa will lead to new interpretations of the classification of the old taxa. In the process, new taxa will be discovered, classified, described, and named; some old taxa recircumscribed or delimited in terms of new and old characters; and the positions and/or ranks of others changed.

Traditional classification for the taxa of a region is an ever-evolving process operating on the basis of using what has gone on before. Most taxonomists conform to the dictum of the "avoidance of changing the rank or position of natural, recognizable groups if no significant change in the concept of their relationships to each other and to other groups results from systematic study" (Cronquist, 1968). Traditional classifiers usually place their phenetically related taxa in a phylogenetic (natural) system of classification (see Tables 6.1 and 6.3).

II. Modern Classification

The classification of plants has rapidly evolved during the past two decades with the application of the principles and methods of numerical taxonomy and the use of modern electronic data-processing techniques. The intuitive and sensory approach of traditional classification is being supplemented and/or replaced by objective, quantitative methods.

Traditional as well as modern classification is based primarily on phenetic resemblances and presumed phylogenetic relationships.

The principles and advantages of numerical taxonomy as presented in Sneath and Sokal (1973) are summarized with slight modification here.*

a. Principles

1. The greater the content of information in the taxa of a classification and the more characters on which it is based, the better a given classification will be.
2. A priori, every character is of equal weight in creating natural taxa.
3. Overall similarity between any two entities is a function of their individual similarities in each of the many characters in which they are being compared.
4. Distinct taxa can be recognized because correlations of characters differ in the groups of organisms under study.
5. Phylogenetic inferences can be made from the taxonomic structures of a group and from character correlations, given certain assumptions about evolutionary pathways and mechanisms.
6. Taxonomy is viewed and practiced as an empirical science.
7. Classifications are based on phenetic similarity.

b. Advantages

1. Numerical taxonomy has the power to integrate data from a variety of sources, such as morphology, physiology, chemistry, affinities between DNA strands, amino acid sequences or proteins, and more. This is very difficult to do by conventional taxonomy.
2. Greater efficiency is promoted through the automation of large portions of the taxonomic process. Thus, much taxonomic work can be done by less highly skilled workers or automata.
3. The data coded in numerical form can be integrated with existing electronic data processing systems in taxonomic institutions and used for the creation of descriptions, keys, catalogs, maps, and other documents.
4. The methods, being quantitative, provide greater discrimination along the spectrum of taxonomic differences and are more sensitive in delimiting taxa. Thus they should give better classifications and keys than can be obtained by conventional methods.
5. The creation of explicit data tables for numerical taxonomy has already forced workers in this field to use more and better-described characters. This necessarily will improve the quality of conventional taxonomy as well.
6. A fundamental advantage of numerical taxonomy has been the reexamination of the principles of taxonomy and of the purposes of classification. This has benefited taxonomy in general, and has led to the posing of some basic questions.
7. Numerical taxonomy has led to the reinterpretation of a number of biological concepts and to the posing of new biological and evolutionary questions.

*Source: Sneath and Sokal, 1973. Used with permission.

c. Goals and Practical Principles

Traditional and modern classifiers of plants have the same fundamental goals. In general, all taxonomists and classifiers try to establish conceptually discrete units with distinctive sets of character correlations as logical, natural groups (taxa) at determined positions and ranks. Most plant classifiers try to develop a natural system of classification for the biotic diversity of their concern (see Table 6.1). Most try

1. To construct taxa that are reliably delimited and circumscribed at their respective positions and ranks.
2. To make taxa that are easily and positively identifiable.
3. To form taxa that are recognizable in nature.
4. To establish taxa that are predictable in character correlations under natural conditions and occurrences.

All classifiers operate essentially on the basic principle that "distinct taxa can be recognized because correlations of characters differ in the groups of organisms under study" (Sneath and Sokal, 1973, principle 4). Most try to use correlative characters that are meaningful, reliable, and distinctive (see Table 5.1). Most try to incorporate characters for the natural groups that are discontinuous among groups of organisms surveyed, noncorrelated (not genetically linked) in the individuals examined, and invariant in the populations sampled. In addition, the characters included in correlations should be measurable or determinable with available resources, present in all the organisms under purview, and germane to the type of classification anticipated.

Most classifiers who are organismal biologists assume that "overall similarity between any two entities is a function of their individual similarities in each of the many characters in which they are being compared" (Sneath and Sokal, 1973, principle 3). They also assume that, to a reasonable point, "the greater content of information in the taxa of a classification and the more characters on which it is based, the better a given classification will be" (Sneath and Sokal, 1973, principle 1). Systematists who are empiricists firmly believe that "phylogenetic inferences can be made from the taxonomic structures of a group and from character correlations, given certain assumptions about evolutionary pathways and mechanisms" (Sneath and Sokal, 1973, principle 5) and that primarily "classifications are based on phenetic similarities" (Sneath and Sokal, 1973, principle 7).

Philosophically, a few taxonomists are convinced that "a priori, every character is of equal weight in creating natural taxa" (Sneath and Sokal, 1973, principle 2), particularly at the lower ranks. In actual practice, however, many taxonomists feel that some characters are more important than others in circumscribing and delimiting higher taxa, for example, number of cotyledons in classification of flowering plants and presence of seeds in delimiting major groups of vascular plants. Taxonomists, now and in the past, have given great weight to characters constant over large groups and usually have made a posteriori decisions (based on experience) as to the value of characters in a classification, particularly for groups at the lower ranks.

The advantages of numerical taxonomy as indicated by Sneath and Sokal above are generally accepted as fundamental assumptions for modern taxonomic and classificatory research.

d. Basic Operations in Modern Classification

Fundamentally, classifiers establish new taxa by selecting groups of organisms (old taxa and possible new taxa) and characters for study. Then a taxon/character data matrix or comparison chart is prepared. The character state(s) of each character for each of the taxa in the matrix is determined (Table 6.2). The resemblances or similarities between the taxa are calculated by determining the coefficients of similarity or resemblance based on the character states. Taxa at determined positions and ranks are recognized on the basis of character state correlations resulting from clustering and ordination procedures (see Appendix A). The character state correlations in the clusters provide the data for the circumscriptions of the taxa. Circumscriptions are used to delimit the old and new taxa at particular positions and ranks. Phylogenetic inferences are made as to relationships between selected taxa based on taxonomic structures and character correlations within the groups studied, frequently using methods developed for cladistic analysis. (See Chapter 10 and Appendix B for methods of phenetic and phylogenetic classification.)

These operations are basic to traditional as well as modern classification. Phylogenetic considerations are frequently omitted from classification devised by numerically oriented taxonomists. Similarly, sophisticated mathematical techniques are not included in the publications of traditional classifiers. The numerical methods and procedures used in classification are usually taught in advanced courses such as numerical taxonomy (see Sneath and Sokal, 1973, Chapters 4 to 6). The operations involved in the establishment and delimitation of taxa at determined positions and ranks require an extensive background in science and mathematics, a willingness to obtain and accept expert help, and the ability to understand and interpret fundamental information and advice.

Table 6.2 CHARACTER COMPARISONS OF THE NORTH AMERICAN TAXA OF *LINDERA*

| Character | Taxon | | |
	L. subcoriacea	*L. benzoin*	*L. melissifolia*
Leaf			
Texture	Subcoriaceous	Membranaceous	Membranaceous
Length × width (cm)	4–7.5 × 2–3.5	6–15 × 3–6	8–16 × 3–6
Shape	Elliptic to oblanceolate	Obovate	Obovate to elliptic
Apex	Obtuse to rounded	Acuminate	Acute
Base	Cuneate	Cuneate	Widely cuneate to rounded
Pubescence (abaxial side)	Present	Present or absent	Present
Orientation	Horizontal to ascending	Horizontal to ascending	Drooping
Fragrance	Faint, piney lemon	Strong, "spicy"	Strong, sassafraslike
Fruit			
Length (mm)	10	10	12
Pedicel	Not thickened at apex; deciduous	Not thickened at apex; deciduous	Thickened at apex; persistent

Source: Wofford, B. E. 1983. A new *Lindera* (Lauraceae) from North America. J. Arnold Arb. 64:325. Used with permission.

e. Training Requirements for Research in Classification

As in all research, a study of what has been done in the past is a requirement for understanding the present and possible future state of a subject. Intelligent people have traditionally benefited from past experiences that were pertinent to their concern. A basic historical perspective of the philosophy, principles, criteria, and general properties pertaining to the development of classification is gained by studying previously published systems. In addition to the history of plant classification, researchers in the subject need an expert knowledge of the principles of evolution and phylogeny. The information on characters provided in traditional courses in botany forms the foundation for character correlations representative of most plant taxa.

Training in multivariate analysis, matrix algebra, and set and graph theory as well as a working acquaintance with computer processing are necessary for understanding and using many methods involved in the process of classification and the development of phylogenetic systems. Numerical techniques and computer manipulations are the modern methods for helping to obtain meaningful classifications while using all types of characters.

A broad background in the physical and biological sciences is a requisite to comprehending the many characters now used in research in systematics. Chemical characters based on data derived from analytical chemistry are standard inclusions in much taxonomic work. Micro- and ultrastructural data from the use of scanning and transmitting electron microscopes provide characteristics of routine use in description, identification, and classification of taxa. Developmental biology and DNA technology will produce characters of fundamental significance for future classifications. The use of a wide variety of evidence is standard procedure for modern research in classification.

Taxonomy truly becomes an integrative discipline as characters from many fields of evidence are used in the development of systems of classification for the biotic diversity of the world. Systematics becomes exceptionally challenging when new techniques and characters are applied to the solutions of problems in phenetic and phylogenetic classifications of diverse taxa.

B. SYSTEMS OF PLANT CLASSIFICATION

Contemporary systems of plant classification* are the result of the labors and insights of numerous workers for more than 2,000 years. The history of plant classification can be divided into the period of ancients, with systems based primarily on habit, from 300 B.C. to about A.D. 1500; the period of herbalists, with systems based on habit and use, from 1500 to 1580; the period of mechanical systems based on a few selected characters, such as sexual, from 1580 to about 1760; the period of natural systems based on a composite of characters from about 1760 to about 1880; and the period of phylogenetic systems presumably based on evolutionary principles from about 1880 to present. The history of classification is really a continuum without sharp lines between the periods (see Figure 6.1 and Table 6.3).

---

*Adapted from Becker, K. M., in Radford, A. E. et al., 1974. Vascular Plant Systematics. Harper & Row, New York. Chapter 28, pp. 583–644.

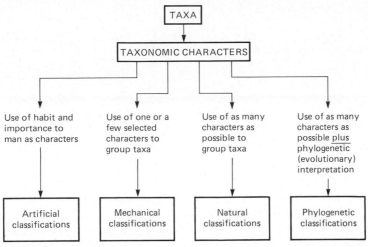

Figure 6.1 From taxa to system in the development of various kinds of general biological classifications.

Several trends in the development of systems of plant classification down through the centuries should be noted: (1) the gradual development of the hierarchy of categories as we know it today; (2) the gradual recognition of the major groups of plants—especially with regard to vascular versus nonvascular plants, cryptogams versus phanerogams, gymnosperms versus angiosperms, monocots versus dicots, and the gradual recognition of more and more angiosperm families as natural groups; (3) the gradual use of an increasing number of characters in classification; and (4) changes in philosophy in the transitions between the various periods in the history of plant classification systems. These trends appear in the summary for the development of classification (Table 6.3) and the relationships of angiosperms (see Figures C.1 and C.2 in Appendix C).

Some contemporary systems take into account, in addition to morphological characters, data from the fields of anatomy, embryology, chemotaxonomy, palynology, and

Table 6.3 SUMMARY FOR THE DEVELOPMENT OF CLASSIFICATION

Period of the Ancients—circa 300 B.C.–A.D. 1500

Theophrastos	Pliny	Wrote to transmit what was generally known about plants. Classification based on habit. A few natural groups recognized.
Dioscorides	Albertus Magnus	

Period of the Herbalists—1500–1580

Brunfels	l'Ecluse	Motivated by medical and commercial considerations—the systems per se were not the primary concern. Classification based primarily on habit.
Bock	l'Obel	
Fuchs	Bankes	
Cordus	Turner	
Mattioli	Gerard	Advent of printing stirred new interest in plants and "mystical respect" for writings of the ancients gave way to a period of original researches into plant structure.
Dodoens		

(continued overleaf)

Table 6.3   *(continued)*

Period of Mechanical Systems—1580 to about 1760

Caesalpino	Magnol	Primary concern was information retrieval—the
J. Bauhin	Tournefort	system itself was considered important. Classification
G. Bauhin	Camerarius	based mainly on form, but included also a number of
Linnaeus		other kinds of criteria. A priori weighting of values
		of individual features for classification purposes was
		the rule (see discussion of Tournefort's system in
		Chapter 2). Hierarchy of categories improved; more
		and more natural groups recognized.

Period of Natural Systems—1760 to about 1880

Adanson	Lindley	Attempted to put seemingly related plants (plants that
B. de Jussieu	Endlicher	look alike) together, using as many characters as
A. de Jussieu	Bentham and	possible (note change in attitude from mechanical
A. de Candolle	Hooker	systems). System of A. deJussieu a turning point in
Ray Brown	Hofmeister	the history of plant classification.
Darwin		Produced no system of his own, but (with A. R.
		Wallace) adopted, recast, expanded, and fostered the
		theory of evolution previously developed, to greater
		or lesser degrees, by Lamarck, Lyell, and others.
		Whereas pre-Darwinian natural-system makers wished
		their systems to reflect the divine plan that was
		thought to underlie all of nature, most post-Darwinian
		phylogenetic system makers thought that, since plants
		and plant groups are the products of evolution, it was
		their task to determine the course of this evolution
		and have their systems reflect it.

Period of Phylogenetic Systems—1880 to present

Eichler	Hutchinson	Attempted to integrate what is known about the
Engler-Prantl	Cronquist	evolutionary history of plant groups into the system.
Bessey	Banks	Most are similar to natural systems, but attempts
Bierhorst		were made to arrange plant groups according to their
		routes of descent. It became obvious to several
		workers in the early days of the phylogenetic period
		that the natural system makers were recognizing
		evolutionarily related groups of plants. Two main
		schools of thought developed as to origin of
		angiosperms and as to which character states were
		considered primitive:

Englerian	**Besseyan-Hallierian**
Primitive flower apetalous, unisexual anemophilous. Ancestors of angiosperms were coniferoid or gnetoid gymnosperms with unisexual strobili.	Primitive flower with perianth of many, free equal parts, bisexual, insect-pollinated. Ancestors of angiosperms were cycadophytes.

cytology. Comprehensive revisions of the classification of vascular and nonvascular plants have appeared in recent years. (See Figures C.1 and C.2 in Appendix C as examples of a modern system of classification for flowering plants.)

SUGGESTED READING

Benson, L. 1962. Plant Taxonomy, Methods and Principles. The Ronald Press Company, New York. Chapter 9, pp. 281–308.

Cronquist, A. 1968. The Evolution and Classification of Flowering Plants. Houghton Mifflin, Boston.

Cronquist, A. 1981. An Integrated System of Classification of Flowering Plants. Columbia University Press, New York. Introduction and Outline of Classification, pp. i–xviii.

Davis, P. H. and V. H. Heywood. 1965. Principles of Angiosperm Taxonomy. D. Van Nostrand Company, Inc., New York. Chapter 1, pp. 1–30; Chapter 2, pp. 31–73; Chapter 3, pp. 74–109.

Jeffrey, C. 1982. An Introduction to Plant Taxonomy, second edition. J & A Churchill Ltd., London. Chapter 1, pp. 1–5; Chapter 2, pp. 6–12; Chapter 3, pp. 13–33.

Radford, A. E. et al. 1974. Vascular Plant Systematics. Harper & Row, New York. Chapter 28, pp. 583–644. (See pp. 641–644 for extensive bibliography and pp. 617–640 for a comparison of selected systems.)

Sneath, P. H. A. and R. R. Sokal. 1973. Numerical Taxonomy. W. H. Freeman and Company, San Francisco. Chapters 4–6.

SUMMARY FOR CLASSIFICATION OF PLANTS

Definition of Classification   *Classification* is the ordering of plants into a hierarchy of taxa in positional compartments with the species being the fundamental unit, rank, or category.

Purpose of Classification   To order organisms into taxa on the basis of their relationships and to provide an orderly arrangement or system, position, and rank that expresses the relationships in a practical or natural way; to produce a system for efficient and effective information storage and use for taxa with particular circumscriptions, positions, and ranks.

Operations in Classification   To note resemblances between organisms, to recognize taxa, and to determine relationships among taxa; to determine positions and ranks for new taxa and establish the correct position and rank of old taxa that have been remodeled, divided, united, transferred, or changed in hierarchy according to an old classification system or in the development of a new system.

Basic Premise for Classification   That in the tremendous variation in the plant world conceptually discontinuous groups exist that can be identified and, via the discontinuity of variation and character correlations, can be ordered into a hierarchy and positional compartments for taxa that express the logical relationships among the groups.

Fundamental Principles of Classification

1. Variation in plants makes possible the establishment of classification systems.
2. Classifications of taxa are changed for the proper reasons of either a more profound knowledge of the facts resulting from adequate taxonomic study or nonconformance to the principles and rules on which the classification system is based.

Guiding Principles for the Development of Classification Systems for Plants

1. No character per se is more important than another in classification, but one character may be more significant within a group.
2. The greater the content of information on the taxon and the more characters on which it is based, the better a given classification will be. The limits of a taxon usually cannot be defined quantitatively or qualitatively by a single characteristic.
3. The totality of similarities and differences of the organisms being studied should be assessed before position and/or rank are determined.
4. In a natural or phyletic classification system, assignments of natural populations or of population systems to taxa and the hierarchical arrangement of those taxa involve phylogenetic inferences based on certain assumptions about evolutionary pathways and mechanisms.

Basic Assumptions for Classification

1. Classification is based on character correlations and discontinuities of variation that are necessary for characterization and delimitation of taxa.
2. A classification system represents the current state of knowledge regarding the relationships of organisms and taxa.
3. A system of classification exists with its principles and rules for the classifying of new taxa and the establishing of the position and rank of old taxa that have been remodeled, united, divided, transferred, or changed in rank.
4. Classifications are based primarily on phenetic similarities.
5. Overall similarity between any two taxa is a function of their individual similarities in each of the many characters in which they are being compared.

QUESTIONS ON CLASSIFICATION

1. What is classification?
2. What are the purposes of plant classification? Why does man classify?
3. What are the basic operations in the development of a system of modern classification?
4. What do classifiers actually do?
5. What is meant by each of the general types of characters below? What is the significance of each to the development of a system of classification?

a. Analytic	*e.* Reliable
b. Synthetic	*f.* Unreliable
c. Weighted	*g.* Diagnostic
d. Nonweighted	*h.* Descriptive

6. How is classification related to characterization? Identification? Nomenclature? What is the logical sequence for the development of a useful classification scheme?
7. How is the establishment of character correlations and discontinuities in variation basic to the development of a classification system?
8. What are the criteria for the establishment of rank? What are the problems in the establishment of a hierarchy for any system of classification?
9. What are the criteria for the choice of kinds and number of characters in the development of a system of classification?
10. What are the procedures for determining the relationships of taxa within and between ranks in a system of classification?

11. What are the desirable properties of any hierarchical system of classification? What are the criteria used in evaluating a good, modern system of classification?

12. Why is a classification system considered an information storage system that is basic to information retrieval?

13. ''Without classification, knowledge would be factual chaos, difficult to retain, and impossible to understand'' is an old statement. How does the application of the principles of classification help people organize factual chaos? Knowledge? Principles? Cite specific examples.

14. How does a study of the principles and systems of classification add to an individual's personal scholastic development in analytical perspicacity, literature expertise, and evaluative ability?

15. Why is the Besseyan-Hallierian system of classification for angiosperms considered an improvement over others?

EXERCISES IN CLASSIFICATION

1. Select a manual pertinent to the vascular flora of your area and determine the following characters for each taxon indicated below:

   a. Diagnostic or circumscriptive characters used in classifying each of the included divisions.

   b. Primary and secondary characters used in distinguishing the families of the Pteridophyta.

   c. Primary and secondary characters used in distinguishing the subdivisions of the Spermatophyta.

   d. Character states or criteria used in distinguishing the classes of Angiospermae.

   e. The primary characters used in separating the families of monocots.

   f. The primary and secondary characters used in distinguishing the families of dicots.

2. Select a manual pertinent to the vascular flora of your region and write a description or circumscription using key characters for each of the following taxa:

   a. Spermatophyta

   b. Selaginellaceae

   c. Gymnospermae

   d. Dicotyledoneae

   e. Alismataceae

   f. Asteraceae

3. Determine the type of classification system used in the manual of your choice—phenetic, alphabetical, phylogenetic, or combination. Give the bases for your decision.

chapter 7

Plant Identification

Taxonomists as identifiers determine the group to which a specimen belongs according to a system of identification; construct systems of identification; and diagnostically distinguish new taxa and determine the diagnostic characteristics for old taxa that have been remodeled, divided, united, transferred, or changed in rank according to a system of classification.

All of us recognize organisms or objects previously seen, smelled, tasted, touched, heard, or otherwise known. We recognize skunk cabbage by its odor; the robin by its song; the dandelion by its color; the zebra by its stripes; the lion by its roar; the prickly pear by its "prickles"; and so on. Recognition presupposes a personal familiarity with the organism or object. On the other hand, identification in the biological sense is the determination of the group to which a specimen belongs, which assumes that organisms have been distinguished from one another, classified and circumscribed, as well as named by professional taxonomists. Recognition involves recall by association; identification usually includes a direct comparison of the unknown specimen with classified, circumscribed, and named taxa or the use of some device such as a key. Recognition is largely intuitive and sensory while identification is cognitive and knowledge-based. Biologically, recognition is an elementary but fundamental type of identification.

Identifiers who establish the identities of new taxa differentiate among similar groups of organisms by determining the characteristics that distinguish one from the other. Identifiers of established taxa assume that these have been delimited, circumscribed, classified, and named and that a workable system of identification has been developed. The identifier using an identification system—a key or polyclave—for the determination of unknown specimens usually procures a name for the taxon for communication or informational-retrieval purposes.

An identification system is composed of named taxa, the useful differentiating characters, and the taxon/character states themselves. A good system provides an efficient, easy, accurate, positive determination for each taxon included. If the application of the system does not result in consistent, positive identification, the differentiating characteristics should be changed or the taxa reclassified.

Identification is a major part of training in systematics. Courses in local flora, woody plants, aquatics, algae, fungi, and bryophytes are identification oriented. Students have to know the names of taxa for communication and informational purposes. The most frequently used method for the determination of the named group (taxon) to which a specimen belongs is the dichotomous key that is found in manuals, floras, and guides for groups such as vascular plants of a region, the woody flora of a state, or the algae of a particular sound. Computer-based methods and a variety of polyclave-type devices have been developed in recent years for identification of taxa in special groups, for example, disease-causing organisms and tropical rain forest seedlings.

A. TRADITIONAL METHODS OF IDENTIFICATION*

The traditional methods of identification include expert determination, recognition, comparison, and the use of keys and similar devices.

In terms of reliability or accuracy the best method of identification is expert determination. In general, the expert will have prepared treatments such as a monograph or revision of the group in question. It is highly probable that the more recent floras or manuals include the expert's concepts of taxa. Although of great reliability, this method presents problems by requiring the valuable time of experts, which is generally not available for routine identification. Recognition can approach expert determination in reliability when based on the identifier's extensive, past experience with the plant group in question.

Another method is the comparison of an unknown with named specimens, photographs, illustrations, or descriptions. Even though this is a reliable method, it may be very time-consuming or virtually impossible to use because of the lack of suitable materials for comparison. Its reliability depends on the accuracy and authenticity of the specimens, illustrations, or descriptions used in the comparison. The most widely employed method of identification is the use of keys or similar devices (synopses, outlines). Keying does not require the time, materials, or experience involved in comparison and recognition.

Keys are devices consisting of a series of contrasting statements or propositions requiring the identifier to make comparisons and decisions based on statements in the key as related to the material to be identified (Figures 7.1 and 7.2). The first dichotomous keys clearly designed for identification were those of Lamarck in his *Flore Francaise* in 1778 (Voss, 1952). The comments of A. P. de Candolle in his dedication to Lamarck in the third edition of this flora concerning keys are equally appropriate today (Voss, 1952):

> As to the artificial method, I have without hesitation, given preference to the one which you have contrived, and which consists of leading the student to the name of the plant

*Adapted from Massey, J. R. in Radford, A. E. et al., 1974, Vascular Plant Systematics. Harper & Row, New York. Chapter 25, pp. 522–536.

A DICHOTOMOUS KEY TO SELECTED GENERA OF SAXIFRAGACEAE

Shrubs or woody vines
Woody vines; petals 7 or more . . . .   3. *Decumaria*
Shrubs; petals 4 or 5.
   Leaves alternate or on short spur branches.
      Leaves pinnately veined; ovary superior;
         fruit a capsule . . . .   1. *Itea*
      Leaves palmately veined; ovary inferior;
         fruit a berry . . . .   2. *Ribes*
   Leaves opposite.
      Petals usually 4; stamens 20–40; fruit
         longitudinally dehiscent, not ribbed . . . . .   4. *Philadelphus*
      Petals usually 5; stamens 8–10; fruit pori-
         cidally dehiscent, 10– to 15–ribbed . . . .   5. *Hydrangea*

Herbs
Staminodia present; petals more than 10 mm long.   6. *Parnassia*
Staminodia absent; petals less than 10 mm long.
   Leaves ternately decompound . . . .   7. *Astilbe*
   Leaves simple.
      Flowers solitary in leaf axils, or in short,
        leafy cymes.
         Sepals 4; carpels 2 . . . .   8. *Chrysosplenium*
         Sepals 5; carpels 3 . . . .   9. *Lepuropetalon*
      Flowers in racemes or panicles.
        Petals pinnatifid or fringed; stem leaves
         opposite . . . .   10. *Mitella*
        Petals not pinnatifid or fringed; stem leaves
         alternate or absent.
         Ovary 1-locular
            Inflorescence paniculate; stamens 5 . . .   11. *Heuchera*
            Inflorescence racemose; stamens 10 . .   12. *Tiarella*
         Ovary 2-locular.
            Stamens 5; leaves palmately lobed . . .   13. *Boykinia*
            Stamens 10; leaves not palmately lobed   14. *Saxifraga*

Figure 7.1 Example of an indented key. (*Source:* Radford, A. E., H. E. Ahles, and C. R. Bell, 1968, Manual of the Vascular Flora of the Carolinas, p. 519, University of North Carolina Press, Chapel Hill. Used with permission.)

by always forcing him to choose between two contradictory characters: in this analytic method I have permitted myself only the slight changes necessitated by the increase in the number of plants described. There, after your example, I have sought to distinguish the plants by the easiest and most apparent characters; and when these characters were not constant, I have tried to foresee their aberrations and to arrive at the same name by different routes; but this ease in the distinguishing of plants is very different in different families: in some, such as the crucifers, it is impossible to distinguish the genera without examination of the fruit. . . . When beginners undergo these difficulties in the use of the analytic method, I beg them, before blaming it, to reflect that the most accomplished botanists meet with the same embarrassment, and that no method can make the work easier to students than it is to the masters. . . . But when the pupil knows the name, let him take care not to think he knows the thing!

A DICHOTOMOUS KEY TO SELECTED GENERA OF SAXIFRAGACEAE

1. Shrubs or woody vines		2.
1. Herbs		6.
2. Woody vines; petals 7 or more	*Decumaria*	
2. Shrubs; petals 4 or 5		3.
3. Leaves alternate or on short spur branches		4.
3. Leaves opposite		5.
4. Leaves pinnately veined; ovary superior; fruit a capsule	*Itea*	
4. Leaves palmately veined; ovary inferior; fruit a berry	*Ribes*	
5. Petals usually 4; stamens 20–40; fruit longitudinally dehiscent, not ribbed	*Philadelphus*	
5. Petals usually 5; stamens 8–10; fruit poricidally dehiscent, 10–15 ribbed	*Hydrangea*	
6. Staminodia present; petals more than 10 mm long	*Parnassia*	
6. Staminodia absent; petals less than 10 mm long		7.
7. Leaves ternately decompound	*Astilbe*	
7. Leaves simple		8.
8. Flowers solitary in leaf axils, or in short, leafy cymes		9.
8. Flowers in racemes or panicles		10.
9. Sepals 4; carpels 2	*Chrysosplenium*	
9. Sepals 5; carpels 3	*Lepuropetalon*	
10. Petals pinnatifid or fringed; stem leaves opposite	*Mitella*	
10. Petals not pinnatifid or fringed; stem leaves alternate or absent		11.
11. Ovary 1-locular		12.
11. Ovary 2-locular		13.
12. Inflorescence paniculate; stamens 5	*Heuchera*	
12. Inflorescence racemose; stamens 10	*Tiarella*	
13. Stamens 5; leaves palmately lobed	*Boykinia*	
13. Stamens 10; leaves not palmately lobed	*Saxifraga*	

Figure 7.2 Example of a bracketed key. (*Source:* Radford, A. E., H. E. Ahles, and C. R. Bell, 1968, Manual of the Vascular Flora of the Carolinas, p. 519, University of North Carolina Press, Chapel Hill. Used with permission.)

Referred by a number in the analytic method to the description, he will find in this second part the details which put together constitute the whole science.

I. Suggestions for the Use of Keys

Select keys appropriate for the materials to be identified. The keys may be in a flora, manual, guide, handbook, monograph, or revision. If the locality of an unknown plant is known, select a general work. If materials to be identified were cultivated, choose one of the manuals treating such plants, since most floras do not include cultivated plants unless naturalized.

Study the introductory comments on format details, abbreviations, and so on before using the key. Read both leads of a couplet before making a choice. Even though the first lead may seem to describe the unknown material, the second lead may be even more appropriate. Try both choices when dichotomies are not clear or when information is insufficient, and make a decision as to which of the two answers better fits the descriptions.

Use a glossary to check the meaning of terms you do not understand. Measure several similar structures when measurements are used in the key, for example, measure several leaves, not a single leaf. Do not base your decisions on a single observation. It is often desirable to examine several specimens.

Verify your results by reading a description and comparing the specimen with an illustration or an authentically named herbarium specimen.

II. Suggestions for Construction of Keys for Floras

Identify all groups to be included in a key. Prepare a description of each taxon. Select key characters with contrasting character states. Use macroscopic, morphological characters and nonvariable character states when possible. Avoid characteristics that can be seen only in the field or on specially prepared specimens, that is, use those characteristics that are generally available to the user. Prepare a comparison chart (Figure 7.3).

Construct strictly dichotomous keys. Use parallel construction and comparative terminology in each lead of a couplet. Use at least two characters per lead when possible. Follow key format (indented or bracketed; see Figures 7.1 and 7.2). Start both leads of a couplet with the same word if at all possible and successive leads with different words. Mention the name of the plant part before descriptive phrases, for example, *leaves alternate* not *alternate leaves* or *flowers blue* not *blue flowers*.

Place those groups with numerous variable character states in a key several times when necessary. Construct separate keys for dioecious plants, for flowering or fruiting materials, and for vegetative materials when pertinent. Test keys for reliability and consistency of identification using specimens previously determined by an expert for the taxa included in the key.

In general, polythetic dichotomous keys are preferable to monothetic. Polythetic keys have multiple contrasting characters used simultaneously in at least some of the couplets. Monothetic keys have single contrasting characters in the couplets requiring a simple either/or answer. A monothetic key with a flower character only will not be of any value to a student trying to identify a fruiting or vegetative specimen. (See the polythetic couplets distinguishing *Itea* from *Ribes* and *Philadelphus* from *Hydrangea* in Figure 7.1.)

COMPARISON CHART
(Character/Taxon Matrix)

	Decumaria	*Itea*	*Ribes*	*Parnassia*	*Heuchera*	*Saxifraga*
Habit	Woody vine	Shrub	Shrub	Herb	Herb	Herb
Leaf arrangement	Opposite	Alternate	Alternate or on spur shoots	Basal (Rosulate)	Basal (Rosulate)	Basal (Rosulate)
Petal number	7–10	5	5	5	5	5
Locule number	7–10	2	1	1	1	2
Stamen number	7+	5	5	5 (staminodia 5)	5	10
Fruit type	Capsule	Capsule	Berry	Capsule	Capsule	Capsule

Figure 7.3 Example of a comparison chart.

B. POLYCLAVES AND OTHER METHODS*

An objective identification system requires a priori information of three kinds: the pertinent taxa, the useful differentiating characters of these taxa, and the taxon/character data themselves.

The hierarchy of natural taxa is the foundation of biological information retrieval as well as systematic synthesis, since data can be stored, retrieved, and studied at any level of generalization. In data processing, we may usually regard a *taxon* as a group of one or more individuals or lower taxa judged sufficiently similar to each other to be treated together formally as a single evolutionary or information unit at a particular level in the taxonomic hierarchy and sufficiently different from other such groups of the same rank to be treated separately from them. Traditionally, each taxon (except the highest) belongs to one and only one taxon of the next higher rank, implying each individual belongs to exactly one species (and has one name) in any particular taxonomic treatment of its group. Taxa are either monothetic or polythetic. For monothetic taxa, possession of a unique set of diagnostic characters is both necessary and sufficient for membership in the group. Polythetic taxa are more loosely circumscribed, since presence of only a large number (none of which is diagnostic alone) of a list of characters is required for membership. In other words, the members of a polythetic taxon exhibit overall similarity. Modern systematists employ the polythetic taxon concept in most dichotomous keys. Other monothetic methods of identification are still used. Theoretically, identification schemes for modern biology ought to assure that no single-character difference, either in population variability or observer error, can result in a misidentification. Polythetic polyclaves offer a solution; here, no possibility is eliminated until several differences have accumulated between the taxon description and the unknown specimen. The appropriate threshold depends on the variability of the taxa, but toleration of just one or two differences often gives a marked improvement in identification success.

Polyclaves of various kinds allow one to select the characteristics for use in identifying each specimen by choosing from some character set and repeating an elimination process until a tentative identification is made. A printed data table, chart, or matrix giving the status of various taxa for useful characteristics is readily used as a polyclave by listing the possible taxa on scratch paper and crossing out those that do not agree with the specimen's characters. Such data tables appear irregularly in the taxonomic literature, often for only the more difficult groups involved but occasionally for all the treated taxa, as done for medical bacteria. For large groups, the diagnostic tables are not only more powerful than the equivalent key, but also take less space to print. Lists of taxa having various characters were among the first nontabular polyclaves. These lists resemble the inverted files common in computerized information systems, where entries are listed according to their characteristics, rather than characteristics by entries. Lists of taxa lacking specified features have also been produced; this modification expedites use as one may then jot down the possible taxa and rapidly cross off those differing from the specimen. Polyclaves are readily mechanized, as shown by the familiar edge-punched cards and the less familiar window keys as well as various mechanical devices. Polyclaves can be computerized with the development of suitable data formats.

*Adapted from Massey, J. R. in Radford, A. E. et al., 1974, Vascular Plant Systematics. Harper & Row, New York. Chapter 25, pp. 522–536.

Polythetic identification, where one difference no longer implies elimination, is available as an option in computer-based systems of identification. Here the program tallies the number of differences and eliminates a taxon only when its tally exceeds a user-determined value, commonly one, two, or three. Although it allows for greater taxon variability or user error, the polythetic polyclave is slower than the monothetic one since more characters must be submitted to assure complete elimination of all other taxa.

When a list of useful characteristics is desired part way through a polyclave procedure, a routine is used to select the best characters to continue dividing the set of remaining possibilities. Actually, a portion of the key-constructing algorithm is used to determine the first such character, and repetition of the procedure (ignoring this character) gives the next-best character, and so forth. An algorithm here is a series of logical steps or instructions by which an identification can be made. With the polyclave, of course, such recommended characters are merely suggestions that need not be employed. No matter which state the specimen shows, the recommended character will eliminate about half the possibilities. Any other character, on the average, will delete fewer taxa because it eliminates a larger number only when in its rarer state. However, if the user happens to notice a specimen displaying a rare character, he or she can eliminate a large number of possibilities at once and identify the specimen much more rapidly. Since the ability to recognize rare characters and realize their power comes only through training and experience, the expert delights in the efficiency and power of polyclave, while the neophyte is lost in its multitude of choices and prefers the supervision and security of the traditional dichotomous key.

If an identification method requires, in general, use of all the characters in some list, it is no longer a polyclave, for no user options remain in selecting characters. Several such character-set methods have been devised; most are statistical. Although well intended, character-set algorithms seem to offer no advantages over polyclaves. However, pattern-recognition methods using no formal characters might be employed effectively for identification in some cases. Such techniques linked with optical scanners, spectroscopy, or chromatography could even offer fully automated identification.

SUGGESTED READING

Benson, L. 1962. Plant Taxonomy, Methods and Principles. The Ronald Press Company, New York. Chapter 14, pp. 391–400.

Davis, P. H. and V. H. Heywood. 1965. Principles of Angiosperm Taxonomy. D. Van Nostrand Company, Inc., New York. Chapter 8, pp. 266–270; Chapter 9, pp. 300–303.

Harrington, H. D. and L. W. Durrell. 1957. How to Identify Plants. The Swallow Press, Chicago.

Jeffery, C. 1981. An Introduction to Plant Taxonomy. J & A Churchill Ltd., London. Chapter 6, pp. 94–105.

Lawrence, G. H. M. 1951. Taxonomy of Vascular Plants. The Macmillan Company, New York. Chapter 10, pp. 223–233.

Pankhurst, R. J. (ed.) 1975. Biological Identification with Computers. Academic Press, London.

Radford, A. E. et al. 1974. Vascular Plant Systematics. Harper & Row, New York. Chapter 25, pp. 522–536.

Voss, E. G. 1952. The history of keys and phylogenetic trees in systematics biology. Journal of the Scientific Laboratories of Denison University 43: 1–25.

SUMMARY FOR IDENTIFICATION

Definitions for Identification   *Identification* is the determination of the similarities or differences between two elements; it is the direct comparison of the characteristics of a specimen in hand with those in keys in order to arrive at a name; and it is the allocation or assignment of an unidentified taxon to the correct class once a system of classification has been established. An *identification system* includes basic information of three kinds: the pertinent taxa, the useful differentiating characters of these taxa, and the taxon/character data themselves. *Keys* are devices consisting of a series of contrasting statements or propositions that require the identifier to make comparisons and decisions based on statements in the key as related to the materials to be identified.

Purpose of the Study of Identification   To determine the group to which a given specimen belongs usually by name so that information can be retrieved from a classification and/or descriptive system; to provide a system of identification designed for ease, accuracy, and certainty of identification.

Operations in Identification   To distinguish new taxa diagnostically and determine the diagnostic characteristics for old taxa that have been remodeled, divided, united, transferred, or changed in rank according to a system of classification; to determine the group to which a specimen belongs according to a system of identification; and to construct keys or polyclaves.

Basic Premise for Plant Identification   That through the use of various techniques (e.g., expert determination, recognition, comparison) and the use of computer programs, keys, or various mechanical devices a given specimen can be properly placed within an established classification system containing appropriate nomenclature allowing for communication and retrieval of information.

Fundamental Principles for Identification

1. Identification is essential to the maintenance and refinement of any classification system and, hence, taxonomic system.
2. Identification directly involves both classification and nomenclature and is related and interrelated with circumscription.
3. Identifications of taxa are changed for the proper reasons of more profound knowledge of the facts resulting from adequate taxonomic study or because the system of identification does not result in consistent, positive identification of the taxon.

Guiding Principles for the Study of Identification

A. Construction of identification devices (keys):
  1. Select characters that are in the usual material to be identified.
  2. Select characters that are easy to observe and interpret.
  3. Select characters that are distinctive between the various states.
  4. Select characters that are independent from other characters used.
  5. Select characters that are tolerant to environmental influences (that is, they do not vary with the environment).

6. Construct strictly dichotomous keys.
7. Use parallel construction and comparative terminology in each lead of a couplet.
8. Use at least two characters per lead when possible.
9. Follow a key format.
10. Start both leads of a couplet with the same word if at all possible and successive leads with different words.
11. Start lead with name of plant part (noun), then follow with descriptor (adjective).
12. Construct separate keys for dioecious plants, for flowering or fruiting material, and for vegetative materials when pertinent.

B. Use of keys or identification devices:
1. Select appropriate key or identification device for materials to be identified, for example, manual appropriate to region, book appropriate to type of material, for example, for cultivars in horticultural treatments.
2. Read introductory comments on format details, abbreviations, and so on before using identification device.
3. Read both leads of key couplet before making choice.
4. Use recommended glossary for terms in key that are not understood.
5. Measure several similar structures when measurements are used in a key.

Basics Assumptions for Identification

1. Taxa have been accurately characterized, diagnostically delimited, hierarchically arranged, and named to make identification schemes possible.
2. A system of identification in which a choice of many characteristics can be used, as in the polyclave method or polythetic key, is more desirable than one in which a rigidly limited number of characteristics, as in a key, is used.
3. Monothetic taxa are basically easier to identify than polythetic taxa; taxa with limited diagnostic characteristics are easier to identify than taxa with numerous combinations or similarities.
4. No more groups can be distinguished than the product of the number of character states.
5. Expert determination precedes recognition, which precedes comparison in identification reliability.

QUESTIONS ON PLANT IDENTIFICATION

1. What is identification from functional, practical, and theoretical standpoints?
2. Why does society need a plant identification service?
3. What do identifiers actually do?
4. Why is an identification system considered an information-retrieval system? A classification system, an information storage system?
5. How do you define each term and distinguish one from the other in each couplet below:
   a. Recognition-identification
   b. Identification-classification
   c. Diagnostic key-synoptic key
   d. Polyclave-comparison chart
   e. Data matrix-comparison chart
   f. Quantitative character-qualitative character
6. What are the methods of identification? Rank as to reliability.

7. How do the suggestions for use of keys help one in the construction of keys?
8. What are the elements of a good synoptic key? A good diagnostic key? Any identification system?
9. What is the significance of rank, order, and sequence in key construction? How are keys hierarchical?
10. What are the three kinds of information required in an objective identification system? Characterize each.
11. Why are identification and classification systems based on characterization that is related to name?
12. How are television commercials, school posters, and cigarette advertisements identification systems?

EXERCISES IN PLANT IDENTIFICATION

1. Identification of an unknown: select an unknown specimen and identify it by keying in an appropriate manual, flora, or monograph. Verify your results by reading a description, comparing with an illustration, or checking with your instructor.
2. Preparation of a comparison chart: select five or more specimens from the group provided by your instructor. Identify each by keying. Verify your results. Prepare a description of each similar to those in a flora or manual. Be sure characters and character states are in the same order. Select contrasting character states and prepare a comparison chart (see Figure 7.3).
3. Construction of keys: construct a dichotomous key to these specimens using the information in the comparison chart above.
4. Select five species (consulting your laboratory instructor) and prepare a comparison chart based on the following characters: leaf arrangement, and so forth (see chart below). Select the diagnostic character states for each taxon in the chart below and write an indented dichotomous key (on a separate sheet) to the five species.

	1.	2.	3.	4.	5.
Leaf arrangement					
Habit					
Petal number					
Locule number					
Stamen number					
Fruit type					

5. Select a key in a manual of your choice and analyze the key for sound construction by using the principles of construction as analytical criteria. Indicate how the construction of the key could be improved.
6. Select a key in a manual of your choice, study it, then indicate how the principles of key use could have been of value in the construction of the key.
7. Preparation of a simple polyclave. Select a family of ten or more genera. Construct a master sheet listing an abbreviation or number for each genus and print in red on a 5 × 8 card. This list

is the *taxa master list underlay*. By examining herbarium specimens or using an appropriate manual or monograph select and compile a list of 50 or more characters and character states for each genus. Duplicate (in black) the taxa master list underlay to equal the number of characters selected. These will become character state overlay cards. Label each card with a specific character state from the character state list. Search each specimen or description for the character state in question and punch out the generic name or number listed on the card for each genus having the character state in question. Punch a card for each character state, for example, if one character state selected is *leaves opposite,* punch out the names of all genera known to have opposite leaves. To use the polyclave, select a specimen identified as belonging in the family in question. Notice character states and begin sorting through overlay cards until you find a card with a characteristic that matches one of the cards you have prepared. Place the character overlay card over the taxa master list. Those names or numbers appearing in red through the holes will represent all genera possessing the character in question. Continue sorting and overlaying until all cards have been used or until a single genus name or number appears.

8. Prepare a taxon delimitation of *Chrysosplenium* from the closely related genera(us) using the characteristics in Figure 7.1 or 7.2.

chapter 8

Taxonomic Evidence

Systematists as researchers select evidentiary characters from many sources and determine character states for a more comprehensive characterization, a more effective delimitation, and the development of a more definitive classification of taxa as well as for a more natural interpretation of the phenetic and phylogenetic relationships of taxa.

A paw print of a cat, an exhibit of a skull, an indicator of waste, a symptom of a disorder, and a deposition for a trial are all different modes of expression for information used as evidence. A fruit salad in a restaurant is information that becomes evidence when used as an example of a diet lunch, as an exhibit in a case of food poisoning, or as the indicator of the quality of the salad bar. *Evidence is information used in context for a purpose.* The results and effects of the use of evidence in systematic studies are manifested in many ways, such as in the proof of a hypothesis, the solution of a problem, the characterization of a taxon, or the classification of a group of organisms.

Intrinsically, evidence takes many forms that can be classified as physical, chemical, or biological (see Table 8.1). Physical evidence occurs as color, shape, size, and arrangement. Chemical evidence may be expressed as structure, process, or product. The biological evidence most frequently used is structural; but functional, developmental, environmental, and geographic types are basic to systematic activities. Physically, wood can be identified by its specific gravity, a fruit by its shape, a flower by its color, and a duckweed by its size. Chemically, the magnolia can be recognized by its fragrance, the potato by its composition, and the stinkhorn fungus by its odor. Biologically, the balsa wood tree is characterized by its structural components, the saltwort by its habitat, and the rare plant by its distribution. Physical, chemical, and biological attributes are traditionally used as evidence in systematic endeavors related to organisms.

173

Evidentiary information can be obtained through our senses, the use of traditional instruments, and the employment of complex tools. The pineapple can be recognized by its taste, the pyramid by its shape, and the dog by its bark. A ruler can be used for measuring the width of a leaf, a balance for the weighing of a seed, and a spectrophotometer for determining the chemical characteristics of a fruit. Some types of evidentiary information can be acquired easily and cheaply, others only through the use of highly involved techniques and expensive instrumentation.

Evidence in its many modes and forms, regardless of how it is obtained, can be used in a wide array of contexts for a great variety of purposes. In the field of systematics, evidence is information procured from any source, whether it is obtained simply and cheaply or by complicated means and expensively, that is used for characterization, identification, and classification of organisms, populations, and taxa as well as for the determination of phenetic, genetic, and phylogenetic relationships.

Evidentiary products such as circumscription, taxon delimitation, characterization, synoptic key, and phylogenetic classification are exemplified and analyzed as to type of evidence and character in A. Taxonomic Evidentiary Analysis. The general types and uses of evidence are treated in B. Uses and Types of Evidence in Systematics.

A. TAXONOMIC EVIDENTIARY ANALYSIS

Evidence is used to show phenetic, genetic, and phylogenetic relationships. Relationships provide the bases for the classification of taxa through shared attributes. Classified taxa are circumscribed and delimited with appropriate character states for selected characters in

Table 8.1   A CLASSIFICATION OF EVIDENCE[a]

A. Physical	B. Chemical	C. Biological
I. Coloration	I. Structure	I. Structural
II. Size	II. Process	II. Functional
III. Number	III. Product	III. Developmental
IV. Fusion		IV. Environmental
V. Shape		V. Geographic
VI. Disposition, etc.		

	C. Biological Evidence	
	I. Structural	
Ia. Morphological	Ib. Anatomical	Ic. Cytological
Id. Palynological	Ie. Embryological	If. Genetical
	II. Functional Evidence	
IIa. Metabolism	IIb. Physical process	IIc. Reproduction
	III. Developmental Evidence	
IIIa. Initiation	IIIb. Patterns	IIIc. Periodicity
	IV. Environmental Evidence	
IVa. Phenology	IVb. Habitat	IVc. Life form
	V. Geographic Evidence	
Va. Origin	Vb. Migration	Vc. Distribution

[a]Physical evidence is classified and described in Chapter 5. Chemical and biological evidence are discussed in B. Uses and Types of Evidence.

one or more types of evidence. Evidence has to be classified to be useful (Table 8.1). Reliable and meaningful evidence used in a taxonomic study is obtained and documented with standardized procedures (see guidelines in B. Uses and Types of Evidence). A description of a species (*Saxifraga michauxii*) in a manual and a taxon delimitation in a key to species are analyzed for evidence, character, and character state in I. Description Characterization, Circumscription and II. Taxon Delimitation, Identification, Key, respectively. Types of evidence used in a classification system and a synoptic key entry are indicated in III. Evidence Used in a Classification System and IV. Evidence Used in a Synoptic Key.

I. Description, Characterization, Circumscription

A very limited description, a limited structural characterization, and a circumscription for the species *Saxifraga michauxii* Britton in the Carolinas, follows. (Radford et al., 1968).*

S. *michauxii*. Plant 1–5 dm tall, pubescent throughout, hirsute to puberulent, usually glandular. Leaves oblanceolate to obovate, to 15 cm long and 4 cm wide, acute, coarsely toothed, base attenuate. Panicle diffuse. Calyx tube absent, sepals 1.5–2.5 mm long, reflexed in fruit; corolla with 3 clawed petals with a yellow spot and 2 spatulate petals; stamens barely exserted, filaments filiform. Capsules nerved, ovoid, 3.5–7 mm long, style 0.5–1 mm long; seeds longitudinally striate, echinate, 0.8–0.9 mm long. June–Aug. Moist rocks and seepage slopes; mts. (VA, GA, TN, WV)

E.	Mor		Phy		Mor		Mor	
C.S.	Plant/	1–5 dm tall,/	pubescent/	throughout,/	hirsute/	to/	puberulent,	
C.	Organism Height		Vestiture		Vestiture			

E.	Mor	Mor	Phy	Phy
C.S.	usually glandular./	Leaves oblanceolate to obovate,/	to 15 cm long	
C.	Vestiture	Organ	Shape	Size

E.	Phy	Phy	Phy	Phy	Mor	Phy
C.S.	and 4 cm wide,/	acute,/	coarsely toothed,/	base attenuate./	Panicle/	diffuse.
C.	Size	Apex	Margin	Base	Organ	Arrangement

E.	Mor	Mor	Phy	Phy	Mor
C.S.	Calyx tube absent,/	sepals/ 1.5–2.5 mm long,/	reflexed in fruit;/	corolla	
C.	Calyx prt	Calyx prt	Length	Orientation	Flower prt

E.	Phy	Phy	Phy	Phy	Phy	Mor	Mor
C.S.	with 3/	clawed petals with a/	yellow spot/	and 2/	spatulate/	petals;/	stamens
C.	#	Shape	Color	#	Shape	Flower prt	Flower prt

E.	Mor	Mor	Phy	Mor	Mor	Phy	Phy
C.S.	barely exserted,/	filaments/	filiform./	Capsules/	nerved,/	ovoid,/	3.5–7 mm
C.	Insertion	Stamen Prt	Shape	Fruit type	Venation	Shape	Length

*E. = evidence; C. = character; C.S. = character state; Mor = morphology; Phy = physical; Phe = phenology; Hab = habitat; Geo = geography, Prt = part; # = number.

E.   Mor   Phy   Mor   Mor   Mor   Phy
C.S. long,/style/ 0.5–1 mm long;/seeds/longitudinally striate,/echinate,/0.8–
C.   Carpel prt   Length Plant prt   Surface   Surface

E.   Phy   Phe   Hab   Geo   Geo
C.S. 0.9 mm long./ June–Aug./ Moist rocks and seepage slopes;/mts./(VA, GA, TN, WV)
C.   Length   Flowering Fruiting   Habitat   Province   State

II. Taxon Delimitation, Identification, Key

S.[*Saxifraga*] *michauxii* delimited from *S. pensylvanica, S. virginiensis,* and *S. pensylvanica* delimited from *S. virginiensis* follow (Radford et al., 1968.)*

Corolla zygomorphic, fruit with distinct longitudinal nerves.
                                                              1. *S. michauxii*
Corolla actinomorphic, fruit without distinct longitudinal nerves.
      Leaves lanceolate to oblanceolate, denticulate to sparsely serrate.
                                                              2. *S. pensylvanica*
      Leaves ovate, coarsely serrate to lobed.
                                                              3. *S. virginiensis*

E.   Mor   Mor   Mor   Mor
C.S. Corolla/zygomorphic;   fruit/with distinct longitudinal nerves. 1. *S. michauxii*
C.   Flower prt   Symmetry Plant prt   Venation

E.   Mor   Mor   Mor   Mor
C.S. Corolla/actinomorphic; fruit/without distinct longitudinal nerves
C.   Flower prt Symmetry   Plant prt   Venation

E.   Mor   Phy   Phy
C.S. Leaves/lanceolate to oblanceolate,/denticulate to sparsely serrate. 2. *S. pensylvanica*
C.   Plant prt   Shape   Margin

E.   Mor Phy   Phy
C.S. Leaves/ ovate,/coarsely serrate to lobed 3. *S. virginiensis*
C.   Plant prt   Shape   Margin

III. Evidence Used in a Classification System

This treatment includes examples of characters used from different fields of evidence for the classification of angiosperms by Cronquist (1981), with the taxa included in parentheses. Modern classification systems developed for plants are based on many types of evidence.

*E. = evidence; C. = character; C.S. = character state; Mor = morphology; Phy = physical; Phe = phenology; Hab = habitat; Geo = geography, Prt = part; # = number.

Anatomical Evidence

Phloem stratified into fibrous and nonfibrous tissue (Sarcolaenaceae/Malvales); internal phloem (Caryophyllales; Malvales, Myrtales); wood parenchyma diffuse versus vasicentric (Umbellales versus Cornales); nodal anatomy (Umbellales versus Cornales); stem with anomalous secondary thickening (Caryophyllidae); schizogenous secretory system (Myrsinaceae; Umbellales versus Cornales); laticifers (Musaceae); secretory canals (Guttiferae, Sapindales); elongate idioblasts (Trochodendrales); ethereal oil cells (Magnoliidae); foliar glands (Leitneriales, Myricales, Juglandales); myrosin cells (Capparales; Tropaeolaceae); mucilage cells (Malvales); raphides, crystal sand (Dilleniaceae); subsidiary cell number (subclasses of monocots); nectary type (Malvales; monocots versus dicots; within Liliidae).

Embryological Evidence

Ovules unitegmic versus bitegmic (subclass level and below); ovules tenuinucellate versus crassinucellate (subclass level and below); endosperm cellular versus nuclear (Polemoniales, Rubiales-Gentianales-Dipsacales); endosperm helobial (Alismatidae, Arecidae); endosperm starchy versus nonstarchy (subclasses of monocots); integumentary tapetum (Ericales/Diapensiales); embryo sac bisporic versus monosporic (Alismatidae); possible free-nuclear stage in embryo development (*Paeonia*).

Cytological Evidence

Chromosome number (Agavaceae/Xanthorrhoeaceae); karyotype (Agavaceae).

Chemical Evidence

Cyanogenetic compounds (within Violales); alkaloids, glucosides (Gentianales); gentiopicrin (Gentianaceae); tryptophane alkaloids (Rubiaceae-Gentianales); betalains (Caryophyllidae); serology (Papaverales/Capparales; Cornales; Caprifoliaceae-Cornaceae).

Palynological Evidence

Pollen binucleate versus trinucleate (subclass level; Sapindales; in Polemoniales); pollen triaperturate versus uniaperturate (subclass Magnoliidae); pollen nonaperturate (Scheuchzeriaceae).

Physiological Evidence

Adaptation to NO_3 deficiency (Sarraceniales); adaptation to water stress (Caryophyllidae; some Violales); mycorrhizal habit (Ericales; Triuridales; Orchidales).

Developmental Evidence

Androecial ontogeny (*Paeonia*); gynoecial ontogeny (Lauraceae/Myristicaceae).

IV. Evidence Used in a Synoptic Key

The following entry is from Cronquist (1981).

Morphological

1. Plants relatively archaic, the flowers typically *apocarpous,* always

Morphological

polypetalous or *petalous* (but sometimes *synsepalous*) and generally with an

Morphological

evident perianth, usually with numerous (sometimes *laminar* or *ribbon-*

Developmental

shaped) stamens initiated in *centripetal sequence,* the pollen grains mostly

Palynological

binucleate and often *uniaperturate* or of a *uniaperturate-derived* type;

Embryological

ovules *bitegmic* and *crassinucellar;* seeds very often with a *tiny* embryo and

Embryological

copious endosperm, but sometimes with a larger embryo and reduced or no

Embryological

endosperm; *cotyledons occasionally more than 2;* plants very often

Chemical

accumulating *benzyl-isoquinoline* or *aporphine alkaloids,* but without

Chemical

betalains, iridoid compounds, or *mustard oils,* and seldom strongly *tanniferous.* . . .
I. Magnoliidae.

B. USES AND TYPES OF EVIDENCE IN SYSTEMATICS

The general guidelines and assumptions pertinent to the use of any type of evidence in systematic research is presented in I. General Guidelines and Assumptions for the Use of Evidence in Systematic Study. The types of evidence used in modern taxonomic studies are treated in II. Types of Evidence Used in Systematics.

I. General Guidelines and Assumptions for the Use of Evidence in Systematic Study

These guidelines and assumptions for uses of evidence are applicable to taxonomic studies in classification, identification, description, and determination of relationships. They are basic to most systematic research. Systematic procedures and processes include the selection and delimitation of characters and character states, the determination of character states for selected characters pertinent to the study of taxa, the establishment of character correlations for circumscriptions of taxa, and so on. Hierarchically, character states are components of characters that are part of character sets comprising evidence. Systematics, to a great degree, is evidentiary study.

a. Guidelines for Use of Evidence in a Systematic Study

1. State the objective(s), premise(s), or hypothesis for the study clearly and succinctly.
2. List the types of evidence to be used in the study and give the reasons for the choice of each type.
3. Relate the proposed use of the evidence to each objective, premise, or hypothesis.
4. Select the characters from each type of evidence to be used in the study and indicate potential use of each.
5. Document the classification, circumscription, and nomenclature of the characters used in the study.
6. Indicate the method(s) of determination and delimitation of the character states to be used in the study. Document the techniques and/or procedures used. (This is particularly pertinent for numerical data used as character states.)
7. Describe and document the procedures used in making character state correlations for the establishment, circumscription, delimitation, and classification of the taxa involved in the study.
8. Summarize and document the treatment of the data and character states as evidence for the proof of a hypothesis, the solution of the problem, or the attainment of the objectives of the study.

b. Basic Assumptions for Use of Evidence in a Systematic Study

1. The more characters and types of evidence on which a taxon is based the better its taxonomy.

2. Evidentiary characters hierarchically arranged and classified for comparability and consistency are relevant and meaningful in circumscriptions, delimitations, and classifications of taxa.
3. No single type of evidence is intrinsically more valuable taxonomically than another.
4. Taxonomic relationships based on more than one type of evidence are usually more significant than those based on only one type of evidence.
5. Evidentiary characters are documented as to definitions, methods of determination, and techniques used.

II. Types of Evidence Used in Systematics

The traditional types of evidence (a) morphological, (b) anatomical, (c) palynological, (d) embryological, (e) cytological, (f) genetic, (g) reproductive biological, (h) chemical, (i) ecological, and (j) geographical are presented in this section from a definition-reasons for use-how used-characters used-reference standpoint. Morphological evidence is classified and described in detail in Appendix D. Physiological characters are included as developmental in anatomical and embryological evidence and as metabolic products and biosynthetic pathways in chemical evidence. Ultrastructural and microstructural evidence, obtained with the use of the transmitting (TEM) and scanning (SEM) electron microscopes respectively, are treated as part of other types of structural evidence, such as palynological and cytological. The electron microscopes have been major tools for providing new evidentiary characters and data of great significance in the resolution of problems of relationships of higher taxa.

a. Morphological Evidence

Morphology is the study of structure and form of plants, usually dealing with the organism and its component organs (see Figures D.1 to D.12). Anatomy is also a study of structure at the cell and tissue levels. Cytology is a study of cell formation, structure, and function with emphasis on the internal structure of the cell.

Some basic evidentiary characters

Plant habit	Perianth structural types
Root structural types	Androecial types
Stem habit	Stamen types
Stem structural types	Gynoecial types
Bud structural types	Carpel types
Leaf structural types	Ovule types
Inflorescence types	Fruit types
Flower types	Seed types

Morphological characters are traditionally useful as evidence at all taxonomic levels, but particularly at the specific and generic ranks (see Appendix D). Morphological evidence provides the basic language for plant characterizations, identification, classification, and relationships. Generally, morphological data are easily observable and obtainable, and thus most frequently used in taxonomic studies.

Evidentiary examples of character/taxon relationships

Members of the Orchidaceae are herbaceous; of the Fagaceae, woody. The bulb is characteristic of the genus, *Allium* (onion); the rhizome, of the genus *Iris;* the stolon, of the genus, *Fragaria* (strawberry). Ventricose or pitcher leaves are typical of *Sarracenia* (pitcher plant); tentacular leaves are typical of *Drosera* (sundew). The ament is characteristic of the Betulaceae; the umbel, of the Apiaceae. The papilionaceous corolla is characteristic of many legumes; the ray corolla, of many composites. The appendicular stamen typifies *Viola;* the petalantherous, some species of *Saxifraga.*

Basic references: Radford, A. E., in Radford et al., 1974, Vascular Plant Systematics, Harper & Row, New York, Chapter 6, pp. 83–166. Stearn, W. T., 1966, Botanical Latin, Thomas Nelson and Sons Ltd. London.

b.   Anatomical Evidence   (Figures 8.1, 8.2)

Anatomy is the study of the structure, organization, and development of plant cells and tissues.

Some basic evidentiary characters

Wood cell type	Epidermal type
Wood cell size	Mesophyll type
Wood cell shape	Stomatal type
Wood cell wall sculpture	Sclereid type
Wood cell pattern	Trichome type
Stelar patterns	Crystal type
Xylem maturation types	Nodal type
Vascular bundle types	Petiole vasculation type
Wood type	Venation type
Ray type	Phloem cell type
Ground tissue type	Periderm origin
Parenchyma type	Specialized cell type

Anatomical characters are most useful at the generic and higher taxonomic category levels in determining relationships. The application of anatomical data to phylogenetic problems has been of great value in elucidating taxonomic relationships.

TRICHOMES

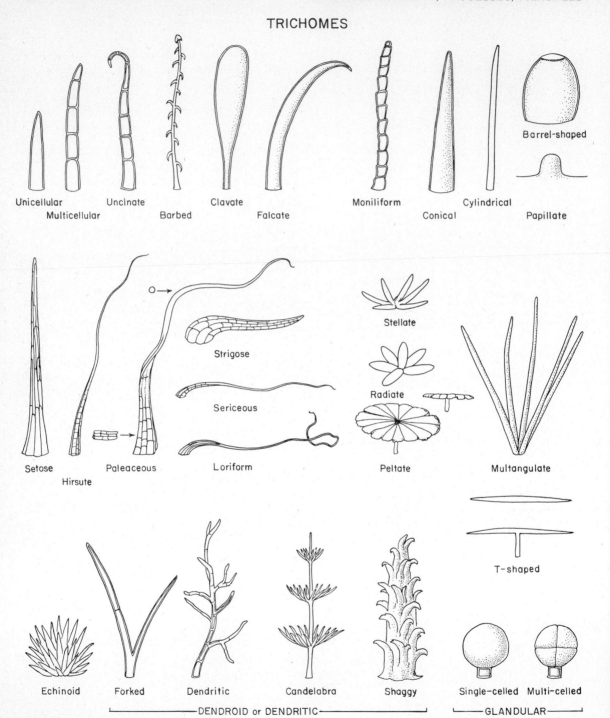

Figure 8.1 Anatomical evidence.

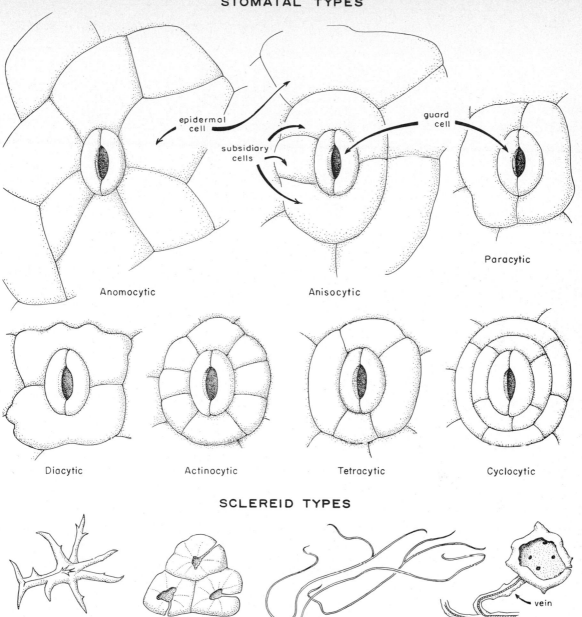

STOMATAL TYPES

epidermal cell

subsidiary cells

guard cell

Anomocytic

Anisocytic

Paracytic

Diacytic

Actinocytic

Tetracytic

Cyclocytic

SCLEREID TYPES

Astrosclereid

Brachysclereid

Filiform sclereid

Terminal sclereid

vein

Macrosclereid

Osteosclereid

Trichosclereid

Figure 8.2 Anatomical evidence.

Evidentiary examples of character/taxon relationships

Nonporous wood is characteristic of gymnosperms; porous, of angiosperms. Scattered vascular bundles are typical of monocots; ring bundles, of dicots. Anomocytic stomata are characteristic of the Ranunculaceae; diacytic, of the Caryophyllaceae (Radford et al., 1974). Uniseriate and homocellular rays are found in *Populus;* multiseriate and heterocellular, in *Eupomatia* (Radford et al., 1974).

Basic references: Dickison, W. C. in Radford et al., 1974, Vascular Plant Systematics, Harper & Row, New York, Chapter 7, pp. 167–210. Esau, K., 1965, Plant Anatomy, second edition, John Wiley & Sons, Inc., New York.

c. **Palynological Evidence**   **(Figure 8.3)**

Palynology is the study of pollen (fossil and modern) and spores, generally focusing on the structure of the wall rather than on living, internal features.

Some basic evidentiary characters

Pollen unit type	Exine structure
Pollen grain polarity	Exine sculpture
Pollen grain shape	Aperture type
Pollen grain symmetry	Aperture number
Pollen grain nuclear state	Aperture position
Pollen wall architecture	Aperture shape
Exine stratification	Aperture structure

Palynological evidence has proven useful at all taxonomic levels, particularly in verifying or refuting relationships in established taxonomic groups. The application of palynological data has proven to be of value in interpreting problems related to origin, migration, and evolution of floras as well as in studies related to stratigraphy, paleoecology, archeology, ethnobotany, distribution, and reproductive biology.

Evidentiary examples of character/taxon relationships

Pollen is binucleate in Magnoliidae, trinucleate in Caryophyllidae (Cronquist, 1981). Pollen is in tetrads in the Ericaceae, in pollinia in Asclepiadaceae (Radford et al., 1974). The pollen wall is echinate in *Taraxacum,* scabrate in *Quercus* (Radford et al., 1974).

Basic references: Dickison, W. C., in Radford et al., 1974, Vascular Plant Systematics, Harper & Row, New York, Chapter 8, pp. 211–222. Erdtman, G., 1969, Handbook of Palynology, Morphology-Taxonomy-Ecology, An Introduction, The Study of Pollen Grains and Spores, Munksgaard, Copenhagen.

POLLEN MORPHOLOGY

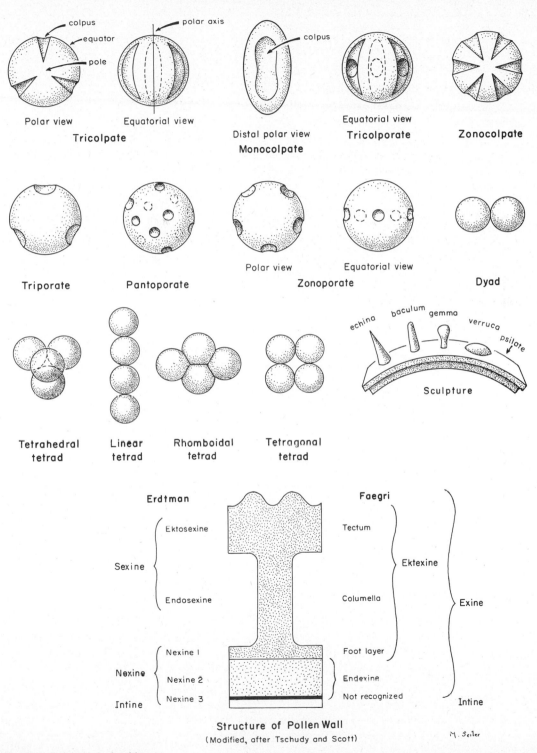

Figure 8.3 Palynological evidence.

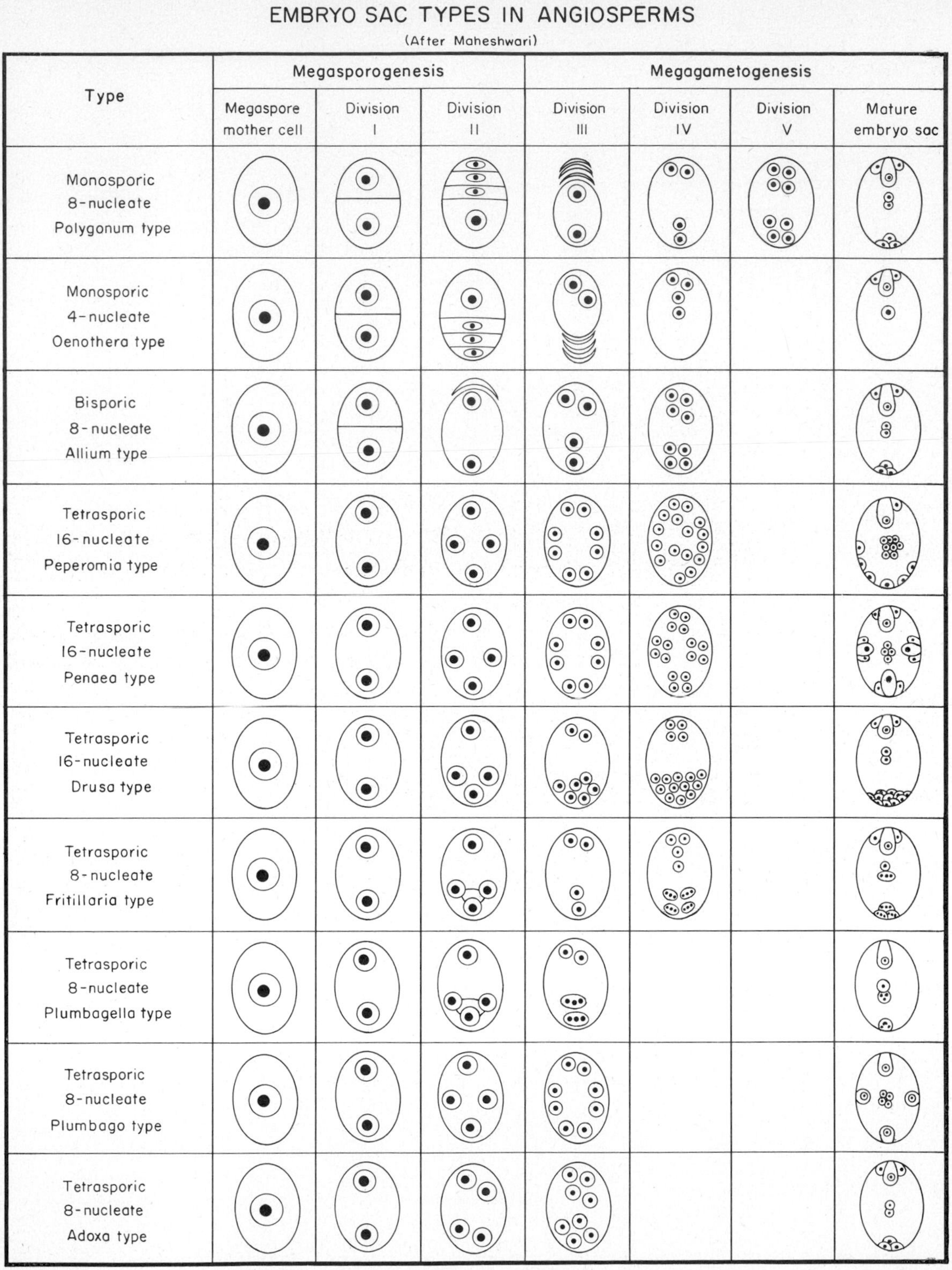

EMBRYO SAC TYPES IN ANGIOSPERMS

(After Maheshwari)

Type	Megasporogenesis			Megagametogenesis			
	Megaspore mother cell	Division I	Division II	Division III	Division IV	Division V	Mature embryo sac
Monosporic 8-nucleate Polygonum type							
Monosporic 4-nucleate Oenothera type							
Bisporic 8-nucleate Allium type							
Tetrasporic 16-nucleate Peperomia type							
Tetrasporic 16-nucleate Penaea type							
Tetrasporic 16-nucleate Drusa type							
Tetrasporic 8-nucleate Fritillaria type							
Tetrasporic 8-nucleate Plumbagella type							
Tetrasporic 8-nucleate Plumbago type							
Tetrasporic 8-nucleate Adoxa type							

Figure 8.4 Embryological evidence.

d. Embryological Evidence   **(Figure 8.4)**

> Embryology is the study of the successive stages of sporogenesis, gametogenesis, and the growth and development of the embryo.

Some basic evidentiary characters

Anther loculi number	Integument number
Anther loculi arrangement	Integument structural type
Anther endothecium type	Ovule orientation type
Anther wall formation type	Ovule position type
Archesporial cell number	Ovule vasculation type
Aril presence	Periplasmodium nature
Embryo sac development type	Perisperm presence
Embryo type	Tapetal type
Embryogeny type	Haustorium formation type
Endosperm formation type	Nucellar feature

> Embryological evidence has been of primary value at the higher category levels, particularly in conjunction with other types of evidence in confirming the systematic positions of taxa. The application of embryological data has also been helpful in answering systematic questions involving two or more positions based on other evidence and in providing new evidence for resolving doubtful systematic positions.

Evidentiary examples of character/taxon relationships

> Ovules are unitegmic and tenuinucellate in Asteridae, bitegmic and crassinucellate in Caryophyllidae. The embryo is embedded in endosperm in Cyperales, peripheral to endosperm in Graminales. The embryo sac is monosporic and eight-nucleate in *Polygonum*, monosporic and four-nucleate in *Oenothera* (Cronquist, 1981).

> Basic references: Dickison, W. C., in Radford et al., 1974, Vascular Plant Systematics, Harper & Row, New York, Chapter 9, pp. 223–236. Davis, G. L., 1966, Systematic Embryology of the Angiosperms, John Wiley & Sons, New York. Maheswari, P., 1964, Embryology in relation to taxonomy, in W. B. Turrill (ed.), Vistas in Botany IV, Macmillan Company. New York.

e. Cytological Evidence   **(Figure 8.5)**

> Cytology is the study of the morphology and physiology of cells. Traditionally, anatomists have dealt with cell shape, size, wall structure, and patterns, whereas cytologists have worked with the internal organelles of the cell and detailed structure of the cell wall.

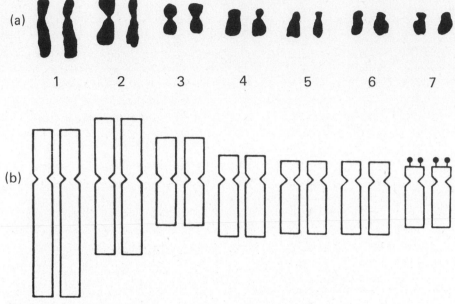

Figure 8.5 Cytological evidence. **a.** Mitotic metaphase chromosomes of *Sullivantia oregana*, which are representative of those of all *Sullivantia* species and of *Boykinia aconitifolia*, arranged by size and centromeric position. **b.** Apparent idiogram representing the same taxa. Scale for mitotic metaphase chromosomes is 5 μm. (*Source:* Soltis, Douglas E. 1980. Karyotypic Relationship among Species of *Boykinia, Heuchera, Mitella, Sullivantia, Tiarella,* and *Tolmiea (Saxifragaceae). Systematic Botany* 5: 17–29. Used with permission.)

Some basic evidentiary characters

Chromosome number	Ploidy level
Chromosome structure	Ploidy type
Chromosome type	Chromosome aberration
Chromosome meiotic behavior	Chromosome meiotic behavior

Cytological evidence is used for distinguishing taxa, determining the possible origin of groups, and for understanding the evolutionary history of related taxa, particularly those at the infraspecific and specific levels. The application of cytological data is very helpful in the resolution of hybridization problems and lineages in phylogenetic studies.

Evidentiary examples of character/taxon relationships

$2N = 26$ is characteristic of Amborellaceae; $2N = 16$, of Trimeniaceae. (Cronquist, 1981). $N = 8$ in *Delphinium ajacis, N = 16* in *Delphinium carolinianum* (Radford et al., 1968).

Basic references: Smith, B. W., in Radford et al., 1974, Vascular Plant Systematics, Harper & Row, New York, Chapter 10, pp. 237–258. Darlington, C. D., Cytology, J. and A. Churchill, London.

f. Genetic Evidence

> Genetics includes the study of variation, its expression in development, its transmission in inheritance, and its relation to breeding systems. Genetic evidence is experimental evidence derived from crosses and analyses of many characters. Experimental genetic data are used primarily to show relationships at the specific, subspecific, variety, and form or race levels. Genetic evidence is used by the taxonomist for a conceptual understanding of the species and infraspecific taxa as biological entities. It also provides evidentiary data as to the speciation that has, and is, occurring within particular populations that can be used in elucidating relationships at the generic and infrageneric levels. Understanding relationships between taxa is fundamental to their classification. Genetics does not provide intrinsic evidentiary characters peculiar to the field of genetics. It provides experimental evidence about characters that may be critical in taxonomic decisions.

> Basic references: Smith, B. W., in Radford et al., 1974, Vascular Plant Systematics, Harper & Row, New York, Chapter 11, pp. 259–268. Stebbins, G. L., 1950, Variation and Evolution in Plants, Columbia University Press. New York.

g. Reproductive Biological Evidence   **(Figure 8.6)**

> Reproductive biology is the study of structures, processes, mechanisms, and significant biotic and abiotic factors involved in the reproduction of an individual, population system, or species.

Some basic evidentiary characters (classes)

Floral biology	Phenology
Pollination pathway	Flowering periodicity
Pollination type	Flowering maturation
Reproductive potential	Plant duration
Reproductive pattern	Seed dispersal
Breeding system	Propagule type

> Evidentiary reproductive biology provides key data to understanding relationships between infraspecific taxa and contributes information for use in interpreting the possible origin, migration, and evolution of taxa.

Evidentiary examples of character/taxon relationships

> Flowers are mostly anemophilous in Plantaginales, mostly entomophilous in related Scrophulariales (Cronquist, 1981). Plants are allautogamous in *Viola,* allogamous in *Hybanthus* (Radford et al., 1974). Plants are hydrophilous in *Zostera,* myrmecophilous (ant-pollinated) in *Asarum.* Plants are dioecious in *Salix, monoecious* in *Betula.* Inflorescence is gynecandrous in *Carex virescens,* androgynous in *Carex rosea.* (Radford, A. E. et al., 1968.)

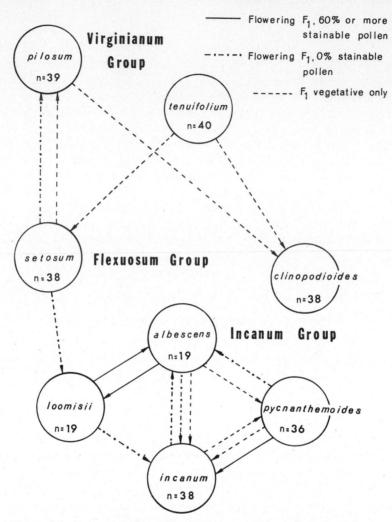

Figure 8.6 Reproductive biological evidence. Crossing diagram of artificial hybridizations. Arrows point from pollen parent to pistillate parent. (*Source:* Chambers, H. L., and K. L. Chambers. 1971. Artificial and natural hybrids in *Pycnanthemum* (Labiatae). Brittonia 23: 71–88. Used with permission.)

Basic references: Bell, C. R., in Radford et al., 1974, Reproductive Biology and Systematics, Chapter 12, pp. 269–284. Faegri, K. and L. van der Pijl, 1966, The Principles of Pollination Ecology, Pergamon Press. Oxford.

h. Chemical Evidence (Figure 8.7)

Chemotaxonomy is the application of natural products chemistry as evidence in taxonomic problems.

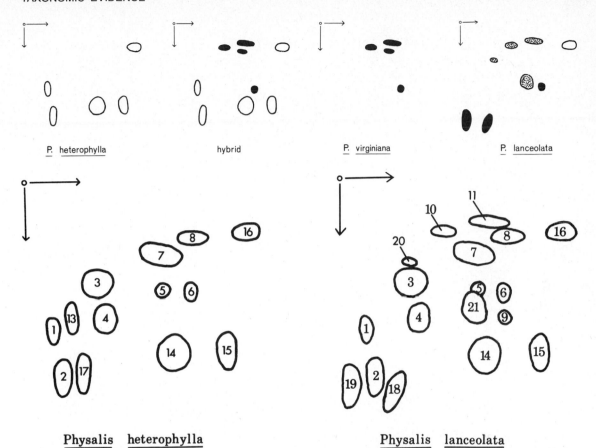

P. heterophylla   hybrid   P. virginiana   P. lanceolata

Physalis   heterophylla   Physalis   lanceolata

Figure 8.7   a, b. Chemical evidence. Chromatographs showing variation in leaf extracts of three species and one hybrid of *Physalis* clearly indicate that *Physalis lanceolata* is not, as once postulated, a hybrid between *P. virginiana* and *P. heterophylla*. Note that in this presentation of chromatographic data, only those spots that are unique to the plants of a species are shown—the many pigment spots the related plants may have in common are omitted from the chromatograph for the sake of clarity. (*Source:* Hinton, W. F. 1970. The taxonomic significance of *Physalis lanceolata* (Solanaceae) in the Carolina Sandhills. Brittonia 22: 14–19. Used with permission.)

Some basic evidentiary characters (classes)

Flavonoids	Amino acids
Terpenoids	Fatty acids
Carotenoids	Aromatic compounds
Polysaccharides	C_3-C_4 photosynthesis
Alkaloids	

Chemical evidence is useful in establishing relationships among taxa and providing clues for alternative interpretations concerning proposed relationships of taxa.

Evidentiary examples of character/taxon relationships

Plants are aromatic in Juglandales, not aromatic in Fagales (Cronquist, 1981). Plants produce betalains but not anthocyanins in the Caryophyllales, anthocyanins but not betalains in Polygonales (Cronquist, 1981). Mustard oils are typical of the Bretschneideraceae, plants are generally tanniferous in Sapindaceae (Cronquist, 1981). Plants generally have alkaloids in Solanaceae (e.g., *Nicotiana, Datura*), plants have highly aromatic compounds in the Lamiaceae (Cronquist, 1981).

Basic references: Parks, C. R. in Radford et al., 1974, Vascular Plant Systematics, Harper & Row, New York, Chapter 13, pp. 285–305. Alston, R. E. and B. L. Turner, 1963, Biochemical Systematics, Prentice-Hall, Englewood Cliffs, NJ.

i. Ecological Evidence

Ecology is concerned with organisms in relationship to their environment. Those relationships may be classified as biotic, abiotic, spatial, and temporal.

Some basic evidentiary characters

Habitat	Spatial Relations
Water	Numercial
Mud	Distributional
Sand	Areal
Rock	Temporal Relations
Bog	Phenology
Marsh	Perodicity
Prairie	Successional
Savannah	Biotic Relations
Scrub	Nutritional
Tundra	Social
Desert	Symbiotic
Abiotic Relations	Reproductive
Climatic	Adaptive Features
Moisture	Life Form
Light	Diaspore
Edaphic	Fidelity
Perodicity	Vitality

Ecological evidence contributes to our understanding of taxa and composition of floras, the genetic and phylogenetic relations of taxa, the variation in populations, and the evolutionary mechanisms involved in speciation. Habitat data are included in most species descriptions and occasionally used in keys. Temporal relations are expressed as flowering-fruiting periods for species and infraspecific taxa. Spatial and biotic relationships are treated in some biosystematic studies.

Basic references: Radford, A. E., in Radford et al., 1974, Vascular Plant Systematics, Harper & Row, New York, Chapter 15, pp. 309–324. Kruckeberg, A., 1969, The implications of ecology for plant systematics, Taxon 18:92–120.

j. Geographic Evidence   (Figure 8.8)

Plant geography is the study of the spatial relationships of past and present plants. Geographic evidence includes data descriptive of the origin, migration, evolution, distribution, and adaptations of taxa as influenced by past events of the earth's history.

Some basic evidentiary characters

Habitat	Flora province
Distribution patterns	Origin and establishment
Disjunction patterns	Migration patterns
Biogeographic region	Distribution center
Biome	

Geographic data correlated with structural features have provided evidence for relationships of taxa at all levels, but particularly for phylogenetic relationships of various taxa.

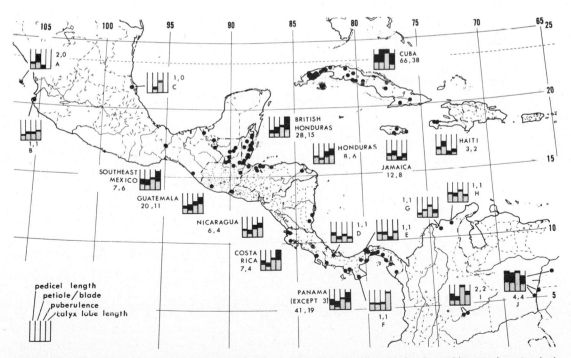

Figure 8.8 Geographical evidence. (*Source:* Morley, T. 1971. Geographic variation in a widespread neotropical species, *Mouriri myrtilloides* (Melastomataceae). Brittonia 23: 413–424. Used with permission.)

Evidentiary examples of character/taxon relationships

> *Amphianthus pusillus,* a monotypic genus, is endemic to granitic flatrock in the southeastern United States; *Solidago notablis* is a localized endemic restricted to limestone. *Hudsonia ericoides* is a northeastern species disjunct in South Carolina; *Hudsonia montana* is a rare species endemic on quartzite in Burke County, North Carolina—closely related to *H. ericoides.* (Radford et al., 1968). *Symplocarpus* and *Tipularia* are relict tertiary genera restricted to eastern North America and eastern Asia but not found in western North America; *Orontium* and *Nestronia* are monotypic endemic genera in the southern Appalachians but not in western North America. (Radford et al., 1974). *Cercidium* and *Bursera* are genera found in the arboreal desert, *Pachycereus* and *Carnegia* in the succulent desert. (Radford et al., 1974).

> Basic references: Radford, A. E., in Radford et al., 1974, Vascular Plant Systematics, Harper & Row, New York, Chapter 16, pp. 325–364. Daubenmire, R., 1978, Plant Geography, Academic Press, New York.

SUGGESTED READING

Cronquist, A. 1981. An Integrated System of Classification of Flowering Plants. Columbia University Press, New York.

Davis, P. H. and V. H. Heywood. 1963. Principles of Angiosperm Taxonomy. D. Van Nostrand Company, Inc., New York. Chapters 5–7, pp. 142–230.

Jones, J. B. and A. E. Luchsinger. 1979. Plant Systematics. McGraw-Hill Book Company, New York. Chapter 5, pp. 60–76.

Radford, A. E. et al. 1968. Manual of the Vascular Flora of the Carolinas. The University of North Carolina Press, Chapel Hill.

Radford, A. E. et al. 1974. Vascular Plant Systematics. Harper & Row, New York. Chapters 6–16, pp. 83–364.

Stace, C. A. 1980. Plant Taxonomy and Biosystematics. University Park Press, Baltimore. Chapters 3–7, pp. 65–190.

SUMMARY FOR EVIDENTIARY STUDY IN SYSTEMATICS

Definition for Evidentiary Study   *Evidence* is information used in context for a purpose.

Purpose of Evidentiary Study   To provide characters from many sources for description, identification, and classification of organisms, populations, and taxa as well as for the determination of phenetic, genetic, and phylogenetic relationships of taxa.

Operations in Evidentiary Study   See beginning of this chapter.

Basic Premise for Evidentiary Study   That data from many sources can be used as evidentiary characters in the characterization, identification, classification, and relationships of taxa.

Fundamental Principles for Evidentiary Study

1. One type of evidence should be used in conjunction with others in establishing definitive taxonomic correlations and relationships.
2. No single type of evidence is more reliable and meaningful than another in determining taxonomic relationships.

Guidelines for Evidentiary Study   See B. I. a. Guidelines for Use of Evidence in a Systematic Study in this chapter.

Basic Assumptions for Evidentiary Study   See B. I. b. Basic Assumptions for Use of Evidence in a Systematic Study in this chapter.

QUESTIONS ON EVIDENTIARY STUDY

1. Under what conditions do data and information become evidentiary characters?
2. What are the advantages and disadvantages of a standardized classification and nomenclature for each type of evidence? What is the relevance of standardization to data banking and systematic research?
3. How is palynological, cytological, chemical, and ultrastructural evidence used in systematic research?
4. How is morphological, anatomical, embryological, and genetic evidence usually analyzed, presented, and documented in systematic studies?
5. Why is ecological and geographic evidence usually included in descriptions of species in floras and manuals?
6. Which types of evidence are frequently used in distinguishing taxa at higher levels in phylogenetically based systems of classification for vascular plants?
7. How is a comparison chart (see Figure 7.3) or a character/taxon matrix of value in studying evidentiary characters for closely related taxa? For a group of objects of any type? How are studies of comparative anatomy, comparative morphology, and comparative physiology examples of good character/taxon matrix study?
8. Why should the guidelines for evidentiary study be considered good practice for any type of study? Poor practice?
9. How are experimentally obtained data intrinsically of more scientific value than observed data as evidentiary characters in systematic studies? Any studies?
10. Under what conditions may characters used in systematic studies become legal evidence? Chemical evidence?

EXERCISES IN EVIDENTIARY STUDY

1. Select a description of a species from a manual of your choice and analyze for evidence, characters, and character states. See A. I. Description, Characterization, Circumscription.
2. Select a key to the species of a genus from a manual of your choice and list the types of evidence, kinds of characters, and character states (for each character) used in the key. Also, list the characters and character states used to delimit two of the species in the genus.
3. Select a paper from a recent taxonomic periodical and indicate the types of evidence used in each evidentiary product in the paper.

4. Select a paper from a recent taxonomic periodical and analyze for evidentiary guidelines used in the study. List the author's guidelines, basic assumptions, and fundamental principles for the use of the evidence described in the work. See B. I. General Guidelines and Assumptions for the Use of Evidence in Systematic Study.

5. Select a recently published system of classification and determine the types of evidence used in distinguishing taxa at each rank, for example, the class, order, and family levels; determine the evidentiary patterns (correlations) for distinguishing families, orders, and classes (see Appendix C).

chapter 9

Variation and Speciation
in Plants

Taxonomists as analyzers of character variation select the object (individual, population, taxon) and character(s) to be analyzed; choose the parameters of the character(s); determine the feasibility of the character variation analysis; analyze; summarize the results of the analysis; and evaluate the results in relation to the original purpose or hypothesis.

Systematists as students of speciation select groups of organisms for analysis of character variation; determine the causal mechanisms for the variations and variants within the populations; ascertain the isolating mechanisms controlling interbreeding between populations; make a critical analysis of the habitat for relationships between variants and environmental factors; and suggest and attempt to prove experimentally the nature of the adaptation of the variant (population) to its environmental stress and natural competition.

Biological systematics, as a unique natural science is a study of individual, population, and taxon relationships for purposes of classification. In this text the study of plant systematics is based upon the premise *that in the tremendous variation in the plant world, there exist conceptually discrete units (usually called species) that can be recognized, classified, described, and named* and on the further premise *that logical relationships developed through evolution exist among these units.* The tremendous *variation* provides the basis for all systematic study. Variation is the foundation for speciation when acted on by natural selection. Variation and change in organisms through the geologic ages is the result of evolution. Species are the products of the basic evolutionary mechanisms—mutation, recombination, selection—acting in different environments through time. The relationships of populations are phenetic, genetic, and phylogenetic. Evidentiary variation is the basis of relationships shown in phenetic and phylogenetic systems of classification.

Table 9.1   A CLASSIFICATION OF TYPES OF VARIATION

Principal types		Integral types
a. Evidentiary types		
General	**Hierarchical level I**	Abiotic, biotic
Abiotic	**Hierarchical level II**	Physical, chemical
Biotic		Biological
Physical	**Hierarchical level III**	Color, size, shape, texture, symmetry, fusion, number, disposition, density
Chemical		Structure, process, product
Biological		Structure, function, development, environment, geography
Structural	**Hierarchical level IV (biological only)**	Morphology, anatomy, embryology, palynology, cytology, genetics
Functional		Metabolism, physical processes, reproduction
Developmental		Initiation, pattern, periodicity
Environmental		Phenology, habitat
Geographic		Origin, migration, distribution
b. Mechanistic types		
Mechanistic	**Hierarchical level I**	Genetic, environmental
Genetic	**Hierarchical level II**	Mutation, recombination, gene interaction
Environmental		Climate, parent material, relief or topography, soils, biology
c. Hierarchical diversity types		
Taxa		Division, class, order, family, genus, species
d. Temporal types		
Temporal		Daily, seasonal, yearly, successional, geologic
e. Group types		
Taxa		Individual, population, taxonomic group

Table 9.1   A CLASSIFICATION OF TYPES OF VARIATION (Continued)

Principal types	Integral types
f. Form types	
Form	Monomorphic, dimorphic, trimorphic, polymorphic, heteromorphic
g. Character types	
Character	Continuous, discontinuous

A. VARIATION IN PLANTS

Variation is a general term covering the act, process, or accident of deviation from a norm or standard in condition, character, or degree. In biological terms, variation is a deviation in structural, functional, or developmental character(s) of an organism from its parents, from others in the same population, or from other populations of the same species or related groups. The variation may include fundamental hereditary changes through which natural selection works to induce evolutionary development as well as purely individual fluctuations that lack evolutionary significance. Genotypic (genetic) variation includes differences in genotypes within a population or species as a result of mutation, recombination, or gene interaction that may be of some evolutionary significance. Phenotypic variation in structural or functional characters resulting from the action of different environments on one or more genotypes may not be of evolutionary significance.

Variety in the plant kingdom is a general term applicable to diversity, variant population, variability, character variation, and many other terms denoting kinds or types of structures, organs, organisms, and so on (Table 9.1)

Diversity refers to the number of types of organisms or taxa in the plant kingdom. Approximately 250,000 living species of vascular plants occur in the world (total diversity), of which there are an estimated 10,000 species of pteridophytes, 600 gymnosperms, and 235,000 angiosperms (total diversity by major group). The species diversity in the genus *Carex* in the flora of the Carolinas is 122; the generic diversity in the family Cyperaceae in the Carolinas is 17. Occasionally, diversity is used in reference to the number of kinds of organs in a taxon. For example, the Rosaceae in the Carolinas is characterized by nine fruit types (a large number for a single family).

A *variant* is an individual or a group of individuals within a population that is definable and recognizable. This is a neutral term generally used without taxonomic significance. It is applied most frequently to an individual or group of individuals having one or only a few distinctive characteristics. It may or may not be the equivalent of the taxonomic category of variety.

Population variability includes three fundamental types: developmental, environmentally induced, and genetic variation. (See A. I. Sources of Genetic Variation for a discussion of the last type.) Observant field students become aware of developmental and environmentally induced variation very early in their careers. The first leaves of the bean plant are opposite and simple, the later ones alternate and pinnately compound. Leaves on shaded sucker shoots (adventitious) of scarlet oak (*Quercus coccinea*) are very similar to normal sun leaves of the red oak (*Q. rubra*); those on shaded sucker shoots of turkey oak (*Q. laevis*) are similar to sun leaves of scarlet oak. The first submersed leaves of water parsnip (*Sium suave*) are pinnately dissected and flaccid, the older emersed leaves are pinnately compound and stiff. The short- and long-day plants of the mermaid weed (*Proserpinaca pectinata*) have pinnatifid leaves. The short-day leaves of *P. palustris* are pinnatifid and the long-day leaves are lanceolate and serrate. Most trees with deciduous leaves exhibit color change with the seasons—some shade of green to yellow, orange, or red and then brown. The number and sizes of roots, leaves, flowers, and fruits are strongly influenced by moisture and nutrient conditions. Late freezes result in loss of flowers but the few surviving fruits may be larger. Insect damage, diseases, and grazing affect shape and size of plant organs.

The systematist has to recognize the causal differences in variability to develop meaningful circumscriptions and delimitations of taxa for effective classification.

I. Sources of Genetic Variation

The sources of genetic variation in natural populations are due to mutation and recombination. Mutation is the occurrence of heritable change in the genotype of an organism that was not inherited from its ancestors. In a theoretical sense, mutation is the ultimate source of all genetic variability. All genetic changes in existing individuals and populations, except those due to recombination, are mutations. A mutation may be as minute as the substitution of a single nucleotide pair in the DNA molecule or as great as a major change in chromosome structure or number (gene and chromosomal mutation, respectively). Chromosomal mutation may be due to deletion, inversion, aneuploidy, or polyploidy (see a general genetics text for definitions).

Recombination is a reassortment of chromosomes, crossover segments, and, therefore, genes that produce new genotypes in sexually reproducing organisms. Recombination is the immediate source of variation observed in populations (such as height of plant and color of corolla) and is the raw material on which the processes of evolution and speciation act. Factors controlling recombination in plants (Grant, 1981) are operative at meiosis (chromosome number, frequency of crossing-over, hybrid sterility) and fertilization (breeding system, pollination system, dispersal potential, population size, crossability barriers, and external isolating mechanisms).

Recombination also results from hybridization and introgression. Natural hybridization provides gene flow between individuals from divergent populations at the local race to taxonomic infraspecific and specific levels (and occasionally at the generic level in plants). Introgression is a process of successive hybridizations that cause the migration of genetic material from one species (or infraspecific taxon) into another. Initial interspecific hybridization is usually followed by a series of backcrosses to one of the parents (or it can occur independently in both directions).

Mutation and genetic recombination produce the genotypic variation in populations basic to the onset of the speciational process. Natural hybridization furnishes the genetic material from established taxa for the possible development of new species under different selective pressures.

II. Analysis of Character Variation

The analysis of character variation in plants is fundamental to taxonomic training and research. This analysis is treated in Appendix A. (Also see the activities and principles involved in the study of variation in the summary of the study later in this chapter.)

B. PLANT SPECIATION

Basically, speciation is the development of populations of freely interbreeding organisms adapted to their environment that are reproductively isolated from other such populations. Present-day species are evolutionary products that have descended from preexisting species. Variation (A. II. Analysis of Character Variation) in organisms provides the raw genetic material that, in concert with environmental variation, makes speciation possible. Variation in natural populations is limited by such factors as self-fertilization, asexual reproduction, genetic fixation, and stabilizing selection. In general, many more individuals are produced by each generation than will survive and reproduce. Variation does occur in each generation and some individuals within the population will have better survival characteristics that are heritable. Organic evolution involves the development of phyletic lines (see evolutionary species in B. IV. Types of Species and Chapter 10) and the transmutation of form and structure within a line that gives rise to morphological (taxonomic) species. The biological species represents the major continuity in ongoing evolution.

This treatment of speciation includes the integral components of population, breeding and asexual reproductive systems, isolating mechanisms, and stabilization and selection, along with a discussion of the many types of species resulting from speciation.

I. Plant Populations

Plant populations are the source of potentially heritable genetic variation, which, in concert with environmentally induced variation, leads to speciation. It is a genetic axiom that the larger the interbreeding population, the greater the size of the gene pool. A large gene pool provides the genetic variation in a population necessary for its adaptation to changes in the selective forces of environment. A small population contains fewer different alleles in its gene pool for adaptive response to changing environmental conditions. The number of individuals at a new site is controlled by the number of propagules per unit area as well as by the suitability and spatial dimensions of the habitat for germination, establishment, and maintenance of the organisms. The fate of the population may well depend on the initial size and heterozygosity of its constituent individuals. A new population started by a single seed in all probability will produce a very limited gene pool of genetically uniform individuals with low speciation potential.

Population size includes more than the gene pool. It involves some form of competition between the individuals of the group. Generally, the greater the number of individuals per unit area (density), the greater the competition between the organisms for moisture, nutrients, and light and the more severe the natural selection. Density directly affects pollination and the reproductive capacity of the population.

The number of populations within an area is determined by the number of suitable habitats available for colonization. Some species are restricted to a very specific soil type, lithology, or set of moisture conditions. The nature and availability of dispersal agents determine the pioneering characteristics of certain species. For the black willow (*Salix nigra*) to produce new populations, low-water sand bars have to be present for germination and establishment almost immediately after dispersal of the seeds from the mature tree. If the sand bars are scattered and small, the black willow populations will be scattered and small, assuming the availability of the propagules. The fireweed (*Epilobium angustifolium*) with windblown seeds has been known to colonize fire-seared areas in

Quebec with literally thousands of individual plants. The availability of suitable habitat and propagules under the proper conditions for germination and establishment control the number, size, and density of the populations.

Population size and density change over time. Among trees and shrubs that live for many years, younger plants get shaded out by older or faster-growing individuals. The first seedlings of annuals in an area usually have a competitive advantage for light, moisture, and nutrients. This leads to self-thinning and the subsequent death of later annuals within the population. Maintenance of populations that leads to the production of different size and age classes is conducive to the survival of the species.

To understand speciation as a process, it is necessary to acquire a knowledge of the functional biology of species as populations. To know the biology of a species, Massey and Whitson (1981) state that "one must adopt a holistic view of the species by studying individuals, *populations,* and *population* systems utilizing the structures, processes, and habitat relations of each major life cycle phase within a particular time reference." The question matrix for major life phases from Massey and Whitson (1981) in Table 9.2 provides an excellent overview of the basic queries for understanding the species biologically and evolutionarily.

To understand how a species is adapted to its environment, it is essential to obtain a knowledge of site-species population relationships. An environmental or habitat analysis of a site occupied by a population is a requirement for the determination of component environmental elements or combinations of elements that affect and/or control the establishment, maintenance, reproduction, and dispersion of the species of concern. An analysis of site-population relationships for several sites can lead to the identification of ele-

Table 9.2   QUESTION MATRIX FOR MAJOR LIFE PHASES

Reproduction	Dispersion	Establishment	Maintenance
Is reproduction occurring?	Are propagules present?	Are new individuals present?	Is there a range of classes?
What types of reproduction are occurring?	What types of viable propagules are present?	What are the origins of the new individuals?	What are the origins of the classes?
What breeding systems are operative?	What dispersal systems are operative?	What establishment processes are operative?	What are the percentages of each class in the population?
What pollination systems are operative?	What are the dispersal units and/or agents?	What are the spatial relationships of establishment processes?	What are the spatial relationships of the classes?
What is the reproductive capacity or status of the population?	What is the dispersal effectiveness of the population?	What are the establishment effectivenesses based on origin?	What is the survivorship of each class progressing to the next class?

Source: Massey and Whitson, 1981. Used with permission.

ments useful in predicting the occurrence or distribution of the species of concern and to the discovery of clues that can be of significance in interpreting the origin, migration, and evolution of the populations under study.

II. Reproductive Biology of Plants

Populations are the gene pool base of genetic systems in plants. The genetic system refers to all of the intrinsic genetic processes that affect genetic recombinations in a population or species. The major components of the genetic system are the chromosomal organization, their behavior in meiosis, and the breeding system. Genetic systems in reproducing populations range from that of the outbreeding diploid with complete pairing and unrestricted crossing-over of homologous chromosomes to that of the obligate apomict without sexual reproduction.

Reproductive biology is a collective term applied to the many mechanisms, agents, and adaptations involved in the reproductive process of a species. In higher plants, reproductive biology includes floral structure, mode and/or agent of pollination, self- or cross-incompatibility, various types of sterility, time and season of flowering, and so on. An understanding of the dynamics (biology and speciation potential) of a natural population depends on an understanding of its reproductive biology.

Many of the evolutionary properties of populations depend on the heterozyosity (dissimilarity in alleles or chromosomes in an individual) maintained by a relatively steady rate of recombination within the population. In outbreeding populations, chromosomal differences (e.g., inversion, translocation), assortive mating, and occasionally varying amounts of infrapopulational sterility provide a degree of restriction to complete or random recombination. Many flowers are normally hermaphroditic, which indicates inbreeding at the level of the individual. Some self-pollinating plants have mechanisms for occasional outcrossing that provide for recombination. Since inbreeding and outcrossing have survival and evolutionary value, it is not surprising to find divergent breeding mechanisms within plant populations.

In general, three breeding systems can be recognized in plants: (1) predominantly outbreeding (outcrossing), (2) predominantly inbreeding (selfing), and (3) mixed out- and inbreeding (outcrossing and selfing). In addition, several highly successful methods of asexual reproduction have evolved among higher plants, with some species having a combination of both.

a. Outbreeding Systems

An *outbreeding system* is one in which sexual reproduction involves the mating and union of gametes of different individuals (resulting in cross-pollination in seed plants). In plants, outbreeding may be obligate or facultative. The necessity or tendency for outbreeding may be reinforced by dioecism, self-incompatibility, heterostyly, and other mechanisms. Outbreeding is usually aided by the wind, insects, or other pollinating agents. Outbreeding may imply mating between individuals that are less closely related than would occur in random mating, as in *panmictic populations,* in which each individual has the same probability of mating with any other individual.

Dioecism, as in *Salix* (willows), in which the sexes (male and female flowers) are on separate plants, makes cross-pollination and cross-fertilization obligatory. *Self-incompatibility,* in which genetically controlled physiological interactions inhibit or prevent self-pollination and/or self-fertilization, occurs in species of *Nicotiana* (tobacco) and *Hemerocallis* (day lily) and thus promotes cross-breeding. *Protandry* and *protogyny,* in which the stamens or carpels mature well ahead of each other, as in *Daucus* (carrot), ensure outcrossing. In some plants (*Oxalis, Lythrum, Houstonia*) with perfect flowers, self-pollination is unlikely due to the arrangement of the anthers and stigmas within the flower. Some *heterostylous* flowers have long styles and short filaments and others short styles and long filaments. The pollinator visiting a long-styled flower will deposit pollen from another flower on the stigma while foraging for nectar and pollen below. However, this is not a foolproof method. Some members of the Asteraceae with the stigma maturing earlier than the nearby anthers will become recurved, thus coming in contact with later maturing anthers, consequently ensuring self-pollination and subsequent fertilization. In this way, seed production within an individual is assured. Many species of plants are predominantly, but not obligate, outbreeders.

Mechanisms that favor or ensure outcrossing or allogamy and generally maximize heterozygosity and variability (high-speciation potential) are dioecism and monoecism, protandry and protogyny, heterostyly, self-sterility, and chasmogamy.

b. Inbreeding Systems

An *inbreeding system* is one in which sexual reproduction involves the mating and union of gametes of individuals that are closely related. Selfing, sib-mating, and back-crossing (e.g., parent-offspring matings) are examples of close inbreeding. Self-pollination is frequent in plants. Many weedy annuals are autogamous (self-pollinating, self-fertilizing inbreeders). Some plants may be obligate inbreeders (e.g., cleistogamous) or facultative and aided by floral adaptations that enhances the opportunities for selfing.

Plants such as *Viola* (violet) have cleistogamous flowers (never opening). The pollen germinates within the anther and the tube grows directly into the ovule. *Viola* also has chasmogamous flowers in which cross-pollination does occur so that the violet has outcrossing and inbreeding. *Myosurus* (mouse tail) normally is self-pollinating with the younger carpels pollinated by the stamens in the same flower. Under certain conditions, carpels are produced within a flower after cessation of stamen production and those carpels are cross-pollinated with pollen from other flowers. Many plant species that are predominantly self-pollinating and inbreeding usually retain the capacity for occasional outcrossing and a renewal of heterozygosity.

Mechanisms that favor or ensure selfing or autogamy and generally maximize homozygosity and genetic uniformity (low-speciation potential) are hermaphroditism, homogamy (stamens and carpels maturing at same time), monomorphy (stamens and carpels same size), self-fertility, and cleistogamy.

c. Vegetative or Asexual Reproduction

Vegetative reproduction or propagation is common among perennial vascular plants. Many species are characterized by having rhizomes, tubers, stolons, or bulbs. The onion has bulbets rising from the bulb that can produce new plants, the Irish potato has tubers that separate from the parent plant and become new ones, and the strawberry gives rise to offsets that become separate plants on breakage or decay of the stolon. Some plants layer (*Forsythia*), others sucker (*Prunus*), and some produce stump sprouts from old root crowns (*Sequoia*). On separation from the parent, the layered shoots, sucker stems, and stump sprouts become new plants. Some of the grasses (*Phragmites*) produce rhizomatous systems covering many square meters with numerous culms appearing as separate plants. With the death of the rhizome or sections thereof, the culms become new plants.

These genetically identical individuals arising through vegetative propagation are considered clones. Some of these clones have survived for hundreds of years with little change in structure or function. The variability within a clonal population is probably phenotypic and caused by slight environmental differences within the habitat. Clonal material is ideal for the study of the morphological and physiological effects of diverse environmental factors on plants from different areas or habitats.

Agamospermy, another form of asexual reproduction, involves the production of viable seeds without fertilization. The embryo in some seeds may develop from an unreduced diploid egg (parthenogenesis) or from a somatic cell in the ovule. If several different genotypes occur in an agamospermous population, these same genotypes will present in the next generation but possibly in different proportions. Although genetic recombination is not possible, a series of populations maintain a small amount of flexibility through time as selection varies the percentage of each genotype in each generation with changing environmental conditions. Agamospermy is typical of the common dandelion (*Taraxacum officinale*) and many kinds of blackberries (*Rubus* spp.).

III. Isolation and Selection

a. Isolating Mechanisms

Interbreeding, hybridization, and gene exchange between species are prevented or reduced by isolating mechanisms of many kinds. An isolating mechanism is any condition that prevents gene exchange between individuals in two or more populations. These mechanisms can be grouped (Grant, 1981) into three main classes: spatial, environmental, and reproductive (Table 9.3). According to Grant (1981), "Spatial or geographic isolation exists between any two allopatric species whose respective geographical areas are separated by gaps greater than the normal radius of dispersal of their pollen or seeds." This would be the case, for example, in *Liriodendron tulipifera* (tulip poplar) of the eastern United States and *L. chinense* of southeastern Asia, which cannot breed under natural conditions but have been crossed artificially (Parks, personal communication). Ecological or environmental isolation exists between two species naturally restricted to different habitats within the same geographic area. *Zygadenus glaucus* is restricted to wooded calcareous habitats and *Z. leimanthoides* to wooded siliceous outcrops in the same region of the Blue Ridge Mountains of North Carolina.

Reproductive isolation comprises blocks to gene exchange between populations that stem from genotypically controlled differences in their reproductive organs, reproductive habits, or fertility relationships (Grant, 1981). Temporal or seasonal isolation exists between two species when their periods of pollination occur at different times of the year, as *Solidago verna* (May–July) and *S. rugosa* (September–October) in the Carolinas. Me-

Table 9.3   CLASSIFICATION OF ISOLATING MECHANISMS IN PLANTS

I. Spatial
   1. Geographical isolation
II. Environmental
   2. Ecological isolation
III. Reproductive (floral isolation)
   A. External
         3. Temporal isolation
              a. Seasonal
              b. Diurnal
         4. Mechanical isolation
         5. Ethological isolation
         6. Isolation due to autogamy
   B. Internal
         7. Incompatibility barriers
              a. Prefertilization
              b. Postfertilization
         8. Hybrid inviability
         9. Hybrid sterility
        10. Hybrid breakdown

Source: Grant, 1981.

chanical isolation results when structural differences between two or more species interfere with or prevent interspecific crosses, for example, as differences in pollinia size between *Asclepias syriaca* and *A. tuberosa*. Ethological isolation of related species due to flower-constant visitation by bees has been observed in *Pedicularis* (Grant, 1981). Autogamous isolation is due to inbreeding, self-pollination, or self-fertility, as in *Arachis hypogaea* (peanut).

Some of the prefertilization incompatibility barriers are due to pollen grains of one species not germinating on a foreign stigma, as *Datura meteloides* on *D. stramonium* (Grant, 1981) or the pollen not developing successfully in a foreign style, as *Heuchera americana* on *H. parviflora* (Wells, 1979). Mechanisms existing in plants that prevent perpetuation of hybrids are invariably due to failure of the hybrid to develop to maturity (frequently dying early) or sterility, in which the hybrid develops to a reproductive stage but does not reproduce, as in the well-known mule resulting from the horse-donkey cross.

b. Natural Selection

Natural selection is the evolutionary force that tends to produce systematic and heritable change between one generation and the next, which may result in adaptation and survival, variation and speciation, or extinction. Natural selection is the differential reproduction of genotypes. Stabilizing selection occurs in every normal population, eliminating extremes in variation and maintaining the adaptation of the group somewhere near its optimum within its habitat. Directional selection occurs when the environment is changing in a systematic fashion, leading to a regular directional change of the adaptive characteristics of the breeding populations. Successional species such as the Miocene *Pinus alvordensis* and recent *Pinus contorta* are morphologically similar species (Axelrod, 1976), with *P. contorta* in all probability being the result of directional selection. Disruptive selection

Table 9.4   COMPARATIVE FEATURES OF ABIOTIC AND BIOTIC SELECTION

Abiotic	Biotic
1. Evolution direction fixed (dependent on unalterable selective force, such as climate).	1. Evolution direction not fixed (dependent on interaction with another evolving organism).
2. May often result in parallel evolution.	2. Rarely results in parallel evolution.
3. Population size not important.	3. Population size important.
4. Biological interactions general.	4. Biological interactions very specific.
5. Does not produce niche subdivision.	5. Produces niche subdivision.
6. Produces low species density.	6. Produces high species density.

Table 9.5   COMPARATIVE FEATURES OF DARWINIAN AND CATASTROPHIC SELECTION

Darwinian	Catastrophic
1. Operates on all individuals in all populations.	1. Operates on marginal individuals in marginal populations.
2. Involves gradual changes in the selective force rarely exceeding range of tolerance of most individuals.	2. Involves relatively sudden changes in the extremes of the selective force that may exceed the tolerances of most individuals.
3. Seldom produces drastic decrease in population density.	3. Often produces drastic decrease in population density.
4. Large gene pool maintained; fixation rare.	4. Gene pool often drastically decreased; fixation likely.
5. Selection homeostatic (multidirectional).	5. Selection often unidirectional.
6. Rate of evolution slow, long cycle.	6. Rate of evolution faster, shorter cycle.

occurs in the breakup of a polymorphic population into local races or taxonomic varieties with normal gene flow interrupted by some local differentiation correlated with environmental differences. The local races of *Hieracium umbellatum* (Turesson, 1922) seem to result from disruptive selection in relation to edaphic differences.

Selection can be considered from causative agent standpoints and intensity. Bell (1974) has provided an excellent summary of the characteristics of biotic-abiotic and Darwinian-catastrophic selection (Tables 9.4 and 9.5).

IV. Types of Species

What are these selected, adapted, stabilized natural products? What is a species as a product of the speciational process? What are these entities called species found in manuals and floras to which so much information is attached? Grant (1981) recognizes five types of species: (1) taxonomic (morphological species, phenetic species), (2) biological (genetical species), (3) microspecies (agamospecies), (4) successional (paleospecies), and (5) biosystematic (ecospecies, coenospecies). These are characterized in Table 9.6, along with the evolutionary species of Simpson (1961). All of these are necessary for a fundamental understanding of the species as taxonomic, biologic, and evolutionary units.

Table 9.6   CHARACTERIZATIONS OF TYPES OF SPECIES

Types of species	Units being classified	Sphere in which applicable
1. Taxonomic	Taxa; groups of morphologically similar individuals.	Groups with morphological differences that are useful in formal classification.
2. Biological	Sexually reproducing population systems.	Sexual organisms on a single time level.
3. Micro	Populations in uniparental organisms.	Uniparental organisms including asexual and parasexual organisms.
4. Successional	Phyletic lineages.	Biological species and microspecies.
5. Biosystematic	Fertility groups.	Sexual plant groups that have been artificially hybridized in the experimental garden.
6. Evolutionary	Combined sexually reproducing populations, uniparental groups, and phyletic lineages.	Composite entity of natural units.

Source: Grant, 1981, for items 1 to 5.

a.   Taxonomic Species

Taxonomic species are the smallest groups that are distinct and distinguishable from all others. These species are the units that are positively identifiable, recognizable in nature by ordinary means (hand lens for higher plants), and are predictable in character correlations under natural conditions and occurrences. These are the basic units that are classified, identified, described, and named in manuals. Presumably, these are the basic natural units in systematics.

The taxonomic species, from a practical standpoint, is an assemblage of morphologically similar individuals that differs from other such groups in one or more structural characteristics. The taxonomic species is the usable unit in phenetic classification and identification. From the standpoint of evolution, the taxonomic species may include populations at every stage of divergence with discontinuities as well as continuities in variation pattern (Figure 9.1). The taxonomic species found in manuals represent a wide variety of types (Table 9.6).

b.   Biological Species

The *biological species* is a group of interbreeding populations composed of biparental organisms reproductively isolated from other breeding populations. Grant (1981) states that this basic populational unit can be viewed as a stage in a more or less continuous series of levels of divergence in population systems. The nodal points in the series are local races, geographical races, allopatric semispecies, allopatric species, and sympatric species (Figure 9.1). If hybridization develops between two biological species, these entities revert to the status of sympatric semispecies, and form a new collective entity, the syngameon!

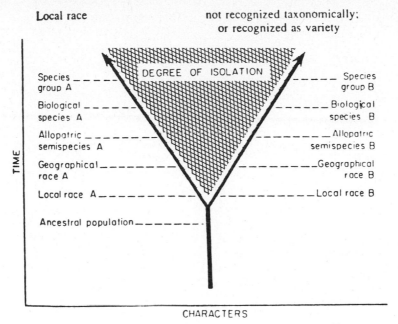

Figure 9.1 Stages in a process of gradual primary divergence. (*Source:* Grant, V. 1981. Plant Speciation. 2nd ed. Columbia University Press, New York. Used with permission.)

Local races are local breeding populations that differ from other conspecific populations in the composition of their gene pools and in their genetically determined phenotypic characters. Geographic races represent local racial variation over a wider range. The geographical variation may be continuous if the habitat is more or less continuous. The geographical variation may be discontinuous and the races disjunct if the habitat is discontinuous, such as alpine races on different mountain peaks. Geographical races are allopatric in that they occupy different geographical areas. The geographic races may also be ecologic. The climatic races of *Potentilla glandulosa* and of the *Achillea millefolium* complex from the alpine zone of the Sierras of California bloom earlier in a uniform garden than the midaltitude races from the same mountains (Clausen, Keck, and Hiesy, 1940, 1948). The edaphic races of *Gilia capitata* and *Streptanthus glandulosus* have been shown by Kruckeberg (1951) to be both serpentine tolerant and serpentine intolerant.

Semispecies are population systems in an intermediate stage of gradual and continuous evolutionary divergence from a geographic race to a sympatric species. Disjunct allopatric populations of *Liriodendron chinense* and *Liriodendron tulipifera* that live in similar habitats in China and the southeastern United States are interfertile when experimentally crossed. These well-recognized taxonomic species from a biological species standpoint are considered allopatric semispecies. Allopatric semispecies theoretically could diverge into allopatric biological species. Natural hybridization and gene flow can take place between biological species if the breeding barriers are not totally effective. Frequently, two widespread sympatric species will hybridize locally, producing sympatric semispecies of hybrid origin. These sympatric semispecies could diverge into biological species with well-established isolating mechanisms.

c. Microspecies

Microspecies, as defined by Grant (1981), are "populations in predominantly uniparental plant groups which are uniform themselves and are slightly differentiated morphologically from one another." They are often restricted to a limited geographic area and frequently of hybrid origin.

Microspecies (Grant, 1981) fall into four main classes according to the mode of reproduction: (1) clonal microspecies, reproducing by vegetative propagation, as in *Phragmites;* (2) agamospermous microspecies, reproducing by agamospermy, as in *Rubus* spp.; (3) heterogamic microspecies, reproducing by the *Oenothera biennis* or *Rosa canina* genetic system; and (4) autogamous microspecies, predominantly autogamous and chromosomally homozygous, as in *Erophila.*

Microspecies are not cryptic species. Cryptic species are those morphologically similar but cytologically or physiologically different. Many tetraploid plants, for example, look very similar morphologically to diploid progenitors but genomic analysis indicates a difference in the number of chromosomes, AABB rather than AA or BB.

d. Successional Species

Successional species (Grant, 1981) are members of the same lineage living in different geologic time periods. They are phyletic species with different sets of characteristics. Phyletic evolution is a continuing process with intergrading forms connecting distinctive species at different points in evolutionary time. Axelrod (1976) in the history of coniferous forests in California and Nevada found several Miocene gymnosperms structurally very similar to recent species and these appear to be related successional species, for example,

Miocene	Recent
Pinus florissanti	*Pinus ponderosa*
Pseudotsuga sonomensis	*Pseudotsuga menziesii*
Abies laticarpus	*Abies magnifica*
Picea lahontense	*Picea engelmannii*
Tsuga mertensioides	*Tsuga mertensiana*

The successional species concept is basic to understanding phylogenetic relationships of species (see Chapter 10).

e. Biosystematic Species

Biosystematic species (Grant, 1981) are classified as categories based on fertility relationships as determined by artificial hybridization experiments. These biosystematic categories form a hierarchy of ascending units: ecotypes, ecospecies, coenospecies, comparium (Turesson, 1922, 1925; Clausen, Keck, and Hiesy, 1939, 1945).

An ecotype consists of "all the members of a species that are fitted to survive in a particular kind of environment within the total range of the species" (Clausen, Keck, and Hiesy, 1945). Two or more ecotypes "are able to exchange genes freely without loss of fertility or vigor in the offspring" (Clausen, Keck, and Hiesy, 1945). The ecospecies is an assemblage of ecotypes. Two or more ecospecies are separated by an incomplete sterility barrier. Coenospecies are groups of related ecospecies in which gene exchange between the assemblages is essentially absent. A comparium is a group of related coenospecies.

Ecotypes are equated with local genetic races that are habitat correlated. Ecospecies and coenospecies are roughly equivalent to biological species; the comparium to a species group.

Turesson (1922) found local races of *Hieracium umbellatum* in southern Sweden with distinctive morphological characteristics correlated with the woodland, sandy field, and dune habitat. He called these habitat-genetic-correlated races ecotypes.

Turesson (1925) stated "Ecotypes . . . do not originate through sporadic variation preserved by chance isolation; they are, on the contrary, to be considered as products arising through the sorting and controlling effect of the habitat factors upon the heterogeneous species-population." The finding by Turesson (1922) and many others of widespread habitat-correlated variation supports the view that natural populations are subject to natural selection, with well-adapted genotypes selected in each habitat.

f. Evolutionary Species

The *evolutionary species* as defined by Simpson (1961) in Grant (1981) is a population that possesses the following characteristics.

1. It is a lineage, an ancestral-descendent sequence of populations existing in space and time.
2. The lineage evolves separately from other such lineages, that is, from other species.
3. It has its own "unitary evolutionary role," which is to say that it fits into its own particular ecological niche in a biotic community.
4. It has its evolutionary tendencies, being susceptible to change in evolutionary role during the course of its history.

The evolutionary species is a basic, discrete biological unit with a distinctive set of correlated characters that is fixed for a moment in evolutionary time. Biological, uniparental, and successional species are integral components of the evolutionary species as a concept. Practically all of the fossil as well as many of the recent species in a phyletic lineage are taxonomic.

Summary Comments

Many morphological/ecological (taxonomic) species have been described. Many remain to be circumscribed and named. The taxonomic species, supposedly, is the type utilized in most floras and manuals for retrieval of information and in systems of classification for storage of that information. The nature of species makes these basic taxonomic units highly variable. A flora is a compilation of species within an area that are treated as taxonomic but, in reality, are combinations of different types. A flora represents the state of research on the groups of species included. In some cases in which apomixis is well known, microspecies (*Rubus*) are treated as taxonomic. In other cases, hybrids and intro-gressants are known with some of the entities recognized as morphological/ecological units (*Quercus*). As stated in the chapter on nomenclature, some organisms are evolving rapidly with intermediates of all sorts, while others represent dead ends with intermediates long since extinct; the remainder are at different stages and conditions of evolutionary development. Primitive taxonomy produces taxonomic species; advanced systematic re-search results in taxonomic species also, but with a better understanding of their intrinsic nature.

A knowledge of the dynamics of speciation leads to an understanding of species and their relationships as natural basic units in systematics.

SUGGESTED READING

Axelrod, D. I. 1976. History of the coniferous forests, California and Nevada. University Calif. Publ. Bot. 70:1–62.

Bell, C. R., in Radford, A. E. et al. 1974. Vascular Plant Systematics. Harper & Row, New York. Chapter 12, pp. 269–284; Chapter 26, pp. 537–552.

Clausen, J., D. D. Keck, and W. M. Hiesy. 1939. The concept of species based on experiment. Amer. Jour. of Bot. 26:103–106.

———. 1940. Experimental studies on the nature of species. I. Effect of varied environments on western American plants. Carnegie Inst. Washington, D.C. Publ. 520.

———. 1945. Experimental studies on the nature of species. II. Plant evolution through am-phiploidy and autoploidy, with examples from the Madiinde. Carnegie Inst. Washington, D.C. Publ. 581.

———. 1948. Experimental studies on the nature of species. III. Experimental responses of cli-matic races of *Achillea*. Carnegie Inst. Washington, D.C. Publ. 581.

Cronquist, A. 1978. Once again, what is a species? Beltsville Symp. Agric. Research 2:3–20.

Davis, P. H. and V. H. Heywood. 1963. Principles of Angiosperm Taxonomy. D. Van Nostrand Co., New York. Chapters 11–14, pp. 350–484.

Grant, V. 1981. Plant Speciation, second edition. Columbia University Press, New York.

Kruckeberg, A. R. 1957. Variation in fertility of hybrids between isolated populations of the ser-pentine species, *Streptanthus glandulosus* Hook. Evolution 11:185–211.

Massey, J. R. and P. D. Whitson, in Radford, A. E. et al., 1981. Natural Heritage: Classification, Inventory and Information. University of North Carolina Press, Chapel Hill, Chapter 4, pp. 111–143.

Mayr, E. 1942. Systematics and the Origin of Species. Columbia University Press, New York.

Simpson, G. G. 1961. Principles of Animal Taxonomy. Columbia University Press, New York.

Smith, B. W., in Radford, A. E. et al. 1974. Vascular Plant Systematics. Harper & Row, New York, Chapter 11, pp. 259–268.

Stebbins, G. L. 1950. Variation and Evolution in Plants. Columbia University Press, New York.

Turesson, G. 1922. The genotypical response of the plant species to the habitat. Hereditas 3:211–350.

———. 1925. The plant species in relation to habitat and climate. Hereditas 6:147–236.

Wells, E. F. 1979. Interspecific hybridization in eastern North American *Heuchera* (Saxifragaceae). Syst. Bot. 4:319–338.

SUMMARY FOR STUDY OF VARIATION

Definitions for Study of Organismic Variation   *Variation* is the deviation of structural, functional, or developmental characteristics of an organism from those typical or standard to the group of which it is a part. *Genotypic variation* is the difference in the genotypes within a population or species as a result of mutation and recombination. *Phenotypic variation* is the morphological or physiological variation resulting from the action of different environments on one or more genotypes.

Purpose of Study of Organismic Variation   To analyze character variation within selected groups for a meaningful correlation of characters in the circumscription, delimitation, and classification of the groups (taxa) as well as for definitive evidence of phenetic and phylogenetic relationships of taxa.

Operations in the Study of Organismic Variation   See beginning of this chapter.

Basic Premise for the Study of Organismic Variation   That in the tremendous variation in the plant world conceptually discrete units occur that can be recognized, classified, described, named, and logically related.

Fundamental Principles for the Study of Organismic Variation

1. Variation provides the raw materials for evolution—without variation natural selection would have nothing on which to act.
2. The current patterns of variation as discrete units are the products of basic evolutionary mechanisms—mutation, recombination, selection—acting in different environments.
3. Variation and change in organisms through time depends on evolution.

Guiding Principles for the Study of Organismic Variation

1. Population variability includes three fundamental components: (1) developmental variation, (2) environmentally induced variation, and (3) genetic variation.
2. Variation may be continuous (without a clear separation of a series into discrete units) or discontinuous (with few or no intermediates in the series and therefore separable into discrete units).
3. Discontinuities in variation result in discrete units that can be recognized, classified, circumscribed, and named.
4. Genotypic variation is limited by: (1) self-fertilization, (2) apomixis, (3) stabilizing selection, (4) fixation, and (5) genetic homeostasis.
5. Phenotypic variation may be limited by environmental constancy.

Basic Assumptions for the Study of Organismic Variation

1. No two individuals are exactly alike.
2. Variation is apparent among species and between individuals of the same species.

3. Variation is not a constant; differences between variation patterns may be large or so small as to be almost imperceptible.
4. An analysis of character variation in selected groups can lead to an understanding of their evolution.

SUMMARY FOR SPECIATION

Definition of Speciation   *Speciation* is composed of evolutionary processes leading to the formation of biological species; the study of speciation includes an attempt to understand the nature and behavior of populations and races, their divergence, and the stabilization of divergent population and races as species of different types.

Purpose of the Study of Speciation   To determine the processes by which species arise; to determine the nature (origin) of existing species; to determine the species biology of present species populations; and to determine the environmental factors that control the stability of existing species populations.

Operations in the Study of Speciation   To determine the species biology of selected populations; to determine the environmental factors that control the evolutionary stability of selected populations; to determine the nature (origin) of existing stable populations; and to determine the processes by which variation is arising in selected populations.

Basic Premise for the Study of Speciation   That speciation involves evolutionary forces that produce variations in existing populations and fixes variations in derivative populations as species of various types.

Fundamental Principles for the Study of Speciation

1. Different types of species exist in the diversity of biological materials on earth (taxonomic, biological, microspecific, and successional—Grant, 1981).
2. The sources of variation in natural populations are a result of mutations, chromosomal recombinations, and genetic interactions.
3. Limitations of variation in natural populations are a result of self-fertilization, asexual reproduction, genetic fixation, and stabilizing selection.
4. Isolating mechanisms that fix variation and stabilize selection in derivative populations are spatial (geographical), environmental (ecological), temporal (seasonal, diurnal), mechanical, and ethological; they include autogamy, incompatibility barriers (prefertilization, postfertilization), hybrid inviability, hybrid sterility, and hybrid breakdown.

Guidelines for the Study of Speciation   See Chapter 8, B. I. a. Guidelines for Use of Evidence in a Systematic Study.

Basic Assumptions in the Study of Speciation

1. The natural or evolutionary species is one of the basic units of organization of living material (Grant, 1981).

2. The evolutionary species possesses certain general properties of its own that can be discovered and elucidated (Grant, 1981).

3. The general properties of evolutionary species can be discovered by the application of the analytical and generalizing methods that have been successful in other branches of theoretical biology (Grant, 1981).

4. Speciation involves the development of new, different gene combinations in separate populations (Grant, 1981).

5. In biological species composed of cross-fertilizing organisms, speciation involves the formation of reproductive isolating mechanisms (Grant, 1981).

QUESTIONS ON VARIATION AND SPECIATION

1. How do you distinguish variation from variant? From variety? From diversity?

2. Why is an understanding of "analysis of character variation" basic to all systematic research? See Appendix A.

3. What are the principles of analysis of character variation in systematics?

4. In data accumulation for your research or any taxonomic problem, what questions and choices arise in choice of (a) system, (b) objects or taxa, (c) characters, (d) parameters of characters, (e) character states, (f) character state codes and filling in and verifying the basic data matrix? See Appendix A.

5. What are the differences in analytical procedures for determination of variation in character states, characters, populations, and taxa?

6. What are the basic types of variation patterns in natural populations and individual organisms?

7. What are the conceptual differences between evolution and speciation? Evolution and phylogeny? Speciation and phylogeny?

8. What are the types of species in plants?

9. What are isolating mechanisms in speciation? Give examples.

10. What are the causative mechanisms for evolutionary divergence? Give examples.

11. What are the breeding systems found in plants? Cite examples of each.

12. How are the studies of populations and species biology related?

13. What is natural selection? How does it occur?

EXERCISES IN VARIATION AND SPECIATION

1. Select a genus such as *Quercus* or *Pinus* with five or more species within the area for a leaf variation study. Prepare a character/taxon matrix for leaf shape, apex, base, margin, l/w ratio, and lower surface vestiture for each species.

2. Select two populations of a variable species within the area for a character variation study. Prepare a character/population matrix for three variable characters with two or more states.

3. Select three individual organisms of different species for ontogenetic variation. List the differences by character states for the first-, middle-, and last developed leaves.

4. Compare the sun and shade leaves of a tree and list the differences; do the same with sucker sprout and normal leaves of the same species.

chapter 10

Phylogeny and Structural Evolution of Plants*

Taxonomists as phylogeneticists analyze the characteristics of a group of organisms by selecting and defining characters and character states, assessing homology, defining morphoclines, and hypothesizing primitive versus derived states; determine the phylogenetic branching pattern of the taxa by grouping monophyletic taxa based on shared derived features; evaluate and reevaluate character state changes and groupings of the cladogram; use the cladogram to devise a classification scheme and to deduce biogeographic history; and use the indicated character state changes to hypothesize past structural evolutionary events.

Phylogeny is the genealogical history of a group of organisms and is a representation of hypothesized ancestor/descendant relationships. *Phylogenetics* (Hennig, 1966; Wiley, 1981; also called *cladistics*) is that branch of systematics concerned with reconstructing phylogeny. Ever since Darwin laid down the fundamental principles of evolutionary theory, one of the major goals of the biological sciences has been the determination of life's history of descent. This phylogeny of organisms, visualized as a branching pattern, can be determined by an analysis of characters from living or fossil organisms, utilizing phylogenetic principles and methodology. The branching pattern can be used as a basis for a system of classification that directly reflects genealogical history (see Appendix B). Additionally, the branching pattern, in conjunction with past or present distributional ranges, can be used to trace the biogeographic history of a group of organisms (see Nelson and Platnick, 1981, for a comprehensive review of biogeographic methods). Ultimately, however, phylogenetic systematics may attempt to deduce the collective genetic changes that have occurred in populations through time. Thus, a knowledge of phylogenetic relation-

*Contributed by Michael G. Simpson, Albright College, Reading, Pennsylvania.

ships may be invaluable in understanding structural evolution as well as in gaining insight into the possible adaptive significance of hypothesized evolutionary changes.

A. PRINCIPLES AND METHODOLOGY

The study of phylogeny begins with the selection of a certain group of taxa to be analyzed. The rationale as to *which* taxa are selected from among thousands of species of land plants rests by necessity on previous classifications or phylogenetic hypotheses. Generally, a group of taxa is chosen for which there are competing or uncertain systems of classification, the objective being to test the bases of those different classification schemes or to provide a new classification system with reference to a rigorous analysis of phylogeny. However, some caution should be taken in choosing which taxa to study. Individual unit taxa must be well circumscribed and delimited from one another, and the entire group must be large enough so that all possible close relatives are treated. (Stated strictly, both unit taxa and the group as a whole must be hypothesized to be monophyletic before the analysis is begun.) Therefore, the initial selection of taxa should always be questioned beforehand to avoid the bias of following past classification systems.

Fundamental in any systematic study is description, the characterization of plants using any number of types of evidence (see Chapters 5 and 8). The phylogeneticist may originally describe aspects of a group of plants or rely partly or entirely on previously published research. In any case, it cannot be overemphasized that the ultimate validity of a phylogenetic study depends on the descriptive accuracy and completeness of the primary investigator. Thorough research and a comprehensive familiarity with the literature on the taxa and characters of concern are prerequisites to the study of phylogeny.

I. Characters

a. Selection and Definition

After taxa are selected and the basic research and literature survey is complete, the next component of a phylogenetic study is the analysis of the characteristics of the group of plants in question. The first step of this analysis is the selection and definition of *characters* (subunits or attributes of the organism) and *character states* (two or more forms or types of a character) to be used in the analysis. Generally, those characters that are heritable, relatively invariable, and denote clear discontinuities from other similar characters and character states should be considered. However, the selection of a finite number of characters from the virtually infinite number that could be used adds an element of subjectivity to the study. Thus, it is important to realize that any analysis is inherently biased simply by the way characters are selected and how the characters and character states are defined. (In some cases, certain characters are weighted over others, a procedure to be discussed.) In addition, because a feature is the manifestation of numerous intercoordinated genes and because evolution occurs by a change in one or more of those genes, the precise definition of a feature in terms of characters and character states is problematical. A structure may be defined broadly as a whole entity with several components, or discrete features of a structure may be defined individually as separate characters and character states. For example, in comparing the evolution of fruit morphology within some taxonomic group, the character, fruit type, may be designated as two character states: berry versus capsule, *or* it may be subdivided into a host of characters with their corresponding states, for example, fruit shape, fruit wall texture, fruit dehiscence, and seed number. In practice, characters are divided only enough to communicate differences in structure between two or more taxa. However, this type of terminological atomization may be misleading with reference to the effect of specific genetic changes in evolution. For example, there is probably not a specific gene that determines fruit shape or seed number. The morphology of a structure is the end product of a host of complex interactions.

b. Homology and Homoplasy

An integral part of the selection and definition of characters and character states is the designation of *homology.* Homology is strictly defined as a hypothesis of common evolutionary origin; characters or character states of two or more taxa are homologous if they were present in the common ancestor of those taxa. These taxa are presumed to share, by common ancestry, gene assemblages that determine the development of the common structure. The determination of homology is one of the most challenging aspects of a phylogenetic study and may involve a variety of criteria (Eldredge and Cracraft, 1980; Wiley, 1981). Generally, homology is hypothesized to be based on some evidence of similarity, either direct similarity (e.g., of structure, position, or development) or similarity via a gradation series (e.g., intermediate forms between character states). Homology should be assessed for each character of all taxa in a study, particularly of those taxa having *similarly termed* character states. For example, if the members of two tribes possess pubescent leaves, the character state, pubescent, for the character, leaf vesture, may be assigned to each of the two tribes. Whether intended or not, the designation in a phylogenetic study of this same character state to the two tribes presupposes that the features are homologous in those taxa and arose by common evolutionary origin. Thus, a careful distinction should be made between terminological similarity and similarity by homology. (Homology may also be defined with reference to similar structures within the *same individual;* two or more structures are homologous if the gene assemblages that determine their similarity share a common evolutionary origin; see B. II. Floral Evolution).

Similarity between organisms can arise not only by common ancestry, but also by independent evolutionary origin. Nonhomologous similarities (known collectively as *homoplasy*) can occur either by *convergence* (synonymous here with *parallelism*), in which a similar feature evolves independently in two or more different lineages, or by a *reversal,* in which a derived feature is lost and replaced by the original ancestral condition. If, in the preceding example, it is determined that leaf trichomes in the two tribes have an entirely different anatomy and ontogeny, it may be hypothesized that pubescence arose independently in the two taxa, that is, by convergence. This hypothesis would necessitate a redefinition of the character states for leaf vesture, such that the two taxa are not ascribed the same character state. In contrast to hypothesizing character convergence, determination of a reversal requires some knowledge of relationships or character distributions of the taxa tested. Phylogenetic analysis allows for the detection of a reversal of character states and may detect previously unrecognized convergences.

c. **Morphoclines**

Once characters and character states are selected and homologies assessed, characters are represented as a sequence of character states, known as *morphoclines* (also called phenoclines or *transformation series*). Morphoclines are generally postulated vis-à-vis some obvious intergradation of character states or stages in the ontogeny of the character, which will be discussed. For example, for the character, ovary position, a morphocline can be designated as ''inferior ↔ half-inferior ↔ superior.'' For a character with only two character states, obviously only one possible morphocline exists (Figure 10.1a). Such a two-state morphocline is more than just a contrast of features between taxa, however. Morphoclines represent past evolutionary changes hypothesized to have occurred within each character, one character state having evolved from another. Characters having several states can have any number of hypothesized morphoclines. For example, for a three-state character, three morphoclines are possible (Figure 10.1b), and for a character with four states, sixteen morphoclines can be constructed (Figure 10.1c).

d. **Polarity**

The next step of character analysis is the assignment of *polarity*. Polarity is the designation of relative ancestry to the character states of a morphocline. A change in character

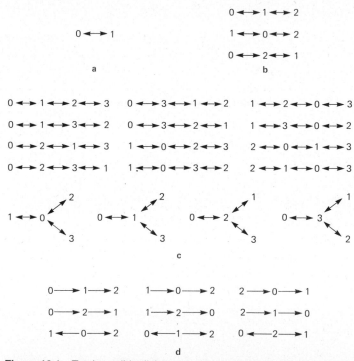

Figure 10.1   Total possible dichotomously branched morphoclines. **a.** A character with two character states (0 and 1). **b.** A character with three character states (0, 1, and 2). **c.** A character with four character states (0, 1, 2, and 3). **d.** Total possible morphoclines for a character with three states (0, 1, and 2), for which the polarity has been determined. Arrows represent directions of evolutionary change.

state represents a heritable evolutionary change from a preexisting structure or feature (termed *plesiomorphic, ancestral,* or *primitive*) to a new structure or feature (*apomorphic, derived,* or *advanced*). From the previous example, if a superior ovary is hypothesized to be primitive, the resultant morphocline would be "superior → half-inferior → inferior." (The possible morphoclines for a three-state character, after polarity is determined, are illustrated in Figure 10.1d). The designation of polarity is often one of the more difficult and uncertain aspects of a phylogenetic analysis. The primary procedure for determining polarity is *out-group comparison,* (which will be discussed).

e. Character × Taxon Matrix

A useful procedure in constructing phylogeny is the assignment of numerical values to the selected characters and character states and the tabulation of these data in the form of a *character × taxon matrix* (e.g., see Figure 10.2a). Character states are assigned nonnegative integer values (beginning with 0) and characters may be designated as a number or letter. The enumerated character states are listed sequentially to correspond with the hypothesized morphocline for that character. For example, for the character, leaf type, the morphocline "simple → pinnately lobed → pinnately compound," in which "simple" is primitive, could be enumerated as "0 → 1 → 2." By convention, the ancestral character state (if known) is designated 0, unless it is intermediate in the morphocline (e.g., 0 ← 1 → 2, in which state 1 is ancestral to both 0 and 2). In the character X taxon matrix, polarity may be indicated by creating a hypothetical ancestor that possesses the most ancestral state of each character (e.g., ANC in Figure 10.2a).

II. The Cladogram

The character X taxon matrix is used to construct a phylogenetic branching diagram, known as a *cladogram* (e.g., Figure 10.2b). A cladogram is a representation of the genealogical history of *groups* of individuals (e.g., populations or species, not of individual organisms). The vertical axis of a cladogram is always an implied, but usually nonabsolute, time scale. Extant taxa (also called terminal taxa or *Operational Taxonomic Units,* abbreviated OTUs; e.g., *W, X, Y, Z* in Figure 10.2b) are placed at the top of the time scale. Each node of the cladogram represents a hypothesized ancestral taxon (termed a *Hypothetical Taxonomic Unit,* abbreviated HTU; e.g., *E, F, G,* in Figure 10.2b). Cladogram internodes (termed lineages or clades) are ancestor-descendant sequences of populations. Thus, each of the bifurcations of a cladogram ultimately represents a past speciation that resulted in two separate lineages. This is true even if the designated OTUs of a study are supraspecific taxa (e.g., tribes, families); a lineage is always theoretically collapsible to a speciation event. Along a given cladogram internode, evolutionary events are portrayed as changes in the states of characters. For example, in Figure 10.2b, numbers in parentheses identify characters, with 0 → 1 representing a change from ancestral character state 0 to derived state 1.

a. Monophyly and Synapomorphy

A primary tenet of phylogenetic systematics is that a cladogram may be constructed from the character × taxon matrix by sequentially arranging, on a branching diagram, sets of

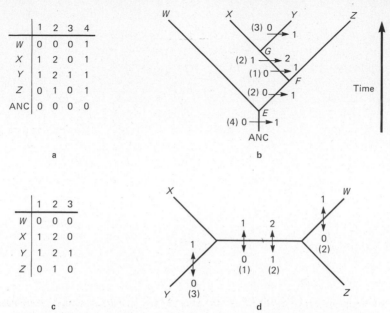

Figure 10.2 **a.** Character × taxon matrix for characters 1 to 4 and taxa W to Z plus hypothetical ancestral taxon ANC (coded for ancestral states of all characters; the morphocline for character #2 is 0 → 1 1 → 2). **b.** Cladogram of taxa (OTUs) W, X, Y, and Z, including common ancestors (HTUs) E, F, and G. Numbers in parentheses designate characters; arrows represent character state changes, which are unique for a given lineage. **c.** Character × taxon matrix in which polarity of states is not indicated, by omission of hypothetical ancestor ANC. Note omission of character #4, which has similar values for all taxa (uninformative without reference to ancestry). **d.** Network, derived from matrix at **c**, portraying relative character state changes between taxa W to Z. Note that the historical direction of character state changes cannot be indicated. In this example, character state values above arrow occur along lineages to the *left* of the arrow.

taxa that together share one or more unique evolutionary events. Each set is called a *monophyletic taxon,* identified and defined by one or more *shared derived character states* (also called *synapomorphies*). Strictly defined, a monophyletic taxon includes a given ancestor and *all and only all* of its descendants. Thus, in Figure 10.2b, the group including W, X, Y, and Z (and common ancestors E, F, and G) is a monophyletic taxon, identified by a synapomorphy of character #4 (0 → 1); the monophyletic taxon of X, Y, and Z (and common ancestors F and G) is identified by the synapomorphy of character #2 (0 → 1); and the monophyletic group including X and Y (and common ancestor G) is identified by synapomorphies of character #1 (0 → 1) and #2 (1 → 2). Note that taxon Z is grouped with X and Y by the synapomorphy 0 → 1 of character #2, even though Z has a different character state value (1) from X and Y (2). The change in character #3 (Figure 10.2b) represents an evolutionary event unique for taxon Y, termed an *autapomorphy*. Two taxa that share a most recent common ancestor are termed *sister groups*. Thus, in Figure 10.2b, taxon X is the sister group of Y (most recent common ancestor G), and X and Y together is the sister group of Z (most recent common ancestor F).

b. Networks

In contrast to a cladogram, a method for the representation of relative character state changes between taxa is the *network*. A network is a branching diagram that minimizes the total number of character state changes among all taxa (Figure 10.2d). Networks are constructed by grouping taxa from a matrix in which polarity is not indicated (e.g., Figure 10.2c, in which no hypothetical ancestor is indicated), perhaps because the polarity of one or more characters cannot be ascertained. Because no assumptions of polarity are made, no genealogical hypotheses are implicit in a network. Note that monophyletic groups cannot be recognized in networks because relative ancestry is not indicated. Thus, in Figure 10.2d, neither *X* and *Y* nor *W* and *Z* can be designated as monophyletic. The character state changes noted on the network simply denote evolutionary changes when going from one group of taxa to another, without reference to direction of change. After a network is constructed, it may be converted into a cladogram. If the relative ancestry of one or more characters can be established, a point on the network may be designated most ancestral, forming the root of the cladogram.

c. Parsimony

In constructing a cladogram, a single branching pattern is selected from among many possibilities. For two taxa, there is obviously only one cladogram (Figure 10.3a); for three taxa, three dichotomously branched cladograms can be constructed (Figure 10.3b); and

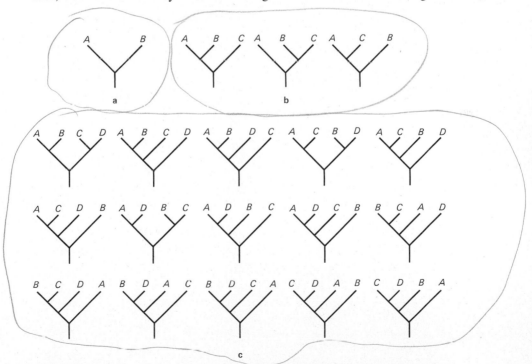

Figure 10.3 All possible dichotomously branched cladograms for a group consisting of the following. **a.** Two taxa (*A* and *B*). **b.** Three taxa (*A*, *B*, and *C*). **c.** Four taxa (*A*, *B*, *C*, *D*).

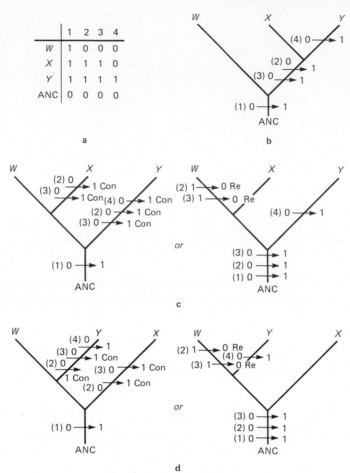

Figure 10.4 **a.** Character × taxon matrix for characters 1 to 4 and taxa *W* to *Z* plus hypothetical ancestor ANC. **b** to **d.** Character changes of three possible cladograms for taxa *W*, *X*, and *Y*. Cladogram at **b,** with a total of four apomorphies, is more parsimonious than that at either **c** or **d,** each with a six apomorphies, two of which are either parallelisms (first cladogram of **c** and **d**) or reversals (second cladogram of **c** and **d**). Con = convergence; Re = reversal.

for four taxa, fifteen dichotomously branched cladograms are possible (Figure 10.3c). Additional possibilities are the occurrence of reticulation or polychotomies, which will be discussed. A major premise of phylogenetic systematics is that hypotheses of genealogical history "minimize requirements for ad hoc hypotheses of homoplasy" (Farris, 1983). In other words, of all possible cladograms for a given group of taxa, the one (or more) implying the least number of reversals or convergences is accepted. This methodology is known as the *principle of parsimony* and is the Ockham's razor of phylogenetics. A consequence of minimizing homoplasy is to minimize the total number of character state changes. For example, for three taxa—*W, X,* and *Y*—three dichotomously branched cladograms can be constructed (e.g., Figure 10.3b). From the character *X* taxon matrix of

Figure 10.4a, the cladogram at Figure 10.4b (with a minimum of four character state changes) is more parsimonious than the cladograms of Figure 10.4c or 10.4d (each having a minimum of six character state changes, two of which are either reversals or pairs of convergences). The principle of parsimony is a valid working hypothesis because it minimizes uncorroborated hypotheses, thus assuming no additional evolutionary events for which there is no evidence (see Farris, 1983).

In practice, construction of a cladogram may be more complex than the preceding example would suggest. The difficulty occurs when taxa have conflicting patterns of synapomorphy. For example, from the character × taxon matrix of Figure 10.5a, taxa X and Y would be grouped by a synapomorphy $(0 \rightarrow 1)$ of character #1; Y and Z would be grouped by a synapomorphy $(0 \rightarrow 1)$ of character #2; and X and Z would be grouped by a synapomorphy $(0 \rightarrow 1)$ of characters #3 and #4. Thus, the distributions of character states are contradictory. The most parsimonious cladogram (with a total of six character state changes) groups together X and Z as a monophyletic taxon (Figure 10.5b). This grouping of taxa X and Z is supported by two synapomorphies. The other two cladograms, grouping X and Y or Y and Z based on a single synapomorphy, would each have a minimum of seven character state changes. Note that by accepting the most parsimonious cladogram, the conflicting character state changes must be hypothesized to be homoplasious (e.g., the four, equally parsimonious combinations of Figure 10.5b to e).

d. Algorithms

Various algorithms are used to determine the most parsimonious cladogram from a given character × taxon matrix. One of the most used methods for cladogram construction is the

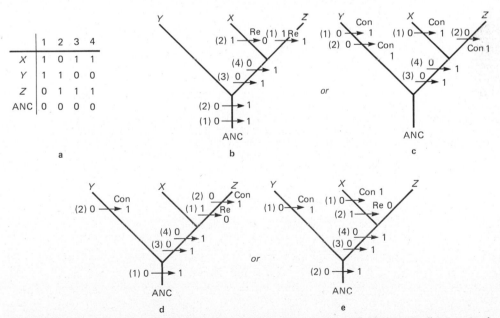

Figure 10.5   a. Character × taxon matrix of taxa X, Y, Z, and ancestor ANC. Note contradictory sets of synapomorphies. **b** to **e.** Most parsimonious cladogram for taxa X, Y, and Z, showing four possible combinations of reversals (Re) and/or convergences (Con).

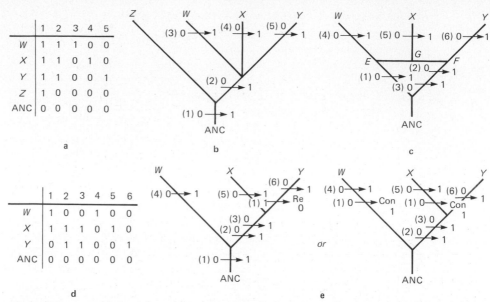

Figure 10.6 **a.** Character × taxon matrix for taxa W, X, Y, Z, and hypothetical ancestor ANC. **b.** Most parsimonious cladogram for matrix at **a,** showing trichotomy of taxa W, X, and Y due to lack of data. Note autapomorphies of characters 3, 4, and 5. **c.** Reticulate cladogram for taxa W, X, and Y, showing hybridization between ancestral taxa E and F, giving rise to taxon G, the immediate common ancestor of X. **d.** Character × taxon matrix of taxa W, X, Y, and ANC, compatible with cladogram at A. **e.** Most parsimonious dichotomous cladogram of taxa W, X, and Y, derived from matrix at **d,** taxa X and Y incorrectly attributed sister-group status. Note that the character state distribution is explained by either a reversal (left) or a pair of convergences (right).

Wagner method, developed by Farris (1970) (see Appendix B). This method is the basis for some phylogeny computer algorithms currently in use. It should be noted that, given the occurrence of homoplasy, more than one equally parsimonious cladogram may exist and that, for a given cladogram, more than one character state change pattern may occur (e.g., as in Figure 10.5b to e).

e. **Polychotomy**

Occasionally, the relationships among taxa cannot be resolved, a situation represented in the cladogram as a polychotomy. A *polychotomy* is a branching diagram in which the lineages of three or more taxa arise from a single hypothetical ancestor. Polychotomies arise either because data are lacking or because two or more of the taxa were actually derived from a single ancestral species. In the first case, there are no derived character states identifying the monophylesis of any two taxa among the group. For example, from the character × taxon matrix of Figure 10.6a, the relationships among taxa W, X, and Y cannot be resolved; synapomorphies do not link W and X, W and Y, and X and Y. Thus, W, X, and Y are grouped as a trichotomy in the most parsimonious cladogram (Figure 10.6b). The other possible reason for the occurrence of a polychotomy is that all of the taxa under consideration diverged independently from a single ancestral species. Thus, no synapomorphic evolutionary event links any two of the taxa as a monophyletic group. The occurrence of a polychotomy in phylogenetic analysis should serve as a signal for the reinvestigation of taxa and characters, perhaps indicating the need for continued research.

f. Reticulation

The methodology of phylogenetic systematics generally presumes the dichotomous or polychotomous splitting of taxa, which ultimately represents an ancestral speciation event. However, another (perhaps likely) possibility in the evolution of plants is *reticulation*, the hybridization of two previously divergent taxa forming a new lineage. A reticulation event between two ancestral taxa (E and F), is exemplified in Figure 10.6c, resulting in the hybrid ancestral taxon G, which is the immediate ancestor of extant taxon X. Most standard phylogenetic analyses do not consider reticulation and would yield an incorrect cladogram if such a process had occurred. For example, the character X taxon matrix of Figure 10.6d is perfectly compatible with the reticulate cladogram of Figure 10.6c. However, the methods of phylogenetic systematics would construct the most parsimonious cladograms of Figure 10.6e, which show homoplasy and require one additional character state change. Reticulation among a group of taxa should always be treated as a possibility. Data, such as from chromosome analysis, may provide evidence for past hybridization among the most recent common ancestors of extant taxa.

g. Selection of Taxa and Polymorphic Characters

As alluded to previously, the initial selection of taxa to be studied may introduce bias in a phylogenetic analysis. Before phylogeny is constructed, each of the smallest unit taxa under study (OTUs) *and* the group as a whole must be hypothesized to be monophyletic before the analysis is begun. Monophyly is ascertained by the recognition of one or more unique shared derived character states that argue for most recent common ancestry of all and only all members of the taxon in question. If such a synapomorphy cannot be identified, any relationships denoted from the phylogenetic analysis may be in doubt. For example, in a cladistic analysis of several angiosperm genera (Figure 10.7a), only if each

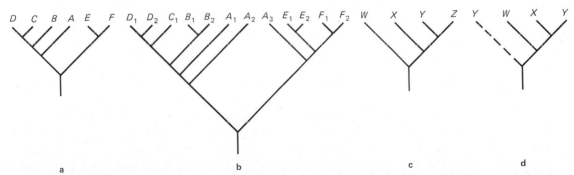

Figure 10.7 **a.** Cladogram of genera A to F, assuming (wrongly) that each genus is a monophyletic unit. **b.** Cladogram of species of genera A to F. Note that genus A is not monophyletic; species A_1 has more recent common ancestry with genera D, C, and B than with species A_2 and A_3; species A_3 has more recent common ancestry with genera E and F than with species A_1 and A_2. **c.** Cladogram of taxa W, X, Y, and Z. Note that Y and Z are sister groups and that Y and Z together are more closely related to X than to W. **d.** Cladogram of taxa W, X, and Z, a nonmonophyletic group because of omission of Y. Sister-group status would be wrongly attributed to taxa X and Z.

of the unit taxa (genera) is monophyletic will the resultant cladogram have meaning. If, however, genus A is not monophyletic, then it may be possible for some species of genus A to be more closely related to (i.e., have more recent common ancestry with) a species of another genus than to the other species of genus A (e.g., Figure 10.7b). Therefore, if any doubt exists as to the monophyly of component taxa to be analyzed, the taxa in question should be subdivided until the monophyly of these subtaxa is reasonably certain. If this is not possible, an exemplar species (selected as representative of a higher taxon and assumed to be monophyletic) may be chosen for a first approximation of relationships. Additionally, if the entire group to be analyzed is not monophyletic, the effect is identical to excluding taxa from the analysis, which could give erroneous results under certain conditions. For example, if Figure 10.7c represents the true genealogy for taxa W, X, Y, and Z, but if only W, X, and Z are included in the analysis, the most that can be deduced is that X and Z are more closely related to one another (i.e., share a more recent common ancestor) than either is to W (Figure 10.7d). It would be erroneous to conclude that X and Z are sister groups (and together comprise a monophyletic taxon) since W, X, and Z are a nonmonophyletic group (a *paraphyletic* group, see Appendix B); taxon Y would be incorrectly presumed to be distantly related to W, X, and Z (Figure 10.7d). The question of monophyly is likely a serious problem for the gymnosperms as a taxon and for many angiosperm tribes, families, and orders that are often not defined by shared derived character states.

Related to the requirement of monophyly is the problem of polymorphic characters. If a unit taxon, determined to be monophyletic, is variable (polymorphic) with respect to the character state values of a character, then it is evident that one of these states is most ancestral and the other state(s) are derived *within that unit taxon*. Any similarity between the derived states in this unit taxon and features in any other taxa must be homoplasious. Therefore, in a polymorphic taxon for which monophyly has been established, only the most ancestral state should be listed in the phylogenetic analysis.

h. Character Weighting

As part of the initial analysis of the features of investigated taxa, the investigator may choose to weight characters. Character weighting is the assignment of greater taxonomic importance to certain characters over other characters in determining phylogenetic relationships. Weighting may be accomplished in cladogram reconstruction by multiple listing of the weighted characters in the character $\times$ taxon matrix to override the unweighted characters. Generally, those characters are weighted for which the designation of homology is considered relatively certain. The expectation is that, by weighting characters for which homoplasy is deemed unlikely, taxa will be grouped by shared derived features. A character that is weighted before the cladogram analysis may be viewed as nonhomoplasious for various reasons. For example, a feature distinctive to two or more taxa may be structurally or developmentally complex, such that the independent evolution of the same character state would seem very unlikely. (It should be realized, however, that if a feature is most likely highly adaptive, parallelism of similar complex features in two or more taxa may not necessarily be ruled out.) Characters may also be weighted because they are correlated (i.e., the corresponding character state values of two or more characters are present in all taxa). The correlation of characters may be viewed as evidence that the derived states of each represent features unique to those taxa that possess them. Alternatively, weighting may be done after the first stage of a phylogenetic analysis. Those characters that exhibit reversals or parallelisms on the cladogram are recognized and either removed or arbitrarily given less weight over those that do not.

Another method of weighting is *character compatibility analysis*. Character compatibility analysis (Meacham, 1981) is a method for phylogeny construction that recognizes incompatibility among any two characters and then identifies the largest set of characters that are mutually compatible. This set of compatible characters alone is used to construct a network; the network may then be rooted by the identification of one or more ancestral character states and the selection of a point on the network as the ultimate ancestor or by including the ancestor as a taxon in the data matrix. If, in this cladogram, the branching pattern of a certain monophyletic subgroup of taxa is unresolved, then further refinement may be made by applying the method to that subgroup alone, utilizing *all* of the original characters. The rationale of this method is that those characters showing numerous incompatibilities are most likely incorrectly assessed with regard to definition, homology, or morphocline designation and may give faulty information (noise) in cladogram construction. Thus, character compatibility analysis is actually a method that selectively weights those characters which are compatible by selectively removing those characters which are not. (Arguments have been presented, however, that character compatibility analysis results in a loss of information; see Farris, 1983).

i. Ontogeny and Paedomorphosis

Phylogeny and the evolution of structure may often be studied only with regard to the mature features of adult individuals. However, a mature plant structure, whether organ, tissue, or cell, is the end product of *ontogeny*, a developmental sequence under the control of a number of genes. A study of this ontogenetic pattern often reveals a series of discrete structural stages or entities, one transforming into the next until the end point (the mature adult structure) is obtained.

One of the major, classical principles of anatomy and morphology is that juvenile stages in the development of a structure may be homologous with adult stages of an ancestor. (Note that juvenile stages of an extant taxon may also be homologous to juvenile features of an ancestor.) This principle (termed *Haeckelian recapitulation* and often summarized by the expression "ontogeny recapitulates phylogeny") may provide a direct method for determining the evolutionary history of a feature by tracing its developmental history. For example, if $S^1 \rightarrow S^2 \rightarrow S^3$ is the developmental sequence of a structure in extant taxon A, with S^3 representing the adult structure, then S^1 and S^2 (which are juvenile structures in the extant taxon) may be homologous to adult structures in respectively older ancestral taxa (e.g., ancestors J and K in Figure 10.8a). These ancestral adult structures may persist and be recognized as adult structures in other extant taxa (e.g., in taxa B and C in Figure 10.8a). Haeckelian recapitulation assumes that evolution occurs only by the sequential addition of new features and that mature ancestral structures will persist as juvenile developmental stages in extant taxa.

However, evolutionary change may result in the modification of mature structures by affecting early developmental stages. For example, if the ontogenetic development of structure S^3 occurs in two discrete steps ($S^1 \rightarrow S^2$ and $S^2 \rightarrow S^3$), each controlled by

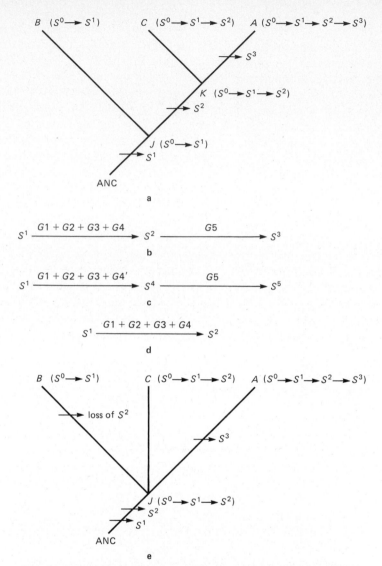

Figure 10.8 **a.** Cladogram of extant taxa A, B, and C and ancestral taxa J and K, showing evolution of developmental sequence $S^1 \rightarrow S^2 \rightarrow S^3$. Note that adult structures of ancestral taxa, S^1 of taxon J and S^2 of taxon K, are homologous to (descended as) adult structures of, respectively, taxa B and C but are juvenile structures of taxon A. **b.** Ontogenetic sequence for mature structure S^3, illustrating developmental stages S^1 and S^2; G1 to G5 represent genes controlling the development of these stages. **c.** Ontogenetic sequence resulting from modification of gene G4 to derived form G4′; note that S^2 and S^3 are not expressed being replaced by S^4 and S^5. **d.** Paedomorphosis by loss or inhibition of gene G5. Note that ancestral structure S^2 is now expressed as the mature adult structure. **e.** Most parsimonious cladogram of taxa A, B, and C, with evidence that ontogenetic sequence $S^0 \rightarrow S^1 \rightarrow S^2$ is ancestral (occurring in common ancestor J). Note that sister-group relationships cannot be determined and that taxon B is paedomorphic by loss of state S^2 (detected as a reversal).

several genes (Figure. 10.8b), then a single alteration of a gene controlling that first developmental sequence may cause a change in both the final structure and the intermediate developmental stage (Figure 10.8c). Subsequent stages would then be recognized as different structural entities (e.g., S^4 and S^5 of Figure 10.8c). This principle (termed von *Baerian recapitulation*) states that structural evolution may occur by modification at any developmental stage and that mature ancestral structures need not be preserved as extant juvenile developmental stages. In addition, the terminal stages of a developmental sequence may be lost (e.g., Figure 10.8d). Such an event, known as *paedomorphosis* (Gould, 1977), results in the transformation of an ancestrally juvenile stage to a mature structure in a derived taxon.

Paedomorphosis has apparently been an important evolutionary mechanism in many groups (Gould, 1977). Paedomorphosis is detected as the reversal of a character state and can only be ascertained from the utilization of other characters in a phylogenetic analysis.

j. Polarity Determination

As mentioned in the discussion on character analysis, knowledge of character polarity is necessary to recognize shared derived character states that define monophyletic taxa. Several criteria may be used to ascertain polarity (Stevens, 1980; Crisci and Stuessy, 1980), including ontogeny, out-group comparison, the doctrine of association, the doctrine of correlation, and in-group comparison.

1. Ontogeny

Ontogeny can provide evidence for the transformation and relative ancestry of character states by the premise that the earlier a feature appears developmentally, the more ancestral is that feature (Haeckelian recapitulation). For example, suppose that taxa B, C, and A possess, respectively, adult character states S^1, S^2, and S^3, the intergradation and polarities of which are unknown. Given that developmental studies show the ontogenetic series $B: S^0 \rightarrow S^1$, $C: S^0 \rightarrow S^1 \rightarrow S^2$, and $A: S^0 \rightarrow S^1 \rightarrow S^2 \rightarrow S^3$, then the relative polarity of the adult structures may be hypothesized as $S^1 \rightarrow S^2 \rightarrow S^3$. The most parsimonious cladogram consistent with these data is that of Figure 10.8a; the other two possible cladograms require a minimum of one more character state change and provide no evidence for the monophyly of any two of the three taxa. (Note that the possession of the developmental stage S^2 is hypothesized as a synapomorphy for taxa C and A and that S^3 is autapomorphic for taxon A.) However, it should be noted that ontogenetic data are not full proof in the assessment of polarity because of the possibility of paedomorphosis. The verification of the polarity of ontogenetic stages, using additional characters, may be necessary. In Figure 10.8a, for example, if it is hypothesized by out-group comparison (see below) that common ancestor J has the ontogenetic series $S^0 \rightarrow S^1 \rightarrow S^2$, then it is apparent that these ontogenetic data provide no evidence for grouping together taxa A and C; the ontogenetic series for taxon B ($S^0 \rightarrow S^1$) would be interpreted as paedomorphic by loss of stage S^2 (Figure 10.8e).

2. Out-group comparison

The method for determining character polarity that requires the minimum number of ad hoc hypotheses is known as *out-group comparison*. Out-group comparison is character

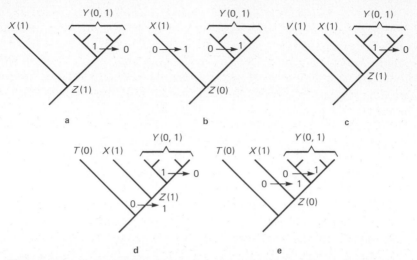

Figure 10.9 Determination of character polarity using out-group comparison. **a.** Most parsimonious assumption in which character state 1 is ancestral and present in ancestor Z. **b.** Alternative, less parsimonious cladogram, in which state 0 is assumed to be ancestral. **c.** Verification of cladogram a by addition of next outgroup V, which also has state 0. **d** to **e.** Cladograms in which additional outgroup T has state 0. Assumption that state 1 is ancestral to group Y (cladogram at D) is equally parsimonious to assumption that state 0 is ancestral (cladogram at E). Taxon X (possessing character state 1) is the nearest outgroup (the sister group) to taxon Y (which possesses both states 0 and 1 among subtaxa).

assessment of the closest relatives to the taxon under study; the character state that is possessed by the closest relatives (particularly by the sister group of the taxon) is considered to be ancestral; states present in the taxon under study, but not occurring in nearest relatives, are derived. For example, given some monophyletic group Y (Figure 10.9) members of which possess both states (0,1) of a character, and given that taxon X (the sister group of Y) possesses only character state 1, then the most parsimonious solution (requiring a single character change: $1 \rightarrow 0$) is that state 1 is ancestral and present in the common ancestor Z; character state 0 is derived within the taxon Y (Figure 10.9a). The alternative, that state 0 is ancestral, requires at least two character state changes (Figure 10.9b). Verification of sister-group comparison is made by considering an additional out-group (e.g., taxon V in Figure 10.9c). If this next out-group possesses only character state 1, then the plesiomorphy of state 1 for taxon Y is substantiated. If, however, out-group V contains only character state 0, then it is equally parsimonious to assume that state 1 is ancestral (Figure 10.9d) rather than derived (Figure 10.9e); in this case, consideration of additional out-groups may resolve polarity. The major problem of out-group comparison is that the cladistic relationships of out-group taxa are often unknown; in such a case, all possible out-groups can be considered.

3.   *Doctrines of association and correlation*

The *doctrine of association* states that the most primitive state of a derived feature will be that which most resembles the structure from which the derived feature evolved (see B.

Structural Evolution). This principle is nothing more than the establishment of a morphocline and the determination of polarity, which would be done in a phylogenetic study. For example, consider the morphocline $A \rightarrow B' \rightarrow B'' \rightarrow B'''$, in which A represents an ancestral structure and B', B'', and B''' are states of a separately defined derived structure B. Because this morphocline represents a hypothesized sequence of evolutionary events, it is apparent that the most primitive state of structure B is state B', which most resembles the ancestral structure A.

The *doctrine of correlation* states that primitive features tend to occur together. For example, assuming that the families of the Magnoliidae have numerous primitive characteristics, then the occurrence of beetle pollination in many of these families may be cited as a correlation in support of the view that pollination by beetles is most ancestral among the flowering plants as a whole. However, the a priori correlation of primitive features (or, for that matter, derived features) is valid only if, in any given lineage, those features have evolved at similar rates, for example, by being genetically related or by having been subject to similar selective pressures. If two or more features are genetically related, then evolutionary change in one feature will presumably bring about a change in the related feature. (In fact, the principle of correlation, in some cases, may be simply a by-product of the division into separate characters and character states of a single structure that is developmentally determined by a number of intercoordinated genes.)

The principle of correlation certainly may be invalid when applied to structures that are genetically unrelated. Different structures may evolve at different rates, dependent, for example, on the selective pressures encountered by the individual. A plant may possess derived features of one structure yet retain ancestral features of another structure; the presence of one primitive feature provides absolutely no evidence that another feature is also primitive. Therefore, the principle of correlation should be used only where there is evidence that separately defined features have evolved in concert.

4. In-group comparison

In-group comparison (also known as the *common ground plan* or *commonality principle*) states that, in a given (presumably monophyletic) group, the primitive structure will tend to be the most common one. The reasoning of this principle is that the evolution of a derived condition will occur in only one of potentially numerous lineages of the group; the ancestral condition will, therefore, tend to be in the majority. However, the fallacy of this principle is readily apparent. In Figure 10.10a, for example, the primitive character state 0 of character #1 is the more common in the monophyletic group that includes *T, U, V, W, X, Y, Z;* in contrast, state 0 is relatively rare in the monophyletic group including *W, X, Y, Z.* Thus, a derived feature may be in the majority in a monophyletic group, dependent on the inclusiveness of that group.

k. Ancestral and Derived Characters

A common point of confusion is seen in the use of the terms ancestral (or primitive) and derived (or advanced). It is advisable that these terms be limited to the description of *characters* (not taxa) and then only relative to monophyletic groups. For example, in the cladogram of Figure 10.10b (constructed from the matrix at Figure 10.10c), state 1 of

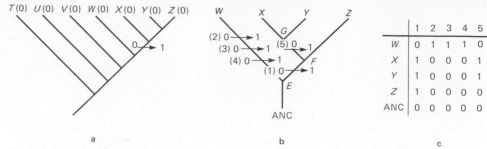

Figure 10.10 **a.** Cladogram of taxa *T* to *Z*, illustrating character #1, showing a change from state 0 to 1. Note that the primitive state 0 is widespread in the monophyletic group including *T* to *Z* but rare in the monophyletic group, including *W* to *Z*. **b.** Most parsimonious cladogram of taxa *W, X, Y,* and *Z* and ancestors *E, F, G,* and ANC. **c.** Character × taxon matrix for taxa *W* to *Z* and ancestor ANC, used to construct cladogram at *b*.

character #1 is *derived* within the group including *W, X, Y,* and *Z* (i.e., state 1 is absent in common ancestor *E*) but is *ancestral* with regard to the monophyletic group *X, Y, Z* (i.e., state 1 is present in *F*, the common ancestor of *X, Y,* and *Z*). (Recall that the original designation of character polarity is made with reference to all taxa included in a study, presumably a monophyletic group.) The use of the terms ancestral and derived to describe *taxa* should be avoided to prevent ambiguity. For example, from Figure 10.10b, a question might be asked as to which taxon is most primitive: *W* or *Z*? Depending on how the term, primitive, is defined, *either W,* which arises as the most basal lineage on the cladogram, *or Z,* which has the fewest number of derived character state changes (one, as opposed to three for taxon *W*), might be designated most primitive. Confusion is avoided by describing taxon *W* as phylogenetically most basal and taxon *Z* as possessing the fewest number of apomorphies.

I. Cladogram Analysis

Once the most parsimonious cladogram is derived, it is important to map out all character state changes. A hypothesis of evolutionary change in a character may be made simply by analysis of character state changes in the cladogram. Monophyletic groupings should be carefully studied in terms of the shared derived states that link them together. If these synapomorphies are questionable (e.g., defined by a single character state change showing convergence), then the monophyly of the group may not be well supported. Homoplasies (convergences or reversals) especially are to be noted. A homoplasy may represent an error in the initial analysis of that character and warrants reconsideration of character state definition, intergradation, homology, or polarity. Thus, cladogram construction should be viewed not only as an end in itself, but as a means of pointing out those areas where additional research is needed to resolve satisfactorily the phylogeny of a group of organisms.

III. A Perspective on Phylogenetic Systematics

The serious student, in constructing cladograms or critically reading the literature, should be aware of several major pitfalls in phylogenetic systematics. Lack of consideration of any of the following renders the study questionable at best and useless at worst. First, what are the sources of the data? The validity of a phylogenetic study is based on the comprehensiveness and accuracy of the original descriptions. Second, which characters are selected and how are they defined? It is important to question the basis for the selection of *these* characters and not others. Third, is homology assessed? Has an effort been made to determine whether similar characters and character states presumably have a common evolutionary origin? Or is the similarity more one of terminology and possibly homoplasious? Fourth, have any characters/character states been weighted a priori? If so, what is the rationale behind it? Fifth, how are polarities determined? Many studies have relied on commonly assumed (and possibly faulty) conceptions of polarity. If out-group comparison was used to assess polarity, the evidence for selection of the out-groups should be thoroughly investigated. Sixth, are unit taxa (OTUs) and the group as a whole monophyletic? If evidence for monophyly is not presented, the study is questionable from the beginning. Finally, is the resultant cladogram analyzed in terms of monophyletic groupings, character state changes, assessment of convergences and reversals, testing of homology, and possible reevaluation of characters and character states? The thorough phylogenetic study critically reviews each step of cladogram construction, considers all alternatives, and evaluates and reevaluates the significance of the phylogenetic analysis in terms of future research that might clarify our understanding of plant evolutionary relationships. Although the determination of phylogeny using the methodology of phylogenetic systematics may be plagued by numerous problems, it has the significant advantage of repeatability; each step of the analysis can be duplicated and unambiguously evaluated and criticized in subsequent investigations.

B. STRUCTURAL EVOLUTION

Structural evolution refers to the historical sequence of changes in the morphology of organs, tissues, or cells. These evolutionary changes in morphology reflect changes in genetic composition and are recognized as a transition from an ancestral feature to a derived feature. Structural evolution may be deduced from fossil evidence and from an analysis of characters and character states of extant taxa, including stages in the development of a structure. A consideration of the structural changes that have occurred during evolutionary history adds insight and meaning to the taxonomic categorization of plants. The study of structural modifications may be important to understanding the possible adaptive significance of such changes in the evolution of organisms within a particular environment and with reference to interactions with other organisms.

The evolution of structure is always studied with regard to a group of organisms and, therefore, can be determined by using the methodology of phylogenetic systematics, including: (1) verification of monophyly for unit taxa and the group as a whole; (2) selection, definition, and coding of characters and character states; (3) assessment of homology; (4) establishment of morphoclines; (5) determination of polarity; (6) cladogram construction from a character × taxon matrix using one or more algorithms; and (7) mapping of character state changes and reevaluation of the data (see A. Principles and Methodology). The resultant cladogram can be used as an analytical device to evaluate the structural evolutionary change of a given character. Examination of the cladogram with reference to the distribution of the states of this character reveals one or more most

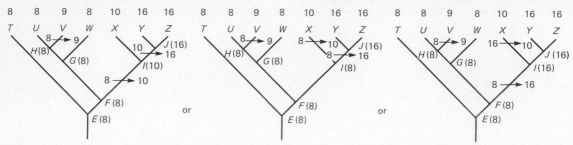

Figure 10.11 Cladogram of taxa *T* to *Z*, illustrating deduction of evolutionary changes in haploid chromosome number, a character not included in the initial analysis. Note three equally parsimonious solutions of character state change in chromosome number with regard to the cladogram. The chromosome numbers of ancestral taxa *E* to *J* are shown in parentheses.

parsimonious explanation(s) for its structural evolution. In addition, certain characters may have been omitted from the original character × taxon matrix because of incomplete data or uncertainty with regard to homology, polarity, or intergradation of character states. However, the sequence of evolutionary changes of these characters may be ascertained by superposing the states of the unincluded character on the terminal taxa of the cladogram. If it is assumed that the cladogram is correct, then the most parsimonious explanation for the distribution of character states may be obtained. For example, Figure 10.11 illustrates the most parsimonious cladogram for taxa *T* to *Z*, in which the character, chromosome number, was not originally included. A superposition of known haploid chromosome numbers may be used to hypothesize three equally parsimonious character state evolutionary patterns (Figure 10.11).

Numerous trends in the evolution of vascular plants have been recognized from comparative studies of extant and fossil plants. Although few broad studies using strict phylogenetic methodology have been made to date (see, however, Bremer and Wanntorp, 1981; Dahlgren and Rasmussen, 1983), recognized directions of structural evolution have generally followed the principles of phylogenetic systematics. Consideration of out-group comparison has been particularly valuable in denoting ancestral versus derived features in the evolutionary history of plants. For example, the common occurrences of embryos and of gametangia and sporangia enveloped by a sterile jacket layer in all members of the land plants (Embryophyta) are believed to constitute (relative to a putative green algal ancestor) shared derived characters, evidence for the monophyly of the land plants as a whole. The possession of specialized conductive cells—tracheary elements and sieve elements— in the vascular plants (Tracheophyta) is generally viewed as a shared derived feature and, therefore, evidence for their monophyly. (Many paleobotanists, however, argue from fossil evidence that lycopods represent a lineage distinct from that of all other vascular plants and, therefore, may have had an independent origin of tracheary elements.) Within the vascular plants, heterospory and secondary growth are considered to be derived structural features, although both apparently evolved independently in different lineages. The seed is usually viewed as a shared uniquely derived feature linking together the seed plants (sometimes called the Spermatophyta, including the gymnosperms and angiosperms) as a monophyletic group, although this is also somewhat controversial.

Several features within the angiosperms, including the distinctive embryo sac, double fertilization with endosperm production, and the carpel, are overwhelmingly accepted as shared derived characters (by out-group comparison, because the other seed plants, that is, gymnosperms, do not possess them), evidence for their monophyly. Finally, within the angiosperms, the occurrence of one cotyledon, the absence of secondary growth, and the possession of numerous multicyclic or scattered stem vascular bundles are widely accepted as shared derived features that link the monocotyledons as a monophyletic taxon. (Dicotyledony, secondary growth, and the eustele are believed to be ancestral within the angiosperms as evidenced by their occurrence in both the dicots and, by out-group comparison, most or all gymnosperms; see, however, Burger, 1983, for an alternative view on the evolution of the monocotyledons.) Among the angiosperms two major features in particular have provided considerable information toward the recognition of broad evolutionary trends: tracheary element anatomy and floral structure. Structural evolutionary trends (and the rationale for their recognition) in these two features will be reviewed in detail, as these trends have greatly influenced angiosperm classification.

I. Tracheary Element Evolution

Tracheary elements are the dead (at maturity), lignified water-conductive cells of vascular plants. Determination of tracheary element evolution has been based on the hypothesis that the two major types of tracheary elements, tracheids and vessels, are homologous as water-conductive cells. Tracheids are elongate, imperforate, water-conductive cells found in almost all lower vascular plants and gymnosperms as well as in some angiosperms. Vessels are perforate elements found in some species of *Equisetum* and *Selaginella*, a few ferns (e.g., *Marsilea, Pteridium*), the Gnetales (sensu lato), and almost all angiosperms.

Within the angiosperms (which are almost certainly a monophyletic group because of their distinctive and presumably derived floral morphology and embryological development), the evolution of tracheary elements can be studied using phylogenetic principles. First, a morphocline is recognized between tracheary element structural types among various angiosperm taxa. This morphocline recognizes the homology and intergradation between imperforate elongate tracheids that have scalariform pitting and various types of vessels, the latter ranging from elongate and narrow cells with numerous scalariform perforation plates, scalariform pitting, and tapering end walls to shorter and wider cells with simple perforation plates, opposite or alternate pitting, and transverse end walls (Figure 10.12). Second, the polarity of the tracheary element morphocline must be determined. Although the early evolutionary history of the angiosperms is largely speculative, fossil evidence supports their derivation from a preexisting gymnosperm (i.e., a noncarpellate seed plant). Because all fossil and almost all extant gymnosperms possess tracheids exclusively (and these with circular-bordered pitting), the use of the gymnosperms as the out-group to the angiosperms supports the hypothesis that the tracheid is the most primitive angiospermous tracheary element. (The Gnetales of the gymnosperms do possess vessels that, because of their distinctive end walls with circular perforations, are generally thought to have been derived independently from those in the angiosperms.) Third, by the doctrine of association, the most ancestral angiospermous vessel type is that one which most resembles a tracheid on the morphocline, that is, a vessel element that is elongate and narrow with tapering end walls (Figure 10.12). In addition, the great majority of angiosperms having elongate, narrow, tapering vessels also have scalariform pitting and perforation plates with numerous, scalariform bars, a correlation that has been estab-

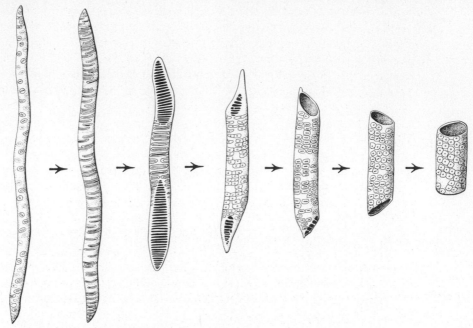

Figure 10.12 Morphocline of tracheary element types, representing a presumed evolutionary series. Note transformation from an imperforate gymnospermous tracheid with circular-bordered pits (left) to an angiospermous tracheid with scalariform pitting to an elongate vessel with scalariform pitting and numerous scalariform bars to progressively shorter, wider vessels with alternate or opposite simple perforation plates and transverse end walls.

lished statistically. Therefore, by the doctrine of correlation, the extrapolation is made that the primitive angiospermous vessel is one also having scalariform pitting and perforation plates with numerous scalariform bars. As mentioned earlier, however, correlation of primitive features may be valid only if those features are presumed to be related genetically, for example, if several related genes interact to determine both general shape (length, width, and end wall angle) and perforation plate morphology and pitting of a vessel member.

The above trends in tracheary element evolution have been utilized in assessing the major features of angiosperm evolution. For example, a major taxonomic controversy has been over which dicotyledonous angiosperm taxa are most primitive (i.e., possess the fewest number of derived features): the Hamamelidae (roughly equivalent to Amentiferae), which have generally small, simple, unisexual flowers with reduced or lacking perianth, *or* the Magnoliidae (previously called Ranales), which possess larger, bisexual flowers often with larger, undifferentiated perianths and numerous stamens and carpels (see below). Studies of tracheary element anatomy demonstrated that taxa of the Magnoliidae have either tracheids only or a primitive vessel type of elongate, narrow cells with tapering end walls and scalariform perforation plates. Because members of the Hamamelidae generally have more advanced vessel structural types, it has been hypothesized by the doctrine of correlation that their smaller, structurally simple, unisexual flowers are derived and therefore evolutionarily reduced from the ancestral Magnoliid floral type. This belief is still largely accepted today and is substantiated by additional evidence. However, as previously discussed, the doctrine of correlation is valid only if the evolutionary modification of one feature affects that of another feature. More extensive studies of carefully defined taxa of the Hamamelidae and Magnoliidae, utilizing a number of characters and phylogenetic methodology, are needed before a full resolution of relative ancestry of these features can be established.

II. Floral Evolution

The evolution of the flower was undoubtedly one of the major factors determining the tremendous success and diversity of the angiosperms. Accordingly, the study of floral structure and evolution has been of primary value in elucidating angiosperm phylogeny and classification. A flower is classically defined as a determinate shoot bearing modified leaves, the latter including perianth parts, male sporophylls (stamens), and female sporophylls (carpels). More accurately stated, the parts of a flower are believed to be homologous with vegetative leaves as exogenous, vascularized appendages. (Also, an intergradation between leaves and bracts is exhibited in many taxa which supports the hypothesis that leaves and bracts are homologous as bifacial vascularized structures.) The similarities between leaves and floral parts are the result of uniquely evolved structural and/or regulatory genes that developmentally cause those similarities. Perhaps the most obvious homology of structure in flowers is that between perianth parts (calyx and corolla) and vegetative leaves. Both arise as exogenous primordia in a specific phyllostachy and both mature as bifacial, vascularized appendages bearing stomata. Sepals are especially leaflike, exhibiting numerous morphological and anatomical resemblances to leaves, including generally being photosynthetic. It is reasonable to hypothesize, therefore, that the similarities between leaves and perianth are the result of a common evolutionary event; many of the same genes, in fact, may function in the development of both structures.

The relative ancestry of various perianth types can be assessed using the doctrine of association. If leaf and perianth are homologous as exogenous, bifacial appendages and if leaves existed before the evolution of the perianth (as is well evidenced from the fossil record), then the most primitive perianth type is that which most resembles the leaves of a vegetative shoot. Among extant angiosperm taxa, however, it is not obvious which perianth type is most ancestral. Classically, the primitive perianth type has been viewed as that which lacks differentiation between calyx and corolla, possesses an irregular number of unfused parts, shows radial symmetry, and has a gradual transition between outer, green, leaflike parts and inner colored parts (Figure 10.13a). This view is supported largely by the doctrine of correlation in that certain angiosperms having a suite of primitive characters (including elongate scalariform vessels and apocarpous carpels with open sutures, see below) often possess such an undifferentiated perianth with numerous parts. However, some angiosperm taxa with otherwise primitive features possess a small, regular number of floral parts. In addition, certain extant taxa have a well-differentiated calyx and corolla but show a gradation between outer petal-like structures and inner fertile stamens; the corolla of these taxa (and their relatives) evolved by a modification of staminodes, not by modification of an ancestrally undifferentiated perianth. Even the most recent fossil evidence provides no conclusive evidence as to the morphology of the most

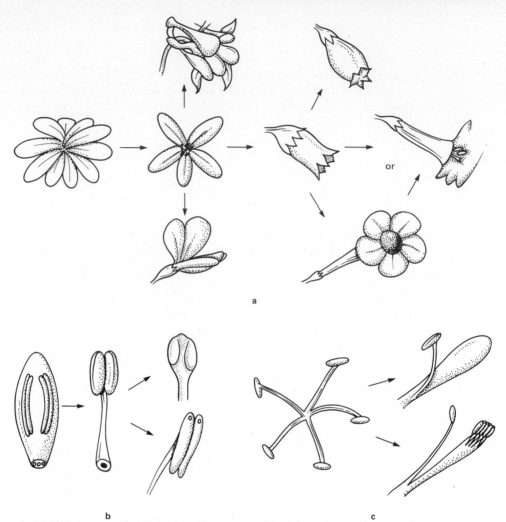

Figure 10.13   a. Morphocline of selected extant angiosperm perianth types. Hypothetical most primitive perianth type (far left) consists of numerous, undifferentiated sepals. Derived features include the evolution of a regular number of parts, differentiated into sepals and petals; zygomorphy of calyx or corolla; and fusion of sepals or petals into tubes or spurs of a great variety of form. **b.** Morphocline of extant stamen types, illustrating the intergradation between the flattened, three-nerved structure with surficial or sunken thecae (left) and the typical stamen composed of well-differentiated anther and filament. Note additional modification in anther dehiscence, from longitudinal to valvular (upper right) or poricidal (lower right). **c.** Various specializations of stamen fusion from the ancestrally distinct and free stamens (left) to fusion of stamens to other floral parts (e.g., epipetaly, upper right) or to one another (e.g., diadelphy, lower right).

ancestral perianth type. Although it may be unclear as to exactly which perianth structural type is most ancestral, derived perianth features are generally accepted to include: (1) the evolution of two distinct whorls (calyx and corolla), (2) the establishment of a regular number of parts, (3) the fusion of sepals or petals, (4) zygomorphy, and (5) the evolution of elaborate calyx or corolla spurs, tubes, and lobes (Figure 10.13a). These derived features apparently evolved in response to selective pressure toward a variety of specialized pollination mechanisms.

Stamens are the male reproductive parts of a flower and generally consist of a supporting structure bearing two pairs of microsporangia (each pair usually developing into a unilocular theca by ontogenetic breakdown of the intervening septum). The great majority of stamens consist of a delimited anther, composed of two longitudinally dehiscing thecae, to which is attached a terete to slightly flattened, uninerved filament. However, the stamens in several taxa consist of a broad, flattened, generally three-nerved supporting structure on one side of which are borne (either surficially or embedded) the two thecae. If it is hypothesized that stamens are homologous to leaves (microsporophylls), then, by the doctrine of association, the most ancestral stamen morphology is that which most resembles a leaf. Therefore, the broad, flattened, three-nerved stamen types with surficial or embedded microsporangia have generally been accepted as the most ancestral stamen type; if true, then evolution has generally proceeded toward a transformation of the stamen-supporting structure into a distinct filament (Figure 10.13b). (However, it has been argued that the flattened three-nerved stamen type is a derived feature, highly specialized in beetle pollination.) Additional derived features in stamen evolution are the development of specialized dehiscence mechanisms (e.g., valvular or poricidal dehiscence; Figure 10.13b) and of various types of stamen fusion, both to other floral parts (e.g., epipetaly; Figure 10.13c) and to one another (e.g., monadelphy, diadelphy; Figure 10.13c).

The carpel is perhaps the major diagnostic feature of flowering plants. It is usually defined as the unit of the gynoecium, consisting of a modified, conduplicate (or revolute) megasporophyll with fused or interlocking margins enclosing one or more ovules. Carpel evolution was apparently an important factor in the success and diversity of the angiosperms, allowing for: (1) effective fertilization by transport of pollen grains to the stigma and growth of a pollen tube; (2) promotion of outbreeding by sporophytic intraspecific and interspecific self-incompatibility; (3) promotion of outbreeding by insect pollinators via the evolution of specialized structural mechanisms (associated with the perianth and stamens); (4) protection of ovules from predation; and (5) seed dissemination via the evolution of numerous mechanical and animal fruit dispersal mechanisms. Because angiosperms appear stratigraphically later than the gymnosperms (the latter including a number of groups, e.g., Pteridosperms, Cycads, Conifers, and Cordaites), angiosperms are presumed to have evolved from a preexisting gymnospermous ancestor. However, the fossil record has to date yielded only a very sketchy notion of the evolution of the angiospermous carpel from a noncarpellate gymnospermous ancestor. Numerous theories (none well substantiated) have been proposed to account for the transition to an enclosed carpel from various hypothetical or known gymnospermous reproductive organs.

Carpels show basic similarities in structure and development to leaves. Both generally are initiated as exogenous primordia (in a particular phyllotactic position) and develop as dorsiventrally flattened, vascularized appendages. The ontogeny of a typical carpel shows a gradual transition from a leaflike structure by fusion of margins and apical extension growth, forming the style and stigma (Figure 10.14a). Thus, it is likely that carpels and leaves are homologous in the sense that the similarities between them are the result of gene assemblages that have a common evolutionary history. (The carpels of some taxa show no evidence of fusion of a dorsiventrally flattened structure; these are usually

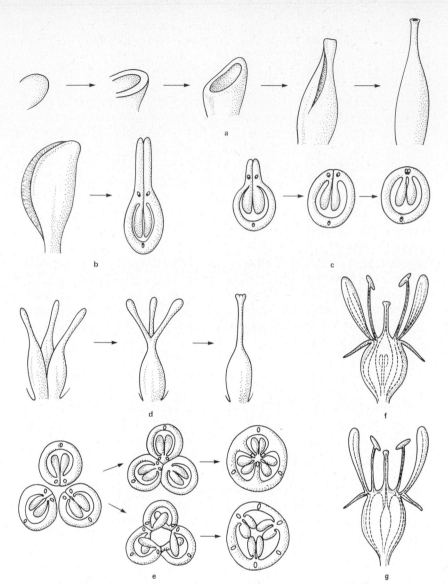

Figure 10.14   a. Ontogenetic sequence of a typical unicarpellate carpel. Note initially bifacial and inrolled structure; mature pistil develops by convolution and fusion of margins and extension of style. **b.** Carpel of *Degeneria* in outer view (left) and cross section (right), showing absence of style and extended flanges of carpellary margins, with interlocking hairs. Note, in cross section, vasculature composed of one median (dorsal) and two lateral (ventral) traces. **c.** Morphocline of carpel structure representing a hypothetical evolutionary series. Note transformation from unfused carpel margins with extended flanges (left) to fused carpel margins with anatomical commissure and absence of flanges (middle) to completely fused carpel margins lacking a commissure and having partially fused lateral traces (right). **d.** Morphocline of intercarpellary fusion types, from basally fused ovaries (left) to fused ovaries but unfused styles (middle) to completely fused ovaries and styles (right). **e.** Morphocline of intercarpellary fusion (cross-sectional view) showing intergradation from unfused ovaries with commissures (left) to central fusion resulting in axile placentation (upper middle) to marginal fusion resulting in parietal placentation (lower middle). Note further loss of lobing, loss of commissures, and fusion of ventral bundles (upper and lower right). **f.** Longitudinal section of a flower with an appendicular inferior ovary, showing vertical traverse of vascular traces peripherally to sepals, petals, and stamens. Note carpellary traces entering style. **g.** Longitudinal section of flower with a receptacular inferior ovary, showing involution of vascular traces leading to carpels.

accepted to be extreme specializations of an ancestrally more leaflike structure.)

The relative ancestry of carpel types among extant taxa can be assessed using two criteria. From the doctrine of association, the primitive carpel type (present in the earliest carpellate angiosperm) is that which most closely resembles the structure from which a carpel is derived, that is, a sporophyll. More strictly stated, the primitive carpel is that most closely reflecting the structural similarities between carpels and leaves, as it is these similarities that are presumed to occur by common ancestry of gene assemblages. In addition, ontogenetic criteria may be used to argue that early developmental stages may be homologous to adult ancestral features. Thus, among the angiosperms, the most primitive mature carpel type may be that which resembles juvenile stages of more derived carpel types (e.g., the intermediate stages in Figure 10.14a). Using either or both of these criteria, the primitive carpel is generally believed to be that which is most leaflike, lacking a differentiated style and having relatively unfused margins (e.g., composed of interlocking hairs as in *Degeneria;* Figure 10.14b). In addition, because taxa with more leaflike carpels possess three major traces—one median (or dorsal) bundle and two lateral (or ventral) bundles—the ancestral carpellary vasculature is generally typified as three-nerved. Derived structural features in carpellary evolution include deviations from the ancestral leaflike condition, such as: (1) complete developmental fusion of carpel margins, resulting in a closed carpel; (2) differentiation of the carpel into ovary, style, and stigma; (3) loss of commissures (evidence of original margins of the carpel); and (4) fusion of vascular bundles, particularly the lateral traces (Figure 10.14c).

Two features of interest in the evolution of carpels are the number of carpels per flower and fusion of carpels, both to one another and to other floral parts. In flowers with more than one carpel, the absence of fusion between them (i.e., an apocarpous gynoecium) is widely thought to be ancestral relative to fusion between carpels (syncarpous gynoecium). This reasoning follows the doctrine of association; each carpel of an apocarpous gynoecium is more similar, developmentally and anatomically, to a leaf than is a carpel in a syncarpous gynoecium. The relative ancestry of the *number* of carpels per flower (ranging from one to hundreds) is unclear. Traditionally, flowers with numerous unfused carpels have been accepted as most ancestral; those with a few or one carpel have been viewed to be derived by reduction. Support for this hypothesis is cited from the doctrine of correlation. Many members of the Magnoliid (Ranalean) complex, which possess a number of presumed ancestral character states (e.g., long, narrow, scalariform vessels and/or carpels with incompletely fused margins), have apocarpous gynoecia with numerous carpels per flower. However, because several other taxa with a suite of primitive features possess only one or a few carpels per flower and because of the paucity of information from the fossil record, the relative ancestry of carpel number is largely speculative and deserving of additional systematic study.

Derived features of carpel evolution include greater degrees of fusion. The ontogenetic fusion between two or more carpels (resulting in a syncarpous gynoecium) is believed to be derived over absence of fusion. The greater the degree of fusion, from synovarious (fusion of ovaries only) to synstylovarious (fusion of ovaries and styles, but not stigmas) to totally syncarpous, the more derived is that state (Figure 10.14d). Initial evolutionary fusion between carpels bearing a number of marginal ovules may theoretically result in either of two placentation types: axile, if fusion of each individual carpel margins occurs prior to fusion between carpels, or parietal, if fusion between carpels occurs by fusion of carpel margins (Figure 10.14e). Derived features of a syncarpous gynoecium include: (1) the evolution of additional placentation types (e.g., from axile or parietal to free central) by loss or ingrowth of septae, elaboration of placentae, or a change in ovule number; (2) the loss of ovary or stylar commissures (boundaries between originally distinct carpels), and (3) fusion of vascular bundles, particularly between adjacent lateral (ventral) bundles (Figure 10.14e).

A final derived feature of carpel evolution is the inferior ovary. An inferior ovary (equivalent to an epigynous androperianth insertion) has been viewed as (and superficially appears to be) a fusion product between a superior ovary and the calyx, corolla, and androecium. The traces leading to sepals, petals, and stamens traverse the outermost tissue of the apparent ovary wall, as if these floral parts were fused to the ovary in layers. (In actuality, the inferior ovary developmentally arises not by congenital fusion, but by the relatively accelerated growth of peripheral meristematic tissue located at and beneath the initiation region of calyx, corolla, and androecium. Thus, because of the precocious growth of this surrounding tissue, the ovary appears fused to the androperianth or embedded within the receptacle.) The inferior ovary is, in general, thought to be derived relative to the superior ovary because virtually all examples of inferior ovaries exhibit few to no ancestral features (such as unfused margins or an undifferentiated style). However, it seems clear that the inferior ovary has evolved independently in a number of angiosperm lineages. Numerous well-defined (and presumably monophyletic) groups include representatives with superior or inferior ovaries; the most parsimonious explanation (assuming that these groups are indeed monophyletic) is that the inferior ovary evolved independently in these groups. Developmental evidence corroborates the independent evolution of the inferior ovary. For example, most inferior ovaries are appendicular, that is, the androperianth traces traverse vertically in the peripheral tissue of the flower (Figure 10.14f). However, some taxa (e.g., the Cactaceae) have a receptacular inferior ovary: The peripheral traces curve inward toward the locules before traces are initiated to the androperianth (Figure 10.14g). The difference between the receptacular versus the appendicular inferior ovary types represents a distinct difference in developmental origin, thus supporting a hypothesis of independent evolutionary origin (i.e., of nonhomology).

III. Conclusions

In conclusion, the principles and methodology of phylogenetic systematics can provide useful insight into the evolution of plant structure. However, the study of structural evolution is important not just for an assessment of what have been the historical changes in cells, tissues, and organs. A knowledge of *what* changes have occurred in the phylogeny of plants is essential to understanding *why* they have occurred in terms of possible adaptive significance. For example, the evolutionary changes in tracheary element anatomy are probably the result of adaptations to changing environmental regimes or habits of growth; changes in corolla structure are in most cases clear adaptions for specialized attractant or pollination systems; the evolution of carpel structure is undoubtedly related to adaptations with regard to self-incompatibility systems, pollination syndromes, ovule nutrition and protection, and fruit and seed dispersal. Therefore, the beginning student should place an emphasis on the supposed function of the structures analyzed. Consideration of functional significance can provide a means of ascertaining the possible selective pressures directing the evolution of structure and may add considerable meaning and insight to the study of phylogeny.

SUGGESTED READING

Bremer, K. and H.-E. Wanntorp. 1981. The cladistic approach to plant classification. In Advances in Cladistics, eds. V. A. Funk and D. R. Brooks. The New York Botanical Garden, New York.

Burger, W. C. 1983. Heresy revived: the monocot theory of angiosperm origin. Evolutionary Theory 5:189–225.

Crisci, J. V. and T. F. Stuessy. 1980. Determining primitive character states for phylogenetic reconstruction. Systematic Botany 5:112–135.

Dahlgren, R. and F. N. Rasmussen. 1983. Monocotyledon evolution: characters and phylogenetic estimation. In Evolutionary Biology, Vol. 16., eds. M. K. Hecht, B. Wallace, and G. T. Prance. Plenum Publishing Corporation, New York.

Eldredge, N. and J. Cracraft. 1980. Phylogenetic Patterns and the Evolutionary Process. Columbia University Press, New York.

Farris, J. S. 1970. Methods for computing Wagner trees. Systematic Zoology 19:83–92.

————. 1983. The logical basis of phylogenetic analysis. In Advances in Cladistics, Vol. 2. eds. N. I. Platnick and V. A. Funk. Columbia University Press, New York.

Gould, S. J. 1977. Ontogeny and Phylogeny. Belknap Press, Cambridge.

Hennig, W. 1966. Phylogenetic Systematics. University of Illinois Press, Urbana.

Meacham, C. A. 1981. A manual method for character compatibility analysis. Taxon 30:591–600.

Nelson, G. and N. I. Platnick. 1981. Systematics and Biogeography: Cladistics and Vicariance. Columbia University Press, New York.

Stevens, P. F. 1980. Evolutionary polarity of character states. Annual Review of Ecology and Systematics 11:333–358.

Wiley, E. O. 1981. Phylogenetics: The Theory and Practice of Phylogenetic Systematics. John Wiley & Sons, New York.

SUMMARY FOR PHYLOGENY AND STRUCTURAL EVOLUTION OF PLANTS

Definitions for Phylogeny   *Phylogeny* is the genealogical history of a group of organisms, representing ancestor/descendant relationships and portrayed as a branching diagram, the *cladogram.*

Purpose of the Study of Phylogeny   To group and classify organisms based on genealogical history; to trace the structural evolutionary changes that have occurred in taxa and assess the possible adaptive significance of those changes; to trace the biogeographical history of a group of taxa.

Operations in the Study of Phylogeny   To verify the monophyly of unit taxa and the group as a whole; to select, define, and code characters and character states; to assess homology of characters and character states; to hypothesize morphoclines for the states of a character and determine polarity; to construct a cladogram, sequentially grouping taxa by shared derived features.

Basic Premise of Phylogeny   That the evolutionary history of taxa, both in terms of genealogical relationships and structural modifications, can be determined by an analysis of the features of those taxa and by identifying monophyletic groups based on their sharing one or more derived character states.

Fundamental Principles of Phylogeny

1. Organisms are conceptually divided into features, defined as characters and character states.
2. Similarly termed features of organisms are studied and assessed for homology, a hypothesis of common ancestry.
3. Intergrading features are arranged in a transformation series, representing a presumed evolutionary sequence.
4. The relative ancestry of features is determined using various criteria.
5. Taxa that share derived features are hypothesized to do so by common ancestry; these taxa together form a monophyletic group, including a common ancestor and all (and only all) descendants of that common ancestor.
6. The phylogeny of a group of organisms can be deduced by sequentially grouping taxa by shared derived features into a branching diagram, the *cladogram*.
7. The classification of organisms may be based directly on their phylogenetic history.
8. The evolution of structural features of organisms can be deduced based on the presumed phylogenetic relationships of those organisms.

Guiding Principles

1. In studying the phylogeny of a group of organisms, both unit taxa and the group as a whole should be assessed for monophyly; if monophyly cannot be ascertained, taxa should be subdivided into entities that are monophyletic.
2. The results of a phylogenetic study are totally dependent on the selection and definition of characters and character states.
3. Characters and character states must be assessed for homology; similarly termed features are, in phylogenetic analyses, treated as homologous, whether or not that is the intention of the investigator.
4. Determining polarity by means of the doctrine of correlation or the common ground plan may be invalid in many cases.
5. In determining polarity by out-group comparison, the choice of the out-group should be carefully substantiated.
6. In the construction of a cladogram, taxa and character state changes should be mapped and assessed; the occurrence of numerous homoplasies may warrant a reevaluation of the data.

Basic Assumptions for Phylogeny

1. A hypothesis of phylogeny can be made only by an analysis of the features of organisms.
2. The possession of a shared derived feature by two or more taxa leads to the assumption, in the absence of contradictory data, that the taxa possess that feature by common evolutionary origin.
3. The most parsimonious cladogram (having the least number of homoplasies) is accepted to portray genealogy most accurately as it assumes the least number of ad hoc hypotheses.

QUESTIONS ON PHYLOGENY AND STRUCTURAL EVOLUTION

1. What is phylogeny? What are the steps in the determination of phylogeny? What is its utility in the biological sciences?
2. What must be considered in selecting the groups of taxa to be analyzed in a phylogenetic study? Why?
3. What problems exist in the selection and definition of characters and character states?
4. What is homology? What problems arise in the determination of homology? How does homology between features in different taxa compare with homology between features in the same individual?
5. What actually is represented by a morphocline? A cladogram? The nodes of a cladogram?
6. What is character polarity? Name and describe five methods for determining polarity, listing the advantages and/or pitfalls of each method.
7. How may cladograms be constructed? Define monophyletic taxon, synapomorphy, autapomorphy, sister group.
8. What is a network? Why are networks constructed?
9. What is the principle of parsimony? How does this principle relate to the grouping of taxa based on shared derived character states?
10. What is a polychotomy? What are two possible reasons for the occurrence of polychotomies in cladogram construction?
11. What is reticulation? What problems arise in the detection of reticulation events?
12. What is character weighting? What criteria may account for weighting characters or character states? How is character compatibility analysis a type of weighting procedure?
13. How is a consideration of ontogeny important in phylogeny construction? What is paedomorphosis? How is it detected in a phylogenetic analysis?
14. What are the major problems in the determination of phylogeny? What are the advantages of these methods?
15. What is structural evolution? How is it determined?
16. Describe the generally accepted trends in the evolution of: (1) tracheary elements; (2) corollas; (3) stamens; and (4) carpels. What is the rationale, in terms of phylogenetic systematics, for the recognition of these trends? What problems exist in assessing large-scale trends of structural evolution?

EXERCISES IN PHYLOGENY

1. Select a subgenus, genus, tribe, subfamily, or family of plants. From available materials (e.g., herbarium, live, or liquid-preserved specimens), study and describe the morphology of all or selected representatives using dissecting microscopes and equipment.
   a. Select and list the characters and character states for each taxon. For any taxa with similarly termed character states, assess their homology. What is the basis for your assessment?
   b. Arrange the character states of each character in an intergradation series (a morphocline), representing a hypothesized evolutionary series.
   c. Select what you believe to be one or more most closely related out-group taxa. What is the basis for your selection? Hypothesize polarity for the morphocline using out group comparison or some other method.
   d. Select a final list of characters and character states and numerically code these in a character × taxon matrix. Review the bases for omitting some characters (e.g., because of uncertainty in the morphocline, homology, or polarity) or for possible weighting of certain characters.
   e. From the character × taxon matrix, attempt to construct a cladogram by sequentially grouping taxa by shared derived features. If a number of homoplasies are apparent, construct the

most parsimonious cladogram by hand or by using the algorithm of Appendix B (or, e.g., a comparable computer program).

  f. Map out all indicated character state changes on the cladogram. Reassess each of these in terms of original hypotheses or assumptions and repeat the procedure of character analysis if needed. Consider: (i) uncertain character assignments warranting further research; (ii) indicated trends of structural evolution and their possible adaptive significance; and (iii) the evolutionary changes of superposed characters not included in the analysis.

  g. Obtain as much data as possible from additional fields of evidence (e.g., anatomy, embryology, karyology, palynology) by consulting the literature or, perhaps, from original studies. Reconstruct the character × taxon matrix and cladogram using at least some of these new data. Was the original cladogram altered? If so, which cladogram is to you most acceptable and why? How has the addition of new characters altered your original suppositions on character weighting?

2. Select a research article on the phylogenetic analysis of a plant group from a botanical journal, such as *Systematic Botany* or *Taxon*. Critically review the article in terms of each step of phylogenetic methodology and principles (see concluding paragraph of A. Principles and Methodology in this chapter). Ask yourself whether you can believe the cladogram presented and why or why not.

3. Construct the most parsimonious cladogram(s) from the following character × taxon matrices (if necessary, use the algorithm of Appendix B):

	1	2	3	4	5
A	0	0	1	1	1
B	1	2	0	1	1
C	0	0	2	0	0
D	0	1	0	1	2
E	0	1	0	1	3
F	1	2	0	1	1
ANC	0	0	1	0	0

	1	2	3	4	5	6
V	1	0	2	1	0	1
W	1	1	1	0	0	0
X	0	0	2	0	0	0
Y	0	0	1	0	1	0
Z	1	0	1	1	1	0
ANC	0	0	0	0	0	0

	1	2	3	4	5	6	7
A	0	1	1	0	0	1	0
B	0	0	1	0	1	1	0
C	1	0	1	0	1	1	0
D	1	0	0	1	0	0	1
E	0	0	0	1	0	0	0
F	1	0	0	0	0	0	1
G	0	0	1	0	0	0	0
H	1	0	0	1	1	0	0
ANC	0	0	0	0	0	0	0

4. A character having three or more states can be coded as two or more two-state characters. (This is done, for example, with all nonlinear morphoclines and in character compatibility analysis.)

  a. Recode characters #1, #2, and #3 into two-state characters, given the following character *X* taxon matrix and morphoclines:

	1	2	3
A	0	0	0
B	1	1	2
C	1	2	3
D	2	3	4
ANC	0	0	1

Character #1: $0 \to 1 \to 2$

Character #2: $0 \to 1 \begin{smallmatrix} \nearrow 2 \\ \searrow 3 \end{smallmatrix}$

Character #3: $0 \leftarrow 1 \begin{smallmatrix} \nwarrow 2 \to 3 \\ \searrow 4 \end{smallmatrix}$

  b. How would the recoding into two-state characters affect the character *X* taxon matrix in terms of character weighting?

part two

SYSTEMATIC INSTITUTIONS

Fundamentally, systematics is a perpetually self-correcting, sporadically evolving endeavor that accepts no axiom as absolute or immutable.

chapter 11

The Botanical Library and Taxonomic Literature*

Systematists use the botanical library as an information source, documentation center, and service facility. The library collections are utilized for obtaining information on nomenclature, characterization, identification, classification, and relationships of taxa of concern; for background perspective on a systematic study; and for keeping abreast of the developments within the field of taxonomy.

The botanical library is an information system† composed of printed materials, electronically stored information, films, drawings, paintings, and other items. The library is a storehouse of books, periodicals, magazines, newspapers, dissertations, movies, tapes, disks, photographs, and so on. This institution is a documentation center of human activities and a data storage system of plant-related subjects. It is also a service facility that enables people to use these information and documentation systems most effectively for the acquisition of pertinent data and information on systematic subjects of interest. Taxonomic literature is a form of documentation and communication for data-information-knowledge, concepts-processes-principles, questions-hypotheses-premises, conclusions-solutions-proofs, and numeric-graphic-pictorial representations as related to the various aspects of systematics.

Through the long course of human history, civilized people have evolved and institutionalized two great, complementary systems for storing and retrieving their accumulated wisdom (their record of history and scientific and cultural achievement). These two

---

*Introduction adapted from Shetler, S. G. in Radford, A. E. et al., 1974, Vascular Plant Systematics, Harper & Row, New York, Chapter 33, pp. 793–801.

†Information system, library or otherwise, may be defined in general terms according to Shetler (1974), "as a total strategy and methodology for gathering, storing, retrieving, and exchanging (interchanging, transferring) information to serve a specified purpose or set of purposes."

colossal systems or institutions, which epitomize civilization and subsume all other lesser systems, are the *library* and the *museum*. People have evolved the first to cope with their written and, more recently, spoken record; and the second to cope with their material record, including artifacts and voucher specimens of nature. The pictorial record—illustrations, photographs, paintings—is divided between these two systems; each documents and supports the other, and each is characterized by a vast data bank.

To the library, in this broad sense, belong all of humanity's written archives, published and unpublished, including the innumerable files of vital statistics such as birth, health, job, property, school, military, bank, and credit records as well as tape and disk recordings and many illustrations, photographs, and paintings. The library, as the repository for the world's literature and the system for exchanging it, symbolizes communication through publication as it has developed in the 500 years since the invention of movable type. It comprises many systems within the total system. Every book or journal in itself can be viewed as a kind of information system, containing selected, organized, and evaluated data displayed and indexed in a particular format for fast, convenient retrieval. Customarily, the printed page is regarded in static terms as being merely a source of information, but any volume also constitutes a means of communicating or transferring information and as such satisfies the criteria of a dynamic system.

Similarly, the museum as an institution is a hierarchy of systems. The functions are storing, curating, displaying, loaning, and exchanging collections of objects and specimens. Apart from mere curiosities, these collections not only vouch for the published record but also provide a vast reservoir of untapped data for future generations to study. There are many classes of museums, for example, art, historical, and technological. Museums also occur within museums, such as herbaria (singular, herbarium) within natural history museums. The system also includes countless informal and private collections beyond the walls of the great formal institutions that constitute the visible museum establishment.

The library and the museum (see Chapter 12) are as generalized as any systems people have been able to conceive. They exist for the benefit of society as a whole and no individual could ever need or be capable of taking advantage of either system in its entirety. The particular needs of individuals are met through specialization within the systems. In the course of history, numerous special-purpose systems have evolved within the larger systems to accommodate the specific needs of the individual disciplines and branches of human thought. Through the years, systematic botany, among the scientific disciplines, has developed its own share of special-purpose information systems. New ones appear constantly as new needs are perceived, but few persist for long as measured by the centuries. The special prominence of historical scholarship in taxonomy gives a special significance to the taxonomic libraries and museums (herbaria) where the essential objects of this scholarship—the literature and the specimens—are preserved and to the systems devised for retrieving information from these repositories.

Of the many types of taxonomic publications that botanists have produced, certain ones stand out in their importance as information retrieval devices: (1) floras and taxonomic monographs, (2) literature indexes, and (3) name indexes and nomenclators. These more durable information systems require further elaboration here.

The flora brings together selected information about the plants of a given geographical region, while the monograph collects the essence of what is known on a worldwide basis about the systematics of a given taxonomic group of plants. In both cases, the information is organized and formatted to answer a time-tested set of predefined questions or to test hypotheses, and effective use is made of a limited vocabulary, as standardized as is ever achieved in written communication. The binomial names are used as the ultimate addresses for the information, and these are grouped hierarchically according to a systematic classification. Diagnostic traits are presented in the form of dichotomous keys to facilitate identification of plants. In short, as a class of publications, floras and taxonomic monographs constitute one of the most effective specialized information systems devised in science. There has been a perennial need for them.

Various systems for indexing and abstracting the taxonomic literature by subject have been introduced through the years, and, although old systems keep giving way to newer ones, some form of literature retrieval continues to survive, out of sheer necessity. Notable among present systems, from the viewpoint of North American vascular plant systematists, are the *Bibliography of Agriculture* and *Biological Abstracts,* which cover the broad areas of agriculture (including botany) and biology, respectively, and have been running for many years, and the various bibliographic indexes being published by the International Association for Plant Taxonomy.

One of the continuing problems that the systematic biologist faces is keeping track of all the new scientific names that are published. For many years, taxonomists of flowering plants have been served by two key name-indexing systems that cite the original publications: *Index Kewensis,* which is produced in book form at the Royal Botanic Gardens, Kew, England, and covers the world; and the *Gray Herbarium Index,* which is produced in card form at Harvard University, Cambridge, Massachusetts, and covers only the western hemisphere. In the realm of nomenclators, special mention should be made of J. C. Willis's *A Dictionary of the Flowering Plants and Ferns,* now in its eighth edition as revised by H. K. Airy Shaw. A nomenclator seeks to establish the acceptable names, which this *Dictionary* does for family and generic names. For a single volume, Willis is a most valuable information retrieval aid for the vascular plant systematist.

Finally, it should be noted that phylogenetic and other taxonomic systems of classification are themselves a type of metaphysical information system, because they enable the systematist to organize knowledge for efficient retrieval, whether in his or her mind, on paper, or in the herbarium, by grouping like organisms and all information pertaining to them according to their intrinsic relationships, that is, their shared characteristics. In another sense, of course, the taxonomic classification is not the system but a part of the system, namely, the scheme whereby the data bank is arranged or structured. Regardless of the emphasis, a taxonomic system of classification is an intellectually thrifty device for storing and retrieving information about plants.

RECENT AND FUTURE INFORMATION SYSTEMS

For 500 years, the printed page has ruled as the medium of information exchange. Now the high-speed electronic computer introduces new media and methods to information handling, making a whole new mode of operation possible. The computer can free the written word from the physical limitations of the printed page. In the data bank of the future, all data can be equally accessible, and virtually any comparison, correlation, or combination can be made at will. Promising new ways of using data are possible, as, for example, use of computer programs to converse directly with the machine to identify plants by matching the characteristics of unknown specimens with the stored values of known species. By taking advantage of the machine's power to permute data and to afford multidimensional querying, the user can procure a completely new functional relationship for his or her data.

Just as scholars first took to writing and then to printing, today's scientists communicate with the aid of machines. As for systematic botanists, surely concern for the future if not the present welfare of their science will impel them to exploit the computer fully as a tool for storing and retrieving literature and original data. Such exploitation must be done in concert with development in other scientific disciplines, of course, because plant systematists' retrieval problems are not unique.

Today's taxonomists can hardly cope with the new literature of their own specialty as it appears; yet they continue to publish, casting their contributions into the swelling sea of literature in the hope that they have truly added to the store of knowledge. With every passing year, however, the chances diminish that their contributions will be found by those who need them. Likewise, no taxonomist can claim realistically that he or she is able to take full advantage of the mass of information locked up among the millions of specimens in the world's herbaria; yet every year taxonomists go into the field to collect more specimens for the stockpile.

The flora or other taxonomic treatise of the future will be a dynamic data bank written on the pages of a computer memory that are available via telecommunications to the entire world and from which can be printed in hard copy any number of subsets and permutations of data for individual or mass consumption. The very process of research and communication will be altered. Today, research culminates in publication. Information retrieval, if any, comes after publication. Tomorrow, the results of research will go directly to the data bank, and conventional publication, if any, will come as output from the data bank. Today, the would-be user of scientific information must wait until the contributor has collected enough data to justify publication. Tomorrow, the user will be able to ask the system at any time to compile the latest information on a given topic, and the contributor will be able to submit data—a datum at a time if he or she so chooses.

Until the electronic library and data bank of the future materialize, systematists will continue to use the present collections in the major university libraries at Harvard, Yale, Illinois, Michigan, California at Berkeley, and Texas, among others. Large botanical libraries associated with museums and botanic gardens such as those at the British Museum, the New York Botanical Garden, Missouri Botanical Garden, and the Gray Herbarium will be focal points of taxonomic activity for years to come. Systematic research and scholarship will long depend on the collections in the great national libraries, for example, the Library of Congress in Washington, the Academy of Sciences in Leningrad, the National Library of Peking, the British Museum in London, and the Bibliothèque Nationale in Paris.

Interlibrary loans, the production of microfiche (photographically reduced material on transparencies) editions, photocopying and microfilming of selected materials, and computer-based bibliographic searching by information specialists will continue as standard library procedures of great value to systematic research and scholarship.

A. BASIC CONCEPTS RELATED TO THE BOTANICAL LIBRARY AND TAXONOMIC LITERATURE

The terms data and information have similar connotations in many contexts and often are used interchangeably, but they are not synonymous. *Data* are the individual measurements and observations expressible as either numbers or words that constitute the factual elements of knowledge. *Information* is the organized and evaluated data—data in context that convey a message. Information is a function of context, thus the same data organized in different ways may convey different information. Data are compiled whereas information must be authored. *Knowledge* is an organized body of information that usually involves comprehension and understanding consequent to the acquisition and organization of the body of information about a subject.

Literature retrieval is the process of identifying and locating by manual or automated means books, journals, articles, or other documents within the total mass of published literature that are of interest to a particular user. *Document retrieval* is a more inclusive concept than literature retrieval, since it covers published as well as unpublished material such as manuscripts, research notes, illustrations, films, tapes, and specimens. *Data retrieval* is the process of identifying, locating, and retrieving by manual or automated means specific facts in stored documents and specimen collections that are of interest to a particular user. Document systems present the user with references to the information desired, whereas modern computerized data systems present the user with the actual data or elements of information desired.

A *document* refers to a written item such as a book, article, manuscript, letter, or other types of objects (specimens, equipment). *Documentation*, in systematics, is the introduction of and/or reference to the written documents, specimens, and other elements used as evidence, source of information, or method of obtaining data in taxonomic studies. The purpose of documentation is the providing of a reference base and depository for the procedures, data, and information used in the systematic research, or, in general, the providing of a means of identification of the source and method for each entry or element in the information system. Documentation of evidence used in a study is necessary for clarity in transmission of information so that the set of evidence and methods used in the investigation can be reassembled and reassessed by subsequent workers in the subject. Documentation in research should be thorough enough so that original observations, analyses, procedures, conclusions, and proofs can be distinguished from previously published ones. Proper use of information and data in interpretation, extrapolation, and comparison depends on knowing the documentary sources and reliability of the information. A *documentation system* is a total strategy for specifying and identifying, referencing and citing, substantiating and authenticating, validating and verifying, and vouchering information used for a specific purpose in a study. Documentation for the included elements in a study should always be sufficiently complete and accurate enough to promote and encourage decision making at a high confidence level.

A *voucher* is a written document or specimen used for attesting, authenticating or verifying some information or justifying a conclusion or act. A *citation* is a quotation from a book or author that is usually a reference to an authority or precedent. A *reference* is a direction to a source of information such as Shetler, S. G. in Radford, A. E. et al, 1974,

Vascular Plant Systematics, Chapter 33, pp. 793–801. A *reference book* is a comprehensive publication consulted for facts or background information, such as an atlas, dictionary, encyclopedia, or glossary. A *bibliography* is a list of source materials used or consulted in writing a text or it may be a complete or selective list of readings on a particular subject.

B. USE OF THE BOTANICAL LIBRARY

Systematists use the botanic library as an information source, documentation center, and service facility. Students using the library as a ready reference source of information, for historical perspective on a taxonomic subject of interest, and/or for the acquisition of knowledge of the current status of their systematic field need to become familiar with the content and use of the facility.

I. Information Source

From this institutional storehouse of botanical materials, taxonomists procure specific information to verify a fact, authenticate the authorship of a paper, validate a procedure, substantiate a citation, ascertain the accuracy of a reference, or simply fill a void in a characterization or classification of a taxon. The acquisition of specific information usually involves a quick perusal of a catalogue, index, or bibliography for a pertinent reference to the information desired or a visit to an appropriate dictionary, glossary, atlas, map case, or encyclopedia for the information required. Students spend additional time searching and reading the literature for background information on a particular problem or for perspective on an honors thesis or research project. Professional systematists regularly examine periodicals and new publications to keep abreast of their field of interest.

II. Documentation Center

The botanical library is a repository for original manuscripts, first editions, rare publications, photographs, illustrations, methods, techniques, and many other archival items. The library is also the reference base for documentary evidence related to terms, concepts, ideas, questions, procedures, and equipment used in taxonomic studies. Collections such as specimens are usually documented in the herbarium. Most documentation in systematic research is based on primary sources of information in articles published in periodicals, since most new taxa, circumscriptions, original observations, successful experiments, and so on are printed therein. Some documentation of evidence is based on secondary sources such as texts, reviews, symposia, and encyclopedias. The documentation for terms, however, is usually from tertiary sources, such as dictionaries and glossaries. Traditionally, the published information in the library is used as documented data for original research in many types of systematic problems. In this sense, the library is a data storehouse for research as well as a documentation center.

III. Service Facility

The botanical library traditionally provides technical, public, and student-staff services for effective utilization of its resources. The botanical library procures new publications pertinent to the field of systematics and fills missing publication voids as needed for taxonomic training and research in active systematic programs within the institution. Library personnel normally catalogue incoming materials, prepare appropriate indexes, and provide the maintenance and record keeping of the collections. Most libraries supply a reference service for assisting patrons in finding needed materials. Some fill public and staff requests for information by photocopying, microfilming, aiding in computer searchings, or referring to some other source. An interlibrary loan service is usually available for materials not locally available.

Modern libraries with appropriate staffing help train students in the use of the library for systematic study by:

1. Introducing them to the most useful reference sources such as dictionaries, glossaries, map files, atlases, gazetteers, handbooks, manuals, encyclopedias, biographical directories, and indexes of various types—the ready references for information.
2. Acquainting them with the primary periodicals, serials, monographs, texts, and most recent bibliographies related to their field—the background perspective for study and keeping abreast of current developments within the subject.
3. Providing instruction on the proper forms and rules for making a bibliography and the methods for keeping up with the literature—mechanics of good research.
4. Analyzing the literature for sound scientific writing and for effective understanding of the material read—methods for sound scholarship.

The nature of plant systematics requires basic training in the use of the library, usually as a cooperative effort between library staff and systematic instructors.

C. TAXONOMIC LITERATURE AND BIBLIOGRAPHIC AIDS*

The descriptive nature of taxonomy, the type and method of documentation, and the principle of priority make it necessary for taxonomists to be bibliographers as well as plant scientists. The degree to which one is able to handle a bibliography will affect, in large measure, one's success as a taxonomist. The resolution of most taxonomic problems will involve some reference to the voluminous and often poorly indexed literature.

I. Classification Used in Libraries

Two systems of classification are currently used in libraries—the Dewey decimal and the Library of Congress (LC). In fact, many libraries are in the process of changing from the older Dewey system to the LC system, and familiarity with both systems is advantageous. The following lists include examples of some of the major categories of each system. For a comparison of the two systems and other pertinent information on library classification systems and theory and botanical bibliography, see L. H. Swift, 1970, *Botanical Bibliographies, A Guide to the Bibliographic Materials Applicable to Botany,* Burgess Publishing Company, Minneapolis, Minnesota.

*This section is adapted from Massey, J. R., in Radford, A. E. et al., 1974, Vascular Plant Systematics, Harper & Row, New York, Chapter 30, pp. 697–750.

Dewey decimal		Library of Congress	
Natural science	500	Natural history	QH
Biology	570	Botany	QK
Botany	580	Zoology	QL
Zoology	590	Agriculture	S
Agriculture	630	Forestry	SD
		Horticulture	SB
		Bibliography	Z

Since the LC system is so widely used, the following examples are given to acquaint the student with the disposition of taxonomic literature and the meaning of call numbers in this system.

a.   Call Number for a Specific Reference

Conrad, Henry S. 1905. The Waterlilies. A monograph of the genus *Nymphaea*.
Carnegie Institution of Washington.
QK 495.N97 C7

Science	Botany	Spermatophyta—Systematics Division	Systematic Group	Author
Q	K	495	.N97	C7

b.   General Groups and Examples

QK 1—PERIODICALS, SOCIETIES
QK 1.C2—California University
Publications in Botany
QK 1.C55—Chronica Botanica
QK 1.F4—Field Museum of
Natural History—Botanical
Series (Chicago Natural History
Museum—Fieldiana: Botany)

QK 11—INDICES,
NOMENCLATORS,
CHECKLISTS
QK 11.L45—Lemee's Dictionnaire
des Generes des Plantes
Phanerogames

QK 31—BIOGRAPHY (Individual)
QK 31.G8 D8—Dupree, A. H.
1959.   Asa Gray.

QK 91—BOTANICAL WORKS OF
LINNAEUS
QK 91.S69—Stafleu, F. A.
Linnaeus and the Linnaeans.

QK 95—CLASSIFICATION
(DISCUSSION AND
COMPARISON OF
SYSTEMS)
QK 95.L3—Lawrence, G. H. M.
1951. An Introduction to Plant
Taxonomy.

QK 96—NOMENCLATURE
QK 96.I58—International Code of
Botanical Nomenclature.

QK 495—SYSTEMATIC DIVISIONS
A–Z
QK 495.A17—Acer
QK 495.Q4—Quercus
QK 495.N97—Nymphaeaceae
QK 495.N97 C7—Conrad, H. S.
The Waterlilies.

QK 97—COMPREHENSIVE
SYSTEMATIC WORKS
QK 97.C2—de Candolle's Mono-
graphiae Phanerogamarum.

QK 97.E62—Engler's Das Pflanzenreich.

QK 101–474—GEOGRAPHIC DIVISIONS (FLORAS AND DISTRIBUTION)
QK 110—North America
  QK 110.A1—North America Flora.
QK 125—Southeastern U. S.
  QK 125.S64—Small, J. K. Manual of the Southeastern Flora.
QK 146—Alaska
  QK 146.H84—Hultén's Flora of Alaska and Yukon.
QK 211—Mexico
  QK 211.M3—Martinez, M. Catalogo de Plantas Mexicana.

QK 281—Europe
  QK 281.H4—Hegi, Flora von Mittel—Europa.
  QK 281.T8—Flora Europaea.
QK 321—Russia
  QK 321.A36—Flora URSS.

QK 520—532—PTERIDOPHYTA
  QK 524—Systematic Divisions
  QK 525—U. S.
  QK 527—Great Britain
  QK 529—Asia

Z—BIBLIOGRAPHY
  Z5356.T8R42—Rehder, A. Bibliography of Cultivated Trees and Shrubs

II. Classification of Taxonomic Literature with Selected References

a. Nomenclature

Nomenclature pertains to references dealing not only with rules of nomenclature (i.e., the Code) and works of value in solving nomenclatural problems but also with those indexes that include the names of species, genera, and families. References cited in this class should assist primarily in locating original descriptions, authors, place of publication, and synonymy and, secondarily, in finding information on monographs, revisions, distribution, and so on.

Engler, A. 1954, 1964. *Syllabus* der Pflanzenfamilien. Ed. 12. Vol. 1. Bakterien bis Gymnospermen (H. Melchior & E. Werdermann, eds.) Vol. 2 Angiospermen (H. Melchior, ed.). Gebruder Borntraeger. Berlin. [This is a valuable source of many kinds of information. In addition to names, the systematic arrangement, description of orders, families, subfamilies, and so on, and notes on the number of genera and the number of species they include, there is an extensive bibliography including information on anatomy and morphology, phylogeny, floras, monographs and revisions; there are also valuable illustrations. This is a reference that should be widely used by students of systematics. Text in German.]

Gray Herbarium Card Index (Gray Card Index). [A card index (also available in volumes), issued quarterly to subscribers, to all new names and new combinations of vascular plants, except fossils, native to the western hemisphere. The starting date for inclusion is 1 January 1886 and the first issue was in 1894. Information includes (1) new taxa—the name, author, reference to place of publication, statement of range; (2) new combinations—new name, bibliographic reference, and basionym; (3) base cards—basionym, bibliographic references (with some exceptions), new name based on it, and other nomenclatural synonyms based on same type; and (4) new issues—including type, collector and herbarium.]

Hooker, J. D. and B. D. Jackson et al. 1893–1895. *Index Kewensis* Plantarum Phanerogamarum. 2 volumes and 16 supplements. Oxford. [An index to the place of publication of generic names or binomials of seed plants published since 1753. Coverage is worldwide. Supplements are issued at five- or ten-year intervals (more frequent recently). Arrangement of information: (1) genus, author, bibliographic reference, (2) specific names arranged alphabetically followed by author, place of publication and double author entry. Indication of synonymy has been discontinued.]

Index Nominum Genericorum. [A project proposed in 1954 to produce a file of names of all genera in *all* plant groups, with each entry the result of painstaking bibliographic research and application of the ICBN. The first issue (1,000 cards), contributed by 14 botanists, was in 1955. Information on each of the cards consists of the validly published generic name, its author, citation of place of publication, as precise a date as possible, type species and its basionym (when appropriate). Numbers on each card are a serial number and a number designation of the person who prepared the card. Class and family in which a genus is placed are also indicated. This index is now computerized and issued as volumes.]

Voss, E. G. (chairman of editorial committee). 1983. International Code of Botanical Nomenclature, adopted by the Thirteenth International Botanical Congress, Sydney, August 1981. Regnum Vegetabile 97. Utrecht. [The Code is published in English, French, and German; in case of inconsistency, the English wording has been arbitrarily decided as correct.]

Willis, J. C. 1973. A Dictionary of the Flowering Plants and Ferns, eighth edition revised by H. K. Airy Shaw. Cambridge University Press. Cambridge. [The seventh edition departed somewhat from previous editions (see preface of seventh edition) in that entries deal strictly with taxonomic matters. The aim of the eighth edition, like the seventh, is to include every published generic name from 1753 to the present and every family name published from the appearance of de Jussieu's *Genera Plantarum* in 1789. This is one of the most useful references for checking the family to which a genus belongs, variant spellings, and synonyms, and is used daily in most herbaria. In addition to the number of corrections, this edition also includes names of intergeneric hybrids of orchids with an indication of their status and a list of family equivalents between this dictionary, the twelfth edition of Engler's *Syllabus* and Bentham and Hooker's *Genera Plantarum*.]

Taxon. [This journal is published for the International Association for Plant Taxonomy by the International Bureau for Plant Taxonomy and Nomenclature, Tweede Transitorium, Uithof, Utrecht, Netherlands, and contains many important papers on nomenclature.]

b.   Terminology and Description of Taxa

References helpful in the translation and/or writing of descriptions (e.g., glossaries and dictionaries) are placed here. It is often important to use the older treatments for changes in usage of terms.

Jackson, B. D. 1928. A Glossary of Botanic Terms, fourth edition. J. B. Lippincott Co. Philadelphia. [A widely used basic glossary, reissued a number of times.]

Lawrence, G. H. M. 1951. Taxonomy of Vascular Plants. The Macmillan Company. New York. [See the illustrated glossary of taxonomic terms, pp. 737–775.]

Radford, A. E. et al. 1974. Vascular Plant Systematics. Harper & Row. New York. Chapter 6. [Illustrated glossary of taxonomic terms.]

Stearn, W. T. 1966. Botanical Latin. Thomas Nelson & Sons. London. [A valuable aid to translation or writing of descriptions.]

c. Identification

Identification is one of the basic aspects of taxonomy. The references cited include a variety of kinds of taxonomic literature useful in identification as well as guides to references for identification. Selected references also include those treating native and cultivated plants; local, state, regional and international floras; and both technical and popular works. Such references are also good sources of other information such as distribution, variation, phenology, ecology, etc.

Bailey, L. H. 1949. Manual of Cultivated Plants, second revised edition. The Macmillan Company. New York. [A convenient manual for identification of cultivated plants of the United States.]

Blake, S. F. and A. C. Atwood. 1942–1961. Geographical Guide to Floras of the World. Parts 1 and 2. U. S. Department of Agriculture Miscellaneous Publications 401 and 797. Washington, D.C. (Part 1 reprinted by Hafner Publishing Co., 1963). [An annotated selected list of floras and floristic works relating to vascular plants, including bibliographies and publications dealing with useful plants and vernacular names. Introduction includes many helpful notes on bibliographies, guides, dictionaries, etc. used in compiling this work. Part I— Africa, Australia, insular floras, North America (including Canada, Mexico, Central America, Greenland, and West Indies) and South America. Part II—Western Europe, European U.S.S.R. and all of Asia including Turkey (not completed). An index to abbreviations and to authors is included in each part.]

Correll, D. S. and M. C. Johnston. 1970. Manual of the Vascular Plants of Texas. Texas Research Foundation. Renner, Texas.

Cronquist, A., A. H. Holmgren, N. H. Holmgren, and J. L. Reveal. 1972. Intermountain Flora. Volume 1. Published for the New York Botanical Garden by Hafner Publishing Company. New York. [Other volumes in progress.]

Davis, P. H. and J. Cullen. 1965. The Identification of Flowering Plant Families, Including a Key to Those Native and Cultivated in North Temperate Regions. Oliver and Boyd. London.

Engler, A. (H. Melchior, ed.). 1964. Syllabus der Pflanzenfamilien. Ed. 12. Gebruder Borntraeger. Berlin. [Text in German.]

Engler, A. and K. Prantl. 1924. Die naturlichen Pflanzenfamilien. Ed. 2. Leipzig.

Fernald, M. L. 1950. Gray's Manual of Botany, eighth edition. American Book Company. New York. [Corrected printing 1970, Van Nostrand Reinhold Company.]

Flora Neotropica. 1968 and forward. Published for the Organization for Flora Neotropica by Hafner Press. Riverside, New Jersey. [A series of monographs of *all* tropical plants growing spontaneously in the western hemisphere. Monographs issued separately without regard to sequence; for example, monograph 1—Cowan, R. S., 1968, Swartzia (Leguminosae, Caesalpinoideae, Swartzieae); monograph 2—Cuatrecasas, J., 1970, Brunelliaceae.]

Gleason, H. A. 1963. The New Britton and Brown Illustrated Flora of the Northeastern United States and Adjacent Canada. 3 volumes, third printing, slightly revised. Published for the New York Botanical Garden by Hafner Publishing Company. New York.

Gleason, H. A. and A. Cronquist. 1963. Manual of Vascular Plants of Northeastern United States and Adjacent Canada. D. Van Nostrand Company, Inc. Princeton, New Jersey.

Hitchcock, C. L., A. Cronquist, M. Ownbey, and J. W. Thompson. 1955–1969. Vascular Plants of the Pacific Northwest. Parts 1–5. University of Washington Publications in Biology Volume 17. Seattle, Washington.

Hultén, E. 1968. Flora of Alaska and Neighboring Territories; A Manual of the Vascular Plants. Stanford University Press. Stanford, California.

Komarov, V. L. and B. K. Schischkin. 1934–1964. Flora U.R.S.S. 30 volumes. Academy of Sciences of the U.S.S.R., Moscow and Leningrad. [Flora of the USSR—English translation in progress; see review in Taxon 18:685–708, 1969.]

Munz, P. A. in collaboration with D. D. Keck. 1959. A California Flora. Supplement, 1968. Published for the Rancho Santa Ana Botanic Garden by the University of California Press. Berkeley, California. [A combined edition of the manual and supplement issued in 1973.]

Tutin, T. G., V. H. Heywood, N. A. Burgess, D. H. Valentine, S. M. Walters, and D. A. Webb (eds.). 1964 and forward. Flora Europaea. Cambridge University Press. Cambridge. [Volumes also include notes on Flora Europaea organization, list of contributors, list of basic and standard floras, appendixes of keys to author abbreviations, book and periodical title abbreviations, glossary and list of English-Latin equivalents, and maps.]

Wood, C. E., Jr. et. al. 1958 and forward. Generic Flora of the Southeastern United States. Journal of the Arnold Arboretum. [Keys to genera, illustrations, and superb bibliographies.]

d. Distribution

The range or geographic distribution of taxa has long been of concern to the taxonomists, but such information has become of increasing importance in a variety of disciplines. The references given not only include major works dealing specifically with distribution but include examples of types of works that should be examined for distribution data.

Willis, J. C. 1973. A Dictionary of the Flowering Plants and Ferns, eighth edition. Revised by H. K. Airy Shaw. Cambridge University Press. Cambridge. [Lists the number of species in each genus, name of family, and general distribution, for example, *Harfordia* Greene and Parry. Polygonaceae. 2 California.]

e. Current Literature

Selected journals of interest to taxonomists.

Aliso. 1948–present. Journal of the Rancho Santa Ana Botanical Garden. Claremont, California.

American Fern Journal. 1940–present. Published by the American Fern Society. Port Richmond, New York.

American Journal of Botany. 1914–present. Published by the Botanical Society of America. Lancaster, Pennsylvania.

American Midland Naturalist. 1909–present. Notre Dame, Indiana.

Annals of the Missouri Botanical Garden. 1914–present. St Louis, Missouri.

Baileya. 1953–present. Journal of horticultural taxonomy. Ithaca, New York.

Bartonia. 1950–present. A botanical annual. Philadelphia, Pennsylvania.

Bauhinia. 1955–present. Zeitschrift der Basler Botanischen Gesellschaft. Basel.

Biota. 1954–present. Magadelena del Mar, Peru.

Blumea. 1934–present. A journal of plant taxonomy and plant geography. Leiden.

Boletim da Sociedade Broteriana II. 1922–present. Coimbra, Portugal.

Botanical Review (interpreting botanical progress). 1935–present. Lancaster, Pennsylvania.

Botanische Jahrbücher fur Systematik, Pflanzengeschichte und Pflanzengeographie. 1881–present. Leipzig.

Botanisk Tidsskrift. 1880–present. Copenhagen. Bruken series.

Bothalia, 1921–present. A record of contributions from the national herbarium, Union of South Africa. Pretoria.

Brittonia. 1931–present. Published by the New York Botanical Garden. New York.

Bulletin of the Botanical Survey of India. 1959–present. Calcutta.

Bulletin of the British Museum, Botany. 1951–present. London.

Bulletin of the Torrey Botanical Club. 1870–present. Lancaster, Pennsylvania.

Canadian Journal of Botany. 1951–present. Ottawa.

Candollea; Organe du conservatoire et du jardin botaniques de la ville de Geneve. 1922–present. Geneva.

Contributions of the Gray Herbarium, Harvard University. 1891–present. Cambridge, Massachusetts.

Folia Geobotanica and Phytotaxonomica. 1966–present. Czechoslovak Academy of Science. Prague.

Journal of the Arnold Arboretum. 1920–present. Harvard Univeristy. Cambridge, Massachusetts.

Kew Bulletin. 1946–present. Royal Botanic Gardens. Kew, England.

Madroño. 1916–present. Published by the California Botanical Society. Berkeley, California.

Memoirs of the Gray Herbarium, Harvard University. 1917–present. Cambridge, Massachusetts.

Memoirs of the New York Botanical Garden. 1900–present. New York.

Memoirs of the Torrey Botanical Club. 1889–present. New York.

Phytologia. 1933–present. (Designed to expedite botanical publication.) New York.

Reinwardtia. 1950–present. Journal on taxonomic botany, plant sociology and ecology. Published by Herbarium Bogoriense. Borgor, Indonesia.

Rhodora. 1899–present. Published by the New England Botanical Club. Lancaster, Pennsylvania.

Sida. 1962–present. (A taxonomic journal started by Lloyd H. Shinners and published privately at Southern Methodist University, W. F. Mahler and John Thieret, eds.) Dallas, Texas.

Svensk Botanisk Tidskrift utgifven Asvenska Botaniska Foreningen. 1907–present. Stockholm.

Systematic Botany. 1976–present. Journal of the American Society of Plant Taxonomists. Lawrence, Kansas.

Taxon. 1951–present. Journal of the International Association for Plant Taxonomy. Utrecht, Netherlands.

Watsonia. 1949–present. Journal of the Botanical Society of the British Isles. Oxford.

f. Abstracts, Guides, and Indexes

Biological Abstracts—BioResearch Index. [Biological Abstracts (BA) is a semimonthly publication consisting of abstracts and indexes published by BioSciences Information Service (BIOSIS), Philadelphia, Pennsylvania. Abstracts of more than 140,000 papers grouped under 623 subject categories, derived from more than 7,600 journals originating in 100 countries are produced annually.]

BioResearch Index (BIOL) is a monthly publication that provides access to more than 100,000 research reports annually, *in addition* to those covered in Biological Abstracts. Types of literature reported in BioResearch Index include: symposia, meetings and congresses, reviews, letters, notes, and bibliographies.

Each issue of both publications contains four indexes: Author, Biosystematic, Cross, and Subject—color coded and keyed to the abstracts and bibliographic references.]

Biological and Agricultural Index. 1964–present. H. W. Wilson and Company. New York. (Formerly Agricultural Index, 1916–1964.) [A subject index to selected lists of agricultural periodicals and bulletins.]

Botany Subject Index. Compiled by USDA library, on index cards, reproduced in 1958 in 15 volumes by Microphotography Company, Boston, Massachusetts, currently available from G. K. Hall and Company, Boston. [Information is worldwide in scope and references are arranged by subject, names of countries, scientific names, and some vernacular names.

Genera are listed under the family, but are also cross-referenced. Each volume has a table of contents.]

Excerpta Botanica. A reference journal to literature of taxonomy and chorology and of phytosociology started in 1959 as a result of cooperation between Gustav Fisher Verlag of Stuttgart and the International Association for Plant Taxonomy, Utrecht.

Holmgren, P. K., W. Keuken, and E. K. Schofield. 1981. Index Herbariorum. Part 1. The Herbaria of the World, seventh edition. Regnum Vegetabile 106. Utrecht. [This valuable guide is divided into four sections: (1) list of herbaria are arranged alphabetically by city in which they are located and entries include such information as name of institution, date of founding, number of specimens, type of collection, important collections, activities (e.g., teaching, research), staff specialties, and availability of loans and exchange; (2) herbarium abbreviations (acronyms); (3) geographical arrangement of herbaria—arranged by country and also city; (4) general index to personal names (curators and staff).

The acronyms for herbaria have become standard and are used in most taxonomic papers when specimens are cited. This is a valuable reference for locating institutions, specific herbaria, and specialists. Although the index is somewhat out of date with respect to the size of collections and staff, it is an important reference.]

Lawrence, G. H. M., A. F. C. Buchhein, G. C. Danniels, and H. Dolezal (eds.). 1968. Botanico-Periodicum-Huntianum. Hunt Botanical Library. Pittsburgh, Pennsylvania. [Botanico-Periodicum-Huntianum (B-P-H) is a compendium of information on all periodical publications that regularly contain (or in some period of their history, included) articles dealing with plant sciences and botanical literature and with the persons who have contributed to botany and its literature. The work includes more than 12,000 titles in 45 languages. Information includes abbreviation, full title as given on title page of Volume 1, city of publication of Volume 1, number of volumes published, dates of publication, report of change in title, and location of entry in Union List of Serials. The two appendixes include a list of abbreviations of words used in title abbreviations (Appendix I) and a list of language equivalents for geographic names found in the titles or imprint of periodicals treated in B-P-H (Appendix II). This is an excellent source for locating correct and complete titles, number of volumes, and standard abbreviations and is extremely helpful in deciphering abbreviations of titles used in other publications. The cross-referencing with the Union List of Serials is an aid in locating the name of the library where a particular serial is on deposit.

Microfiche editions of herbaria. The Inter Documentation Company, Zug, Switzerland, has prepared microfiches of a number of herbaria (e.g., The Willdenow Herbarium and a Systematic Index to the Willdenow Herbarium) as well as books, catalogues, and lists. New editions are announced in Taxon and a catalogue may be requested from IDC, Postrasse 14, 6300 Zug, Switzerland. This method of dissemination of information is gaining favor and makes it possible for the contents of many herbaria and books to be consulted by those working outside the large botanical centers.

SUGGESTED READINGS

Lawrence, G. H. M. 1951. Taxonomy of Vascular Plants. The Macmillan Company, New York. Chapter 14, pp. 284–331.

Radford, A. E. et al. 1974. Vascular Plant Systematics. Harper & Row, New York. Chapter 30, pp. 697–748 and Chapter 33, pp. 793–801.

Smith, R. C. et al. 1980. Smith's Guide to the Literature of the Life Sciences, ninth edition. Burgess Publishing Company, Minneapolis, Minnesota. 223 pp.

SUMMARY FOR THE STUDY OF TAXONOMIC LITERATURE

Definitions for Study of Taxonomic Literature: The *botanical library*—a depository for taxonomic literature—is a research, training, and service institution that functions as a source of information, center for documentation, and facility for storing, curating, loaning, exchanging, and retrieving archival collections. *Taxonomic literature* refers to all inclusive writings (published and unpublished) and numeric and graphic representations that relate to classification, identification, nomenclature, description, and relationships of organisms and taxa. *Collections* consist of books, manuscripts, magazines, journals, films, tapes, drawings, illustrations, and paintings.

Purpose of Study of Taxonomic Literature To develop scholarship in systematics through acquiring a knowledge of fundamental references pertaining to the subject for general information by obtaining a background in the literature of particular studies and by keeping abreast of developments within the field.

Operations in the Study of Taxonomic Literature To become acquainted with the classification, identification, and names of the available systematic references for efficient retrieval and effective use.

Basic Premise for the Study of Taxonomic Literature That archival documents are available and can be located, examined, analyzed, and categorized for effective study of a taxonomic subject of interest.

Fundamental Principles for Study of Taxonomic Literature

1. Each archival collection has intrinsic informational value.
2. Variation in the archival collections makes possible the classification, identification, and nomenclature of taxonomic literature for effective use on topics of concern.

Guiding Principles for the Use of Taxonomic Literature

1. The user will conform to the rules established for the use of the botanical library.
2. The user will conform to the general guidelines established for the handling of different types of archival documents.
3. The user will conform to the guidelines for borrowing collections and using loans—including handling, protecting, photocopying, and removal of materials.

Basic Assumptions for the Study of Taxonomic Literature

1. Guides and procedures are available for the use of collections in botanical libraries.
2. Taxonomic literature has been properly classified, identified, named, and catalogued for effective storage, retrieval, and use.
3. Taxonomic literature is a form of documentation for the data, information, knowledge, research, and scholarship of the past.

QUESTIONS ON TAXONOMIC LITERATURE

1. Why do taxonomists have to be excellent bibliographers to be successful in their field?

2. What are the purposes of documentation in a systematic publication?

3. How is a botanical library an information storage-retrieval system? A documentation center? A service facility? A data bank?

4. What types of taxonomic indexes are used in systematic studies? How? Why?

5. How are books, periodicals, reprints, and indexes usually arranged in your local library?

6. How do you distinguish the terms in each of the following couplets:

    a. Manual-flora

    b. Monograph-revision

    c. Taxonomic guide-catalogue

    d. Journal-periodical

    e. Documentation-reference

    f. Bibliography-literature citation

    g. Library-museum

    h. Name index-nomenclator

    i. Biological Abstracts-Excerpta Botanica

    j. Kew Index-Gray Card Index

    k. Data-information

    l. Primary-secondary sources of documentation

7. What should be included in basic training for a student of systematics in the use of a botanical library?

8. What are the basic references for solution of a problem in nomenclature? For identification of an unknown vascular plant specimen from Spain, the Ukraine, or Texas?

EXERCISES IN TAXONOMIC LITERATURE

1. Select any species of vascular plant described 25 or more years ago and give the following information:

    a. Citation of the original publication.

    b. List of synonyms (to ten) and references.

    c. Copy and translate (if necessary) the original description.

    d. List of references (to ten) to illustrations.

    e. Location of type specimen.

    f. Collector of type and type locality.

    g. Correct generic name, authority, and bibliographic reference.

    h. Type species of the genus.

    i. Number of species in genus.

    j. Infraspecific taxa described since 1885.

    k. Distribution of this species in the United States and Canada and general world distribution for the genus.

    l. Most recent monograph of the group.

    m. List the key characters used in at least three manuals to separate this species (or genus) from closely related ones.

    n. An economically or horticulturally important member of the group.

    o. Chromosome number.

    p. List current papers or research projects published within the last three years treating this species (or genus).

    q. Current authority or worker on the group.

SAMPLE FORM FOR RECORDING INFORMATION

Number	Indexes and guides consulted (e.g., *Index Kewensis*)	LC or Dewey number	Answer	Documentation (citation of specific references)
a.				
b.				

2. Select a paper from a taxonomic periodical such as Taxon or Systematic Botany and determine the documentation for (a) descriptive terms, (b) nomenclature, (c) system of classification, (d) background information, and (e) procedures or methods.

3. Select a text in systematics and give a few examples of documentation from primary, secondary, and tertiary sources as well as original data and information.

4. Develop a constituent element summary for the study of documentation in taxonomic research.

chapter 12

The Herbarium

Herbarium collections are examined by taxonomists to determine character states of selected characters within taxa for a more comprehensive characterization and more effective delimitation of taxa, for the elucidation of pertinent intra- and intertaxon relationships, and for a better understanding of evolutionary trends and distribution patterns within taxa of concern.

In the simplest sense, the herbarium is a collection of pressed and dried specimens usually arranged according to a classification system. A modern herbarium includes diverse collections of flowering plants, gymnosperms, ferns, mosses, liverworts, lichens, fungi, algae, and fossils. Other resources found in this institution are microscope slides, photographs, photomicrographs, wood specimens, cellulose acetate peels, camera lucida drawings, field notebooks, diaries, letters, unpublished reports, manuscripts, lists of exsiccatae, reprints, and botanical illustrations and paintings. The herbarium of today is a research, training, and service institution that serves as a reference center, documentation facility, and data storehouse.

As a reference center, an herbarium is a fundamental resource for identification of plants by practicing taxonomists, field ecologists, workers with endangered and threatened species, natural heritage employees, conservationists, naturalists, and environmentalists. The plant collections, reprints, exsiccatae, manuscripts, and so on—the documents—are fundamental references for basic and applied research in botany, biology, agriculture, medicine, pharmacy, and genetics.

As a documentation facility, the herbarium is the repository for collections of historic significance such as types of new taxa, representatives of new discoveries, economically important introductions, and geographic disjuncts; sets of specimens providing the bases for floristic, revisionary, and monographic studies; and annotated collections showing the interpretations of different authorities down through the years. Many herbaria also have cytological and cytogenetic voucher specimens; ecological schemes and data attached as notes to herbarium sheets; photomicrographs of anatomical sections and pollen grains affixed to documentary collections; photographs showing morphological detail on separate sheets; and many reference notes on individually mounted vouchers that indicate the precise data use of that particular specimen. The herbarium is the institutional facility for documentation of research in systematics.

As a data storehouse, its collections are used by geneticists searching for new sources of DNA material, chemists studying alkaloids, pharmacists and other researchers systematically seeking new antitumor compounds, and individuals looking for new natural food and energy sources. Data for determining the environmental impact of various industrial installations and different land uses also may be found in herbaria. Today, herbaria are major sources of information on habitat, ecology, distribution, and taxonomy of rare species of plants. Traditionally, the herbarium has been a natural data resource for comparative morphological and phylogenetic studies; a special collection for anatomical and palynological research; a warehouse of material for working out ranges and ecological distributions; and a primary source of information on humanity's explorations and observations of flora and vegetation. Herbaria are data centers for all types of research in systematics.

The herbarium, as an educational resource, routinely develops and maintains collections for courses in local flora, horticulture, general taxonomy, advanced systematics, and special groups of plants such as aquatics, fungi, grasses, and trees. Collections serve as a standard resource for graduate students working on problem selection and feasibility in advanced degree programs. Many professional staff members actively participate in undergraduate and graduate teaching. Occasionally, herbaria sponsor short professional and public interest courses such as spring flora, trees in winter, and culinary herbs. The herbarium collections provide a foundation for training in plant diversity.

The herbarium, like a library, is a service institution. The collections constitute a vast reservoir of facts about plants—some of these collections are rare and no longer replaceable. Many valid requests for advice concerning plant-related questions and problems come from public school students, teachers, and individuals working for state and federal agencies and environmental firms. Herbaria also receive a wide variety of queries about plants from scientists, administrators, corporate personnel, doctors, lawyers, writers, and amateur as well as professional naturalists. Knowledgeable staff members make an effort to respond to reasonable requests for information on plants and plant materials associated with herbaria.

Herbaria are the laboratories for much systematic research. The plant collections are basic to the scholarly research of taxonomists and the training of students in systematics. The curatorial staffs provide a significant service to the public on botanical questions and problems. The collections and staff expertise found in herbaria throughout the world form an international plant resources system of irreplaceable and inestimable value to science and humanity.

Figure 12.1 A herbarium in use. (Courtesy Allen Rokach, The New York Botanical Garden.)

A. THE HERBARIUM—PAST, PRESENT, FUTURE

I. Historical Development

The beginning of the herbarium as a collection of dried specimens* affixed to paper is attributed to Luca Ghini (1490?–1556). Ghini seems to have been the sole initiator of the art of herbarium making and this art was disseminated over Europe by his students. Gherards Cibo, a pupil of Ghini, began collecting and preserving specimens as early as 1532 and his herbarium is extant today in Florence, Italy. John Falconer, an Englishman, is mentioned in the writings of Lusitanus in 1553 and William Turner in 1569 as possessing an herbarium, and it is believed that he also learned of herbarium making either directly or indirectly from Ghini. Although the herbarium technique was a well-known botanical practice at the time of Linnaeus, he departed from the convention of the day (mounting specimens and binding them into volumes) by mounting his specimens on single sheets and storing them horizontally, much as is the practice today (Stearn, 1957). Although this method became general during the second half of the eighteenth century, it was by no means universal. In fact, as late as 1833 Asa Gray was offering bound volumes of grasses and sedges for sale (DeWolf, 1968).

In addition to changes in mounting, early workers also began depositing specimens in established collections, as well as exchanging specimens. It is indeed fortunate for botanists of today that such practices developed so early. Many herbaria have been de-

*See Appendix E for modern field collection and preparation of specimens.

stroyed by fire, insects, war, and ignorance; and all that remains of some collections are the duplicates sent to other institutions on exchange.

In the United States several herbaria are known to have existed in the mid-1700s. Many of these and later ones found their way to Europe, where they are preserved. According to Holmgren, Keuken, and Schofield (1981), the oldest institutional herbarium in the United States was started at Salem College in 1772. Other early institutions include the Academy of Natural Sciences of Philadelphia, Amherst College, Charleston Museum, and others, all founded before 1860 (Holmgren, Keuken, and Schofield, 1981).

During the first 400 years of the existence of herbaria, they have gone from the personal portfolio of a few pressed and dried specimens to small collections of local or regional significance to large institutions of national and international stature with millions of specimens. Most of the systematic research institutions with worldwide collections such as those at Kew, Leningrad, Paris, Geneva, and Cambridge (United States) were founded in the nineteenth century. The great herbarium in Paris was founded in 1635 (Holmgren, Keuken, and Schofield, 1981).

II. Present Size and State of Herbaria

According to Holmgren, Keuken, and Schofield (1981), three herbaria reported collections of five million specimens or more, for example, Kew in Great Britain, Leningrad in the Soviet Union, and Paris in France. The largest herbaria in the United States are those at Harvard University, the New York Botanical Garden, the Smithsonian Institution, the Missouri Botanical Garden, and the Field Museum.

In 1969, Shetler estimated the herbarium resources of the world included as many as 250 million specimens, although the total number actually reported by the various institutions was 148 million. Of this estimate, about 78 million were in European collections and 36 million in North America. Since that time millions of specimens have been added to the major herbaria, and doubtless also to many private and small public herbaria that oftentimes do not report their holdings.

Today, collections are more accessible than ever before due to the efforts of many individuals and national and international organizations. Two extremely useful indexes for locating collections are *Index Herbariorum* (Holmgren, Keuken, and Schofield, 1981), which includes location and curators of herbaria of the world, size and type of collections, type of institution, publications (serials), and important historical collections; and *Index Xylariorum* (Stern, 1978), an index to wood collections of the world. The principal wood collections of the United States includes those of the Arnold Arboretum (Aw), Cambridge, Massachusetts; Rancho Santa Ana Botanic Garden (RSAw), Claremont, California; U. S. Forest Products Laboratory (MADw, SJRw), Madison, Wisconsin; and Smithsonian Institution (Sw), Washington, D. C. Various herbaria, particularly those with an abundance of types and other historical collections, often issue accounts of their collections: the Sloane Herbarium in the British Museum (Dandy, 1958), Herbaria of the Department of Botany in the University of Oxford (Clokie, 1964), and the Linnaean Herbaria (Stearn, 1957). A growing number of institutions (e.g., New York Botanical Garden, Gray Herbarium, Philadelphia, Smithsonian) are compiling type registries and/or issuing microfiche editions of their type collections. Such projects are tremendous aids to systematic research and also to specimen preservation.

During the past 50 years, the use made by systematists of characters and techniques from the fields of anatomy, palynology, cytology, ultrastructure, biochemistry, and so on in the solution of their problems has added diversity to the collections. Modern herbaria now have bulk, liquid, and frozen specimens as well as photographs and illustrations of all types. Pressed and dried specimens mounted on high-quality paper remain the heart of the herbarium, however.

Insects in the collections are controlled by fumigating, heating, freezing, and other techniques appropriate to the conditions and nature of the pests. Most dried collections are now stored in insect- and fire-proof steel cases. Specimen storage space is more effectively used with the recent development of compactors that permit the elimination of aisles between the rows of cases. Management, protection, and storage of the diverse collections continue to present many problems.

III. The Herbarium and the Future

Herbaria have been established for a variety of purposes. Some institutions have complete but limited collections of species of vascular plants from their immediate areas for local flora classes. Others have collections restricted to aquatics, trees, lichens, and so on for courses covering each of these groups. Some scholars have private collections of specialty groups. Small teaching and research herbaria have the common problems of pest control and adequate storage. Both have ongoing maintenance chores for specimens due to handling—the teaching collections much more so due to inexperienced users. The larger regional and national research institutions have these same basic problems plus others. Financial, spatial, and staffing constraints are forcing the larger herbaria to reassess their roles in science and society.

The fundamental principle for the use of herbarium collections, that each specimen has intrinsic informational value, has generated many questions on content and significance of collections. The modern principle for the use of herbarium collections, that the data in the plant storehouse be efficiently retrievable by electronic methods, has stimulated a great deal of discussion on the entire problem of efficient and cost-effective data storage and retrieval.

The researcher using the major herbaria assumes—since the institutions are reference centers, documentation facilities, and data storehouses—that the collections are adequate or at least somewhat useful for the comprehensive characterization and effective delimitation of taxa, the elucidation of intra- and intertaxon relationships, and an understanding of evolutionary trends and distribution patterns within taxa of concern. The environmentalist using the herbarium as a data storehouse for information on an unusual community reported from an impact area expects the species that made the community unique are represented by voucher collections. A taxonomist studying the habitat of a species typically occurring on granitic outcrops anticipates seeing the specimens studied and collected by a student from a diabase dike. An evolutionist interested in the genetic variation within populations of a rare species hopes to see the documentation for a new variety described by a natural heritage worker. The professional users' assumptions of most systematists are not always valid and the expectations of many workers are occasionally not realized.

Collections do have intrinsic informational value but financial reality dictates asking and answering the questions of how much information and of what value to whom. Management, preparation, maintenance, and accessioning of collections depend on the resources and interests of the institution and the systematic community. Herbarium policy, for example, determines the number of collections of a taxon accessioned from a county, province, or state; the number of varieties within a taxon processed; the number of historical collections retained; and the types and quality of specimens added.

Documentation for new taxa and research by staff and doctoral students are well represented in herbaria. Specimens resulting from ecological, natural heritage, species biology of rare plants, and beginning graduate research frequently accumulate in temporary storage awaiting curatorial decision on disposition. Some of the collections in temporary storage can be of real legal consequence in ecological impact controversies and of great significance in the protection of rare elements in our natural heritage. Not all research in systematics is documented in institutional herbaria and some collections are not readily available to the user. What to add, whose to keep, and which to maintain are continuing questions for administrators of systematic collections.

The herbarium is a storehouse of data that have to be retrieved to be used. The handling of specimens creates maintenance problems and costs. Efficient means of data retrieval are sorely needed. The computer and data bank seem to be logical solutions to a number of problems associated with collections. Numerous programs are becoming available at reasonable prices. Cost effectiveness is, however, a constant problem in all decisions on the use of electronic as well as traditional methods. User time and convenience remain considerations for research and service institutions.

The establishment of the Association of Systematics Collections in 1972 was a major step toward the improvement of the condition of systematic collections and their services. The condition of plant collections as a national resource and the efficiency of services associated with these systematic collections will slowly but surely improve. A significant national resource of systematic collections demands national responsibility, from local to federal levels, for its development and maintenance.

B. HERBARIUM USE*

Herbaria exist only for use regardless of the type of collection—research, identification (synoptic), teaching, or other. Every effort of ensuring maximum efficiency in locating material in a collection (information retrieval) should be made without endangering the collection. (For operation and maintenance of herbaria, see Massey in Vascular Plant Systematics, pp. 766–768).

1. *User Instructions*. Every herbarium user should be provided with a guide or some introductory instructions to a collection. These instructions should include the systematic and geographic arrangement of the collections; color code of folders; a list of families with numbers and case number (including segregate and combined families; e.g., Ericaceae versus Pyrolaceae, Monotropaceae); a map of case arrangement; a list,

---

*Adapted from Massey, J. R. in Radford, A. E. et al., 1974, Vascular Plant Systematics, Chapter 31, pp. 763–768.

location, and arrangement of special collections; and a generic catalog. Herbarium policy concerning reshelving of materials, handling of specimens, use of equipment and other facilities, loan procedures, and any other pertinent information should be included. Printed guides or instruction sheets can save considerable time on the part of visitors as well as of the herbarium staff.

2. *Specimen Use.* Specimens are preserved for use and should be handled with extreme care, since they are of scientific value and generally irreplaceable. The following suggestions are for aiding users in the proper handling of specimens.

   a. Keep sheets flat, do not shuffle or leaf through a folder like a book. Specimens are brittle and easily damaged.

   b. Store properly. Preferably, specimens should be stored in herbarium cases or at least on shelves. Do not crowd too many specimens into a box or pigeon hole. Keep materials in folders or containers when not in use.

   c. Do not lay books or other heavy objects on specimens.

   d. Place loose fragments in packets or envelopes, if the specimens to which the fragment belongs can be ascertained.

   e. Use materials sparingly. It is often necessary to dissect flowers or fruits. Use no more than is required and place all fragments in a fragment envelope on the sheets. For dissection, flowers and fruits may be softened by boiling or treating with a softening or wetting agent. Pohl's softening agent (Pohl, 1965) is an excellent wetting material. This solution will not stain or discolor sheets but is best used in a watch glass. Place materials to be softened in a watch glass and add several drops of the solution. Many materials are sufficiently softened in a few minutes. Fragments may be blotted dry and returned to the fragment envelope.

   f. Use a long-armed dissecting microscope. An entire specimen can be studied with this instrument without bending the sheet.

   g. Support the specimens with a ventilator or other stiff board when carrying specimens, even for short distances.

   h. Call the attention of curators to those specimens in need of repair.

   i. Do not write on the herbarium sheets unless permission to do so has been given by someone in charge.

   j. Reshelve specimens only with permission and always with extreme care.

   k. For additional information on annotations, borrowing specimens, and so on, see item 4 below.

3. *Visitors.* Visiting the herbarium is, of course, one of the best possible ways to study materials. An investigator can examine the exact materials desired as well as search undetermined folders, consult types and other species collections (some not available on loan), and use the library and botanic garden as well as exchange ideas with fellow botanists. Visitors would do well to follow common courtesy if they wish to have access to collections, particularly at unusual hours, such as evenings, weekends, or holidays. A note in advance can mean the difference in having or not having a comfortable place to work and access to a microscope, as well as gaining admittance to special collections and libraries.

4. *Loans.* Most herbaria lend materials to other institutions (rarely to individuals). Reciprocal agreements between many institutions often result in transit costs only one way. One should, however, be sure that such agreements exist before requesting a loan. Otherwise, the borrower pays all transit costs. For the most part, loans should be requested after some preliminary work has been done so that unduly long-term loans are avoided (loan periods are generally for six months). Requests should be for reasonable amounts of material and for species as taxa rather than species in communities or

floristic provinces. Loans should be requested by the curator or director of the herbarium for study by a specific person, and should be addressed to the director or curator of the lending institution.

Special permission should be obtained for removal of fragments, pollen, or other materials from specimens for study. Even with permission, the utmost discretion should be exercised. Many herbaria ask that duplicate slides, photographs, and other preparations made from their materials be included when the specimens are returned. When in doubt, the borrower should consult the curator of the lending institution for special instructions.

A commonly accepted procedure is that all specimens be annotated whether they are used in ecological, floristic, monographic, anatomical, biosystematic, or other studies. This procedure should include even those specimens correctly determined. A convenient way to agree is by using an exclamation point (!) followed by the name of the investigator and date (at least the year). Printed annotation labels affixed above the original label are the most desirable. The original label should not be altered. Some institutions allow the use of a rubber stamp with small, yet readable print (preferably using black ink), for example, ''Examined in a survey of parasitic angiosperms of Southwestern U. S., John Doe, 1971.'' The name of the institution in which the work is being conducted is also quite desirable. Annotations should reflect the history of use of individual specimens.

Materials should be returned promptly on the date due or an extension should be requested. Most institutions honor requests as soon as staff, time, and funds permit. Loan forms to be returned are generally sent out before or at the time the loans are shipped. On receipt of a loan, the borrower should examine, count, and acknowledge the shipment. Any damage occurring in transit due to mishandling or insects should be noted on the loan acknowledgment form. Each sheet should be checked for the stamp of the lending institution. Since sheets may have the seal of more than one institution, careful notes should be made, particularly if the borrower has material from several institutions. Specimens packed by the lender in a certain order should be returned the same way.

Borrowed materials should be housed in fireproof, and dust- and pest-free cases. Materials may be fumigated and repellents used, provided no damage is done to specimens, labels, and photographs. Great care should be used in all cases. Plastic strapping is softened by strong fumigants and repellents such as paradichlorobenzene. This and other information is often included on loan forms. Most specimens are sent library rate by parcel post at considerable savings to all; however, the lending institution determines the mode of transit and this should be followed unless permission to do otherwise is obtained.

In addition to the annotations, the borrower can perform many other services: determining duplication, supplying additional data (obtained from duplicates from other institutions), correcting locality data (by annotation), pointing out mixed labels, determining specimens as to type (isotype, paratype, etc.), and indicating a mixed collection. In most cases, both parties should profit from a loan.

SUGGESTED READINGS

Arber, A. 1938. Herbals, Their Origin and Evolution, second ed. University Press, Cambridge.
Clokie, H. N. 1964. An Account of the Herbaria of the Department of Botany in University of Oxford. Oxford University Press, London.

Croat, T. B. 1978. Survey of Herbarium Problems. Taxon 27:203–218.

Dalla Torre, C. G. and H. Harms. 1900–1907. Genera Siphonogamarum ad systema Englerianum conscripta. Leipzig.

Dandy, J. E. (ed. and compiler). 1958. The Sloane Herbarium. British Museum (Natural History). London.

DeWolf, G. P., Jr. 1968. Notes on making an herbarium. Arnoldia 28:69–111.

Fosberg, F. R. and M. Sachet. 1965. Manual for tropical herbaria. Regnum Vegetabile 39:5–132.

Franks, J. W. 1965. A Guide to Herbarium Practice. Handbook for Museum Curators, Part E, Section 3. The Museums Association. London.

Hicks, A. J. and P. M. Hicks. 1978. A Selected Bibliography of Plant Collection and Herbarium Curation. Taxon 27:63–99.

Holmgren, P. K., W. Keuken, and E. K. Schofield. 1981. Index Herbariorum, Part I. The Herbaria of the World, seventh ed. Regnum Vegetabile 106. Bohn, Scheltema & Holkema, Utrecht.

Irwin, H. O., W. Payne, D. Bates, and P. Humphrey (eds. and compilers). 1973. America's Systematics Collections: A National Plan (Prepublication Draft).

Massey, J. R. 1974. The Herbarium. In Radford, A. E. et al., Vascular Plant Systematics. Harper & Row, New York. Chapter 31, pp. 751–774.

Payne, W. W. et al. (compilers). 1974. Systematic Botany Resources in America Part I. Survey and Preliminary Ranking. A Report to the National Science Foundation and to the Botanical Systematics Collection Community. New York Botanical Garden, Carey Arboretum, Millbrook, New York.

————. 1979. Systematic Botany Resources in America Part II. Costs of Services. A Report to the Public and Botanical Systematic Community. The New York Botanical Garden, Carey Arboretum, Millbrook, New York.

Pohl, R. W. 1965. Dissecting equipment and materials for the study of minute plant structures. Rhodora 67:95–96.

Rollins, R. C. 1955. The Archer method for mounting herbarium specimens. Rhodora 57:294–299.

Shetler, S. G. 1969. The herbarium: past, present and future. Proceedings of the Biological Society of Washington 86:687–758.

Stearn, W. T. 1957. An introduction to the "Species Plantarum" and cognate botanical works of Carl Linnaeus. Prefixed to the Royal Society facsimile of Linnaeus' Species Plantarum, 1. London.

Stern, W. L. 1978. Index Xylariorum. Institutional wood collections of the world. Taxon 27(2/3):233–269.

SUMMARY FOR TAXONOMIC STUDY
OF PLANT COLLECTIONS IN THE HERBARIUM

Definitions for Taxonomic Study of Plant Collections The *herbarium*—a repository for plant collections—is a research, training, and service institution that functions as a reference center, documentation facility, and data storehouse. *Collections* consist of specimens that are samples of populations and taxa from nature, experimental garden, or laboratory.

Purpose of Taxonomic Study of Plant Collections To develop meaningful classification by providing a more comprehensive characterization and effective delimitation of taxa, elucidating pertinent intra- and intertaxon relationships, and producing a sharper discernment of phylogenetic trends and distribution patterns within taxa of concern.

Operations in Taxonomic Study of Plant Collections To circumscribe accurately, to delimit diagnostically, to determine position and rank, to name correctly, and to determine relationships for taxa of concern.

Basic Premise for the Taxonomic Study of Plant Collections That in the tremendous variation in the plant world as represented by collection (samples of populations and taxa), conceptually discontinuous groups with character correlations exist that can be recognized, classified, circumscribed, named, and logically related.

Fundamental Principles for Taxonomic Study of Plant Collections

1. Each specimen or collection has intrinsic informational value.
2. Variation in plants and plant collections makes possible the establishment of a taxonomic system for taxa of concern.
3. The data in the storehouse of plant collections are most effectively and efficiently stored and retrieved by electronic means.

Guiding Principles for the Use of Plant Collections

1. The user will procure a guide or receive instructions for the use of the collections within an herbarium and conform to the policies therein while using the herbarium.
2. The user will conform with the general guidelines for handling specimens (see B. Herbarium Use).
3. The user will conform with the general guidelines for borrowing collections and using loans—including shipping, handling, protecting, annotating, and removal of materials (see B. Herbarium Use).
4. The user will add to the historical, scientific, and societal resource values of the collections.

Basic Assumptions in the Taxonomic Study of Plant Collections

1. Guides and procedures for the use of plant collections are well established by research herbaria.
2. Plant collections are representative samples of the genetic and ecological diversity of the individuals, populations, and species of concern.

QUESTIONS ON THE HERBARIUM

1. How has the role of the herbarium changed since its inception in 1532?
2. How is the herbarium used as (a) documentation center, (b) data source, (c) information storage system, (d) basic reference collection, and (e) data retrieval system?
3. What are the purposes, goals, and practices of a general herbarium?
4. What are the traditional sources of materials found in herbaria?
5. How do collections increase in value to science? To society? Why are certain collections of more value to science than others?
6. What is an herbarium specimen? How are the various types of specimens prepared?
7. What precautions should be taken in handling of specimens by the user? By any herbarium worker? (Dried, wet, fossil, pollen, type specimens, etc.)

8. What are the obligations of a borrower in the handling, care, and use of a herbarium loan?
9. What is the relevance of data banking and data retrieval to the development or demise of herbaria and botanical libraries?
10. What should the future role of herbaria be in taxonomic research, conservation, and teaching?
11. What are the accepted procedures for obtaining herbarium materials for use in anatomical, palynological, embryological, chemical, ecological, and geographical studies? What are the obligations of the investigator for the use of these materials?
12. What are the documentation procedures for the use of herbarium materials in anatomical, palynological, chemical, and ecological studies?

EXERCISES IN THE HERBARIUM

1. Assume that you are going to start a small institutional herbarium (teaching, special research, or other type of your choosing) of some 10,000 to 20,000 specimens. Outline and document, where possible, your goals and organization; costs; manpower, materials, and space needs; accession, loan, and exchange policies; system of filing and basic references to be housed in the herbarium; record-keeping and data-retrieval systems.
2. Consult the basic references for the herbarium. Evaluate and annotate each as to history, concepts, principles, terminology, organization, operation, administration, and general significance.
3. Review the herbarium literature, then characterize it and indicate the basis of its separation from other taxonomic literature.

chapter 13

Botanic Gardens and Arboreta*

Living collections are used by systematists in studies on character variation in local populations, reproductive biology of selected organisms, breeding experiments with related taxa, demographics of groups of concern, and critical habitat analysis of ecotypic variants for understanding relationships between taxa and comprehending the biology of species of interest.

Botanic gardens and arboreta are institutions that maintain living plant collections representing a great number of species, genera, and families. Generally, the botanic garden includes plants of all kinds, while the arboretum is often restricted to woody plants (trees, shrubs, vines, and ground covers). Large gardens and arboreta,† for example, the New York Botanical Garden and the U. S. National Arboretum, include libraries and herbaria as integral parts of their facilities. Miscellaneous resources available in gardens are color slides, photographs, field notebooks, reprints, botanical illustrations, paintings, and specimens of such types as wood, pollen slides, pickled duckweeds, and palm fronds.

The modern botanic garden is a biological facility that serves as a center for research, data storage, documentation, and reference, as well as for educational, interpretive, conservation, and public service activities. Its living collections and other resources are used for

Research in organismal biology by undergraduates, graduates, and postdoctoral students.

*Adapted from Moore, J. K., in Radford, A. E. et al., 1974, Vascular Plant Systematics, Harper & Row, New York, Chapter 32, pp. 775–790.
†Although the phrase ''botanic gardens'' appears more frequently than ''arboreta'' in this chapter, the discussion refers to both, because of similarities in organization and use.

Graduate and undergraduate training in organismal biology.

Workshops and short courses for professional and advanced teacher training in organismal biology.

Education in biology and natural history for teachers of primary and secondary grades.

Interpretive training and public instruction in natural heritage for rangers, park personnel, recreation leaders, outdoor specialists, natural area preservation managers, and other interested individuals.

Storage and dissemination of botanical and horticultural information to professionals and the general public.

The collections and staff expertise found at botanic gardens throughout the world comprise an international botanical resources system of inestimable value to science and mankind.

A. HISTORICAL DEVELOPMENT OF BOTANIC GARDENS

The following summary of the history of the development of botanic gardens is based primarily on Hill (1915), Stafleu (1969), Stearn (1971), and Moore (1974).*

I. The Earliest Botanic Gardens

Although gardens existed in ancient Egypt and Mesopotamia for the growing of herbs, food plants, and ornamentals, or for pleasure and/or as status symbols, they cannot properly be designated as botanic gardens in which plants are collected and maintained for scientific purposes. The first such botanic garden with the primary function of science and education may have been that of Theophrastos, the father of botany. This garden, attached to his school (the Lyceum near Athens), had been bequeathed to him by his teacher, Aristotle.

The Romans maintained small gardens as sources of medicine and as aids to medical studies. Similar gardens were also established in Europe. The medieval monastic gardens appeared in the late eighth century during the reign of Charlemagne, who assigned the task of medical training to the monasteries. The typical monastic garden included the hortus for vegetables and fruit, and the herbularis for various herbs; the latter was the precursor of the physic gardens that were often established in affiliation with the medical faculties of universities during the sixteenth and seventeenth centuries.

The credit for the establishment of the first modern botanic garden belongs to the Italian, Luca Ghini (ca. 1490–1556), a professor of botany called from Bologna to Pisa in 1543 or 1544, where he almost immediately began planting a botanic garden connected with the university. Two other university botanic gardens were begun at Padua and Florence in 1545; all three of these were aided by the patronage of the Medici family. The establishment of other important gardens followed in succession: Bologna, Italy (1567);

*I. The Earliest Botanic Gardens is based on Hill (1915); II. The European Period to VIII. Subsequent Periods (item 3) are based on Stafleu (1969) and Stearn (1971); VIII Subsequent Periods (item 4) and IX. American Gardens are based on Moore (1974).

Leyden, Netherlands (1587); Montpellier, France (1593); Heidelburg, Germany (1593); Strasbourg, France (1619); Oxford, England (1621); Paris, France (1653); Groningen, Netherlands (1642); Berlin, Germany (1646); Uppsala, Sweden (1655); Edinburgh, Scotland (1670); Chelsea, England (1673); Amsterdam, Netherlands (1682); Vienna, Austria (1754); Kew, England (1760); Cambridge, England (1762); and Coimbra, Portugal (1773). Luca Ghini is usually credited with having established the first herbarium. He also taught scientific courses in plant taxonomy that were not limited simply to medicinal herbs.

Though the Pisa Botanic Garden does not exist today, records of its design demonstrate the geometric patterns of plantings originating with the monastic gardens that are characteristic of many continental gardens even today.

II. The European Period (to 1560)

Plant collections of this time were principally of indigenous European species and those collected from southern and southeastern Europe and the adjacent Mediterranean lands, including Egypt. Some species had to be maintained in tubs as potted plants, which were kept in cubicula tepida during the winter; these glass houses were simple rooms with windows facing south, another important innovation attributed to Luca Ghini.

III. The Near East Period (1560–1620)

This oriental period of plant introductions from southeastern Europe and adjacent Asia may be characterized by a few groups outstanding for fragrance and brilliant color, such as hyacinths, tulips, and lilies. The Flemish-Austrian botanist and father of the bulb industry, Carolus Clusius, was very interested in this group of introduced plants. He was an inquisitive natural scientist, botanical traveler, horticulturist, and plant taxonomist. While studying and traveling through the Netherlands, Vienna, Frankfurt, Leiden, Hungary, Italy, and Spain, he was busy obtaining and describing new plants. Clusius became a dominant figure in botany during the last half of the sixteenth century, and as director of the botanic gardens at Vienna and Leiden, he greatly influenced the development of such gardens in Europe. By the end of the Near East period, the center of plant activity had shifted from Italy to Austria, Germany, and the Netherlands.

IV. The Period of Canadian and Virginian Herbaceous Plants (1620–1687)

With the decline of Spanish sea power, French exploration expanded, and France became the new center of botanical activity. While the Dutch were involved in the tropics and the African cape and the English with their Virginia territories, the French exploited the Canadian wilderness. Introductions into the botanic garden in Paris included such familiar plants as arborvitae, sumac, poison ivy, black locust, and perennials such as black-eyed susan, dutchman's breeches, and goldenrod. The botanic garden at Paris, known as the Jardin du Roi or Jardin des Plantes, was in no way a palace garden; it was an independent educational and scientific institution with a royal endowment, originally founded as an establishment to promote the teaching of pharmaceutical botany. At present, its collections, particularly in its herbarium and paleobotanical and paleontological departments, are among the world's best; it is the oldest and most important nonuniversity botanic garden still in existence. Some of the private gardens of England received their first North American plants by way of French introductions. A few decades later, private gardens of England, such as that of the Tradescants at Lambeth, had early Virginian introductions that included red maple and tulip poplar, thus marking the second phase of this botanical period. The introduction of North American species to the Netherlands from Dutch settlements began a third period.

V. The Cape Period (1687–1772)

Dutch sea traffic around the Cape of Good Hope during the golden age of Holland is responsible for the most characteristic plant introductions of this period. By now, glass houses and conservatories had become much more sophisticated so that the introductions of geraniums and numerous succulents could be better accommodated. After travel in South Africa and India, Paul Hermann, a German immigrant, was appointed professor of botany at Leiden where he was instrumental in the great increase in the plant collection at that botanic garden. In Amsterdam at about the same time, 1682, Jan Commelin was appointed to establish a botanic garden. The efforts of these two botanists stimulated a strong interest in plants among the Dutch sea merchants, resulting in vigorous and enthusiastic plant collecting. Linnaeus completed his botanical training in Holland, utilizing the latest collections that often contained plants representative of families previously unknown to him. From 1735 to 1737, he was employed by George Clifford, a wealthy Dutch banker, to describe the collections in his own private botanic garden, De Hartecamp, which included a permanent herbarium (now at the British Museum). With access to Holland's excellent botanical libraries and printers, Linnaeus was able to complete and publish *Systema naturae* (1735), *Bibliotheca* and *Fundamenta botanica* (1736), *Genera plantarum* (1737), and *Hortus Cliffortianus* (1737). In the latter work the collections of Clifford's garden, many of which were introductions from the cape, were described. In 1741, Linnaeus became director of the botanic garden at Uppsala, Sweden. Prior to this time, the garden had been allowed to deteriorate to such an extent that plant collections were down to fewer than 300 species. Through Linnaeus' recommendation, the garden obtained the services of the chief gardener of Clifford's estate, so that by the time Linnaeus was stationed at Uppsala, the garden had begun to improve. Under Linnaeus's supervision, plant collections grew to over 3,000 species within the next seven years (Blunt, 1971).

At first, cape plants reached England almost exclusively through Holland, but during the course of the eighteenth century, direct introductions into England rapidly increased. By the end of this period the English gardens began to rival those of Holland; the gardens of both nations had surpassed the important German, French, and Italian gardens of earlier periods.

VI. The Period of North American Trees and Shrubs (1687–1772)

Through the efforts of Lancelot Brown, English gardening experienced a revolution in the development of an open landscape style, free from the rigid geometrical formality that had persisted in the continental gardens due to the monastic influence. Introductions of new woody species from North America gained prominence. Many of the introductions reached England through the efforts of Peter Collinson, whose private gardens at Packam and Mill Hill were significant botanical collections. Collinson corresponded extensively with the American naturalist, John Bartram, who is credited with having established the first botanic garden in America in 1731 near Philadelphia, on his estate on the Schuylkill River. Bartram studied and propagated the native plants prior to sending them to his friend Collinson. Through Collinson, many of the American introductions reached other gardens such as those at Chelsea and Kew.

The great botanic gardens of England were preceded by the establishment of several private gardens devoted to the cultivation of medicinal herbs and other plants of botanical interest. The establishment of the botanic garden at Oxford in 1621 was specifically for the purpose of cultivating "physic" plants and for the demonstration of principles in the study of botany. The Chelsea Physic Garden was founded in London in 1673 as the garden of the Society of Apothecaries. Sir Hans Sloane, who deeded the land for the Chelsea garden, served botanical study well when he stipulated that 50 preserved specimens of diverse plants from the garden be presented to the Royal Society of London each year until 2,000 distinctly different types of plants had been donated. The appointment of Philip Miller as head gardener in 1723 was an important event in the life of the institution. Under his practical skill and botanical knowledge, many plants recently acquired from all over the world were being grown and propagated for the first time in cultivation. Kew owes its establishment, in 1760, to the interest of Princess Augusta, Princess Dowager of Wales, who set aside a portion of the Royal Garden at Kew House for a physic garden. William Aiton, the first curator and a pupil of Miller, was in charge of this garden for 34 years. After the death of the princess in 1772, George III inherited the Kew property and united the gardens with those lying contiguously that had formed the gardens of the Palace of Richmond, thus producing the extensive area known as the Royal Botanic Gardens, Kew. Kew differs from other botanic gardens in that it has no connection with any university or educational organization as such, thus its usefulness has largely been with the economic aspects of botany and with the assisting and encouraging of botanists, travelers, merchants, and manufacturers in various botanical investigations.

VII. The Australian Period (1772-1820)

Sir Joseph Banks of Kew Gardens accompanied Captain Cook on his first voyage (1768–1771), marking the beginning of the Australian period of development of botanical gardens. Species of numerous representatives of families of plants from the southern hemisphere were discovered and introduced into England and from there to the continent. The tropical botanic gardens were developed during the Australian period. Great Britain is credited with establishing the first of these economic gardens (for the cultivation of spices and other commercial plants) on the island of St. Vincent, British West Indies, in 1764. The Calcutta Botanic Garden, founded in 1786 for the cultivation of spices, is noted for potato cultivation and the introduction of tea, mahogany, jute, sugar cane, and the quinine-yielding cinchonas. The Botanic Garden at Buitenzorg in Java, Indonesia, founded in 1817, is noted for the cultivation of rubber and coffee. Tropical gardens were later established in Malaya and Ceylon during this period.

VIII. Subsequent Periods

1. Period of tropical glasshouse plants and hardy plants from Japan and North America (1820–1900).*

2. Period of west Chinese plants (1900–1930).

3. Period of hybrids (1930–1960).

4. Period of emphasis on regional native floras and recognition of the need to preserve natural vegetation on a global scale (1960–present).

IX. American Gardens (1801–   )

Botanic gardens lost the restrictive character of physic gardens and became more extensive in size and function, with the result that many well-known universities without a botanic garden began making plans for one. It was near the end of the Australian period that true botanic gardens began to appear in North America. The Elgin Botanic Garden, established in 1801, accumulated a valuable plant collection and became the Botanic Garden of the State of New York in 1801. It was subsequently granted to Columbia College. Although it does not exist today, several of this garden's associates, Thomas Eaton, John Torrey, and Asa Gray, will be remembered as some of America's leading botanists.

Before 1900, the Missouri Botanical Garden (Shaw's Garden) in St. Louis (1859), the Arnold Arboretum at Harvard University (1872), and the New York Botanical Garden

*Items 1 to 3 are based on Stearn (1971) and item 4 is based on Moore (1974).

Figure 13.1   A conservatory. (Courtesy Allen Rokach, The New York Botanical Garden.)

in New York City (1891) were the major botanic gardens established in America. Not until after 1920 did botanic gardens and arboreta begin to appear in the nation in significant numbers.

A number of private plant collections, attesting to the avid interest of amateur botanists and horticulturists, are scattered over the world. These private collections often rival or surpass similar collections of professional institutions. They hold enormous potential for the research of plant systematists.

B. BOTANIC GARDENS OF TODAY

Historically, the role of the botanic garden has been the providing of information on food plants, medicinal herbs, and ornamentals as well as the propagation of plants introduced from various parts of the world for horticultural purposes. The botanic garden of today is an institution established to supply living plant resources for basic and applied research in systematics, genetics, ecology, horticulture, and other associated disciplines. It also serves as a professional training and public instructional facility. Conservation activities now associated with many gardens and arboreta are the protection of endangered and threatened species, the propagation of rare plants, and the preservation of plant diversity in unique communities. In general, the botanic garden is a multipurpose facility that serves as a center for research, education, conservation, and public service.

I. Research

The living collections in botanic gardens provide an invaluable reservoir of material for systematic research. These living materials are used in variation analyses within populations, in reproductive biology of selected organisms, in breeding and hybridization studies of related taxa, and in demographic studies of many kinds. These collections are also used by geneticists seeking new sources of DNA material, by medical researchers making allergy studies, by horticulturists searching for more effective pest- or disease-free clones, and by ecologists managing and manipulating a preserve under garden control for maintenance of populations of rare species or unique communities.

Botanic gardens are conservatories of living collections representative of newly described species, geographic disjuncts, economically important plant introductions, historically significant taxa, and products of breeding and hybridization studies. Gene pools and germ plasm of rare species are maintained in living populations for their commercial potential and natural heritage values. Reference collections are kept for professional identification purposes as well as for sources of horticultural propagative materials. Occasionally, excellent collections used in monographic or revisionary studies are maintained as living documentation of the biological diversity used in the research.

Well-run botanic gardens are ever-growing resources of genetic and organismic diversity. They also provide physical facilities for professional researchers in many fields, with special value for systematists, geneticists, and horticulturists.

II. Education

As centers for educational training, botanic gardens provide classroom, laboratory, nursery, greenhouse, and field resources for teaching and learning experiences. Gardens associated with academic institutions supply facilities for courses in local flora, horticulture, plant propagation, introductory taxonomy, pollination ecology, hybridization, dendrology, agrostology, and so on. Institutions with natural areas have plant communities for vegetation studies; those with experimental plots have space for studies in the biology of species. Some gardens have extensive collections and family beds that are excellent resources for courses in phylogeny and classification. Living populations are frequently available for field analysis of characters and of reproductive biology studies in biosystematics.

Educational activities are usually provided for the public through regularly scheduled interpretive classes, lectures, and field trips for groups of all ages. Special-interest courses, such as culinary and medicinal herbs and trees in winter, are made available to the general public. Workshops and short training sessions for teachers, naturalists, and garden club members are standard features of educational programs in botanic gardens. Symposia and conferences on rare species, the protection of natural areas, the design of gardens, and many other related subjects are part of the ongoing activities of progressive institutions. As educational centers, botanic gardens in general make good use of their plant resources by providing meaningful learning experiences that are related to actual work with living organisms.

III. Conservation

In recent years, botanic gardens have assumed a leading role in the conservation of rare species and have become more active in the protection of nature preserves and habitats. Gardens are doing yeoman's service in making the public aware of the disappearance of our native species and the destruction of our natural areas. For example, *Extinction is Forever,* the published proceedings of the Symposium on Threatened and Endangered Species of Plants in the Americas, sponsored in 1976 by the New York Botanical Garden, has had a salutary effect on the academic community by stimulating action on conservation. And *Survival or Extinction,* the published proceedings of the Conference on the Practical Role of Botanic Gardens in the Conservation of Rare and Threatened Plants, sponsored in 1978 by the Royal Botanic Gardens, Kew, has elicited similar positive action on conservation. Today, many gardens actively support the natural heritage programs sponsored by The Nature Conservancy, the only national organization with the sole purpose of protecting critical habitats to save natural diversity.

Many botanic gardens are now involved in the propagation and cultivation of rare species, the discovery and documentation of wild populations of concern, the surveillance and monitoring of rare plants of interest, and the development of cloned and cultivated material for study of the biology of endangered or threatened taxa. Gardens are now protecting native habitats of rare plants and manipulating plant communities for the maintenance of wild populations nearing extinction. Rare plants in threatened sites are transplanted to protected areas controlled by gardens. These organizations now protect unique plant communities by acquisition or easement. Seed banks presently act as a genetic resource for reintroduction of rare species into natural habitats or for establishment in appropriate preserves. Gardens actively teach and practice conservation. Botanic gardens are an integral part of the cosmopolitan effort to protect, maintain, and propagate our biological diversity for the benefit of humanity.

IV. Public Service

Botanic gardens provide information to the general public on identification of native and exotic species of plants, methods of propagation, pest and disease control of cultivars, when and how to plant certain species, and many other related problems. Professional botanical services are supplied to conservation groups, federal and state agency personnel, nursery workers, garden club members, and people in educational consortia of many types. Much information on garden topics, such as home gardening and care of plants in winter, is transmitted through radio and television programs. Gardening facts, announcements, and advice are regularly released to interested citizens in leaflets, pamphlets, newsletters, booklets, and newspaper articles prepared by institutional staff members and friends.

Plant materials, such as flowering shrubs, various herbs, and seeds, are supplied to the public through annual sales and/or complimentary distribution. Under staff supervision, native plants of horticultural or conservation interest are often rescued from sites of imminent destruction and made available to the private sector.

Some botanic gardens now play a significant role in programs for mental health, hortitherapy, and disturbed children. Today, gardens provide an aesthetically pleasing environment in a relaxed atmosphere for the enjoyment and benefit of large segments of our society.

C. BOTANIC GARDENS OF TOMORROW

Botanic gardens have provided formal plots and pleasant landscapes for the pleasure of many. Most gardens have produced culinary and medicinal herbs, many have introduced new plants into cultivation, others have developed new horticultural varieties. A few gardens that have been centers of botanical research and training have contributed much to our knowledge of plants. In recent years, the research and training institutions have become focal points for preservation of endangered and threatened species and habitats. Leadership in much of the conservation effort of today comes from individuals in these research and training organizations.

The traditional aesthetic values of botanic gardens need to be augmented. More pleasant walkways through well-designed landscapes and natural areas should be constructed for the enjoyment of our citizens. Small garden plots and flower beds should be made available to interested people for their therapeutic values as well as for food and flowers. Trails through the woods and fields are basic to passive recreation. Botanic gardens, now and in the future, will have the opportunities and challenges of developing the aesthetic potential of their establishments for the welfare of many in our increasingly complex civilization. Many garden institutions will become more proficient in their public service endeavors in the practical uses of plants.

Botanic gardens have a societal responsibility for the preservation and utilization of plant diversity through their biological resources, staffing, facilities, skills, and practices. These organizations should have ever-expanding roles in the development of gene pools in living populations, in the improvement of seed banks, in the promotion and establishment of germ plasm centers, and in the management of genetic resources in natural communities. Much of their research activities will be directed toward propagation and cultivation of rare species, the experimental manipulation of threatened communities, the monitoring of disappearing populations, the analyzing of unique habitats, and the understanding of the reproductive biology of taxa of concern. Much time will be spent on organisms with economic potential or conservation value. The inherent nature of these institutions indicates their potential and responsibility for becoming excellent centers of systematic, genetic, ecological, physiological, horticultural, and evolutionary research and professional training in organismal biology.

Botanic gardens and arboreta, herbaria and museums, and nature sanctuaries and preserves will be the organizations providing the expertise for the wise use of our natural resources. These institutions will train future generations in understanding and appreciating the roles of plants and animals in our environment, economy, everyday lives, and biological heritage. These research centers will acquire knowledge about these organisms that are basic to our existence. Many of these biological institutions will continue to combine their talents and resources to realize, as economically as possible, their collective potential for the best care and use of our natural heritage and biological resources.*

Botanic gardens, in concert with herbaria and conservation organizations, have a responsibility for leadership in educating the public and developing a political constitutency in our citizenry for effective stewardship of our genetic and ecological resources for the enjoyment, use, and betterment of present and future generations.

SUGGESTED READING

Avery, G. S., Jr. 1957. Botanic gardens—what role today? American Journal of Botany 44(3):268–271.

Bethel, J. S. 1973. Botanical gardens—who needs them? Arboretum and Botanical Garden Bulletin 7(1):22–26.

Blunt, W. 1971. The Compleat Naturalist—A Life of Linnaeus. Viking Press, New York.

Brown, R. A. and R. D. MacDonald. 1971. The A. H. S. plant records center. Arboretum and Botanical Garden Bulletin 5(4):108–117.

Bryan, J. E. 1972. Education in botanical gardens and arboreta. Arboretum and Botanical Garden Bulletin 6(3):74–76.

Bunce, F. H. 1971. Arboreta, Botanical Gardens, Special Gardens. Bulletin No. 90. National Recreation and Park Association, Washington, D.C.

Chan, A. (ed.). 1968. Arboretum and Botanical Garden Bulletin 2(4):86–97. (A special issue on research programs.)

Fletcher, H. R., D. M. Henderson, and H. T. Prentice. 1969. International Directory of Botanical Gardens. International Bureau for Plant Taxonomy and Nomenclature, Utrecht, Netherlands.

Fogg, J. M., Jr. 1970. The nature and functions of an arboretum. Arboretum and Botanical Garden Bulletin 4(2):47–54.

*An outstanding example of collective effort for the common good in recent years is the formation of the Plant Sciences Data Center (PSDC) at Mt. Vernon, Virginia. The PSDC became a reality in 1968 through the combined efforts and support of The American Association of Botanical Gardens and Arboreta, the American Horticultural Society, and The Longwood Foundation (MacDonald, 1971). The PSDC provides computerized references to living plant resources, in cultivation or under natural areas management of major botanic gardens in the United States and Canada. The data for each collection include the scientific name, garden, accession number, original source, and year of accession. The PSDC has been a major force in standardizing plant records of living collections and is now the basic data system for information on living collections in gardens and arboreta. In the *Master Inventory of Botanical Taxa,* second edition, published in 1979 by PSDC in microfiche format, data were given on 173,000 living collections in more than 49,000 taxa representing 4,000 genera.

Hill, A. W. 1915. The history and functions of botanic gardens. Annals of the Missouri Botanic Garden 2:185–223.

Howard, R. A. 1969. The botanical garden—an unexploited source of information. Boissiera 14:109–117.

MacDonald, R. D. 1971. A history of the Plant Records Center. Arboretum and Botanical Garden Bulletin 5(4):107–108.

Moore, J. K., in Radford, A. E. et al. 1974. Vascular Plant Systematics. Harper & Row, New York. Chapter 32, pp. 775–790.

Niering, W. A. 1973. Arboreta—their environmental role. Arboretum and Botanical Garden Bulletin 7(2):55–57.

Prance, G. T. and T. S. Elias. 1977. Extinction Is Forever. The New York Botanical Garden, New York, New York.

Pratt, C. B. 1974. Education and arboreta. The Bulletin, American Association of Botanical Gardens and Arboreta 8(3):95–96.

Pyle, R. 1937. Report on the Committee on Botanical Gardens and Arboreta. American Association of Nurserymen, Chicago.

Stafleu, F. A. 1969. Botanical gardens before 1818. Boissiera 14:31–46.

Stearn, W. T. 1971. Sources of information about botanic gardens and herbaria. Biological Journal of the Linnean Society 3(3):225–233.

Synge, H. and H. Townsend. 1979. Survival or Extinction: The Practical Role of Botanic Gardens in the Conservation of Rare and Threatened Plants. The Bentham-Moxon Trust. Royal Botanic Gardens, Kew, England.

Teuscher, H. 1933. The botanical garden of the future. Journal of the New York Botanical Garden 34:49–62.

Thompson, P. A. 1972. The role of the botanic garden. Taxon 21(1):115–119.

Tsitsin, N. V. 1969. The system of botanical gardens of the USSR and its role in investigating vegetable resources of the country. Boissiera 14:135–140.

Wyman, D. 1959. The Arboretums and Botanical Gardens of North America. Arnold Arboretum of Harvard University. Jamaica Plain, Massachusetts.

——. 1970. How to establish an arboretum or botanic garden. Arboretum and Botanical Garden Bulletin 4(52):52–60.

SUMMARY FOR SYSTEMATIC STUDY OF LIVING COLLECTIONS IN BOTANIC GARDENS AND ARBORETA

Definitions for the Systematic Study of Living Collections The *botanic garden* (and *arboretum*)—a conservatory, sanctuary, and experimental laboratory for living plant collections—is a biological resource center that provides data storage and distribution, documentation, education and interpretation, conservation, and public services. *Collections* consist of living specimens that are samples of natural populations and taxa maintained in cultivation or managed in the natural state.

Purpose of the Systematic Study of Living Plant Collections To develop knowledge of the biology and taxonomy of taxa by providing a more thorough characterization and efficacious delimitation of taxonomic groups by comprehending relevant intra- and inter-taxon relationships and by generating a better understanding of speciation and variation within organismal groups of interest; and also to formulate recommendations for preservation of rare species and effective management of wild populations and unique communities.

Operations in the Systematic Study of Living Plant Collections To circumscribe accurately, to delimit diagnostically, to determine position and rank, to name correctly, and to determine genetic and phenetic relationships for taxa of concern; to determine the species biology of existing populations; to determine the isolating mechanisms for stable populations; to determine the environmental factors that control selected populations; and to determine the type and extent of selected populations.

Basic Premise for the Systematic Study of Living Plant Collections That in the tremendous variation in the plant world, as represented by living collections (samples of populations and taxa), conceptually discontinuous groups with character correlations exist that can be recognized, classified, circumscribed, named, and logically related.

Fundamental Principles for the Systematic Study of Living Plant Collections

1. Each living plant specimen, population, or community has intrinsic informational value.
2. Variation in plants makes possible the establishment of a taxonomic system for taxa of concern.
3. The data in the storehouse of living plant collections are most effectively and efficiently stored and retrieved by electronic means.

Guiding Principles for the Use of Living Plant Collections

1. The user will procure a guide or receive instructions for the use of the collections within a botanic garden and conform to the policies therein while using the facilities and collections.
2. The user will conform with the general guidelines (which vary among institutions) regarding research and other uses of collections, acquisition of plants or plant parts, or introduction of plants into the living collections or facilities of the institution.
3. The user will add to the historical, scientific, and societal resource values of the collections.

Basic Assumptions in the Systematic Study of Living Plant Collections

1. Guides and procedures for the use of living plant collections are well established by each botanical garden.
2. Living plant collections are representative samples of the genetic and ecological diversity of the individuals, populations, and species of concern.

QUESTIONS ON BOTANIC GARDENS AND ARBORETA

1. How is the historical development of botanic gardens parallel and related to the development of plant taxonomy?
2. What are the objectives of the Plant Sciences Data Center? What is the significance of this center to the future development of botanic gardens and systematics?

3. What is the relevance of botanic gardens and arboreta to monographic, floristic, and biosystematic research? Document the relevance from published research in books and periodicals for each type of study.

4. How is your local or regional botanic garden or arboretum (a) an information storage system, (b) a data-retrieval system, (c) a data source, (d) a documentation center, (e) a basic reference center, (f) an education center, (g) a conservation center, and (h) a public service center?

5. What is the relevance of data banking and computer-based procedures to the development of a botanic garden or arboretum?

6. What are the standard procedures for (a) handling incoming specimens, (b) for maintaining living family, habitat, herb and drug collections, and (c) research collections?

7. What standard precautions should be taken for (a) educational use of collections, (b) research use, and (c) conservation use?

8. What are the obligations of (a) a researcher in the handling and use of living collections and materials and (b) of a teacher?

9. What are the contributions of your local or regional botanic garden arboretum to (a) public education, (b) conservation, and (c) professional research?

10. How has the expanded conservation role of gardens and arboreta affected the routine functions of (a) collections acquisitions and maintenance, (b) academic teaching and research, and (c) public education?

EXERCISES IN THE BOTANIC GARDEN

1. Assume that you are going to start a small institutional botanic garden (teaching, special research, or other type of your choosing) of some 10,000 to 20,000 specimens. Outline and document, where possible, your goals and organization; costs; manpower, materials, and space needs; accession policies; basic references to be maintained; record-keeping and data-retrieval systems.

2. Consult the basic references for the botanic garden. Evaluate and annotate each as to history, concepts, principles, terminology, organization, operation, administration, and general significance.

3. Review the botanic garden literature, then characterize it and indicate the basis of its separation from other taxonomic literature.

part three

TAXONOMIC RESOURCE INFORMATION

appendix *A*

Data Analysis in Systematics*

A discussion of data analysis must by necessity begin with an understanding of some of the principles of data collection. Data collection itself reflects the investigator's perception of the problem. A taxonomist does not choose to study a particular group by running his or her finger down a list of taxa and randomly stopping at a taxon. The choice of a particular taxonomic group is made because the taxonomist has perceived a problem with the traditional treatment of the group. Perhaps, in using the manual for an area, plants that he or she believes belong to the same species will key out as two or more species. The taxonomist may have noticed phenological or ecological factors that appear to affect the identification of the individual (e.g., juvenile plants may key out as one species, while the adults key out as a different species). On the other hand, he or she may believe that what is being treated as a single species may actually be two or more species. Geographical variation may have been found that is not being accounted for in the traditional treatment. He or she may have observed that different manuals were treating the taxa in a different manner. Or perhaps the taxonomist is simply aware of the variation that exists within a species and wishes to study this variation in more detail.

Each of these problems has different implications for data collection. For example, if the taxonomist has observed greater variation in a taxon than is accounted for in standard treatments, there are several possible explanations: (1) two or more species are present instead of a single one; (2) only one species occurs, but with a greater range of variation than has been previously noted; or (3) there exist hybrids and/or introgressive individuals between two species. Each of the other observations listed in the preceding paragraph would have its own set of hypotheses that could explain what the taxonomist had observed.

---

*Contributed by James Doyle, University of North Carolina at Chapel Hill.

The problem that is observed and the hypotheses that are advanced to explain the problem will in turn affect the techniques that are used to analyze the problem. If, for example, the taxonomist believed that what was being considered as one species was really two species separated ecologically, with one species being found on dry sites and the other species in more mesic habitats, then techniques should be used that could help separate environmental plasticity from the genetic component of the variation. The same would hold true for geographically isolated taxa. In these cases, he or she must distinguish phenotypic plasticity from genetic fixity. Techniques such as reciprocal transplants or electrophoresis of allozymes can help differentiate between these two causes of variation. If the two taxa are observed growing in mixed populations, it is less likely that phenotypic plasticity is an important factor in explaining the difference between the taxa.

After recognizing a problem, a taxonomist typically consults the literature to see if other workers had observed or studied the problem. He or she would examine revisions, prior studies, and manuals for other regions in which the taxa in question were found. The taxonomist would consult with peers who might be familiar with the group. He or she would obtain and examine herbarium specimens over the range of the species. If the problem had not been satisfactorily resolved by other workers, the taxonomist would then proceed to study the taxa in greater depth.

The classical approach to studying variation involves extensive and intensive knowledge of the taxa. The taxonomist studies the plants in the field as well as from herbarium specimens over long periods of time. He or she might take many measurements, but these data would not be analyzed using modern statistical methods. The analysis of the variation was accomplished using the taxonomist's intuition. What this meant was that the taxonomist, based on first-hand knowledge of the group, would become aware of the correlations of characters and the discontinuities in the range of variation that indicate the separation of the taxa. He or she might separate two taxa based on the shape of the calyx lobes and the pubescence of the pedicels, but often could not explain why he or she believed these characters were the diagnostic characters. Other taxonomists would judge his or her work based on their evaluation of taxonomic sense and familiarity with the group, as well as their own observations of the taxa. This classical approach lacked repeatability—other researchers might examine the same specimens but arrive at different conclusions. The taxonomic community's judgment of a particular study was largely a reflection of their judgment of the ability of the researcher. Although this method lacked repeatability, it is important to note that many of the conclusions made by well-regarded taxonomists were also accepted by other researchers studying the same groups, and these same conclusions are often verified using the objective, statistical methods that are in use today.

Modern statistical methods of data analysis assure repeatability—if another researcher follows the same procedure, he or she will arrive at the same conclusion. However, this does not imply that the results are necessarily correct. There is still a significant amount of taxonomic judgment that must be exercised in studying biological variation. It is doubtful that this subjectivity will ever be eliminated, and it is not clear that this would be desirable. Taxonomic judgment is necessary in the choice of characters to be measured, the coding of the characters, and the types of statistical tests that are employed (not to mention even the very choice of taxa to be studied). Statistical methodology does not assure that all researchers will come to the same taxonomic conclusions, but if proper precautions are taken concerning the choice of characters, type of tests employed, and so on, it is likely that taxonomic treatments will not differ significantly.

A. NUMBER AND CHOICE OF CHARACTERS

Once a taxonomist has chosen a group for study and has formulated hypotheses to explain the observed variation, the next step is to measure the variation. Potentially an infinite number of characters could be measured, and these characters could be taken from many fields of evidence, such as morphology, chemistry, cytology, anatomy, and ecology. How does one decide how many characters and which ones to measure?

In regard to the number of characters, clearly the more characters that are measured, the more likely it will be that overall similarity and differences are uncovered. Likewise, the more fields of evidence that are employed, the greater the likelihood of finding overall similarities. Since different characters and groups of characters will evolve at different rates, the choice of too few characters, or characters from a single field of evidence, may produce distorted results. Sneath and Sokal (1973) have suggested 60 characters to be the minimum number used whenever possible. One possibility is to compare the results when a few characters are eliminated. If the same patterns appear, it is likely that the resulting classification is stable with respect to the addition of further characters. If results differ depending on which characters are eliminated, this indicates that more characters must be incorporated into the data. As many fields of evidence as possible should be used in addition. Many examples have been reported in which a classification based on morphology will differ from a classification based on isozymes, which differs from a classification based on chemistry. *Homo sapiens* relies heavily on his or her sense of vision, so it seems imperative that morphological characters be included in any classification. It is not practical to arrive at a classification that requires complex chemical tests or electron microscopy to identify an individual. However these other fields of evidence are important for producing a classification based on overall resemblance.

Since the number of characters that are to be measured is obviously limited, the choice of which characters to measure is clearly of utmost importance. Diagnostic and key characters that have been used by other researchers studying the group should always be included. Characters that are important in differentiating related members should likewise be included. Characters that are invariant should not be included. Obviously, the characters that introduced the problem to the taxonomist would also be measured. The hypotheses advanced may suggest the use of other characters and fields of evidence. If, for example, one hypothesis is that hybrids between two species exist and are involved in the variation pattern that is observed, one potentially significant field of evidence could be crossing experiments with the presumed parental species. If geographical variation had been observed and it was necessary to separate phenotypic plasticity from genetic fixity, evidence from electrophoresis of allozymes or reciprocal transplant experiments would be appropriate. Familiarity with the taxonomic literature is extremely important. Taxonomic conclusions based on data that did not include characters that other researchers had considered important would be suspect. On the other hand, if the characters used include only those characters that other researchers found important, the resulting conclusions would be biased.

B. CODING OF CHARACTERS

To analyze the data statistically, all characters must be numerical. This is not a problem for characters such as leaf length, or the length/width ratio, which are already numerical. A character such as flower color can be numerically coded in the following way: if there are only two colors, red and white, plants could be given a value of 1 for red and 0 for white. If there were three flower colors, red, pink, and white, individuals could have a value of 0 for white, 1 for pink, and 2 for red. The value of 1 is intermediate between 0 and 2 as the color pink is intermediate between white and red. If, however, the three flower colors were red, yellow, and blue, it would not be correct to assign values of 0, 1, and 2, since this would be equivalent to stating that one of these colors was intermediate between the other two. In this case, two variables (or characters) are created for flower color:

	Red	Yellow
Red	1	0
Yellow	0	1
Blue	0	0

The blue flower has values of zero for both characters. In general, characters that cannot be ranked (such as the three colors above) can be expressed using a number of characters equal to one less than the number of states of the original character (three flower colors expressed in two characters).

Some characters that may appear unable to be ranked, such as leaf shape, are actually better expressed directly as measurements. It would be impossible to rank shapes such as ovate, lanceolate, elliptic, and obovate. However, awareness of the basis of these terms points out a solution. Leaf shape is a combination of the length/width ratio of the leaf, and the ratio of the leaf length divided by the distance from the leaf base to the widest point of the leaf. The measurement of these two characters will allow a more accurate and complete description of the leaf shape than any coding of characters could produce. Likewise, leaf apices and bases are best determined by measuring the angle of the apex or base rather than coding terms such as acute, obtuse, and so forth. Leaf margins can be assessed by counting the number of teeth per cm, and the depth and angle of the teeth. Using these techniques, the characters measured should all be able to be expressed in mathematical units.

C. THE CHARACTER × TAXON MATRIX

The data that have been collected can be presented in a character-taxon matrix. The characters are the rows and the taxa are the columns. In many cases, the taxa represent individuals, so the character × taxon matrix is a listing of the character states for each of the individuals. In some cases, the taxa may represent groups of individuals, such as populations, varieties, or species. In these situations, the values for a particular character may represent averages over the taxon. The individual taxa are called operational taxonomic units (OTUs), whether they represent individuals or larger groupings. Table A.1 is an abbreviated character × taxon matrix with six OTUs and five characters.

Table A.1 CHARACTER × TAXON MATRIX

Character	A	B	C	D	E	F
			Taxon			
Leaf length/width	.55	.63	.78	.27	.19	.69
Petiole length/leaf length	.42	.47	.29	.72	.68	.35
Calyx apex (degrees)	78	82	55	58	57	68
Stem pubescence (trichomes/mm²)	173	219	233	107	123	218
Bract length (mm)	16.1	15.2	12.3	14.2	13.8	13.9

D. STATISTICS

I. Means

One basic statistical measurement is the mean of a series of numbers. The mean is the average, which is found by adding the series of numbers, and dividing the total by the number of values. The mean of the leaf length/width ratio in Table A.1 is (.55 + .63 + .78 + .27 + .19 + .69)/6 = .518

Mathematically, this formula can be written as follows:

$$\bar{X} = \frac{\Sigma X_i}{n}$$

where $\bar{X}$ represents the mean; Σ indicates a summation of all the values of X; X_i represents the individual values; and n represents the number of observations (or taxa in this case). $\bar{X}$ is the mean leaf length/width ratio for the six individuals measured only. If these individuals all belong to the same species, we cannot say that $\bar{X}$ is the mean ratio for the species. However, it is our best estimate for the mean value for the species.

II. Variance and Standard Deviation

The mean is an indication of the central tendency of a set of numbers, but it gives no indication how dispersed the numbers are around the mean value. In one situation, we could have a set of numbers that were all identical; the mean would be that value, and there would be no variance from the mean. In another case, we might have a set of numbers that ranged from very small to very large. The mean would be somewhere in between, and there would be a high variance—some numbers would be much greater or smaller than the mean. Two different formulas are used for determining the variance: the first is used to determine the variance of a population (where all individuals have been measured), and the second is used to determine the variance of a sample from the population. The statistical meaning of the word population as used here does not refer to a biological population, but rather to the total group of objects under study. In our example in Table A.1, if the six individuals are all from the same species, they are a sample from the total number of individuals of that species (unless these six individuals are all of the individuals of that species!) and the second variance formula would be used. The formula for the variance of a population, written as σ^2 (sigma squared), is given by:

$$\sigma^2 = \frac{\Sigma (X_i - \overline{X})^2}{n}$$

The difference between each value of X and the mean $\overline{X}$ is squared, and these squares are added together (this is the sum of squared deviations); the sum of these squares is divided by n, the number of observations (taxa). The formula for the sample variance, written as s^2 to distinguish it from the population variance, is the same, except that the sum of squared deviations is divided by $n - 1$ instead of n:

$$s^2 = \frac{\Sigma (X_i - \overline{X})^2}{(n - 1)}$$

The standard deviation is simply the square root of the variance; since there are two variance formulas, there are also two standard deviation values, depending on whether we are determining the standard deviation of the total population or only of a sample from the population.

$$s = \sqrt{s^2} \quad \text{or} \quad \sigma = \sqrt{\sigma^2}$$

The variance and the standard deviation are both indicators of the variation of the data around the mean. In a set of numbers that are normally distributed (i.e., the values form a bell-shaped curve), approximately 66 percent of the values will lie within 1 standard deviation of the mean; 95 percent of the values will be within 2 standard deviations; and 99 percent will be within 3 standard deviations. In our example from Table A.1, the sample variance of the leaf length/width ratio is

$$[(.55 - .518)^2 + (.63 - .518)^2 + (.78 - .518)^2$$
$$+ (.27 - .518)^2 + (.19 - .518)^2 + (.69 - .518)^2]/5 = .0562$$

The sample standard deviation is the square root of the variance or

$$\sqrt{.0562} = .237$$

The mean, sample variance, and sample standard deviation for each of the five characters in Table A.1 are shown below.

	Mean	Variance	Std deviation
Leaf length/width	.518	.0562	.237
Petiole length/leaf length	.488	.0308	.1754
Calyx apex (degrees)	66.3	133.9	11.57
Stem pubescence	178.8	2878.6	53.65
Bract length	14.25	1.691	1.300

III. Correlation

Species are not differentiated from each other based on single characters, but it is the correlation of characters that enables us to define species. Thus, character correlations are extremely important to taxonomists. At the same time, correlations among characters are potentially suspect. If two characters are necessarily correlated, including both characters has the effect of doubling the importance of the character. For example, if leaf color is one character and chlorophyll concentration is another, these two characters are basically measuring the same quality. Including both of these in the analysis is equivalent to saying that leaf pigmentation is twice as important as other characters. In some cases, it is not clear that two or more characters are necessarily correlated. If leaf length and leaf width are measured and then the length/width ratio is determined, these three values may show low correlation with each other. However, many times these values can be highly correlated with each other. In such cases it is best not to include all of these in the analysis. If two characters are correlated but are not necessarily correlated, it seems best to include both, since these correlations are what is being investigated. For example, if stem pubescence and leaf length are correlated, both should be included, since there is no necessary correlation between these characters.

The most common measure of the correlation between variables is the Pearson product moment correlation coefficient. The formula for this correlation coefficient is given by:

$$\text{Corr} = \frac{\Sigma \, (X_i - \bar{X})(Y_i - \bar{Y})}{\sqrt{\Sigma \, (X_i - \bar{X})^2 \, \Sigma \, (Y_i - \bar{Y})^2}}$$

where X_i represents the individual values of the first variable and Y_i represents the corresponding values for the second variable.

The value of this coefficient ranges from $+1$ to -1. A value of 0 indicates no correlation between the two variables; a large positive value (near $+1$) indicates a strong positive correlation; a negative value near -1 indicates a strong negative correlation. If there is a positive correlation, the numerator will tend to be positive. This is true since whenever X is large (relative to its mean value), Y will also tend to be large (relative to its mean value); the value of $X_i - \bar{X}$ will be positive, $Y_i - \bar{Y}$ will be positive, and the product will be positive. When X is small, Y will also tend to be small; the value of $X_i - \bar{X}$ will be negative, $Y_i - \bar{Y}$ will be negative, and the product will again be positive. The sum of these positive numbers will be a positive correlation. Likewise, if X and Y are negatively correlated, whenever $X_i - \bar{X}$ is positive, $Y_i - \bar{Y}$ will tend to be negative, and vice versa. The products will tend to be negative, and the correlation value will also be negative. In all cases, the denominator will never be negative. The formula shown above might look intimidating, and if one had to work out the correlation values among all the variables by hand, it would certainly take a great deal of time.

Fortunately, rarely does one need to calculate correlations by hand; there are computer programs that will run on mainframe computers (such as SAS and SPSS) that will determine correlation among all the variables simply by giving on-line commands (for example, in SAS, the command *PROC CORR;* will produce a matrix of all the correlations among the variables). There are also several software programs that will run on microcomputers such as the Apple or IBM PC that can also determine correlations. Most of the programs can also easily determine the other statistics discussed here, as well as many other important statistical measurements. Memorizing these formulas is unnecessary—what is important is to know which formulas are used for which purposes. Table A.2 is a matrix of the correlations among all the characters listed in Table A.1. Note that this matrix is symmetrical—the correlation between any two characters A and B is the same as the correlation between B and A. Note also, as expected, the correlation of any character with itself must equal the number 1.

Table A.2   MATRIX OF THE CORRELATIONS AMONG THE VARIABLES

Characters*	1	2	3	4	5
1	1	−.9635	.3335	.9652	−.1591
2	−.9635	1	−.2746	−.9316	.1696
3	.3335	−.2746	1	.3648	.8321
4	.9652	−.9316	.3648	1	−.1813
5	−.1591	.1696	.8321	−.1813	1

*The characters are represented by the following numbers: (1) leaf length/width ratio; (2) petiole length/leaf length; (3) calyx apex; (4) stem pubescence; (5) bract length.

IV.   Standardizing the Data

In our example, the first two characters are in unitless dimensions, some of the data are in degrees (calyx apex), some in mm (bract length), and some in numbers per millimeters squared (stem pubescence). The mean values for all these variables, as well as their variances, differ markedly. If we used the data as they exist at present to try to classify these individuals, we would find that those characters that had high mean values or high variance would be more important in determining the classification than characters with low values. This is clearly not satisfactory, since we could change the importance of a character simply by changing the units we measured it in (bract length would have become less important as a character, had we measured it in centimeters rather than millimeters). To avoid this problem, the data are standardized; as a result of standardization, all characters will have a mean value of 0, and a variance of 1. To achieve this result, the mean for the character is subtracted from each number, and this value is divided by the standard deviation:

$$Z = (X - \bar{X})/s$$

where Z represents the standardized variable. Table A.3 is identical to Table A.1, except each of the values has been standardized.

If we calculate the mean value for the leaf length/width ratio, we will see that it is equal to 0; in addition, the variance is equal to 1. This is true for each of the other characters as well.

E.   SIMILARITY MATRICES

After calculating the character × taxon matrix, the next step is to create a similarity matrix, which will have the taxa for both the rows and the columns. The values in the similarity matrix will indicate how similar each of the taxa are to each other based on the

Table A.3   STANDARDIZED CHARACTER × TAXON MATRIX

Character	A	B	C	D	E	F
Leaf length/width	.135	.473	1.105	−1.046	−1.384	.726
Petiole length/leaf length	−.388	−.103	−1.129	1.323	1.095	−.787
Calyx apex (degrees)	1.011	1.357	−.977	−.717	−.804	.147
Stem pubescence	−.108	.749	1.010	−1.338	−1.040	.731
Bract length	1.423	.731	−1.500	−.038	−.346	−.269

Table A.4 DISTANCE MATRIX

Taxa	A	B	C	D	E	F
A	0	1.236	3.903	3.310	3.435	2.196
B	1.236	0	3.456	3.686	3.730	1.731
C	3.903	3.456	0	4.285	4.087	1.766
D	3.310	3.686	4.285	0	.598	3.560
E	3.435	3.730	4.087	.598	0	3.470
F	2.196	1.731	1.766	3.560	3.470	0

characters measured. There are several similarity indexes, each with its own advantages and disadvantages. The easiest to understand is the distance measure, which actually produces a dissimilarity matrix. We will use the standardized character × taxon matrix (Table A.3) for this example.

Looking at the leaf length/width ratio, the distance between taxon A and taxon B is .338. Similarly, the distance between A and B for the petiole length ratio is .285. The distances between each of the taxa for each of the characters can be easily determined. The Pythagorean theorem can be used to determine the total distance between each of the taxa. The general Pythagorean formula is given by:

$$\text{Distance} = \sqrt{\Sigma \ (A_i - B_i)^2}$$

where A_i is the value of character i for taxon A, and B_i is the value of character i for taxon B.

Solving this to compute the distance between A and B yields 1.236. In this same manner, we can fill in this distance matrix for each of the taxa (Table A.4).

Another commonly used similarity index is the correlation coefficient discussed earlier. This coefficient allowed us to determine the correlations among all the variables (see Table A.2). If we imagine turning the character × taxon matrix on its side, we could use the same coefficient to determine the correlations among each of the taxa. Taxa that tended to have high and low values for the same variables would have a high positive correlation. Taxa that differed greatly from each other based on the characters measured would have a large negative correlation. A matrix of these values would also be a similarity matrix. Many multivariate statistical methods are based on the correlation × coefficient similarity matrix. Table A.5 is the similarity matrix based on the correlation coefficient for our example (using the standardized data from Table A.3).

There are other similarity indexes that can be used to create the similarity matrices, but the two shown above are the most common ones. Sneath and Sokal (1973) present a basic description of the most commonly used similarity indexes.

Table A.5 SIMILARITY MATRIX BASED ON CORRELATION COEFFICIENT

Taxa	A	B	C	D	E	F
A	1	.761	−.568	−.316	−.451	.037
B	.761	1	−.222	−.652	−.729	.411
C	−.568	−.222	1	−.551	−.435	.779
D	−.316	−.652	−.551	1	.965	−.950
E	−.451	−.729	−.435	.965	1	−.893
F	.037	.411	.779	−.950	−.893	1

F. REDUCING DIMENSIONS; CLUSTER ANALYSIS AND PRINCIPAL COMPONENTS ANALYSIS

If we tried to map the position of each of the taxa in our example as a point in space based on our similarity matrix, we would find that it would take up to $n - 1$ dimensions to place each point accurately. If, for example, we had only taxa A and B, we could separate them on a single line to represent their distance from each other. If we add taxon C, we need a two-dimensional plane so that each of the taxa is the proper distance from each other. When we add taxon D, we now need to plot our points in three-dimensional space. With more and more taxa, it clearly becomes impossible to visualize their relative positions in space. To reach an understanding of the relationships among the taxa, it is necessary to reduce the dimensions of the space. This reduction is achieved at the expense of a loss of information. There are two major approaches to reducing dimensions: cluster analysis and principal components analysis. Many variants exist for each of these approaches. Because of the widespread availability of computers, it is easily possible to compare the results of classification based on different cluster techniques and principal components techniques, as well as different similarity indexes. Comparisons among these methods will point out which aspects of the resulting classification are consistent across the different methods, and which may be merely a product of a single statistical assumption. The result is a classification that does not depend on limiting and possibly erroneous assumptions.

I. Cluster Analysis

Cluster analysis produces a hierarchical classification of entities (taxa) based on the similarity matrix. There are different types of cluster analyses. Some of these types are divisive, that is, they start with all the taxa in one large group and then break this up into the best two cluster separation, then the best three clusters, and so on. Agglomerative techniques start with each OTU in its own group, and then combine these into larger and larger groupings until all the OTUs are in a single group. The types of cluster analysis differ in several other characteristics as well, perhaps the most important being whether they are weighted or unweighted methods. There are various ways to weight the different associations—for example, the importance of the different members of a cluster may depend on the level of the clustering process at which each entered the group. Another important distinction among the cluster methods is whether they are pair-group methods that combine groups a single step at a time, or whether they are variable-group methods that can join together several OTUs or groups in a single step. Sneath and Sokal (1973) present an excellent introduction to the different types of clustering techniques and the advantages and disadvantages of each.

The results of the cluster analysis are presented in a phenogram (Figure A.1). The phenogram illustrates the order in which the groups have been formed. The ordinate indicates the distance scale at which the clusters are formed. The taxonomist must decide at what level he or she wishes to recognize these groups. Thus, in the example in Figure A.1, if the taxonomist believed that groups more than three units apart were different species, he or she would recognize two species, the first formed of OTUs A, B, C, and F; the second formed of D and E. If the taxonomist felt that groups two units apart were different species, he or she would accept three species in the example—A and B, C and F, and D and E.

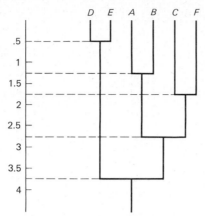

Figure A.1 Phenogram of taxonomic relationship.

The most commonly used method of cluster analysis, which also appears to produce the best results (Sneath and Sokal, 1973), is the unweighted pair-group method using arithmetic averages (UPGMA). The following is an example of this technique using the data from Table A.4. Clearly, this process can be tedious when done by hand, particularly with large data sets. There are several programs for both mainframe and microcomputers that can readily calculate the phenograms from the original data.

The first step is to find the two OTUs that are most similar. Taxa D and E are the most similar, since these two taxa have the lowest dissimilarity values (.598). We combine D and E into a new taxon, which we will name P, and create a new matrix, substituting P for D and E. To determine the distance from A, B, and C to the new taxon P, we take the average of the original distances. Thus, the distance from A to D was 3.310, and the distance from A to E was 3.435. The new distance from A to P is $(3.310 + 3.435)/2 = 3.3725$. The new matrix formed is given in Table A.6.

This procedure can be repeated a second time, again looking for the most similar taxa. In this case, A and B are the most similar with a dissimilarity value of 1.236, so a new taxon Q is formed. The new matrix is produced in the same manner (Table A.7).

The same process is repeated again, this time combining taxa C and F into the new taxon R (Table A.8). From this matrix, it can be seen that taxa Q and R will join to form S, and then finally S and P will join (Table A.9).

These results are summarized in the phenogram in Figure A.1. Note that taxa D and E are joined at .598, the dissimilarity value from the original matrix. Every pair of taxa is joined at the corresponding level from the matrices created above.

II. Principal Components Analysis

A second method for reducing the dimensions of the original data is principal components analysis (PCA). Although the mathematics required to solve this type of analysis are beyond the scope of this presentation, a basic understanding of the concept is possible. If a straight line represented a single character, all the OTUs could be placed along the line according to their value for that character. If two characters were used, a two-dimensional

Table A.6   DISTANCE MATRIX, TAXA D AND E COMBINED INTO P

Taxa	A	B	C	P	F
A	0	1.236	3.903	3.3725	2.196
B	1.236	0	3.456	3.708	1.731
C	3.903	3.456	0	4.186	1.766
P	3.3725	3.708	4.186	0	3.515
F	2.196	1.731	1.766	3.515	0

Table A.7   DISTANCE MATRIX, TAXA A AND B COMBINED INTO Q

Taxa	Q	C	P	F
Q	0	3.6795	3.54025	1.9635
C	3.6795	0	4.186	1.766
P	3.54025	4.186	0	3.515
F	1.9635	1.766	3.515	0

Table A.8   DISTANCE MATRIX, TAXA C AND F COMBINED INTO R

Taxa	Q	P	R
Q	0	3.54025	2.8215
P	3.54025	0	3.8505
R	2.8215	3.8505	0

Table A.9   DISTANCE MATRIX, TAXA Q AND R COMBINED INTO S

Taxa	P	S
P	0	3.6954
S	3.6954	0

graph would suffice to locate all OTUs. With three characters, the OTUs would be points in space, their location determined by their values for all three characters. With n characters, an n-dimensional space is required to locate all OTUs.

Principal components analysis determines the line through the cloud of points that accounts for the greatest amount of variation. This is the first principal components axis. Next, a second axis is produced, perpendicular to the first, that accounts for the next greatest amount of variation. This process continues with each succeeding axis being perpendicular to the others until all the variation is accounted for (the final number of axes is the lesser of the number of characters or one less than the number of OTUs).

It is now possible to display the data in various forms. In two dimensions, the OTUs are presented as points, and the X and Y axes are any two principal component axes. Typically, the first and the second or the first and the third axes are displayed, since these account for the greatest amount of variation. The position of the points relative to each other is an indication of their taxonomic relationship. Several programs can also produce three-dimensional graphs that display the first three PCA axes. Unlike cluster analysis, PCA allows visual interpretation of the relationships among the taxa. Ideally, a hybrid taxon would be located intermediate between the parental taxa. The closeness of the

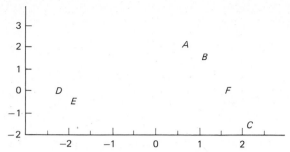

Figure A.2 First and second principal components analysis axes.

relationship can also be inferred from the PCA positions. It is important to remember that OTUs that appear close to each other on the first two or three PCA axes may actually be distant when all axes are considered together. However, since the amount of variation accounted for decreases with each succeeding axis, the later axes usually represent only the noise in the original data. Character correlations and discontinuities in the variation will weigh heavily in the first axes. These are the true indicators of taxonomic relationship.

Figure A.2 presents the results of a principal components analysis using the data from Table A.1. The first two PCA axes are shown. Note that taxa D and E, which were the first to be joined in the cluster analysis, are the closest together in the PCA. A and B are the next closest pair, which also matches favorably with the cluster analysis. In this analysis, 61 percent of the total variation in the data was accounted for by the first PCA axis; 97 percent by the first two axes together. It is clear that the first two axes in this case are sufficient to describe the data with little loss of information. Of course, in a typical data set with many characters, the first axes would account for much less of the total variation.

Table A.10 lists the eigenvectors for the five PCA axes. The eigenvectors indicate the importance of a character for a particular axis. The larger the eigenvector in absolute value, the more important is that character. For example, in the first axis, the length/width ratio, petiole length/leaf length ratio, and the stem pubescence are most important. These are the characters that are separating taxa D and E from the other four OTUs. Thus, the PCA not only produces a visual representation of the relationship among the taxa, but also indicates which characters are the most important in defining that relationship.

Table A.10   EIGENVECTORS OF PRINCIPAL COMPONENTS ANALYSIS

Char/axes[*]	PRIN 1	PRIN 2	PRIN 3	PRIN 4	PRIN 5
1	.563	−.075	−.094	.817	−.028
2	−.553	.100	.626	.472	.265
3	.248	.662	.415	−.082	−.566
4	.560	−.070	.500	−.316	.576
5	−.038	.735	−.421	.063	.526

[*]The characters are represented by the following numbers: (1) length/width ratio; (2) petiole length/leaf length; (3) calyx apex; (4) stem pubescence; (5) bract length.

G. GENERAL CONSIDERATIONS

With the large number of tests that are available, students are often concerned about which tests to use for which purposes, and which tests to perform first. Some general advice follows.

First, before any statistical tests are performed, verify that the data you have entered into the computer (assuming the tests will be performed by computer) are identical to the original data. Nothing is more frustrating than to have invested a great deal of time and perhaps money performing various statistical tests only to realize that a decimal point has been misplaced in one or more data points, and thus all the tests must be repeated. In any large data set, chances are that there will be several errors in the data entry process. The time spent double-checking all the data is invested wisely.

The use of simple statistics such as means and standard deviations is helpful for determining the distribution of the data. Calculating the correlations among the variables should be one of the first tests performed. With most statistical programs, it is also relatively easy to calculate the significance of the correlation values—that is, the probability that two variables would be as highly correlated by chance alone. A close examination of the correlation values and their significance level and an understanding of the basis of the characters will allow the proper exclusion of logically or necessarily correlated characters, as discussed above.

It is best to perform both cluster analysis and principal components analysis on the data. PCA produces a visual representation of the relative locations of the OTUs. Various pairs of axes can be plotted. It is most common to plot the first and second, and the first and third axes. One drawback is that taxa that appear close together based on the first three axes may actually be far apart when all axes are taken into account. Cluster analysis overcomes this handicap by creating clusters based on the total distance between the taxa. The technique of cluster analysis, however, will produce clusters even if the distribution of the OTUs is totally random. A combination of both techniques will help avoid the problems inherent in each of them and assure that biologically realistic results are achieved. It is best to perform several versions of both the cluster analysis and principal components analysis, as discussed above. If the result of the principal components analysis is two or more distinct groups, it is advisable to perform separate PCAs on each of the groups. This will help clarify the relationships within each group.

Many other techniques have been used by taxonomists to reduce the dimensions of the original data. Most of these are variants of cluster analysis or principal components analysis. Nearest neighbor analysis and discriminant analysis are related to cluster analysis. Principal coordinate analysis, factor analysis, canonical variate analysis, and multidimensional scaling are all related to principal components analysis.

This has been a brief introduction to the use of statistics in analyzing and interpreting taxonomic data. For a more detailed description of the assumptions, limitations, and explanations of these and other statistical techniques, consult the Suggested Reading.

SUGGESTED READING

Anonymous. 1982. SAS User's Guide: Basics. SAS Institute, Cary, North Carolina.
———. 1982. SAS User's Guide: Statistics. SAS Institute, Cary, North Carolina.
Bhattacharyya, G. K., and R. A. Johnson. 1977. Statistical Concepts and Methods. John Wiley &

Son, New York.

Cole, A. J., ed. 1969. Numerical Taxonomy. Proceedings of the Colloquium in Numerical Taxonomy held in the University of St. Andrews, September 1968. Academic Press, New York.

Cooley, W. W., and P. R. Lohnes. 1971. Multivariate Data Analysis. John Wiley & Son, New York.

Everitt, B. S. 1980. Cluster Analysis, second edition. Heineman Educational Books, Ltd, London.

Felsenstein, J, ed. 1983. Numerical Taxonomy. Springer-Verlag, New York.

Gower, J. C. 1967. A comparison of some methods of cluster analysis. Biometrics 23:623–637.

———. 1969. A survey of numerical methods useful in taxonomy. Acarologia 11:357–375.

Jardine, N., and R. Sibson. 1971. Mathematical Taxonomy. John Wiley & Son, New York.

Kshirsasger, A. M. 1972. Multivariate Analysis. Marcel Dekker, New York.

Mardia, K. V., J. T. Kent, and J. M. Bibby. 1979. Multivariate Analysis. Academic Press, New York.

Mezzich, J. E., and H. Solomon. 1980. Taxonomy and Behavioral Science. Academic Press, New York.

Morrison, D. F. 1976. Multivariate Statistical Methods, second edition. McGraw-Hill, New York.

Rao, C. R. 1964. The use and interpretation of principal component analysis in applied research. Sankya A 26:329–358.

Rohlf, F. J., and R. R. Sokal. 1981. Comparing numerical taxonomic studies. Syst. Zool. 30:459–490.

Sneath, P. H. A., and R. R. Sokal. 1973. Numerical Taxonomy. The Principles and Practice of Numerical Classification. W. H. Freeman & Co, San Francisco.

Sokal, R. R., and P. H. A. Sneath. 1963. Principles of Numerical Taxonomy. W. H. Freeman & Co, San Francisco.

Phenetic and Phylogenetic Classification*

Taxonomic entities can be classified in different fashions, dependent on the choice of features that are analyzed and also on the methodology used in their analysis. Two major schools of classification are generally recognized: *phenetics* and *phylogenetics*. (A third, rather disparate means of classification is that of *evolutionary taxonomy,* which also will be discussed.) The supposed utility of either type of classification has been a popular topic of discussion (or, at times, of vehement argument) in recent years. However, it is important to understand the basic assumptions and operations of each methodology, as both are valid means of classification. The decision of which method to utilize as a reference system depends on the ultimate goals of the classifier and may be affected by the particular taxa and characters selected for study.

A. PHYLOGENETIC CLASSIFICATION

Phylogenetic classification is the grouping of taxa by genealogical descent, as denoted by a *cladogram*. As discussed in Chapter 10, constructing a cladogram from a character × taxon matrix may be performed by inspecting the character state values of each character and sequentially grouping, in lineages, those taxa that possess shared derived features. If, however, there are many taxa, many characters, or numerous character incompatibilities (indicated as homoplasies), a more rigorous method may be needed to determine that cladogram having the least number of evolutionary steps (character state changes). Determining the most parsimonious cladogram utilizes an algorithm suggested by Kluge and Farris (1969) and elaborated upon by Farris (1970, 1972). The following example summarizes this technique, which is the basis for many available phylogenetic computer packages.

Phylogeny Construction

1. Construct the *character × taxon matrix* containing a hypothetical ancestor (ANC) coded for the ancestral states of all characters (by convention given the value 0 unless the ancestral state is intermediate in a morphocline). If the data of the matrix are incomplete, one or more characters or one or more taxa may be omitted from the analysis. An unrooted network may be constructed by deleting ancestor ANC and proceeding as in cladogram construction. For example,

	1	2	3	4	5
A	1	2	2	1	0
B	1	2	1	0	0
C	0	1	0	1	0
D	0	0	0	0	1
ANC	0	0	0	0	0

Character × Taxon Matrix

2. From the character × taxon matrix, calculate the distance between each pair of taxa (including ANC) and tabulate the data in a *distance matrix*. *Distance* is defined as the total number of character state differences between any two taxa, calculated by the formula:

$$d(X,Y) = \sum_{i=1}^{n} |V_{X_i} - V_{Y_i}|$$

where $d(X,Y)$ = the distance between taxa X and Y
n = the total number of characters
V_{X_i} = the character state value of taxon X for character i
V_{Y_i} = the character state value of taxon Y for character i

For example, from the above working matrix,

$$d(B,C) = |1 - 0| + |2 - 1| + |1 - 0| + |0 - 1| + |0 - 0|$$
$$= 4$$

	A	B	C	D	ANC
A	—	2	4	7	6
B		—	4	5	4
C			—	3	2
D				—	1
ANC					—

Distance Matrix

3. Draw a lineage connecting the ancestor ANC to that OTU that is least distant. (In this example, taxon D is least distant from ANC; $d(D,\text{ANC}) = 1$.) If two or more OTUs are equally distant from ANC, select one of these arbitrarily. Character state changes may be indicated on this lineage, but this is not necessary for the cladogram construction.

4. Determine which unplaced OTU is connected as a side branch of the original lineage; a hypothetical ancestral taxon (termed HTU1) is placed at the intersection of this side branch with the lineage. Of all unplaced OTUs, that one for which the fewest number of character state changes occur between HTU1 and the OTU is selected for the side branch. The procedure for determining this is as follows:

a. Calculate the distance between each unplaced OTU and HTU1. Because the characteristics of HTU1 (in terms of character state values) are not as yet known, these distance values must be calculated with reference to ancestor ANC, the first placed OTU, and the unplaced OTU according to the formula:

$$d(X,\text{HTU1}) = \frac{d(X,Y) + d(X,\text{ANC}) - d(Y,\text{ANC})}{2}$$

where
$\qquad X$ = an unplaced OTU
$\qquad \text{HTU1}$ = the hypothetical ancestor
$\qquad Y$ = the first placed OTU
$\qquad \text{ANC}$ = the original ancestor

For example,

$$d(A,\text{HTU1}) = \frac{d(A,D) + d(A,\text{ANC}) - d(D,\text{ANC})}{2}$$

$$= \frac{7 + 6 - 1}{2} = 6$$

$$d(B,\text{HTU1}) = \frac{d(B,D) + d(B,\text{ANC}) - d(D,\text{ANC})}{2}$$

$$= \frac{5 + 4 - 1}{2} = 4$$

$$d(C,\text{HTU1}) = \frac{d(C,D) + d(C,\text{ANC}) - d(D,\text{ANC})}{2}$$

$$= \frac{3 + 2 - 1}{2} = 2$$

Cladogram illustrating calculation of distance between X and HTU1

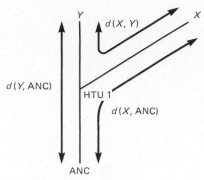

Cladogram Illustrating Calculation
of Distance Between X and HTU1

b. Connect that OTU having the least distance to HTU1 as a side branch to the original lineage. If two or more OTUs have the same minimum distance, both must be considered separately; that is, there may be two or more equally parsimonious cladograms. In this example, taxon C is minimally distant to HTU1; $d(C, \text{HTU1}) = 2$.

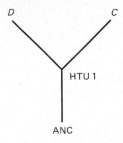

5. Determine the character state values of HTU1. The states of an HTU depend on the states of the three nearest taxonomic units (whether OTUs, other HTUs, or ANC). HTUs are constructed such that the number of evolutionary steps is minimally reduced. The values of an HTU are derived as follows:

a. If all three nearest taxa have the same state, the HTU is assigned that state. For example, for characters #1 and #3,

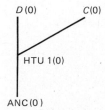

b. If two of the three nearest taxa have the same state, the HTU is assigned that state. For example, for characters #2 and #4,

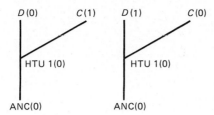

c. If all three nearest taxa have different states, the HTU is assigned the intermediate state. (In the present example, this does not occur.) For example,

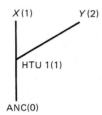

6. Add HTU1 to the character × taxon matrix. Note that, in this example, HTU1 is identical to ANC. For example,

	1	2	3	4	5
A	1	2	2	1	0
B	1	2	1	0	0
C	0	1	0	1	0
D	0	0	0	0	1
ANC	0	0	0	0	0
HTU1	0	0	0	0	0

7. Calculate the distance between HTU1 and every remaining taxon, including ANC. Add HTU1 to the distance matrix. For example,

	A	B	C	D	ANC	HTU1
A	—	2	4	7	6	6
B		—	4	5	4	4
C			—	3	2	2
D				—	1	1
ANC					—	0
HTU1						—

8. Repeat the procedure for each remaining taxon and for each internode in the clado-gram as it is constructed. In this example:

   a. Determine the next branch by testing taxa *A* and *B* relative to internodes ANC—HTU1, D—HTU1, and C—HTU1.

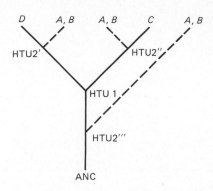

$$d(A,\text{HTU2}') = \frac{d(A,D) + d(A,\text{HTU1}) - d(D,\text{HTU1})}{2} = 6$$

$$d(B,\text{HTU2}') = \frac{d(B,D) + d(B,\text{HTU1}) - d(D,\text{HTU1})}{2} = 4$$

$$d(A,\text{HTU2}'') = \frac{d(A,C) + d(A,\text{HTU1}) - d(C,\text{HTU1})}{2} = 4$$

$$d(B,\text{HTU2}'') = \frac{d(B,C) + d(B,\text{HTU1}) - d(C,\text{HTU1})}{2} = 3$$

$$d(A,\text{HTU2}''') = \frac{d(A,\text{HTU1}) + d(A,\text{ANC}) - d(\text{ANC},\text{HTU1})}{2} = 6$$

$$d(B,\text{HTU2}''') = \frac{d(B,\text{HTU1}) + d(B,\text{ANC}) - d(\text{ANC},\text{HTU1})}{2} = 4$$

   b. Add taxon *B* to the cladogram at HTU2″ [$d(B,\text{HTU2}'') = 3$, the minimal distance]. Determine the character state values of HTU2 (=HTU2″); add to the character *X* taxon matrix and to the distance matrix (not illustrated).

   c. Add the final branch to the cladogram.

9. Map out final, most parsimonious cladogram(s), including the values of all HTUs and the character state changes along each lineage. Note, in this example, the indication of convergence (Con) for character #4.

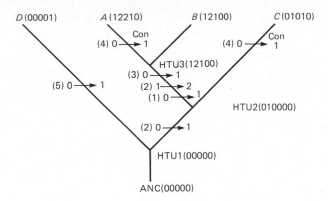

10. If homoplasy is indicated in the final cladogram, consider all possible combinations of reversals and parallelisms. For example, an equally parsimonious cladogram to the one above (below) shows an apomorphy and a reversal (Re), instead of a convergence, for character #4.

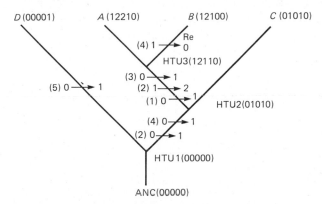

From the arrangement of taxa on the cladogram, a hierarchical classification system can be derived that directly reflects genealogical history. The procedure is simple. First, monophyletic groups are recognized as taxa. For example, in Figure B.1a, the monophyletic groups consisting of genera W and X; Y and Z; T and U; W, X, Y, and Z; V, W, X, Y, and Z; and T, U, W, X, Y, and Z are defined as taxa. (Each genus and the group as a whole are hypothesized to be monophyletic a priori.) Second, respective sister groups are given equivalent hierarchical ranks. In Figure B.1a, taxa WX and YZ are of equal rank; $WXYZ$ and V are of equal rank; TU and $VWXYZ$ are of equal rank; and S and $TUVWXYZ$ are of equal rank. Third, the more inclusive the monophyletic group, the higher its taxonomic rank. For example, from Figure B.1a, either of two classification systems, differing only in rank, might be designated (Figure B.1b or B.1c). Note that both classification schemes are not only hierarchical, they are also dichotomous; each change in rank reflects a dichotomy (ultimately a speciation) in the phylogenetic tree. Which of the two classifications is selected is arbitrary and may depend on usefulness and tradition. The important point is that the classification directly reflect phylogenetic relationship; the original cladogram can be reconstructed given the hierarchical rankings.

An obvious difficulty with the assignment of rank as directly reflecting phylogeny is that of utility. The number of available taxonomic rankings is totally insufficient for the phylogenetic classification of the over 250,000 species of plants; such a classification, if possible to construct, would be incredibly unwieldy. One possible solution is the annotated classification as proposed by Wiley (1979). An annotated system allows for the nondichotomous assignment of rank, without loss of phylogenetic information. For example, from the cladogram of Figure B.1a, tribal rank might be assigned to the monophyletic groups *S*, *TU*, *V*, and *WXYZ* (Figure B.1d). Each tribe is listed in sequence by the order of divergence of that tribal lineage from the common ancestor for the entire group. Thus, by using the *sequence* of tribes listed in the annotated classification of Figure B.1d, the cladogram of Figure B.1a can be directly reconstructed. Such an annotated classification has the advantage of flexibility. Traditional taxa can more readily be retained. In addition, taxa that have a large number of independently evolved distinctions (autapomorphies) can be distinguished. For example, genera *W*, *X*, *Y*, and *Z* are included in a single tribe in the classification of Figure B.1d. If, however, a large number of autapomorphies distinguish *WX* from *YZ* (illustrated at Figure B.1e), then the two taxa might be distinguished as separate tribes (Figure B.1f). As long as each group is monophyletic in a strict sense and sister-group relationships are apparent, the classification of taxa based in part on autapomorphic separation is not inconsistent with the methodology of phylogenetic systematics.

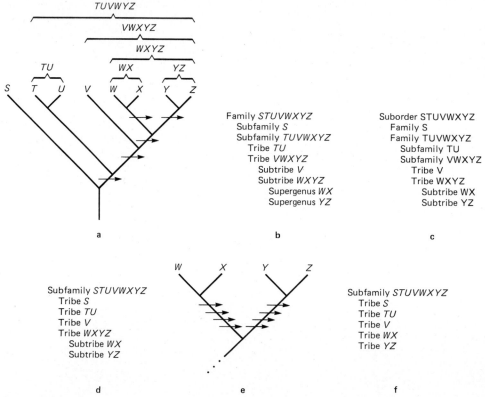

Figure B.1

B. PHENETIC CLASSIFICATION

As opposed to phylogenetic classification, in which taxa are grouped by shared derived features, *phenetic classification* is the grouping of taxa by overall similarity, regardless of whether these similarities arc symplesiomorphous or synapomorphous in a phylogenetic sense. Of the many methodologies that fall in the realm of phenetics, (including quantitative phylogenetic analysis, multivariate statistical analysis, and nonhierarchical classification; Sokal and Sneath, 1963; Sneath and Sokal, 1973), *cluster analysis* (see Appendix A) is most often utilized in devising a phenetic classification scheme and will be reiterated here. An algorithm for phenetic cluster analysis is as follows.

Phenetic Construction

1. Select taxa, characters, and character states and code them in a character $\times$ taxon matrix. Characters are often only two-state, essentially representing either the presence (usually denoted by 1) or absence (usually denoted by 0) of a particular character. Multistate characters, where present, do not strictly represent a presumed evolutionary sequence (a morphocline as in phylogenetic systematics), but denote *classes* of a continuum (either arbitrarily defined or defined by discontinuities of taxonomic characters). Because no a priori hypotheses of evolutionary direction (polarity) are made, the character $\times$ taxon matrix does not include a hypothetical ancestor.

	1	2	3	4	5
A	1	2	2	1	0
B	1	2	1	0	0
C	0	1	0	1	0
D	0	0	0	0	1

2. Calculate the similarity between each pair of taxa by means of the *coefficient of overall similarity* (s_{jk}). The coefficient of overall similarity is variously defined (see Sneath and Sokal, 1973), one of the more commonly used formulas being:

$$s_{jk} = 1 - \frac{\sum\limits_{i=1}^{n} \dfrac{|x_{ij} - x_{ik}|}{R_i}}{n}$$

where x_{ij} = the ith character state of taxon j
x_{ik} = the ith character state of taxon k
R_i = the range of the ith character, the highest minus the lowest values
n = the number of characters

For example, for the above character $\times$ taxon matrix:

$$s_{AB} = 1 - \frac{\frac{0}{1} + \frac{0}{2} + \frac{1}{2} + \frac{1}{1} + \frac{0}{1}}{5} = .7$$

$$s_{AC} = 1 - \frac{\frac{1}{1} + \frac{1}{2} + \frac{2}{2} + \frac{0}{1} + \frac{0}{1}}{5} = .5$$

$$s_{AD} = 1 - \frac{\frac{1}{1} + \frac{2}{2} + \frac{2}{2} + \frac{1}{1} + \frac{1}{1}}{5} = 0$$

$$s_{BC} = 1 - \frac{\frac{1}{1} + \frac{1}{2} + \frac{1}{2} + \frac{1}{1} + \frac{0}{1}}{5} = .4$$

$$s_{BD} = 1 - \frac{\frac{1}{1} + \frac{2}{2} + \frac{1}{2} + \frac{0}{1} + \frac{1}{1}}{5} = .3$$

$$s_{CD} = 1 - \frac{\frac{0}{1} + \frac{1}{2} + \frac{0}{2} + \frac{1}{1} + \frac{1}{1}}{5} = .5$$

3. Arrange the coefficients of overall similarities in a *similarity matrix*. (Note that s_{jk} between a taxon and itself = 1.) In the above example, the similarity matrix is:

	A	B	C	D
A	1	.7	.5	0
B		1	.4	.3
C			1	.5
D				1

4. Apply a *hierarchical clustering strategy* to the similarity matrix (to devise a hierarchical classification, one having sequentially inclusive groupings). The steps of this process are:
   a. Identify the pair of taxa, j and k, that are mutually most similar. In the above example, taxa A and B, in which $s_{AB} = .7$, are most similar.
   b. Form a new taxon P from most similar taxa j and k (in the above example, A and B become taxon P) and delete the rows and columns for j and k in the similarity matrix and replace with P; for example:

	P	C	D
P	1		
C		1	.5
D			1

   c. Calculate the coefficient of overall similarity between P and all remaining taxa. This may be performed by a variety of formulas and is a major difference between different clustering methods. The formula used in UPGMA (unweighted pair group analysis; Sneath and Sokal, 1973) is:

$$s_{Pm} = \frac{N_j s_{jm} + N_k s_{km}}{N_j + N_k}$$

   where
   P = new taxon formed from j and k
   m = all remaining taxa
   s_{jm} = similarity coefficient between taxon j and taxon m (from original similarity matrix)
   N_j = the number of original taxa contained in the jth taxon (by creation of new taxa)
   s_{km} and N_k = as above, but for taxon k

For example, in the above case,

$$s_{PC} = \frac{(1).5 + (1).4}{1 + 1} = \frac{.9}{2} = .45$$

$$s_{PD} = \frac{(1)0 + (1).3}{1 + 1} = \frac{.3}{2} = .15$$

	P	C	D
P	1	.45	.15
C		1	.50
D			1

    d. Repeat steps a to c for remaining taxa. For example, from the above:
      i. Form new taxon Q from C and D ($s_{CD} = .5$).
      ii. Construct a new similarity matrix, replacing C and D with Q:

	P	Q
P	1	
Q		1

      iii. Calculate s_{QP}

$$s_{QP} = \frac{(1).45 + (1).15}{1 + 1} = \frac{.6}{2} = .3$$

5. Construct a hierarchical classification by the order in which taxa are united to form a new taxon (based on the maximum s_{jk}). This classification can be represented as a *phenogram,* having a scale of similarity (from 0 to 1), denoted as *phenetic distance.* The connection between any two taxa occurs at that phenetic distance equal to their s_{jk}.

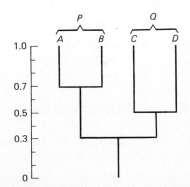

The phenogram, like the cladogram, can be used to construct a hierarchical system of classification directly, achieved by sequentially grouping most similar pairs of taxa (see below). An arbitrary similarity value (phenetic distance) may be chosen to delimit taxonomic ranks (e.g., in defining species, genera, tribes, etc.). As with a phylogenetic classification, traditional taxonomic categories may be preserved as long as they fit the basic framework of the phenogram.

C. PHENETIC AND PHYLOGENETIC CLASSIFICATION

In the above examples of cladogram and phenogram construction, the resultant groupings are quite different, even though the same data matrix was used for taxa A to D. From these examples, the hierarchical classifications of taxa A to D, using phylogenetic versus phenetic methodology, is:

Phylogenetic Classification	*Phenetic Classification*
Taxon $ABCD$	Taxon $ABCD$
Taxon D	Taxon AB
Taxon ABC	Taxon A
Taxon AB	Taxon B
Taxon A	Taxon CD
Taxon B	Taxon C
Taxon C	Taxon D

The possible discrepancy between phenetic and phylogenetic classifications is the result of the different premises of their methodologies. Phenetic classification groups taxa based on overall similarity, the object being to measure relative similarity between pairs of taxa mathematically. In contrast, a major emphasis of phylogenetic systematics is that shared primitive character states (*symplesiomorphies*) reveal nothing about genealogy; thus, similarities between taxa that are symplesiomorphic are not utilized in phylogeny construction and may, in fact, be deceptive as to supposed genealogical relationships. For example, from the character × taxon matrix of Figure B.2a, taxa X and Y would be grouped together in a phenogram (Figure B.2b) because X and Y share the states of more characters (state 0 of characters 2 and 3) than either does with taxon W. However, with the addition of a hypothetical ancestor (Figure B.2c), it is apparent that taxa W and X share a derived feature (state 1 of character 1), constituting evidence that these taxa share a more recent common ancestor with one another than either does to taxon Y; taxa X and X, therefore, would be grouped together in a cladogram (Figure B.2d). The similarities between X and Y are all symplesiomorphic and yield no information with regard to phylogeny. Thus, according to phylogenetic principles, only shared derived features can be used to detect the true genealogical history. The common practice in traditional taxonomy of defining natural groups based on overall similarity may be invalid in a phylogenetic classification if those similarities are symplesiomorphic.

The validity of phylogenetic versus phenetic classifications has been a controversial topic of debate in recent years. Those in the phylogenetic camp (the cladists) have argued

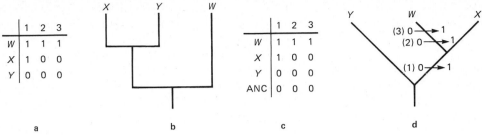

Figure B.2

that grouping taxa based on evolutionary descent is itself sufficient reason in support of phylogenetic classification and that a phylogenetic classification is more stable (Michevich, 1978) and more natural (Farris, 1976) than a phenetic one. Pheneticists have countered that a phylogenetic system of classification is difficult to construct because of general uncertainties in devising morphoclines and in determining polarity (e.g., out-group taxa being difficult to recognize). In addition, the methodology of phylogenetic systematics relies on the principle of parsimony and assumes the existence of past evolutionary events that can be detected today as a synapomorphy in defining a particular monophyletic group; the occurrence of numerous reversals and convergences (homoplasy) in a cladogram may likely be an indication that such synapomorphies cannot be confidently recognized. However, rather than blindly follow one or the other methodology in classification, it is important to understand the assumptions and purposes of each. Utility is always a major impetus in classification. A phylogenetic system, directly reflecting inferred genealogy, may be preferable when there is a high degree of confidence in that genealogy (e.g., regarding hypotheses of monophyly, homology, or character polarities). A phenetic cluster analysis may be preferable in classifying taxa of low rank, for example, species that are difficult to delimit because of rampant hybridization; in addition, the polarities of certain types of quantitative data may be virtually impossible to assess; the only meaningful information obtainable from these data may be via an approach portraying overall similarity. The thorough taxonomic study should include several approaches of classification, perhaps including phenetic clustering, principal components analysis, Wagner tree cladogram construction, and character compatibility analysis. The resultant groupings can then be compared, and the bases for their differences can be analyzed with respect to the data and methodologies used, undoubtedly providing insight as to the original problems and controversies of classification.

D. EVOLUTIONARY TAXONOMIC CLASSIFICATION

Evolutionary taxonomy is a disparate philosophy of classification characterized by the acceptance of paraphyletic groups as valid taxa. *Paraphyletic groups* (also called grades) are those in which not all descendants of a common ancestor are included (see Farris, 1974, for a necessarily more strict definition of paraphyly and polyphyly). For example, in Figure B.3a, the group containing taxa *T*, *U*, *V*, and *W* is paraphyletic because taxon *X* is omitted. Various justifications for the acceptance of paraphyletic groups have been made, particularly the occurrence of a large morphological gap (numerous character state changes) separating the unincluded group from the paraphyletic one (see Wiley, 1981, for a review of the purported justifications for accepting paraphyletic taxa). For example, in Figure B.3a, taxon *X* may possess numerous autapomorphies, so distinguishing it from the other taxa as to presumably warrant its classification equal in rank to *T*, *U*, *V*, and *W*. An example of a group illustrating paraphyly is the group gymnosperm. If, as is commonly accepted, the angiosperms arose from a gymnosperm ancestor, then the gymnosperms are not a monophyletic taxon because that group does not include *all* descendants of a common ancestor. The gymnosperms are a paraphyletic group (analogous to taxon *TUVW* in Figure B.3a); the many species of angiosperms are a monophyletic group, defined by numerous synapomorphies (analogous to taxon *X* in Figure B.3a).

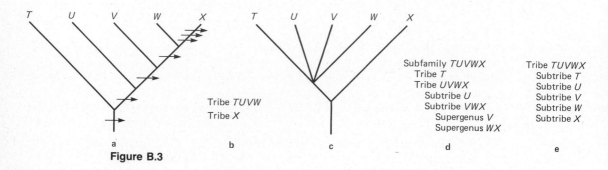

Figure B.3

Proponents of evolutionary taxonomic classification may or may not accept the methodology of phylogenetic systematics in the construction of genealogical branching patterns. However, the net effect is that the paraphyletic group (as recognized in a cladogram) is classified as a discrete taxon, usually of equal or lower rank than the unincluded group. For example, an evolutionary taxonomic classification might group taxa *T*, *U*, *V*, and *W* as one tribe and genus *X* as a separate tribe (Figure B.3b). This grouping might in itself be used to reconstruct a presumed genealogical branching pattern (as in Figure B.3c), one that is, however, nonrigorous and does not consider character polarity. Alternatively, the classification of Figure B.3b could be devised even though the cladogram of Figure B.3a is treated as valid by the classifier. The only difference, then, between the phylogenetic and the evolutionary taxonomic classification is that, in the latter, monophyletic taxa are not necessarily accepted as taxonomic entities. However, the classification of taxa *T*, *U*, *V*, and *W* as a tribe and *X* as another tribe reveals nothing regarding the true genealogical history of taxa *T*, *U*, *V*, *W*, and *X* and is misleading because the two tribes are *not* sister groups. In contrast, a phylogenetic classification, whether dichotomous (Figure B.3d) or annotated (Figure B.3e), is derived directly from the cladogram and can be used to reconstruct it directly. Thus the advantage of a phylogenetic classification is its greater information content with regard to constructing that genealogical history.

SUGGESTED READING

Farris, J. S. 1970. Methods for computing Wagner trees. Systematic Zoology 19:83–92.

———. 1972. Estimating phylogenetic trees from distance matrices. American Naturalist 106:645–668.

———. 1974. Formal definitions of paraphyly and polyphyly. Systematic Zoology 23:548–554.

———. 1976. On the phenetic approach to vertebrate classification. In Hecht, Goody, and Hecht (eds.), Major Patterns in Vertebrate Evolution, pp. 823–850. Plenum Press, New York.

Gilmour, J. S. L. 1961. Taxonomy. In Macleod and Cobley (eds.), Contemporary Botanical Thought. Quadrangle Books, Chicago.

Kluge, A., and J. S. Farns. 1969. Quantitative phyletics and the evolution of anurans. Systematic Zoology 18:1–32.

Michevich, M. G. 1978. Taxonomic congruence. Systematic Zoology 27:112–118.

Rohlf, F. S. 1970. Adaptive hierarchic clustering schemes. Systematic Zoology 19:58–82.

Sneath, P. H. A. and Sokal, R. R. 1973. Numerical Taxonomy. W. H. Freeman, San Francisco, California.

Sokal, R. R. and Sneath, P. H. A. 1963. Principles of Numerical Taxonomy. W. H. Freeman, San Francisco.

Wiley, E. O. 1979. An annotated Linnean hierarchy, with comments on natural taxa and competing systems. Systematic Zoology 28:308–337.

———. 1981. Phylogenetics, the Theory and Practice of Phylogenetic Systematics. John Wiley & Sons, New York.

appendix C

Families of Flowering Plants

Twenty-four of the 27 plates used to illustrate families in this appendix are part of a series that Carroll E. Wood, Jr., has selected for use in teaching from among some 490 illustrations prepared for A Generic Flora of the Southeastern United States, *which is currently directed by him at Harvard University and by Dr. Norton G. Miller at the New York Biological Survey, Albany, New York. One-hundred twenty of these specially labeled illustrations including representatives of 70 families have been published in* A Student's Atlas of Flowering Plants: Some Dicotyledons of Eastern North America *(Harper & Row, 1974), and all of those used here were also published in* Vascular Plant Systematics *(Harper & Row, 1974).*

These illustrations have been intended to show as many structural and biological details as possible, and not to be merely "recognition" drawings. For teaching purposes these plates have been modified by placing the labels directly on the illustration, rather than in a conventional legend below it. The labels were prepared by Professor Wood with the collaboration of Dr. Elizabeth A. Shaw, and the help of Dr. Kenneth R. Robertson. Although originally prepared for reduction to one-half, in the present format it has not been feasible to give the magnification of each component figure. Of the 24 plates, 11 were originally published (with a greater reduction and with conventional legends) in the Journal of the Arnold Arboretum, *in which the extensive series of papers (108 to date) treating the genera of flowering plants in the southeastern United States has been appearing since 1958 (vol. 39). Additional information about these illustrations and the artists who drew them can be found in the two books referred to above.*

The family is a rank between order and genus in the taxonomic hierarchy. The family as a taxon is a group of plants with a particular circumscription, position, and name at the rank of family. The family as a taxon includes one or more genera. No set number of characteristics, no special types of characters, and no particular number of included taxa are required for the designation of any group as a family.

Natural groups now known as families, such as the mints, umbels, legumes, and composites, have been recognized for centuries by layman and taxonomist alike. Most of our major groups, now termed families, have been known since the time of Antoine de Jussieu, a Frenchman of the eighteenth century. Pierre Magnol, another Frenchman, understood the family as a concept prior to 1700.

With the recent advent of numerical taxonomy and the use of characters from many fields of evidence, more objectivity has been introduced into the delimitation and classification of families. Some of the major groups previously known as families are now being more accurately circumscribed and effectively delimited. The family is a fundamental unit in the taxonomy of plants.

Traditionally, the family has been a study unit in systematic teaching and training, with most basic training occurring in local flora courses (see Family Study Worksheet, p. 405). Expedience in identification of species comes with the recognition of families. Comprehension of present classification systems depends on a fundamental understanding of diagnostic features for the family. Three approaches are used in this presentation for understanding the flowering plant family conceptually and identifiably: (1) synoptic keys to classes and subclasses and descriptions of individual families are presented as conceptual aids for the fundamental understanding of higher taxa and selected families, (2) diagnostic characters are given as aids in field identification, and (3) plates illustrating features of the structurally more difficult families are included for visual help.

Since most individuals gain a high percentage of their experience with families while using floras or manuals, a few remarks are in order on the treatments of families in those publications.

A. FAMILIES IN FLORAS AND MANUALS

Summary charts of the number of divisions, classes, orders, *families,* genera, and species are found in the prefaces of most manuals. The largest families, based on the number of genera and/or species included, are frequently noted in the introduction to the book. The families are indexed alphabetically and frequently given a numerical code number on the inside front cover of some floras or manuals. The indexed and coded families provide convenience for retrieval of information. The nature of family descriptions and the descriptive format are traditionally placed in the preface.

The author of each family treated in the book usually is listed in the preface or in a footnote on the first page of the description. Family adaptations from previous manuals or publications are also documented in footnotes accompanying the family treatment. Basic references are included at the end of the family description. The herbarium locations for the specimens used in the descriptions generally are indicated in the preface.

Synoptical as well as artificial keys to the families are placed in the introductory portions of the book or toward the end of the prefatory material. Occasionally, descriptions of orders are included in the manual along with family keys.

Family descriptions vary in length and complexity. Monotypic families have brief diagnostic descriptions or none at all—with the family characteristics encompassed in the generic or specific descriptions. Structurally complex families such as the Asteraceae, Poaceae, and Asclepiadaceae require detailed accounts of their morphology. Families with a diversity of characteristics—for example, apocarpy and syncarpy; epigyny, perigyny, and hypogyny; capsules, berries, drupes, and follicles—definitely require lengthier descriptions. Families with few genera usually have intermediate length treatments.

Some manuals have the families, genera, and species numbered so that the taxa can be encoded easily for data-banking purposes. For example in the *Manual of the Vascular Flora of the Carolinas,* 179.85.1 refers to Asteraceae (Family 179.), *Tanacetum* (Genus 85.), *T. vulgare* (Species 1), and 38.2.2 indicates Commelinaceae (Family 38.), *Aneilema* (Genus 2.), *A. keisak* (Species 2). The family number is frequently used as a running head. In the flora of the Carolinas, for example, 179 is in boldface type at the top of all pages dealing with the Asteraceae, and 38 is at the top of the pages treating the Commelinaceae.

A conceptual understanding of a family treated in a flora or manual includes the following:

A study of the family description in the manual for circumscriptive features.

An analysis of the characters used in the synoptic and artificial keys to families found in the manual for diagnostic characteristics.

A visual examination of the illustrations in the family treatment for structural variation in the family.

A perusal of the flowering/fruiting data in the family treatment for a phenologic characterization of the family.

A reading of habitat data in the family treatment for general, specialized, or unique habitats within the family.

A study of the distribution maps in the family treatment for indications of geographic patterns within the family.

B. SYNOPSES AND SYNOPTICAL ARRANGEMENTS

I. Synopsis of Characteristics of the Divisions of Vascular Plants and Classes of Flowering Plants*

Polypodiophyta (Ferns) [Pteridophyta]

Seedless plants with alternate, generally large and compound leaves with a branching vein system. Sporangia borne on ordinary foliage leaves, occasionally on highly modified leaves, in clusters known as sori on the lower surface or margin.

Lycopodiophyta (Club Mosses) [Pteridophyta]

Seedless plants with alternate or opposite small and simple leaves with an unbranched midvein (rarely 2). Sporangia borne singly on ordinary foliage leaves or reduced leaves grouped into a terminal cone known as a strobilus.

*Cronquist, A., 1982, Basic Botany, second edition. Harper & Row, Publishers, New York. Chapter 27, p. 494. Used with permission.

Equisetophyta (Horsetails) [Pteridophyta]

Seedless plants with whorled, small, and simple leaves with a single unbranched midvein. Sporangia borne in groups of 5 to 10 on a sporangiophore with many sporangiophores grouped into a terminal cone known as a strobilus.

Pinophyta (Gymnosperms) [Spermatophyta]

Naked seed-bearing plants without flowers and with generally alternate needle- or scalelike leaves with simple venation patterns. Sporangia (male and female) borne in groups of 2 or more on sporophylls grouped into cones known as strobili(us).

Magnoliophyta (Angiosperms) [Spermatophyta]

Enclosed seed-bearing plants with flowers and leaves of varied shapes, sizes, and venation patterns. Male sporangia borne in groups of 1 to 4 in the anther; female sporangia borne in groups of 1 to many enclosed in the ovary.

Class: Magnoliopsida (Dicots)

Cotyledons 2 (seldom 1, 3, or 4)

Leaves mostly net-veined

Intrafascicular cambium usually present

Vascular bundles usually borne in a ring that encloses a pith

Floral parts, when of definite number, typically borne in sets of 5, less often 4, seldom 3 (carpels often fewer)

Pollen typically triaperturate, or of triaperturate-derived type, except in a few of the more primitive families

Mature root system either primary or adventitious, or both

Class: Liliopsida (Monocots)

Cotyledon 1 (or the embryo sometimes undifferentiated)

Leaves mostly parallel-veined

Intrafascicular cambium absent; usually no cambium of any sort

Vascular bundles generally scattered or in 2 or more rings

Floral parts, when of definite number, typically borne in sets of 3, seldom 4, never 5 (carpels often fewer)

Pollen of uniaperturate or uniaperturate-derived type

Mature root system wholly adventitious

II. Synoptical Arrangements of the Subclasses of Magnoliopsida and Liliopsida
Synoptical Arrangement of the Subclasses of Magnoliopsida†

(Figure C.1e)

1. Plants relatively archaic, the flowers typically apocarpous, always polypetalous or apetalous (but sometimes synsepalous) and generally with an evident perianth, usually with numerous (sometimes laminar or ribbon-shaped) stamens initiated in centripetal sequence, the pollen grains mostly binucleate and often uniaperturate or of a uniaperturate-derived type; ovules bitegmic and crassinucellar; seeds very often with a tiny embryo and copious endosperm, but sometimes with a larger embryo and reduced or no endosperm; cotyledons occasionally more than 2; plants very often accumulating benzylisoquinoline or aporphine alkaloids, but without betalains, iridoid compounds, or mustard oils, and seldom strongly tanniferous. I. Magnoliidae. (Figure C.1d)

1. Plants more advanced in one or more respects than the Magnoliidae; pollen grains triaperturate or of a triaperturate-derived type; cotyledons not more than 2; stamens not laminar, generally with well-defined filament and anther; plants only rarely producing benzylisoquinoline or aporphine alkaloids, but often with other kinds of alkaloids, or tannins, or betalains, or mustard oils, or iridoid compounds. 2.

2. Flowers more or less strongly reduced and often unisexual, the perianth poorly developed or wanting, the flowers often borne in catkins, but never forming bisexual pseudanthia, and never with numerous seeds on parietal placentas; pollen-grains often porate and with a granular rather than columellar infratectal structure, but also often of ordinary type. II. Hamamelidae. (Figure C.1b)

2. Flowers usually more or less well developed and with an evident perianth, but if not so, then usually either grouped into bisexual pseudanthia or with numerous seeds on parietal placentas, only rarely with all of the characters of the Hamamelidae as listed above; pollen grains of various architecture, but rarely if ever both porate and with a granular infratectal structure. 3.

3. Flowers polypetalous or less often apetalous or sympetalous, if sympetalous then usually either with more stamens than corolla lobes, or with the stamens opposite the corolla lobes, or with bitegmic or crassinucellar ovules; ovules only rather seldom with an integumentary tapetum; carpels 1 to many, distinct or more often united to form a compound pistil; plants often tanniferous or with betalains or mustard oils. 4.

4. Stamens, when numerous, usually initiated in centrifugal (seldom centripetal) sequence; placentation various, often parietal or free-central or basal, but also often axile; species with few stamens and axile placentation usually either bearing several or many ovules per locule, or with a sympetalous corolla, or both. 5.

5. Plants usually either with betalains, or with free-central to basal placentation (in a compound ovary), or both, but lacking both mustard oils and iridoid compounds, and tanniferous only in the two smaller orders; pollen grains trinucleate or seldom binucleate; ovules bitegmic, crassinucellar, most often campylotropous or amphitropous; plants most commonly herbaceous or nearly so,

†These arrangements are keylike for placing the subclasses in perspective; they are not intended for identification. From Cronquist, A., 1981, An Integrated System of Classification of Flowering Plants, Columbia University Press, New York. Used with permission.

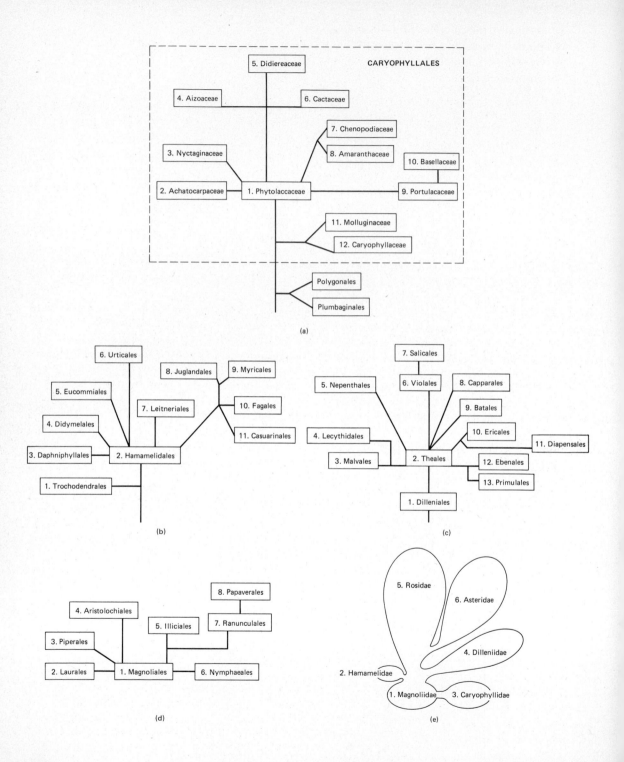

CARYOPHYLLALES

5. Didiereaceae

4. Aizoaceae   6. Cactaceae

3. Nyctaginaceae   7. Chenopodiaceae

8. Amaranthaceae

10. Basellaceae

2. Achatocarpaceae   1. Phytolaccaceae   9. Portulacaceae

11. Molluginaceae

12. Caryophyllaceae

Polygonales

Plumbaginales

(a)

6. Urticales

5. Eucommiales   8. Juglandales   9. Myricales

4. Didymelales   7. Leitneriales   10. Fagales

3. Daphniphyllales   2. Hamamelidales   11. Casuarinales

1. Trochodendrales

(b)

7. Salicales

5. Nepenthales   6. Violales   8. Capparales

9. Batales

4. Lecythidales   10. Ericales   11. Diapensales

3. Malvales   2. Theales   12. Ebenales

13. Primulales

1. Dilleniales

(c)

8. Papaverales

4. Aristolochiales   5. Illiciales   7. Ranunculales

3. Piperales

2. Laurales   1. Magnoliales   6. Nymphaeales

(d)

5. Rosidae   6. Asteridae

4. Dilleniidae

2. Hamamelidae

1. Magnoliidae   3. Caryophyllidae

(e)

334

with woody species usually with anomalous secondary growth or otherwise anomalous stem structure; petals distinct or wanting, except in the Plumbaginales; in the largest order (Caryophyllales) the sieve tubes with a unique sort of P-type plastid, and the seeds usually with perisperm instead of endosperm. III. Caryophyllidae. (Figure C.1a)

5. Plants without betalains, and the placentation only rarely (except in the Primulales) free-central or basal; plants often with mustard oils or iridoid compounds or tannins; pollen-grains usually binucleate (notable exception: Brassicaceae); ovules various, but seldom campylotropous or amphitropous except in the Capparales; plants variously woody or herbaceous, many species ordinary trees; petals distinct or less often connate to form a sympetalous corolla, seldom wanting; seeds only seldom with perisperm; plastids of the sieve tubes usually of S-type, but in any case not as in the Caryophyllales. IV. Dilleniidae. (Figure C.1c)

4. Stamens, when numerous, usually initiated in centripetal (seldom centrifugal) sequence; flowers seldom with parietal placentation (notable exception: Saxifragaceae) and also seldom (except in parasitic species) with free-central or basal placentation in a unilocular, compound ovary, but very often (especially in species with few stamens) with 2 to several locules that have only 1 or 2 ovules of each; flowers polypetalous or less often apetalous, only rarely sympetalous; plants often tanniferous, and sometimes with iridoid compounds, but only rarely with mustard oils, and never with betalains. V. Rosidae. (Figure C.2h)

3. Flowers sympetalous (rarely polypetalous or apetalous); stamens generally isomerous with the corolla lobes or fewer, never opposite the lobes; ovules unitegmic and tenuinucellar, very often with an integumentary tapetum; carpels most commonly 2, occasionally 3 to 5 or more; plants only seldom tanniferous, and never with betalains or mustard oils, but often with iridoid compounds or various other sorts of repellents. VI. Asteridae. (Figure C.2g)

III. Synoptical Arrangement of the Subclasses of Liliopsida   (Figure C.2f)

1. Plants either with apocarpous (sometimes monocarpous) flowers or more or less aquatic or very often both, always herbaceous, but never thalloid; vascular system generally not strongly lignified, often much reduced, the vessels confined to the roots, or wanting; endosperm mostly wanting, not starchy when present; subsidiary cells mostly 2, pollen trinucleate. I. Alismatidae. (Figure C.2a)

1. Plants with syncarpous (or seldom pseudomonomerous) flowers except in some arborescent taxa, usually terrestrial (or epiphytic), much less often more or less aquatic, although sometimes not only aquatic but also thalloid and free-floating; vessels, endosperm, stomates, and pollen various, but not combined as in the Alismatidae. 2.

2. Flowers usually numerous, usually small, and subtended by a prominent spathe (or several spathes), often aggregated into a spadix (but the inflorescence much reduced in the Lemnaceae); plants very often either arborescent or with relatively broad leaves that do not have typical parallel venation or both (but sometimes lacking both of these features, and in the Lemnaceae even thalloid and free-floating); septal nectaries wanting except in many Arecaceae; subsidiary cells typically 4, less often 2 or more than 4; vessels generally present in all vegetative organs, except among the Arales; endosperm not starchy except in some Arales. II. Arecidae. (Figure C.2d)

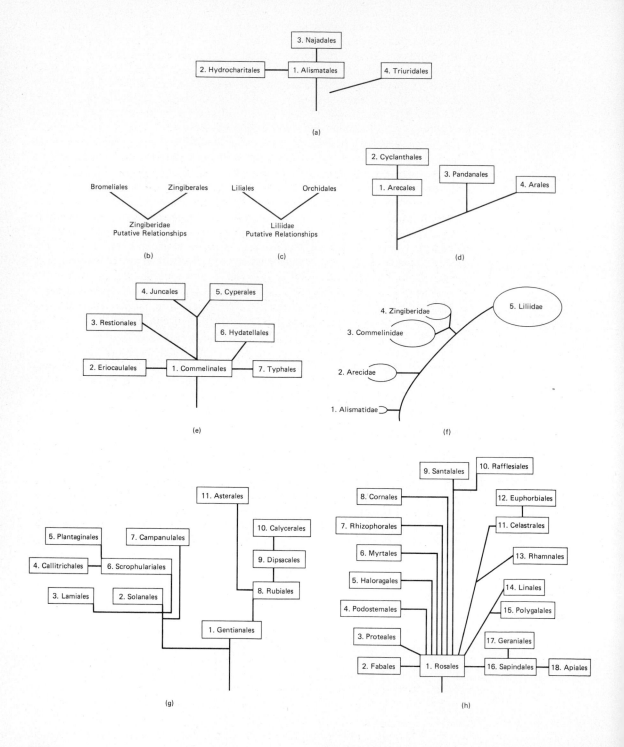

2. Flowers few to numerous and small to large, but never aggregated into a spadix, and usually without a distinct spathe (inflorescence with 1 or more spathelike bracts in many Zingiberidae, but the flowers then larger and showy); plants herbaceous or much less often arborescent; leaves narrow and parallel-veined or less often broader and more or less net-veined or with a special type of pinnate-parallel venation; nectaries, stomates, vessels, and endosperm various but not combined as in typical Arecidae. 3.

3. Nectar and nectaries mostly wanting; perianth in the more archaic families trimerous and with well-differentiated sepals and petals, in the more advanced families reduced and chaffy and often not obviously trimerous, or wanting, the families with reduced perianth typically adapted to wind pollination; ovary always superior (or nude); vessels generally present in all vegetative organs; endosperm wholly or in large part starchy, commonly mealy and with compound starch grains, without significant reserves of hemicellulose and usually also without significant reserves of oil or rarely the endosperm wanting; stomates mostly with 2 subsidiary cells, seldom without subsidiary cells or with more than 2. III. Commelinidae. (Figure C.2e)

3. Nectar and nectaries of one or another sort (often septal nectaries) generally present; perianth generally well developed, not reduced and chaffy, the flowers typically adapted to pollination by insects or other kinds of animals; ovary superior or very often inferior; vessels most often confined to the roots, but sometimes occurring also in the stem or in all vegetative organs; endosperm and stomates various. 4.

4. Sepals usually well differentiated from the petals, often green and herbaceous, sometimes petaloid in texture but still unlike the petals; endosperm typically starchy and mealy, with compound starch grains, only seldom hard and presumably with reserves of hemicellulose; subsidiary cells (2) 4 or more; plants not obviously mycotrophic; leaves narrow and parallel-veined, or equally often with a broad, expanded, petiolate blade and a characteristic pinnate-parallel venation. IV. Zingiberidae. (Figure C.2b)

4. Sepals usually petaloid in form and texture, sometimes differentiated from the petals but still petaloid in appearance, only rarely green and herbaceous; endosperm when present typically very hard, with reserves of hemicellulose, protein, and oil, less commonly starchy, but not mealy in texture, the starch grains typically simple rather than compound, or very often the endosperm wanting; subsidiary cells 2 or often none, only seldom four; plants often strongly mycotrophic; leaves typically narrow and parallel-veined, but without the characteristic pinnate-parallel venation of many Zingiberidae. V. Liliidae. (Figure C.2c)

C. FAMILY DESCRIPTIONS AND ILLUSTRATIONS*

I. Index to Families of Flowering Plants

Aceraceae	Aquifoliaceae
Alismataceae	Araceae
Amaranthaceae	Asclepiadaceae
Apiaceae	Asteraceae

This treatment of families except for the illustrations is from Massey, J. R. et al., 1979, Laboratory Guide for An Introduction to Plant Taxonomy, Burgess Publishing Company, Minneapolis, Minn. Used with Permission.

Betulaceae	Juncaceae
Boraginaceae	Labiatae
Brassicaceae	Lamiaceae
Cactaceae	Leguminosae
Campanulaceae	Liliaceae
Caryophyllaceae	Magnoliaceae
Chenopodiaceae	Malvaceae
Clusiaceae	Nymphaeaceae
Commelinaceae	Oleaceae
Compositae	Onagraceae
Convolvulaceae	Orchidaceae
Cornaceae	Plantaginaceae
Crassulaceae	Poaceae
Cruciferae	Polemoniaceae
Cucurbitaceae	Polygonaceae
Cyperaceae	Potamogetonaceae
Ericaceae	Primulaceae
Euphorbiaceae	Ranunculaceae
Fabaceae	Rosaceae
Fagaceae	Rubiaceae
Geraniaceae	Salicaceae
Gramineae	Scrophulariaceae
Guttiferae	Solanaceae
Hamamelidaceae	Typhaceae
Hydrocharitaceae	Ulmaceae
Hypericaceae	Umbelliferae
Iridaceae	Verbenaceae
Juglandaceae	Violaceae

II. Systematic Arrangement of Selected Families of Flowering Plants (after Cronquist, 1981)

Class	Subclass	Order	Family
Liliopsida	Alismatidae	Alismatales	Alismataceae
Liliopsida	Alismatidae	Hydrocharitales	Hydrocharitaceae
Liliopsida	Alismatidae	Najadales	Potamogetonaceae
Liliopsida	Commelinidae	Commelinales	Commelinaceae

Class	Subclass	Order	Family
Liliopsida	Commelinidae	Juncales	Juncaceae
Liliopsida	Commelinidae	Cyperales	Cyperaceae
Liliopsida	Commelinidae	Cyperales	Poaceae
Liliopsida	Commelinidae	Typhales	Typhaceae
Liliopsida	Arecidae	Arales	Araceae
Liliopsida	Liliidae	Liliales	Liliaceae
Liliopsida	Liliidae	Liliales	Iridaceae
Liliopsida	Liliidae	Orchidales	Orchidaceae
Magnoliopsida	Magnoliidae	Magnoliales	Magnoliaceae
Magnoliopsida	Magnoliidae	Nymphaeales	Nymphaeaceae
Magnoliopsida	Magnoliidae	Ranunculales	Ranunculaceae
Magnoliopsida	Hamamelidae	Hamamelidales	Hamamelidaceae
Magnoliopsida	Hamamelidae	Urticales	Ulmaceae
Magnoliopsida	Hamamelidae	Juglandales	Juglandaceae
Magnoliopsida	Hamamelidae	Fagales	Fagaceae
Magnoliopsida	Hamamelidae	Fagales	Betulaceae
Magnoliopsida	Caryophyllidae	Caryophyllales	Cactaceae
Magnoliopsida	Caryophyllidae	Caryophyllales	Caryophyllaceae
Magnoliopsida	Caryophyllidae	Caryophyllales	Chenopodiaceae
Magnoliopsida	Caryophyllidae	Caryophyllales	Amaranthaceae
Magnoliopsida	Caryophyllidae	Polygonales	Polygonaceae
Magnoliopsida	Dilleniidae	Theales	Clusiaceae
Magnoliopsida	Dilleniidae	Malvales	Malvaccae
Magnoliopsida	Dilleniidae	Violales	Violaceae
Magnoliopsida	Dilleniidae	Violales	Curcurbitaceae
Magnoliopsida	Dilleniidae	Salicales	Salicaccac
Magnoliopsida	Dilleniidae	Capparales	Brassicaceae
Magnoliopsida	Dilleniidae	Ericales	Ericaceae
Magnoliopsida	Dilleniidae	Primulales	Primulaceae
Magnoliopsida	Rosidae	Rosales	Crassulaceae
Magnoliopsida	Rosidae	Rosales	Rosaceae
Magnoliopsida	Rosidae	Rosales	Fabaceae
Magnoliopsida	Rosidae	Myrtales	Onagraceae
Magnoliopsida	Rosidae	Cornales	Cornaceae
Magnoliopsida	Rosidae	Celastrales	Aquifoliaceae
Magnoliopsida	Rosidae	Euphorbiales	Euphorbiaceae
Magnoliopsida	Rosidae	Sapindales	Aceraceae
Magnoliopsida	Rosidae	Geraniales	Geraniaceae
Magnoliopsida	Rosidae	Umbellales	Apiaceae
Magnoliopsida	Asteridae	Gentianales	Asclepiadaceae
Magnoliopsida	Asteridae	Polemoniales	Solanaceae
Magnoliopsida	Asteridae	Polemoniales	Convolvulaceae
Magnoliopsida	Asteridae	Polemoniales	Polemoniaceae
Magnoliopsida	Asteridae	Lamiales	Boraginaceae
Magnoliopsida	Asteridae	Lamiales	Verbenaceae
Magnoliopsida	Asteridae	Lamiales	Lamiaceae
Magnoliopsida	Asteridae	Plantaginales	Plantaginaceae
Magnoliopsida	Asteridae	Scrophulariales	Oleaceae

(*continued overleaf*)

Class	Subclass	Order	Family
Magnoliopsida	Asteridae	Scrophulariales	Scrophulariaceae
Magnoliopsida	Asteridae	Campanulales	Campanulaceae
Magnoliopsida	Asteridae	Rubiales	Rubiaceae
Magnoliopsida	Asteridae	Asterales	Asteraceae

III. References for Families of Flowering Plants

The following references have been used in the preparation of the descriptions of the selected families of flowering plants. Numbers 8, 20, 21, 25, 29, 30, 31, 36, and 40 were used as basic references in all family descriptions. The remaining references refer to specific families. The numbers below correspond to numbers on the individual family descriptions under the section entitled References. The numbers for the Illustrations correspond with the reference numbers.

1. Benson, L. 1957. Plant Classification. D. C. Heath and Co., Boston.
2. Brizicky, G. K. 1961. The genera of Violaceae in the Southeastern United States. J. Arnold Arbor. 42:321–333.
3. ———. 1963. The genera of Sapindales in the Southeastern United States. J. Arnold Arbor. 44:462–501.
4. ———. 1964. The genera of Celastrales in the Southeastern United States. J. Arnold Arbor. 45:206–234.
5. Channell, R. B. and C. E. Wood, Jr. 1959. The genera of Primulales in the Southeastern United States. J. Arnold Arbor. 40:268–288.
6. Chase, A. 1968. First Book of Grasses, third edition. Smithsonian Institution Press, Washington, D. C.
7. Cronquist, A. 1955. Phylogeny and taxonomy of the Compositae. Amer. Midl. Nat. 53:476–511.
8. ———. 1968. The Evolution and Classification of Flowering Plants. Houghton Mifflin Co., Boston.
9. Elias, T. S. 1970. The genera of Ulmaceae in the Southeastern United States. J. Arnold Arbor. 51:18–40.
10. ———. 1971. The genera of Fagaceae in the Southeastern United States. J. Arnold Arbor. 52:159–195.
11. ———. 1974. The genera of Mimosoideae (Leguminosae) in the Southeastern United States. J. Arnold Arbor. 55:67–118.
12. Ernst, W. R. 1963. The genera of Hamamelidaceae and Platanaceae in the Southeastern United States. J. Arnold Arbor. 44:193–210.
13. ———. 1964. The genera of Berberidaceae, Lardizabalaceae and Menispermaceae in the Southeastern United States. J. Arnold Arbor. 45:1–35.
14. Ferguson, I. K. 1966. The Cornaceae in the Southeastern United States. J. Arnold Arbor. 47:106–116.
15. Gould, F. W. 1968. Grass Systematics. McGraw-Hill, New York.
16. Graham, S. and C. E. Wood, Jr. 1965. The genera of Polygonaceae in the Southeastern United States. J. Arnold Arbor. 46:91–121.
17. Hitchcock, A. S. 1951. Manual of Grasses of the United States, second edition. Revised by A. Chase. U. S. Department of Agriculture Miscellaneous Publications. Washington, D. C.

18. Hunt, D. R. 1967. Cactaceae. In J. Hutchinson (ed.), The Genera of Flowering Plants (II), Clarendon Press, Oxford, pp. 427–467.

19. Hutchinson, J. 1967. The Genera of Flowering Plants. Clarendon Press, Oxford. 2 volumes.

20. ———. 1969. Evolution and Phylogeny of Flowering Plants. Academic Press, New York.

21. ———. 1973. The Families of Flowering Plants, third edition. Clarendon Press, Oxford.

22. Isely, D. 1957. Leguminosae: Papilionoideae. Iowa State University (mimeograph), Ames.

23. ———. 1973. Leguminosae of the United States. I. Subfamily Mimosoideae. Mem. New York Bot. Gard. 25:1–152.

24. ———. 1975. Leguminosae of the United States. II. Subfamily Caesalpinoideae. Mem. New York Bot. Gard. 25:1–228.

25. Lawrence, G. H. M. 1951. Taxonomy of Vascular Plants. Macmillan Co., New York.

26. Melchior, H. (ed.). 1964. A. Engler's Syllabus der Pflanzenfamilien (II), twelfth edition. Gebrüder Borntraeger, Berlin.

27. Munz, P. A. 1965. Onagraceae. In C. T. Rogerson (ed.), North American Flora Series II, Part 5, New York Botanical Garden, New York, pp. 1–278.

28. Pijl, L. van der and C. H. Dodson. 1966. Orchid Flowers: Their Pollination and Evolution. The Fairchild Tropical Garden and University of Miami Press, Coral Gables.

29. Pool, R. J. 1941. Flowers and Flowering Plants, second edition. McGraw-Hill, New York.

30. Porter, C. L. 1967. Taxonomy of Flowering Plants, second edition. W. H. Freeman, San Francisco.

31. Radford, A. E. et al. 1974. Vascular Plant Systematics. Harper & Row, New York.

32. Rendle, A. B. 1956. The Classification of Flowering Plants (II). University Press, Cambridge.

33. Robertson, K. R. 1972. The genera of Geraniaceae in the Southeastern United States. J. Arnold Arbor. 53:182–201.

34. ———. 1974. The genera of Rosaceae in the Southeastern United States. J. Arnold Arbor. 55:303–332, 344–401, 611–662.

35. ——— and Y. Lee. 1976. The genera of Caesalpinoideae (Leguminosae) in the Southeastern United States. J. Arnold Arbor. 57:1–53.

36. Smith, J. P., Jr. 1974. Introduction to the Families of Vascular Plants. Mad River Press, Eureka, California.

37. Solbrig, O. T. 1963. The tribes of Compositae in the Southeastern United States. J. Arnold Arbor. 54:436–461.

38. Spongberg, S. A. 1976. Magnoliaceae Hardy in Temperate North America. J. Arnold Arbor. 57:250–312.

39. Webster, G. L. 1967. The genera of Euphorbiaceae in the Southeastern United States. J. Arnold Arbor. 48:303–361, 363–430.

40. Willis, J. C. 1973. A Dictionary of the Flowering Plants and Ferns, eighth edition. Revised by H. K. Airy Shaw. University Press, Cambridge.

41. Wilson, K. A. 1960. The genera of the Arales in the Southeastern United States. J. Arnold Arbor. 41:47–72.

42. ———. 1960. The genera of Hydrophyllaceae and Polemoniaceae in the Southeastern United States. J. Arnold Arbor. 41:197–212.

43. ———. 1960. The genera of Convolvulaceae in the Southeastern United States. J. Arnold Arbor. 41:298–317.

44. ——— and C. E. Wood, Jr. 1959. The genera of Oleaceae in the Southeastern United States. J. Arnold Arbor. 40:369–384.

45. Wood, C. E., Jr. 1958. The genera of the woody Ranales in the Southeastern United States. J. Arnold Arbor. 39:296–346.

46. ———. 1959. The genera of Nymphaeaceae and Ceratophyllaceae in the Southeastern United States. J. Arnold Arbor. 40:94–112.

47. ————. 1961. The genera of Ericaceae in the Southeastern United States. J. Arnold Arbor. 42:10–80.

48. ————. 1974. A Student's Atlas of Flowering Plants: Some Dicotyledons of Eastern North America. Harper & Row, New York.

49. ———— and P. Adams. 1976. The genera of Guttiferae (Clusiaceae) in the Southeastern United States. J. Arnold Arbor. 57:74–90.

IV. Family Descriptions and Illustrations*

ACERACEAE   (Maple family)

Magnoliopsida (Dicotyledoneae). Two or three genera and approximately 150 species, primarily North Temperate Zone and tropical mountains. Genera: *Acer (Negundo)* Ca. 148 species, *Dipteronia* 2 species.

Trees and shrubs, sap occasionally milky. Leaves simple or pinnately compound, opposite, exstipulate. Inflorescence corymbose, racemose, paniculate or flowers in fascicles. Flowers bisexual or more frequently unisexual (species monoecious, dioecious, or polygamodioecious), actinomorphic; perianth mostly biseriate (rarely apetalous); calyx of 4 to 5 distinct or partially connate sepals; corolla of 4 to 5 (rarely 0) distinct petals; nectariferous disc usually present, extrastaminal or intrastaminal, lobed or sometimes deeply 5-parted; androecium of 4 to 10 distinct stamens, usually 8, mostly inserted on the edge of the disc; pistillodes often present in staminate flowers; gynoecium of 1 compound pistil, ovary superior, carpels 2, locules 2, placentation axile, styles 2 (distinct or basally connate), stigmas terminal; staminodes often present in pistillate flowers. Fruit a schizocarp (samara). Seeds often compressed; embryo with elongate radicle, cotyledons flat or plicate; endosperm absent.

Aids for Field Identification: Trees or shrubs; species usually dioecious or polygamodioecious; leaves usually palmately veined and palmately lobed, opposite; fruit a schizocarp (samara).

Illustrations: 21, 25, 48;   *References:* 3, 8, 21, 25, 29, 30, 31, 36, 40.

ALISMATACEAE (Figure C.3) [ALISMACEAE]   (Arrowhead family)

Liliopsida (Monocotyledoneae). Approximately 13 genera and 90 species, widely distributed but mainly temperate and tropical regions of northern hemisphere. Genera: *Alisma, Echinodorus, Sagittaria.*

Annual or perennial marsh or aquatic herbs. Roots fibrous. Leaves simple, basal, whorled or clustered, exstipulate, blades linear to ovate, often with sagittate or hastate bases, principal veins parallel with margins but converging at apex, transverse veins parallel. Inflorescence racemose or paniculate, flowers often whorled. Flowers bisexual or unisexual, actinomorphic; perianth biseriate, imbricate; calyx of three distinct segments; corolla of three distinct segments (deciduous); androecium of (3) 6 to many distinct stamens, anthers extrorse, 2-locular; gynoecium of numerous apocarpous pistils (rarely

---

*The plates for the 24 of the families included in this appendix were prepared for *A Generic Flora of the Southeastern United States* under the supervision of Carroll E. Wood, Jr., of Harvard University. Plates for 3 families (Alismataceae, Cyperaceae, Poaceae) were prepared by Susan Sizemore Whitfield for *An Atlas and Illustrated Guide to the Threatened and Endangered Vascular Plants of the Mountains of North Carolina and Virginia* under the supervision of J. R. Massey. Used with permission.

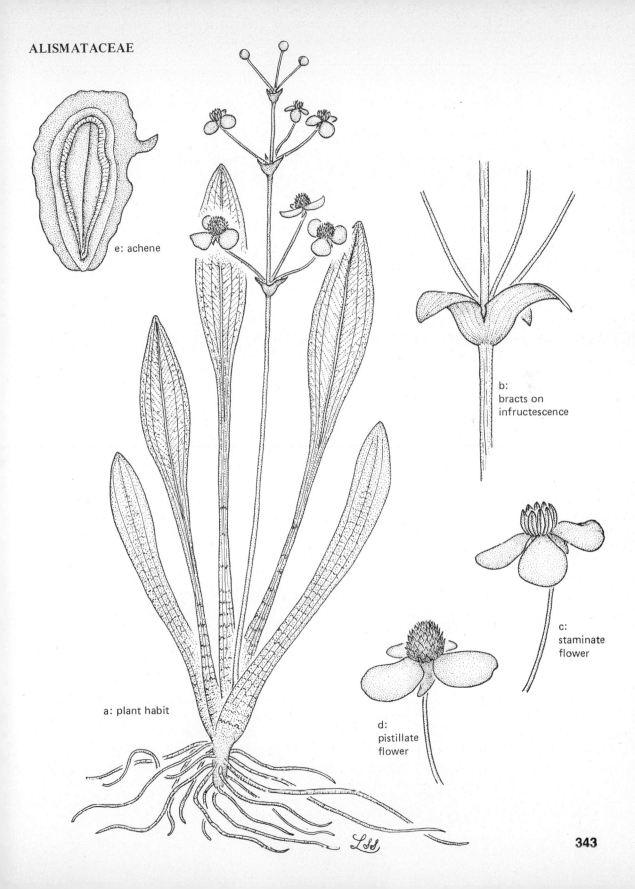

ALISMATACEAE

e: achene

b:
bracts on
infructescence

c:
staminate
flower

d:
pistillate
flower

a: plant habit

343

united at base), ovary superior, unicarpellate, unilocular, placentation basal, style and stigma one. Fruit an achene (rarely follicular). Seeds solitary (infrequently several); embryo hippocrepiform; endosperm absent.

Aids for Field Identification: Perianth present; floral parts in 3s; carpels form a disk or a ring on receptacle; fruit an achene.

Illustrations: 21, 25;   *References:* 8, 21, 25, 29, 30, 31, 36, 40.

AMARANTHACEAE (Figure C.4)   (Pigweed family)

     Magnoliopsida (Dicotyledoneae). Sixty-five genera and circa 850 species, mostly tropical but also in temperate regions. Selected Genera: *Amaranthus, Celosia, Alternanthera, Gomphrena, Iresine, Froelichia, Tidestromia.*

     Annual or perennial herbs or rarely woody shrubs or vines, mostly weedy herbs. Leaves simple, alternate or opposite, exstipulate. Inflorescence a spike, raceme, head or the flowers solitary. Flowers usually bisexual or when unisexual the species mostly dioecious or polygamodioecious, small, each flower often with scarious bract(s) and bracteoles; perianth uniseriate; calyx of 3 to 5 distinct or basally connate, mostly dry and membranous sepals; corolla absent; androecium of mostly 5 stamens, opposite the sepals, monadelphous (separate in *Amaranthus*), the tube often with lobes, ennations, or staminodes alternating with the anthers; anthers 2- to 4- celled, longitudinally dehiscent; gynoecium of 1 pistil, ovary superior, carpels 2 to 3, unilocular, placentation basal, styles 1 to 3, stigmas 1 to 3. Fruit a utricle, capsule, pyxis or achene or rarely a berry or drupe. Seeds 1 to few; embryo curved; endosperm mealy.

Aids for Field Identification: Mostly herbs; flowers small with dry scarious bracts; stamens connate, at least at the base; fruit a utricle or capsule.

Illustrations: 21, 25, 48:   *References:* 8, 21, 25, 29, 30, 31, 36, 40.

APIACEAE (Figure C.5) [Umbelliferae, alternative name; Ammiaceae, includes Hydrocotylaceae]   (Parsley family)

     Magnoliopsida (Dicotyledoneae). Approximately 275 genera and 3,000 species, cosmopolitan but primarily North Temperate Zone.

Selected Genera (after Engler, in reference 40):

     Subfamily I.   HYDROCOTYLOIDEAE
          Tribe Hydrocotyleae: *Azorella, Hydrocotyle*
          Tribe Mulineae: *Bowlesia*

     Subfamily II.   SANICULOIDEAE
          Tribe Saniculeae: *Eryngium, Sanicula*
          Tribe Lagoecieae: *Lagoecia*

     Subfamily III.   APIOIDEAE
          Tribe Echinophoreae: *Echinophora*
          Tribe Scandiceae: *Anthriscus, Chaerophyllum, Torilis*
          Tribe Coriandreae: *Coriandrum*
          Tribe Smyrnieae: *Conium*
          Tribe Ammieae: *Apium, Bupleurum, Carum, Foeniculum, Ligusticum*

AMARANTHACEAE: Amaranthus. a-h, A. spinosus; i-j, A. retroflexus

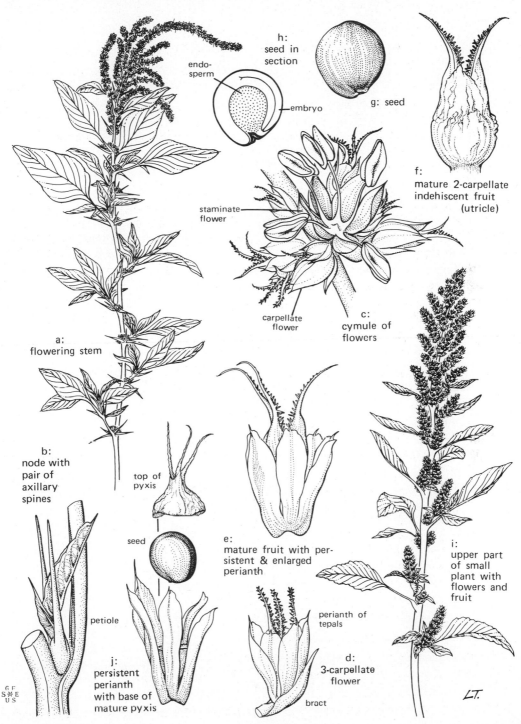

h:
seed in
section

endo-
sperm

embryo

g: seed

f:
mature 2-carpellate
indehiscent fruit
(utricle)

staminate
flower

carpellate
flower

c:
cymule of
flowers

a:
flowering stem

b:
node with
pair of
axillary
spines

top of
pyxis

seed

e:
mature fruit with per-
sistent & enlarged
perianth

i:
upper part
of small
plant with
flowers and
fruit

petiole

perianth of
tepals

j:
persistent
perianth
with base of
mature pyxis

d:
3-carpellate
flower

bract

L.T.

APIACEAE (UMBELLIFERAE): Daucus. a-m, D. carota; n, D. pusillus

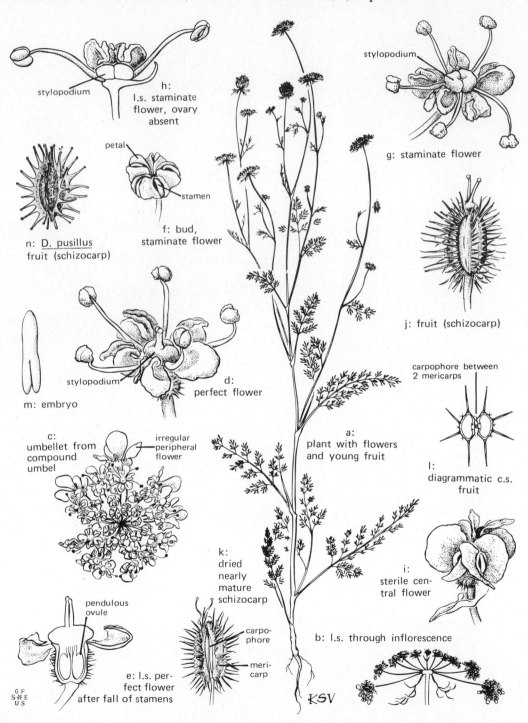

h: l.s. staminate flower, ovary absent

stylopodium

stylopodium

g: staminate flower

petal

stamen

f: bud, staminate flower

n: D. pusillus fruit (schizocarp)

j: fruit (schizocarp)

stylopodium

m: embryo

d: perfect flower

carpophore between 2 mericarps

l: diagrammatic c.s. fruit

c: umbellet from compound umbel

irregular peripheral flower

a: plant with flowers and young fruit

i: sterile central flower

k: dried nearly mature schizocarp

pendulous ovule

carpophore

mericarp

b: l.s. through inflorescence

e: l.s. perfect flower after fall of stamens

KSV

Tribe Peucedaneae: *Angelica, Ferula, Pastinaca*
Tribe Laserpitieae: *Laserpitium, Thapsia*
Tribe Dauceae: *Daucus*

Mostly biennial or perennial herbs with stout stems, internodes hollow. Leaves usually pinnately or palmately compound, rarely simple, alternate, stipulate to exstipulate, petioles with sheathing bases (pericladial). Inflorescence a simple or compound umbel or capitate, often bracteate or involucrate. Flowers bisexual or unisexual (last umbellets to develop are sometimes staminate; species monoecious, dioecious, or polygamodioecious), actinomorphic; perianth biseriate; calyx adnate to the ovary, sepals 5, evident, reduced to small lobes, or absent; corolla of 5 distinct petals; androecium of 5 distinct stamens alternate with the petals, inserted on an epigynous disc, anthers bilocular with longitudinal dehiscence; gynoecium of 1 compound pistil, ovary inferior, carpels 2, locules 2, placentation axile, styles 2, often with a thickened or swollen base (stylopodium), stigmas 2. Fruit a schizocarp (2 mericarps), divided along the commissure at maturity by the central axis (carpophore), mericarps usually longitudinally ribbed. Seeds with a minute embryo; endosperm copious.

Aids for Field Identification: Herbs; leaf bases pericladial; inflorescence umbellate; ovary inferior; stylopodium present; fruit a schizocarp with a carpophore.

Illustrations: 21, 25, 48; *References:* 8, 21, 25, 29, 30, 31, 36, 40.

AQUIFOLIACEAE [including Phellineaceae]   (Holly family)

Magnoliopsida (Dicotyledoneae). About 3 or 4 genera and 450 species, tropical and temperate zones. Genera: *Ilex (Byronia), Nemopanthus, Phelline*.

Trees or shrubs. Leaves simple, evergreen or deciduous, alternate, often coriaceous, exstipulate or with small stipules. Inflorescence cymose or fascicled, flowers seldom solitary. Flowers bisexual or unisexual and the species either dioecious or polygamodioecious, actinomorphic; perianth biseriate; calyx of 3 to 6 imbricate sepals, basally connate; corolla of 4 to 9 distinct or basally connate petals; androecium of 4 to 9 distinct stamens that are epipetalous or alternate with the petals, anthers 2-celled with longitudinal dehiscence (pistillodes often present in staminate flowers); gynoecium of 1 compound pistil, ovary superior, carpels and locules 3 to many, placentation axile, style 1, stigma lobed or capitate (staminodes often present in pistillate flowers). Fruit drupaceous. Seeds with a straight embryo; endosperm fleshy and copious.

Aids for Field Identification: Trees or shrubs; inflorescence axillary; flowers perfect or imperfect (plants dioecious or polygamodioecious); fruit drupaceous.

Illustrations: 21, 25, 48. *References:* 4, 8, 21, 25, 29, 30, 31, 36, 40.

ARACEAE (Figure C.6)   (Arum family)

Liliopsida (Monocotyledoneae). About 115 genera and 2,000 species, mostly tropical (92 percent) and temperate. Selected Genera: *Acorus, Aglaonema, Anthurium, Arisaema, Calla, Colocasia, Dieffenbachia, Lysichiton, Orontium, Peltandra, Philodendron, Pistia, Spathiphyllum, Symplocarpus.*

Rhizomatous or tuberous, terrestrial herbs (rarely aquatic or woody-climbing), sap milky, watery or pungent, calcium oxalate crystals generally present in tissues; plant

ARACEAE: Arisaema. a-j, A. atrorubens; k-m, A. stewardsonii; n, A. dracontium

h: ovule (orthotropous)

i: unusually large infructescence

g: carpellate flower, l.s.

f: carpellate inflorescence with most of spathe removed

e: staminate flower

j: fruit, l.s., showing seed

n: A. dracontium, habit

sterile part of spadix (appendix)

c: staminate inflorescence, most of spathe removed

d: portion of staminate spadix

l: A. stewardsonii, spathe in back view

m: spathe, A. stewardsonii, from above

a: A. atrorubens, habit

k: spathe in side view

b: spathe, A. atrorubens, from above

occasionally epiphytic with aerial roots. Leaves simple or compound, basal or cauline, alternate, petioled with membranous sheathing base, blades ensiform (sword-shaped) and parallel-veined or variously lamellate with pinnately or palmately netted venation. Inflorescence a spadix, sometimes subtended by a spathe. Flowers bisexual or unisexual (species then monoecious or rarely dioecious); perianth of 4 to 6 distinct or connate segments (present in bisexual flowers and rare in unisexual flowers); androecium of 2, 4, or 8 distinct or connate stamens, anthers 2-celled with dehiscence by pores or slits; gynoecium of 1 pistil, ovary superior or inferior (then embedded in spadix axis), locules 1 to many, placentation variable (axile, apical, basal, or parietal), style various (sometimes absent), stigma various. Fruit usually a berry. Seeds 1 to many; embryo large; endosperm usually present.

Aids for Field Identification: Inflorescence a spadix; perianth absent or inconspicuous.

Illustrations: 21, 25; *Reference:* 8, 21, 25, 29, 30, 31, 36, 40, 41.

ASCLEPIADACEAE (Figure C.7) [including Periplocaceae, Stapeliaceae]
(Milkweed family)

Magnoliopsida (Dicotyledoneae). Approximately 130 genera and 2,000 species, mostly tropics and subtropics. Selected Genera: *Asclepias, Ceropegia, Cryptostegia, Cynanchum, Gonolobus, Gymnema, Hoya, Huernia, Matelea, Oxypetalum, Sarcostemma, Stapelia.*

Perennial herbs, shrubs, or rarely small trees, sometimes climbing, often fleshy or cactuslike (*Stapelia et al.*), sap usually milky. Leaves simple, reduced or absent in some succulent species, usually opposite or whorled, stipules absent or if present minute, margins generally entire. Inflorescence determinate and a dichasial or monochasial cyme or racemose or umbellate. Flowers bisexual, actinomorphic; perianth biseriate; calyx of 5 sepals, distinct or basally connate, imbricate or valvate; corolla of 5 petals, sympetalous, the tube often short and with lobes, contorted or valvate. A nectariferous corona of variable structure may be present (typically consisting of 5 hoods and their associated horns) and adorns the ring of short connate filaments. Androecium of 5 stamens, rarely distinct, typically the filaments connate into a short tube and anthers usually adnate or adherent to the stigmatic area of the gynoecium to form a gynostegium (gynandrium); the anthers 2-celled, pollen often agglutinated within each anther sac into a pollinium; 2 pollinia, one each from adjacent anthers, are joined together by 2 arms (translators, retinacula) attached to a central gland (corpusculum); the pollinia usually occur behind the hoods, with the translators and corpusculum visible between the corona segments. The gynoecium consists of one bicarpellate pistil composed of 2 distinct, unicarpellate, unilocular ovaries, each superior to slightly inferior, placentation marginal, styles 2, stigma 1 and 5-lobed, often much enlarged, nonreceptive except for 5 longitudinal strips of glandular stigmatic surface (stigmatic slits) on the edge of the stigma between the adjacent anthers. Fruit of 2 separate carpels (sometimes one by abortion) which are follicular. Seeds numerous, flattened, and comose with long, silky hairs; embryo large, nearly as long as the seed, and straight; endosperm thin and small.

Aids to Field Identification: Herbs or vines; milky sap present; leaves opposite or whorled; flowers 5-merous, gynoecium bicarpellate with distinct ovaries and styles and 1

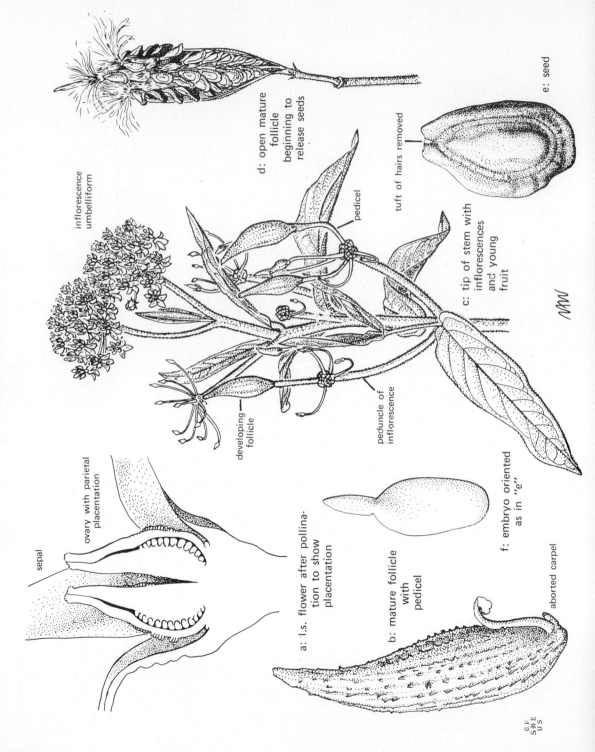

ASCLEPIADACEAE: Asclepias. a, b, A. syriaca; c-f, A. incarnata

inflorescence
umbelliform

d: open mature
follicle
beginning to
release seeds

e: seed

tuft of hairs removed

pedicel

c: tip of stem with
inflorescences
and young
fruit

peduncle of
inflorescence

developing
follicle

ovary with parietal
placentation

sepal

a: l.s. flower after pollination to show
placentation

b: mature follicle
with
pedicel

f: embryo oriented
as in "e"

aborted carpel

G F
S✳E
U S

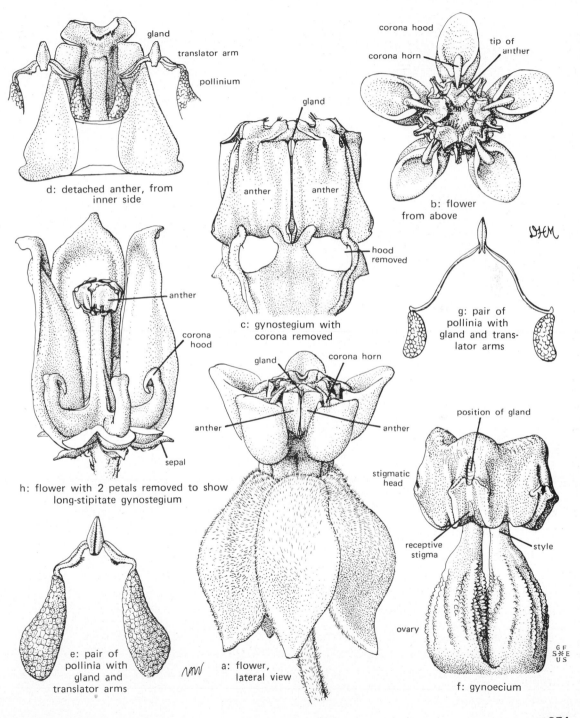

ASCLEPIADACEAE: Asclepias. a-f, A. syriaca; g, A. connivens; h, A. pedicellata

d: detached anther, from inner side

gland
translator arm
pollinium

corona hood
corona horn
tip of anther

b: flower from above

gland
anther
anther
hood removed

c: gynostegium with corona removed

g: pair of pollinia with gland and translator arms

anther
corona hood
sepal

h: flower with 2 petals removed to show long-stipitate gynostegium

gland
corona horn
anther
anther
stigmatic head

position of gland
receptive stigma
style
ovary

f: gynoecium

e: pair of pollinia with gland and translator arms

a: flower, lateral view

stigma; corona, gynostegium, and pollinia present; fruit 2 follicles with comose seeds. The family is closely related and similar to Apocynaceae.

Illustrations: 21, 25, 48;   *References:* 8, 21, 25, 29, 30, 31, 36, 40.

ASTERACEAE (Figure C.8)   (Composite family)

   Magnoliopsida (Dicotyledoneae). One thousand to 2,000 genera and circa 20,000 species (one of the largest families of flowering plants), cosmopolitan.

Selected Genera:

   Subfamily I.   TUBULIFLORAE (some florets of head with tubular or bilabiate corollas; sap not milky):
      Tribe 1.   Heliantheae: *Ambrosia, Bidens, Borrichia, Coreopsis, Echinacea, Eclipta, Helenium, Helianthus, Iva, Parthenium, Rudbeckia, Silphium, Verbesina, Xanthium, Zinnia*
      Tribe 2.   Astereae: *Aphanostephus, Aster, Astranthium, Baccharis, Bellis, Bigelowia, Boltonia, Brintonia, Chrysoma, Chrysopsis, Conyza, Erigeron, Euthamia, Grindelia, Gutierrezia, Haplopappus, Heterotheca, Solidago*
      Tribe 3.   Anthemideae: *Achillea, Anthemis, Artemisia, Chrysanthemum, Hymenopappus, Matricaria, Soliva, Tanacetum*
      Tribe 4.   Arctotideae (chiefly South Africa—*Arctotis calendulacea* an escape in South Carolina and Florida): *Arctotis*
      Tribe 5.   Inuleae: *Anaphalis, Antennaria, Evax, Facelis, Filago, Gnaphalium, Inula, Pluchea, Pterocaulon, Sachsia*
      Tribe 6.   Senecioneae: *Arnica, Cacalia, Emilia, Erechtites, Gynura, Petasites, Senecio, Tussilago*
      Tribe 7.   Calenduleae (chiefly African and Mediterranean): *Calendula*
      Tribe 8.   Eupatorieae: *Ageratum, Brickellia, Carphephorus, Eupatorium, Garberia, Hartwrightia, Kuhnia, Liatris, Mikania, Sclerolepis, Trilisa*
      Tribe 9.   Vernonieae: *Elephantopus, Stokesia, Vernonia*
      Tribe 10.   Cynareae (Cardueae): *Arctium, Centaurea, Cirsium, Cnicus, Silybum*
      Tribe 11.   Mutisieae: *Chaptalia, Mutisia, Gerbera*

   Subfamily II.   LIGULIFLORAE (all florets of head ligulate; sap milky):
      Tribe 12.   Cichorieae: *Cichorium, Crepis, Hieracium, Krigia, Lactuca, Prenanthes, Pyrrhopappus, Sonchus, Taraxacum, Tragopogon*

   Annual or perennial herbs, shrubs, rarely trees or vines; sap watery or milky; often with stolons, rhizomes, tubers, or fleshy roots; often spiny or with various types of vestiture, such as tomentose, pannose, strigose, lanate, floccose. Leaves mostly simple, sometimes needlelike or reduced to scales, often rosulate and/or alternate, opposite, fascicled, rarely whorled, radical and/or cauline, exstipulate, margins often pinnately or palmately lobed to divided. The primary inflorescence a capitulum (head) with many (rarely 1) flowers (florets) borne on a flat, conical, convex, or concave receptacle, each floret often subtended by a receptacular bract (chaff, pale) that is often scarious, the receptacle usually subtended by an involucre of bracts (phyllaries) of various sizes, shapes, and textures disposed in one or more series. The heads may be arranged in various

ASTERACEAE (COMPOSITAE): Inflorescence, florets, and stamens

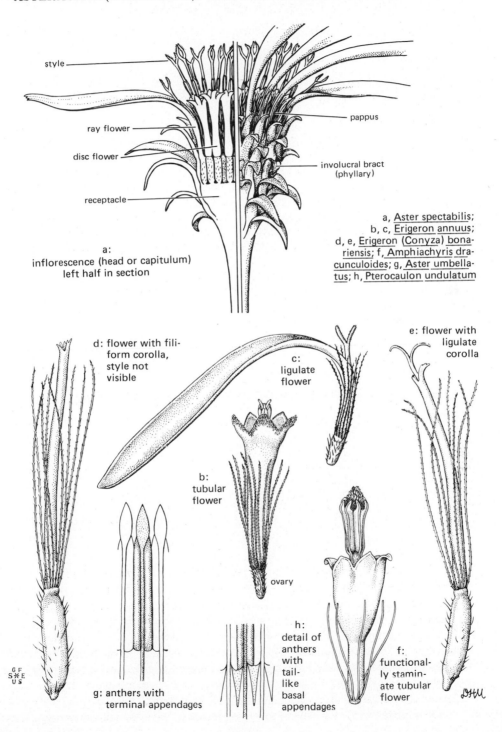

style

pappus

ray flower

disc flower

involucral bract
(phyllary)

receptacle

a:
inflorescence (head or capitulum)
left half in section

a, <u>Aster spectabilis</u>;
b, c, <u>Erigeron annuus</u>;
d, e, <u>Erigeron</u> (<u>Conyza</u>) <u>bona-
riensis</u>; f, <u>Amphiachyris dra-
cunculoides</u>; g, <u>Aster umbella-
tus</u>; h, <u>Pterocaulon undulatum</u>

d: flower with fili-
form corolla,
style not
visible

e: flower with
ligulate
corolla

c:
ligulate
flower

b:
tubular
flower

ovary

h:
detail of
anthers
with
tail-
like
basal
appendages

f:
functional-
ly stamin-
ate tubular
flower

g: anthers with
terminal appendages

G F
S ✳ E
U S

DHM

353

ways, often in racemes, panicles, or corymbs. Florets bisexual, unisexual, or neuter, actinomorphic or zygomorphic. Calyx represented by a pappus of awns, bristles, or scales or absent, epigynous. Corolla of 5 petals, sympetalous, represented by three basic types: (1) tubular (discoid)—cylindrical in shape and forming a conspicuous tube and short limb, (4)5-lobed; (2) ligulate (ray)—strap-shaped and with a very short tube, often with 2 to 5 apical teeth; and (3) bilabiate—modified from the tubular corolla, with 3-lobed upper lip and 2 slender recurved lower lips. Based on the arrangement and position of these 3 types of corollas, 3 types of capitula exist: (1) ligulate—only ray florets present, these usually bisexual, sometimes unisexual and the species dioecious; (2) discoid—only disk florets present, these usually bisexual, sometimes unisexual and the species then monoecious or dioecious; and (3) radiate—ray florets present on the periphery (usually neutral or pistillate) and disk florets present on the rest of the receptacle (usually bisexual). Androecium typically of (4)5 stamens, syngenesious (forming a cylinder around the style), epipetalous, the anthers 2-celled, longitudinally dehiscent, introrse, and apical appendage (rarely absent) and connective present, a rounded auriculate, caudate, or sagittate basal appendage often present on each theca. Gynoecium of 1 compound pistil, ovary inferior, carpels 2, locule 1, ovule 1, anatropous, placentation basal, style 1, usually 2-branched with receptive stigmatic tissue inside, of diverse forms. Fruit a cypsela (achene of most authors) of various shapes, often crowned by the persistent pappus, sometimes enclosed by persistent bracts. Seed 1 per fruit; embryo large and straight; endosperm absent.

Aids for Field Identification: Herbs, vines, or shrubs; inflorescence an involucrate capitulum; pappus present; carpels 2, ovary inferior, ovule 1; stamens syngenesious; fruit a cypsela.

Illustrations: 21, 25, 48.   *References:* 7, 8, 21, 25, 29, 30, 31, 36, 37, 40.

BETULACEAE [including Corylaceae and Carpinaceae]   (Birch family)

Magnoliopsida (Dicotyledoneae). Six genera and circa 157 species, mostly in north temperate regions, some in tropical mountains, the Andes, and Argentina. Selected Genera: *Carpinus, Ostrya, Corylus, Alnus, Betula.*

Trees and shrubs. Leaves deciduous, simple, alternate, stipulate (often caducous), margins serrate, venation strongly penni-parallel. Inflorescence almost always compound, erect or pendulous staminate and pistillate catkins composed of a primary axis with condensed cymules (''flowers''), each subtended by a bract and spirally arranged; in the ancestral theoretical cymule a group of three flowers forms the dichasium in the axil of each bract and includes primary, secondary, and tertiary bracts; modifications exist in which some flowers are absent or in which the secondary and tertiary axes and/or bracts are absent. Flowers unisexual and species monoecious; perianth of 4 rudimentary tepals or absent, staminate inflorescence composed of 3-flowered cymules, androecium in each cymule of 2 to 20 stamens, distinct or basally connate, the anthers 2-celled with cells separate or connate, longitudinally dehiscent; pistillate inflorescence composed of 2-flowered cymules (3-flowered in *Betula*), gynoecium of 1 compound pistil, the ovary inferior or nude, carpels 2, locules 2 below and 1 above the placental septum, each locule with 2 ovules or reduced to 1 by abortion, placentation apical, styles 2 or style 1 and deeply divided into 2 parts, stigmas 2. Fruit a nut, often with accessory wings, or a samara. Seed solitary; embryo large and straight; endosperm absent.

Aids for Field Identification: Trees and shrubs; monoecious; pistillate cymules in catkins; 2-carpellate.

Illustrations: 21, 25, 48.   *References:* 8, 19, 21, 25, 29, 30, 31, 32, 36, 40.

BORAGINACEAE [including Ehretiaceae, Heliotropiaceae]   (Borage family)

Magnoliopsida (Dicotyledoneae). Approximately 100 genera and 2,000 species, tropics and temperate regions, especially the Mediterranean. Selected Genera: *Amsinckia, Borago, Cryptantha, Cynoglossum, Echium, Hackelia, Heliotropium, Lithospermum, Mertensia, Myosotis, Plagiobothrys, Pulmonaria, Symphytum.*

Mostly herbs, a few shrubs or trees, sometimes climbing, vesture often bristly, hirsute, or hispid. Leaves simple, alternate (rarely opposite), exstipulate, margins usually entire. Inflorescence usually a helicoid or scorpioid cyme. Flowers mostly bisexual, actinomorphic (rarely zygomorphic); perianth biseriate; calyx of 5 sepals, distinct or basally connate, imbricate (rarely valvate); corolla of 5 petals, sympetalous, often lobed, imbricate or contorted in bud, rotate, salverform, infundibular, or campanulate; androecium of 5 stamens, distinct, epipetalous, antisepalous, the anthers 2-celled, longitudinally dehiscent, introrse; nectariferous disk sometimes present; gynoecium of 1 compound pistil, ovary superior and deeply 4-lobed, carpels 2, locules 2 (seemingly 4 by a false septum), placentation axile but seemingly basal, the ovules usually 2 in each carpel, style 1 and usually gynobasic, stigma typically 1. Fruit a schizocarp (nutlets 2 to 4), drupe, or a nut. Seeds with or without endosperm, when present fleshy and scant; embryo straight or curved.

Aids for Field Identification: Bristly or hispid herbs; stems round; leaves alternate; inflorescence often a helicoid cyme; flowers actinomorphic, 5-merous, ovary 2-locular; ovary 4-lobed and developing into a schizocarp (nutlets), style gynobasic. The Boraginaceae are closely related and similar to the Lamiaceae and Verbenaceae. The main differences lie in its terete stems, alternate leaves, and mostly actinomorphic corolla.

Illustrations: 21, 25, 48.   *References:* 8, 21, 25, 29, 30, 31, 36, 40.

BRASSICACEAE (Figure C.9) [CRUCIFERAE]   (Mustard family)

Magnoliopsida (Dicotyledoneae). Approximately 375 genera and 3,200 species, cosmopolitan but mostly northern hemisphere. Selected Genera: *Alyssum, Arabis, Armoracia, Barbarea, Capsella, Cardamine, Dentaria, Draba, Erysimum, Iberis, Isatis, Lepidium, Lunaria, Nasturtium, Rhaphanus.*

Annual, biennial, or perennial herbs (rarely subshrubs), sap watery, herbage commonly with stellate or branched unicellular hairs. Leaves mostly simple, alternate (rarely opposite or subopposite), exstipulate. Inflorescence typically racemose. Flowers perfect, mostly actinomorphic; perianth triseriate; calyx of 4 sepals, distinct, in two whorls; corolla of (0) 4 petals, distinct, often clawed; androecium of 6 tetradynamous (rarely fewer or more than 6) stamens, anthers 2-locular (rarely 1-locular), longitudinally dehiscent; gynoecium of 1 pistil, ovary superior, carpels 2 (rarely 4-carpellate), locules 2 by means of a false but complete septum (sometimes transversely divided into many locules), plac-

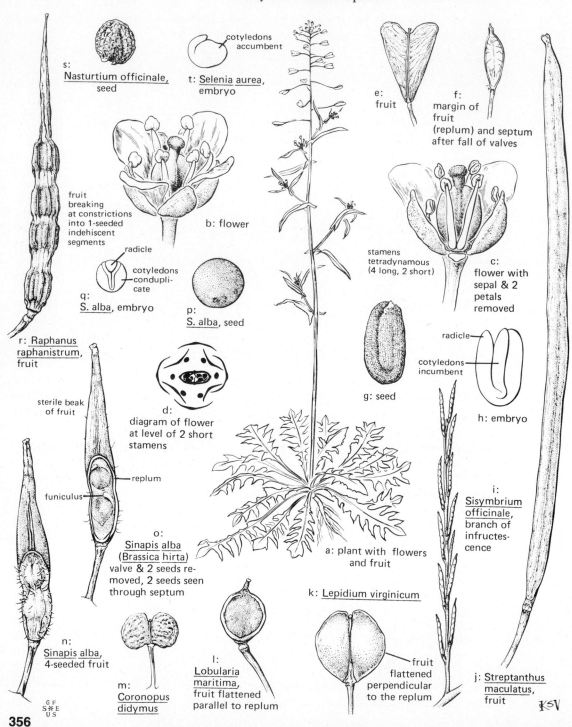

BRASSICACEAE (CRUCIFERAE): a-h, Capsella bursa-pastoris; i-t, fruits, seeds, & embryos of various species

s: Nasturtium officinale, seed

t: Selenia aurea, embryo

cotyledons accumbent

e: fruit

f: margin of fruit (replum) and septum after fall of valves

fruit breaking at constrictions into 1-seeded indehiscent segments

b: flower

radicle

cotyledons conduplicate

q: S. alba, embryo

p: S. alba, seed

stamens tetradynamous (4 long, 2 short)

c: flower with sepal & 2 petals removed

r: Raphanus raphanistrum, fruit

radicle

cotyledons incumbent

g: seed

h: embryo

sterile beak of fruit

d: diagram of flower at level of 2 short stamens

replum

funiculus

o: Sinapis alba (Brassica hirta) valve & 2 seeds removed, 2 seeds seen through septum

a: plant with flowers and fruit

i: Sisymbrium officinale, branch of infructescence

n: Sinapis alba, 4-seeded fruit

m: Coronopus didymus

l: Lobularia maritima, fruit flattened parallel to replum

k: Lepidium virginicum

fruit flattened perpendicular to the replum

j: Streptanthus maculatus, fruit

GF
S*E
US

356

entation parietal, style 1 or obsolete, stigmas 2 or connate. Fruit a silique or silicle (occasionally a nut). Seeds 1 to few; embryo large; endosperm scant to absent.

Aids for Field Identification: Herbs, apopetalous; floral parts in 4s; stamens tetradynamous; fruit commonly a silique or silicle.

Illustrations: 21, 25, 48.   *References:* 8, 21, 25, 29, 30, 31, 36, 40.

CACTACEAE   (Cactus family)

Magnoliopsida (Dicotyledoneae). Approximately 84 genera (50 to 220 depending on authority) and 2,000 species, American with *Rhipsalis* occurring in Africa, Madagascar, Seychelles, Ceylon, and Mauritius but doubtfully native.

Key to Tribes (see reference 18):   (Selected Genera)

Leaves broad and flat, more or less persistent, or small, awl-shaped (*Maihuenia*), glochids absent; seeds not arillate, black . . . . .Pereskieae
Leaves and glochids present, the former sometimes very small, early deciduous, the latter usually persistent; seeds encased in a bony white aril, rarely winged or with matted hairs . . . . .Opuntieae
Leaves vestigial or more commonly absent; glochids absent; seeds not encased in a bony aril, brown or black . . . . .Cacteae

Tribe 1. Pereskieae (2 genera): *Pereskia, Maihuenia*
Tribe 2. Opuntieae (5 genera): *Opuntia, Pereskiopsis*
Tribe 3. Cacteae (Cereeae of some authors) (77 genera): *Cereus, Echinocactus, Mammillaria, Gymnocalycium, Ferocactus, Lophophora, Astrophytum, Rhipsalis, Zygocactus, Epiphyllum*

Perennial herbs, shrubs, or trees of diverse forms, succulent, often spiny, stems simple to branched or arborescent. Leaves simple, alternate, flat, scalelike or absent in many genera; spines, tufts of bristles (glochids) and flowers arising from cushionlike axillary pads or areoles. Flowers mostly solitary or clustered, rarely corymbose or paniculate in *Pereskia,* sessile, mostly bisexual and actinomorphic; perianth segments numerous and weakly differentiated into calyx and corolla, often fused to form a perianth tube or hypanthium; calyx often petaloid, corolla mostly in several series; androecium of numerous stamens in several series or groups and often fused to the perianth tube, anthers 2-celled, longitudinally dehiscent; gynoecium of 1 pistil, ovary mostly inferior, syncarpous, carpels 2 to many, ovary unilocular, placentation parietal (rarely basal), styles typically 1, stigmas often as many as the carpels. Fruit mostly a many-seeded berry, often glochidiate, spiny, scaly or bristly, fleshy or dry. Seeds often arillate or strophiolate; embryo straight or curved; endosperm little or absent, food storage mainly perispermous.

Aids for Field Identification: Fleshy and mostly prickly or spiny plants; flower parts numerous; ovary inferior, unilocular; placentation parietal.

Illustrations: 21, 25, 48.   *References:* 8, 18, 21, 25, 26, 30, 31, 36, 40.

CAMPANULACEAE [including Lobeliaceae]   (Bellflower family)

Magnoliopsida (Dicotyledoneae). Sixty to 70 genera and circa 2,000 species, temperate and subtropical regions and tropical mountains.

Key to Subfamilies: (Selected Genera)

Corolla actinomorphic; anthers distinct . . . . CAMPANULOIDEAE
Corolla zygomorphic; anthers syngenesious, very rarely distinct . . . . . LOBELIOIDEAE

> Subfamily I.   CAMPANULOIDEAE: *Campanula, Platycodon, Triodanis, Wahlenbergia, Heterocodon, Githopsis*
>
> Subfamily II.   LOBELIOIDEAE: *Nemacladus,   Parishella,   Downingia, Legenere, Lobelia*

Annual or perennial herbs or subshrubs, rarely trees, sap often milky. Leaves simple, alternate (rarely opposite), exstipulate. Inflorescence basically a determinate dichasial or monochasial cyme and seemingly racemose or thyrsiform and resembling a panicle, sometimes a head or flowers solitary in leaf axils, usually bracteate and often bracteolate. Flowers bisexual, actinomorphic or zygomorphic; perianth biseriate; calyx usually of 5(3 to 10) sepals, synsepalous or basally connate; corolla usually of 5 petals (variable in number, sometimes absent), sympetalous in most genera, actinomorphic and campanulate or tubular or zygomorphic and strongly bilabiate (often split down one side); androecium of as many stamens as petals, distinct or variously coherent or connate (syngenesious in Lobelioideae) filament bases often expanded and forming a dome shaped chamber over the nectariferous epigynous disk, antisepalous, generally epipetalous at the extreme base of corolla, the anthers 2-celled (locular), longitudinally dehiscent, introrse; gynoecium of 1 compound pistil, ovary inferior (half-inferior in some; e.g., *Lobelia* spp.), carpels and locules typically 2 to 5, placentation axile (sometimes parietal, rarely basal or apical), the ovules usually numerous and anatropous, the style 1 and slender, sometimes with 2 to 5 branches, the stigmas or stigma lobes usually 2 to 5. Fruit typically a capsule dehiscing by various means, sometimes a berry. Seeds numerous; embryo small and usually straight; endosperm copious and fleshy.

Aids for Field Identification: Herbs or shrubs; corolla 5-merous, usually sympetalous, actinomorphic or zygomorphic; stamens sometimes united into a tube; ovary inferior; placentation axile; ovules numerous.

Illustrations: 21, 25, 48.   *References:* 8, 21, 25, 29, 30, 31, 36, 40.

CARYOPHYLLACEAE (Figure C.10) [including Illecebraceae and Paronychiaceae]

(Pink family)

Magnoliopsida (Dicotyledoneae). Approximately 70 genera and 1,750 species, cosmopolitan.

Selected Genera: (see reference 26)

> Subfamily I.   PARONYCHIOIDEAE: *Paronychia,   Spergula,   Spergularia, Polycarpon*
>
> Subfamily II.   ALSINOIDEAE: *Scleranthus, Minuartia, Sagina, Arenaria, Stellaria, Cerastium*
>
> Subfamily III.   SILENOIDEAE: *Silene, Dianthus, Lychnis, Saponaria, Vaccaria, Gypsophila*

Annual or perennial herbs or infrequently suffrutescent shrubs. Stems characteristically with swollen nodes. Leaves simple, opposite (rarely alternate), often connected at base by a transverse line or by a shortly connate-perfoliate base, decussate, stipules when

CARYOPHYLLACEAE: Silene: a-j, S. virginica; k, S. caroliniana; l, S. ovata; m, n, S. antirrhina

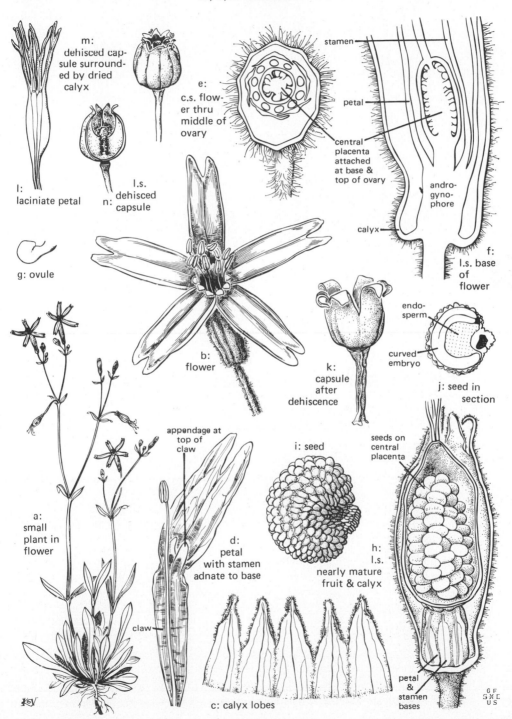

m: dehisced capsule surrounded by dried calyx

e: c.s. flower thru middle of ovary

stamen

petal

central placenta attached at base & top of ovary

l: laciniate petal

n: l.s. dehisced capsule

andro-gyno-phore

calyx

f: l.s. base of flower

g: ovule

endo-sperm

curved embryo

j: seed in section

b: flower

k: capsule after dehiscence

a: small plant in flower

appendage at top of claw

i: seed

seeds on central placenta

d: petal with stamen adnate to base

h: l.s. nearly mature fruit & calyx

claw

petal & stamen bases

c: calyx lobes

359

present scarious. Inflorescence simple or compound dichasia or the flowers solitary and terminal. Flowers typically bisexual (when unisexual the species dioecious), actinomorphic; perianth biseriate; calyx of 5 (rarely 4) sepals, distinct or connate; corolla typically of as many petals as sepals, distinct, imbricate and often cleft and/or differentiated into a claw and limb, rarely minute or absent; androecium of as many as or twice as many stamens as petals (typically 5 or 10) in 1 to 2 whorls, distinct or basally connate, staminodia sometimes present and petaloid, anthers 2-celled, longitudinally dehiscent; gynoecium of 1 pistil, ovary superior, syncarpous, carpels 2 to 5, mostly unilocular and placentation free-central or basally 3 to 5 locular with axile placentation or axile below and free central above or ovules solitary and basal, styles and stigmas mostly 2 to 5, rarely 1. Fruit a many-seeded capsule and variously dehiscent or an achene or utricle. Seeds with curved or straight embryo; endosperm hard or soft.

Aids for Field Identification: Herbs; leaves opposite; nodes often swollen; inflorescence cymose or flowers solitary; placentation free-central.

Illustrations: 21, 25, 48.   *References:* 8, 21, 25, 26, 29, 30, 31, 36, 40.

CHENOPODIACEAE   (Goosefoot family)

Magnoliopsida (Dicotyledoneae). Approximately 102 genera and 1,400 species; cosmopolitan but with chief centers in Australia, the pampas, the prairies, the Mediterranean coasts, the Karoo (South Africa), the Red Sea shores, the southwest Caspian coast, Central Asia (Caspian to Himalayas—deserts), salt steppes of East Asia. Selected Genera: *Atriplex, Beta, Chenopodium, Salicornia, Salsola, Spinacia, Suaeda.*

Predominantly halophytic annual or perennial herbs or shrubs (rarely small trees). Stems often fleshy, jointed, and nearly leafless. Leaves simple, fleshy in some or reduced to scales, alternate (rarely opposite, in *Salicornia*), exstipulate. Inflorescence usually small, dense, dichasial or unilateral cymes. Flowers often bracteate, minute, greenish, bisexual or unisexual and the species monoecious or dioecious, actinomorphic; perianth uniseriate; calyx typically of 5 (2 to 5) connate sepals (rarely absent), often accrescent in fruit; corolla absent; androecium of as many stamens as the sepals and opposite them, usually distinct, anthers 2-celled (locular), incurved in bud, longitudinally dehiscent; gynoecium of 1 compound pistil, ovary superior (inferior in *Beta*), carpels 2 to 3, locule 1, ovule solitary, erect, or suspended from a basal funiculus, styles 1 to 3, stigmas mostly 2 to 3. Fruit a utricle. Seed with a peripheral or coiled embryo surrounding the endosperm (sometimes absent).

Aids for Field Identification: Halophytic, fleshy, farinaceous herbs or shrubs; flowers minute and green; gynoecium 1-locular, 2- to 3- carpellate, and 1-ovulate. The absence of scarious bracts and the presence of distinct stamens distinguish the family from the Amaranthaceae.

Illustrations: 21, 25, 48.   *References:* 8, 21, 25, 29, 30, 31, 36, 40.

CLUSIACEAE [= Guttiferae, including the Hypericaceae]   (St. John's wort family)

Magnoliopsida (Dicotyledoneae). Approximately 40 genera and 1,000 species, mostly tropical. Selected Genera: *Ascyrum, Clusia, Garcinia, Hypericum, Mammea.*

Trees, shrubs, and herbs (rarely vines) with resinous sap. Leaves simple, usually opposite or whorled, exstipulate, often glandular dotted. Inflorescence cymose. Flowers bisexual or unisexual, species often polygamodioecious, actinomorphic; perianth biseriate; calyx of 2 to 10 imbricate sepals, sometimes basally connate; corolla of 4 to 12 distinct petals, may be clawed, sessile, imbricate or contorted; androecium of few to many distinct or variously connate stamens that are often fascicled (rarely monadelphous), anthers 2-celled with longitudinal dehiscence (pistillodes often present in male flowers); gynoecium of 1 compound pistil, ovary superior, carpels 3 or 5 (or as many of the locules), locules 1 to many, placentation axile, basal or pendulous, styles 3 to 5 (or as many as the locules) and either distinct or basally connate, stigmas 3 to 5 (or as many as the locules). Fruit a berry or capsule, rarely a drupe. Seeds with a large, straight, or curved embryo; endosperm absent.

Aids for Field Identification: Trees, shrubs, or herbs with resinous sap; leaves opposite or whorled, often glandular-dotted; stamens usually in fascicles; carpels usually 3 to 5.

Illustrations: 21, 25, 48.   *References:* 8, 21, 25, 29, 30, 31, 36, 40, 49.

COMMELINACEAE   (Spiderwort family)

Liliopsida (Monocotyledoneae). Approximately 38 genera and 500 species, mostly tropical and subtropical. Selected Genera: *Aneilema, Callisia, Commelina, Cyanotis, Rhoeo, Tinantia, Tradescantia, Zebrina.*

Succulent annual or perennial herbs, acaulescent or with nodose stems, roots fibrous or tuberlike. Leaves simple, alternate, flat or troughlike, often with tubular sheath at base, margins entire. Inflorescence a cyme, thyrse, panicle, or flowers solitary. Flowers bisexual, usually actinomorphic, sometimes subtended by a cymbiform (boat-shaped) spathe or foliaceous bract; perianth biseriate; calyx of 3 distinct sepals (rarely fused or petaloid); corolla of 3 petals, distinct, rarely united into a tube, the third petal sometimes much reduced; androecium typically of 6 distinct (rarely fused) stamens, often reduced to 3 (by abortion of 3 to staminodes, rarely aborted to 1 functional stamen with no staminodes), filaments often bearded, anthers 2-celled (parallel or divergent), dehiscence longitudinal or poricidal; gynoecium of 1 pistil, ovary superior, sessile or stipitate, carpels 3, locules 3 (2), placentation axile, style 1, stigma 1 (capitate or trifid). Fruit a loculicidal capsule, rarely fleshy and indehiscent. Seeds few to solitary in each loculus; embryo small; endosperm abundant, mealy.

Aids for Field Identification: Herbs; leaves with tubular sheath; floral parts in 3s; calyx green; corolla colored; ovary superior; fruit a capsule.

Illustrations: 21, 25.   *References:* 8, 21, 25, 29, 31, 36, 40.

CONVOLVULACEAE [including Dichondraceae, Humbertiaceae; excluding Cuscutaceae]

(Morning glory family)

Magnoliopsida (Dicotyledoneae). Approximately 55 genera and 1,650 species, tropics and temperate regions. Selected Genera: *Breweria, Calonyction, Calystegia, Convolvulus, Cressa, Dichondra, Evolvulus, Ipomoea, Jacquemontia.*

Annual or perennial herbs, shrubs, or trees, often climbing or prostrate; sap often milky. Leaves simple, alternate, exstipulate, usually petiolate, margins entire, lobed, or pinnatifid. Inflorescence typically an axillary cyme or the flowers solitary and axillary. Flowers usually bisexual (if unisexual the species then dioecious), actinomorphic, bracteate, the bracts often composing an involucre; perianth biseriate; calyx of 5 sepals, distinct, imbricate, persistent; corolla of 5 petals, sympetalous, often lobed, usually contorted in bud, often infundibular or salverform; androecium of 5 stamens, distinct, epipetalous, antisepalous, the anthers 2-celled, longitudinally dehiscent, introrse, dorsifixed; intrastaminal disk usually present; gynoecium of 1 compound pistil, ovary superior, carpels typically 2 (rarely 3 to 5), locules typically 2 (rarely 1 to 4, the latter by false septation), placentation axile, the ovules 1 to 2 in each locule, style usually 1, stigmas 1 or 2. Fruit a capsule, berry, or nut. Seeds sometimes hairy; embryo large, straight, cotyledons folded or spirally coiled and emarginate or bilobed; endosperm scanty and cartilaginous.

Aids for Field Identification: Ours mostly twining vines, with milky sap; flowers 5-merous, carpels typically 2 or 3; corolla plicate, often infundibular or salverform; flowers axillary.

Illustrations: 21, 25, 48.   *References:* 8, 21, 25, 29, 30, 31, 36, 40, 43.

CORNACEAE   (Dogwood family)

Magnoliopsida (Dicotyledoneae). Approximately 12 genera and 100 species, temperate zones and tropical mountains. Selected Genera: *Aucuba, Cornus.*

Trees, shrubs, rarely perennial herbs or woody vines. Leaves simple, alternate or opposite, exstipulate, usually petiolate, margins entire. Inflorescence a head, cymose or paniculate, occasionally with large showy bracts. Flowers bisexual or unisexual (species monoecious, dioecious, or polygamodioecious), actinomorphic; perianth biseriate (adnate to ovary); calyx of 4 to 5 lobes (often minute, sometimes absent); corolla of 4 to 5 (0) distinct petals; androecium of 4 to 5 distinct stamens alternate with the petals, anthers 2-celled with longitudinal dehiscence; gynoecium of 1 compound pistil, ovary inferior, carpels and locules 2 to 4 (seldom unilocular), placentation usually axile, styles 1 to several (arising from glandular disk), stigmas 1 to several. Fruit a berry or drupe. Seed 1 per locule; embryo straight, small; endosperm copious.

Aids for Field Identification: Mostly trees and shrubs; leaves usually opposite; flowers usually 4- to 5-merous; ovary inferior; fruit drupaceous.

Illustrations: 21, 25, 48.   *References:* 8, 14, 21, 25, 29, 30, 31, 36, 40.

CRASSULACEAE [excluding Penthoraceae]   (Stonecrop family)

Magnoliopsida (Dicotyledoneae). Approximately 35 genera and 1,500 species, cosmopolitan but chiefly South Africa. Selected Genera: *Sedum, Kalanchoe, Echeveria, Crassula, Sempervivum.*

Annual or perennial herbs or shrubs, sometimes scandent and mostly succulent. Leaves mostly simple, alternate or opposite (rarely whorled), exstipulate, succulent and mostly entire. Inflorescence usually cymose, bracteate. Flowers bisexual (rarely unisexual and the species dioecious), actinomorphic; perianth biseriate; calyx commonly 4 or 5 (but ranging from 3 to 30) sepals, distinct or united; petals usually the same number as sepals, distinct or variously connate; androecium of as many as or twice as many stamens as the petals and typically in 2 whorls, distinct or basally connate, generally hypogynous except where petals are connate then epipetalous, anthers 2-celled, introrse, longitudinally dehiscent; gynoecium of 3 or more distinct or basally connate pistils (carpels), the pistils typically as many as the petals, each subtended by a nectariferous gland or scale, ovaries superior, 1-carpellate, unilocular, placentation marginal, style 1, stigma 1. Fruit typically a follicle. Seeds minute, mostly numerous; embryo straight; endosperm fleshy and scant or rarely absent.

Aids for Field Identification: Herbs; apopetalous; pistils mostly 4 to 5, simple, distinct (apocarpous) or basally connate and each with a scalelike gland; fruit a follicle or follicetum.

Illustrations: 21, 25, 48.   *References:* 8, 21, 25, 29, 30, 31, 36, 40.

CUCURBITACEAE   (Gourd family)

Magnoliopsida (Dicotyledoneae). Between 100 and 110 genera and 650 to 850 species, mainly tropics and subtropics but with some temperate members in both hemispheres.

Selected Genera: (after Jeffrey in reference 21)

Subfamily I. CUCURBITOIDEAE:	*Citrullus, Ecballium, Lagenaria, Luffa, Momordica, Echinocystis, Cyclanthera, Sicyos, Cucurbita, Cayaponia, Melothria, Cucumis*
Subfamily II. ZANONIOIDEAE:	*Fevillea*

Annual or perennial prostrate herbs or vines (rarely shrubby or arborescent) usually with spirally coiled tendrils. Leaves simple or compound, palmately veined, alternate, tendrils present (variously interpreted, perhaps stipular in origin), exstipulate. Inflorescence axillary and basically cymose or the flowers solitary. Flowers unisexual (rarely bisexual), the species monoecious or dioecious, actinomorphic; perianth biseriate; calyx of 5 sepals, united or distinct, imbricate or open; corolla typically of 5 petals, sympetalous to apopetalous, lobes imbricate to valvate; androecium of staminate flowers of 1 to 5 stamens (typically 5 but modified into a number of types of by cohesion and twisting of filaments and anthers), actinomorphic or zygomorphic, pistillodes often present in staminate flowers, anthers 1- or 2-celled; gynoecium of 1 pistil, ovary inferior, syncarpous, carpels 3 to 5, typically 3, locules 1 to 3, placentation mostly parietal, sometimes the placentae meeting in the middle and appearing axile or rarely apical, styles 1 to 3, stigmas as many as carpels and each often bifid or notched. Fruit a pepo or capsule. Seeds often flat, numerous; embryo straight; endosperm absent.

Aids for Field Identification: Climbing or prostrate herbs with tendrils; flowers 5-merous; ovary inferior; fruit a pepo.

Illustrations: 21, 25, 48.   *References:* 8, 21, 25, 29, 30, 31, 36, 40.

CYPERACEAE

b: perigynium

c: pistillate scale

d:
pistillate
infructescence

e: staminate scale

f:
staminate
inflorescence

a: plant habit

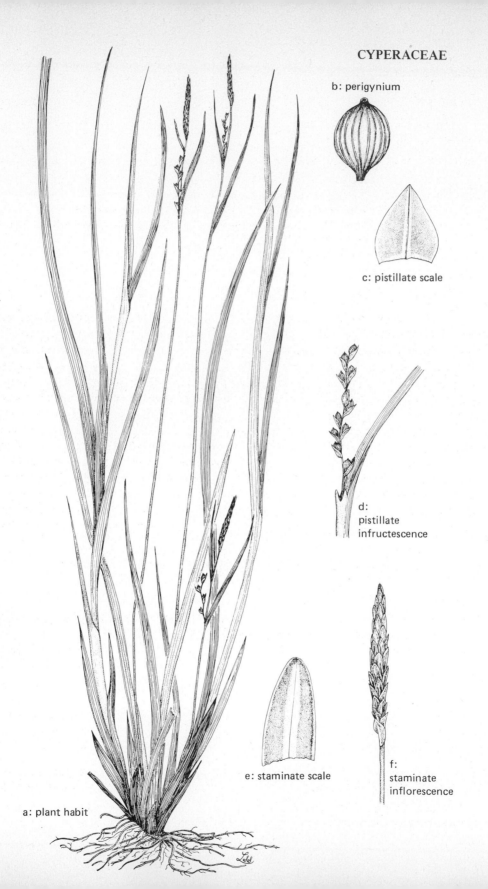

CYPERACEAE (Figure C.11)   (Sedge family)

Liliopsida (Monocotyledoneae). Approximately 90 genera and 4,000 species, cosmopolitan in damp or wet habitats.

Selected Genera: (see reference 21)

Tribe 1. Cypereae: *Cyperus, Kyllinga*
Tribe 2. Dulichieae: *Dulichium*
Tribe 3. Rhynchosporeae: *Cladium, Dichromena, Rhynchospora*
Tribe 4. Scirpeae: *Eleocharis, Eriophorum, Fuirena, Lipocarpha, Hemicarpha, Bulbostylis, Fimbristylis, Scirpus*
Tribe 5. Hypolytreae: *Hypolytrum, Mapania*
Tribe 6. Sclerieae: *Scleria*
Tribe 7. Cryptangieae: *Cryptangium, Lagenocarpus, Didymiandrum*
Tribe 8. Cariceae: *Cymophyllus, Carex, Uncinia, Kobresia*

Annual or perennial herbs. Stems (culms) solid (rarely hollow), often triangular, generally unbranched below the inflorescence, creeping rhizomes common. Leaves in basal tufts or if cauline, crowded near base of the stem, often 3-ranked, blade often grasslike, sheath closed (rarely open), ligule rarely present. Inflorescence—basic unit a spikelet, these variously arranged in spikes, racemes, panicles or umbels, well-developed foliaceous bracts often subtend the inflorescence (compound) of those of the umbellate type. Flowers (florets) small, bisexual or unisexual (the species monoecious or dioecious), each borne in the axil of a glumelike bract, the bracts often distichous or spiral and imbricate; perianth absent or represented by hypogynous scales, bristles, or hairs; androecium of typically 3 but ranging from 1 to 6 stamens, hypogynous, filaments distinct, anthers basifixed, 2-celled, dehiscence longitudinal; gynoecium of 1 pistil, ovary superior (subtended and enclosed in a perigynium in *Carex*), carpels 2 to 3, syncarpous, locule 1, placentation basal, styles 2 to 3. Fruit a lenticular or trigonous, indehiscent achene. Seed one per fruit; embryo small and embedded in the mealy or fleshy endosperm.

Aids for Field Identification: Herbs; culms solid, triangular; leaves 3-ranked; fruit an achene.

Key to Poaceae and Cyperaceae:

Florets usually subtended by a single bract or scale; seed coat free from the pericarp; embryo embedded in the endosperm; leaf sheaths mostly closed; perianth of bristles or scales or absent; fruit an achene; stems mostly 3-angled and solid; leaves 3-ranked . . . . CYPERACEAE

Florets usually subtended by a pair of bracts (lemma and palea); seed coat adnate to pericarp; embryo peripheral to the endosperm; leaf sheaths mostly open; perianth absent or of 1 to 3 lodicules; fruit a caryopsis; stems mostly terete and internodes hollow; leaves 2-ranked . . . . POACEAE

Illustrations: 21, 25, 48.   *References:* 1, 8, 21, 25, 29, 30, 31, 36, 40.

ERICACEAE (Figure C.12) [including Rhododendraceae, Rhodoraceae, Menziesiaceae, Arctostaphylaceae, Vacciniaceae; excluding Pyrolaceae, Monotropaceae](Heath family)

Magnoliopsida (Dicotyledoneae). Between 50 and 80 genera and 1,350 to 2,000 species, cosmopolitan, except in deserts, almost absent from Australasia.

ERICACEAE: Oxydendrum. a–g, O. arboreum

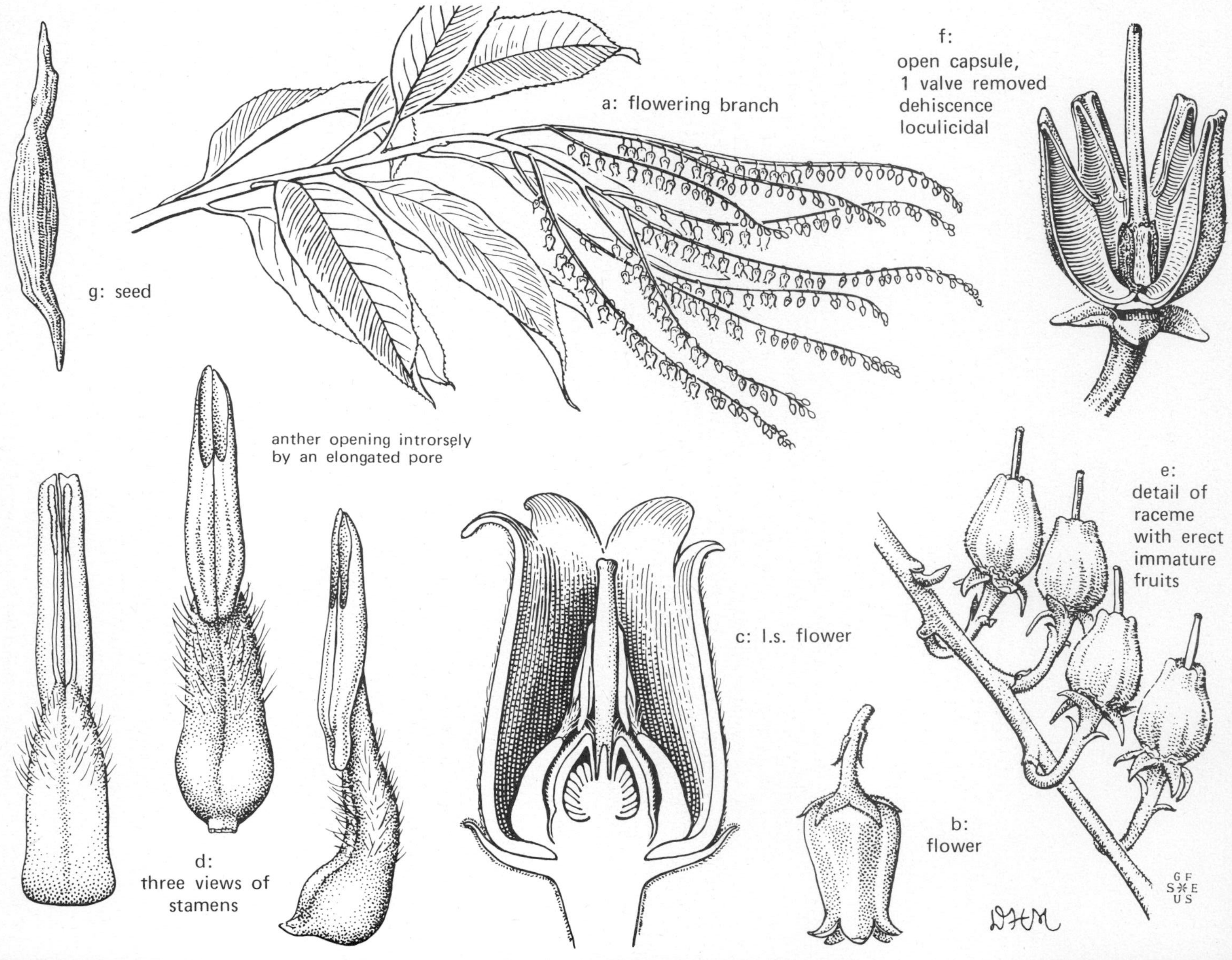

a: flowering branch

f:
open capsule,
1 valve removed
dehiscence
loculicidal

e:
detail of
raceme
with erect
immature
fruits

b:
flower

c: l.s. flower

anther opening introrsely
by an elongated pore

d:
three views of
stamens

g: seed

366

Selected Genera: (see reference 47)

Subfamily I.   ERICOIDEAE (ovary superior):   *Arbutus, Arctostaphylos, Befaria, Calluna, Cassandra, Cassiope, Elliottia, Epigaea, Erica, Gaultheria, Kalmia, Ledum, Leiophyllum, Leucothoe, Lyonia, Menziesia, Oxydendrum, Phyllodoce, Pieris, Rhododendron, Zenobia*

Subfamily II.   VACCINIOIDEAE (ovary inferior):   *Gaylussacia, Vaccinium*

Predominantly shrubs, occasionally suffrutescent perennial herbs, rarely trees or vines. Leaves simple, usually alternate, sometimes opposite or whorled, exstipulate, often coriaceous and evergreen. Inflorescence a corymb, raceme, or panicle or flowers solitary and axillary. Flowers bisexual, actinomorphic or slightly zygomorphic (as in *Rhododendron* spp.); perianth biseriate; calyx of 4 to 7 sepals, usually synsepalous and lobed, often persistent; corolla of 4 to 7 petals, usually sympetalous and lobed, often campanulate, infundibular, or urceolate; androecium typically of twice as many stamens as petals, distinct (rarely somewhat connate), arising from the base of a disk, obdiplostemonous, the anthers 2-celled, poricidally dehiscent, introrse, thecae often saccate and basally gibbous, frequently with appendages, pollen grains in tetrads; gynoecium of 1 compound pistil, ovary superior (e.g., Rhododendreae, Ericeae, Epigaeeae) or inferior (Vaccinioideae), carpels and locules typically 5(4 to 10), placentation axile, ovules usually numerous in each locule, anatropous, style and stigma typically 1. Fruit a capsule, berry, or drupe. Seeds usually small, sometimes winged; embryo straight and cylindrical; endosperm copious and fleshy.

Aids for Field Identification: Predominantly shrubs; leaves often coriaceous and evergreen; flowers often 5-merous and campanulate or urceolate; anthers poricidally dehiscent; fruit a capsule or berry.

Illustrations: 21, 25, 48.   *References:*   8, 21, 25, 29, 30, 31, 36, 40, 47.

EUPHORBIACEAE (Figure C.13)   (Spurge family)

Magnoliopsida (Dicotyledoneae). Approximately 300 genera and 5,000 species, cosmopolitan, except the Arctic. Selected Genera: *Acalypha, Codiaeum, Croton, Euphorbia, Hevea, Jatropha, Manihot, Pedilanthus, Ricinus, Sapium, Tragia.*

Trees, shrubs, or herbs (rarely vines), mostly with milky latex, a diverse group ranging from succulent and cactuslike, cricoid, or phyllocladous to nettlelike forms with stinging hairs. Leaves simple or variously compound, alternate but sometimes opposite or whorled, mostly stipulate, the stipules sometimes modified into glands, hairs, or spines. Inflorescence variable but mostly determinate. Flowers unisexual (species mostly monoecious), actinomorphic, sometimes greatly reduced; perianth of both calyx and corolla or the latter or both absent, valvate or imbricate; calyx when present 5-merous, sepals distinct; corolla of 0 to 5 distinct or united petals; androecium of 1 to 100 stamens but usually as many or twice as many as petals, distinct or monadelphous, anthers 2-locular (rarely 3 to 4), longitudinally or transversely (rarely poricidally) dehiscent; gynoecium of 1 pistil, sometimes present as a pistillode in staminate flowers, ovary superior, syncarpous, mostly 3-carpellate (rarely 2 to 4) and 3-locular, placentation axile, styles 3, distinct or variously united, each often 2-lobed, stigmas 3 or 6 or sometimes dissected into numerous filiform segments. Fruit a capsule (often schizocarpic) or drupe. Seeds solitary or paired in each

EUPHORBIACEAE: a-i, Euphorbia corollata; j, E. inundata; k, l, E. commutata; m, E. dentata; n-p, Chamaesyce maculata

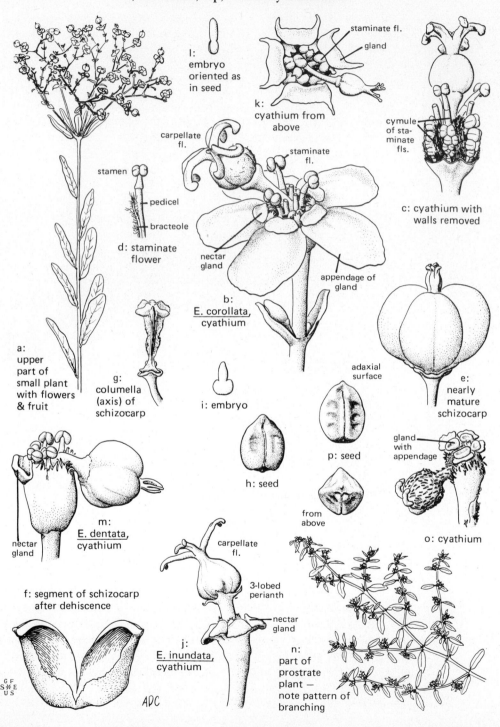

l: embryo oriented as in seed

staminate fl.

gland

k: cyathium from above

cymule of staminate fls.

c: cyathium with walls removed

carpellate fl.

stamen

pedicel

bracteole

d: staminate flower

staminate fl.

nectar gland

appendage of gland

b: E. corollata, cyathium

g: columella (axis) of schizocarp

a: upper part of small plant with flowers & fruit

i: embryo

adaxial surface

h: seed

p: seed

from above

e: nearly mature schizocarp

gland with appendage

o: cyathium

m: E. dentata, cyathium

nectar gland

f: segment of schizocarp after dehiscence

carpellate fl.

3-lobed perianth

nectar gland

j: E. inundata, cyathium

n: part of prostrate plant — note pattern of branching

GFE
S✳E
US

ADC

368

locule, often with a conspicuous caruncle; embryo straight; endosperm copious and fleshy.

Aids for Field Identification: Monoecious or dioecious herbs, shrubs, or small trees; often apetalous; carpels 3; inflorescence a cyathium in *Euphorbia*.

Illustrations: 21, 25, 48.   *References:* 8, 21, 25, 29, 30, 31, 36, 39, 40.

FABACEAE (Figure C.14) [including Caesalpinaceae, Leguminosae, Mimosaceae, Papilionaceae]   (Legume family)

Magnoliopsida (Dicotyledoneae). Approximately 600 genera and 12,000 species (the third largest family of flowering plants), cosmopolitan.

Key to Subfamilies: (Selected Genera)

Flowers actinomorphic; perianth aestivation valvate or rarely imbricate; stamens 10 or more, anthers sometimes with terminal gland . . . . MIMOSOIDEAE

Flowers zygomorphic; perianth aestivation imbricate; stamens 5 to 10

Corolla aestivation imbricate-ascending, posterior petal innermost; petals typically 5 and distinct; stamens distinct or monadelphous . . . . CAESALPINIOIDEAE

Corolla aestivation imbricate-descending, posterior petal outermost; petals 5, the anterior petals basally connate; stamens monadelphous or diadelphous, rarely distinct . . . .
. . . . PAPILIONOIDEAE (Lotoideae, Faboideae)

Subfamily I.	MIMOSOIDEAE:	*Acacia, Albizzia, Mimosa, Prosopis, Schrankia*
Subfamily II.	CAESALPINIOIDEAE:	*Bauhinia, Cassia, Cercis, Gleditsia, Gymnocladus*
Subfamily III.	PAPILIONOIDEAE (see reference 22)	
Tribe 1.	Sophoreae:	*Cladrastis, Sophora*
Tribe 2.	Podalyrieae:	*Baptisia*
Tribe 3.	Genisteae:	*Crotalaria, Lupinus*
Tribe 4.	Trifolieae:	*Medicago, Melilotus, Trifolum*
Tribe 5.	Loteae:	*Lotus*
Tribe 6.	Galegeneae:	*Astragalus, Oxytropis, Robinia, Tephrosia*
Tribe 7.	Psoraleae:	*Dalea, Psoralea*
Tribe 8.	Indigofereae:	*Indigofera*
Tribe 9.	Hedysareae:	*Arachis, Desmodium, Lespedeza*
Tribe 10.	Dalbergieae:	*Dalbergia*
Tribe 11.	Vicieae:	*Lathyrus, Lens, Pisum, Vicia*
Tribe 12.	Phaseoleae:	*Glycine, Phaseolus, Pueraria, Rhynchosia, Vigna*

Herbs, shrubs, trees, or vines. Leaves typically pinnately compound, often trifoliolate, sometimes palmately compound, unifoliolate by suppression of leaflets, represented by phyllodes (many *Acacia* spp.), or rarely reduced to scales, alternate, usually stipulate, sometimes stipellate. Inflorescence a spike, raceme, panicle, head, or flowers solitary. Flowers bisexual, zygomorphic in Caesalpinioideae and Papilionoideae and actinomorphic in Mimosoideae; perianth biseriate; calyx of 5 sepals, synsepalous and lobed; corolla of 5 petals (rarely absent or reduced to a single petal), distinct or the 2 anterior petals connate along their lower margins to form the keel; androecium typically of 10 stamens

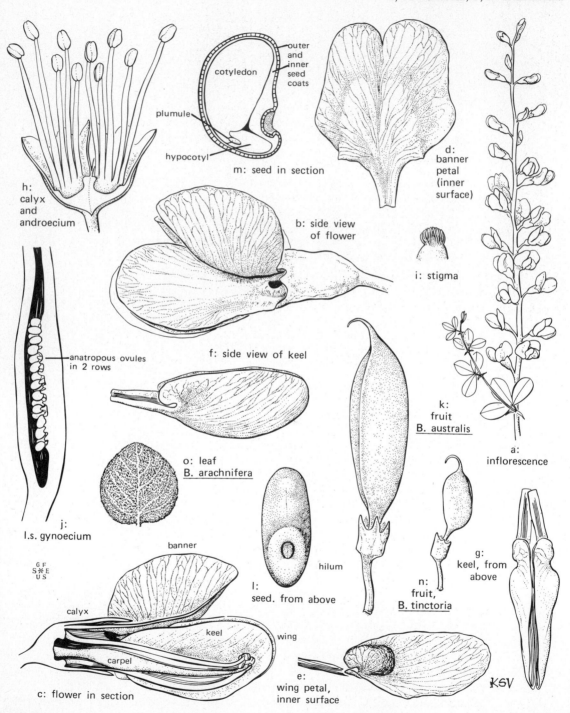

FABACEAE (LEGUMINOSAE) subfamily FABOIDEAE: Baptisia. a-m, B. australis; n, B. tinctoria; o, B. arachnifera

h: calyx and androecium

cotyledon

outer and inner seed coats

plumule

hypocotyl

m: seed in section

d: banner petal (inner surface)

b: side view of flower

i: stigma

anatropous ovules in 2 rows

f: side view of keel

k: fruit B. australis

o: leaf B. arachnifera

a: inflorescence

j: l.s. gynoecium

hilum

l: seed, from above

g: keel, from above

banner

n: fruit, B. tinctoria

GF S✳E US

calyx

keel

carpel

wing

c: flower in section

e: wing petal, inner surface

KSV

370

(numerous in some Mimosoideae), distinct, monadelphous, or diadelphous, the anthers 2-celled, longitudinally dehiscent, sometimes with a deciduous gland at apex; gynoecium of 1 simple pistil, ovary superior, carpel and locule 1, placentation marginal, style and stigma 1. Fruit typically a legume or loment, sometimes follicular, rarely indehiscent. Seeds usually with a leathery testa, some with funiculus modified into a fleshy aril or a callosity; embryo large, cotyledons usually with a large food reserve; endosperm generally absent or very scant.

Aids for Field Identification: Herbs, shrubs, trees, or vines; leaves often trifoliolate or pinnately compound; flowers 5-merous, carpel 1; corollas usually papilionaceous; stamens monadelphous or diadelphous; fruit a legume or loment.

Illustrations: 21, 25, 48.　　*References:* 8, 11, 21, 22, 23, 24, 25, 29, 30, 31, 35, 36, 40.

FAGACEAE (Figure C.15)　　　　　　　　　　　　　　　　　　　　　(Beech family)

　　　Magnoliopsida (Dicotyledoneae). Approximately 8 genera and 900 species, cosmopolitan, but mostly temperate and tropical regions of the northern hemisphere.

Genera: *Castanea, Castanopsis, Chrysolepis, Fagus, Lithocarpus, Nothofagus, Quercus, Trigonobalanus.*

Synoptic Key to Genera (see reference 40):

Flowers in axillary clusters, rarely solitary.
　　Male inflorescence many-flowered, globular; female inflorescence with 2 flowers
　　. . . . .*Fagus*
　　Male inflorescence 1- to 3-flowered; female inflorescence with 1, 3, or 7 flowers
　　. . . . .*Nothofagus*
Flowers in catkins.
　　Male catkins rigid; stamens usually 12.
　　　　Cupules not lobed, or showing signs of vertical divisions . . . .*Lithocarpus*
　　　　Cupules lobed or divided into segments (sometimes segments remain fused), vertical divisions evident in young cupules.
　　　　　　Female flowers on separate inflorescences. . . . .*Castanopsis*
　　　　　　Female flowers on androgynous inflorescences.
　　　　　　　　Styles 6 to 9; fruits not separated by cupule lobes . . . .*Castanea*
　　　　　　　　Styles 3; fruits separated by inner cupule lobes . . . .*Chrysolepis*
　　Male catkins rigid or flexed; stamens usually 6.
　　　　Fruit round; cupule margin entire. . . . .*Quercus*
　　　　Fruit 3-angled; cupule lobed . . . .*Trigonobalanus*

　　Trees and shrubs. Leaves deciduous or evergreen, simple, alternate, stipulate. Staminate inflorescences pendulous heads or catkinlike racemes or flowers solitary; pistillate inflorescence a cymule or flowers solitary. Flowers unisexual (rarely bisexual), and species monoecious (dioecious in *Nothofagus*); staminate flowers with perianth of 4, 6, or 7 distinct tepals; androecium of 4 to 40 distinct stamens, anthers basifixed, 2-celled; pistillate flowers with or without involucre, perianth of 4 to 6 distinct tepals, adnate to ovary, gynoecium of one pistil, ovary inferior, carpels 3 to 6, locules 3 to 6, placentation axile, styles 3 to 6, ours with mostly 3 to 6 stigmas. Fruit a 1-seeded nut subtended by a

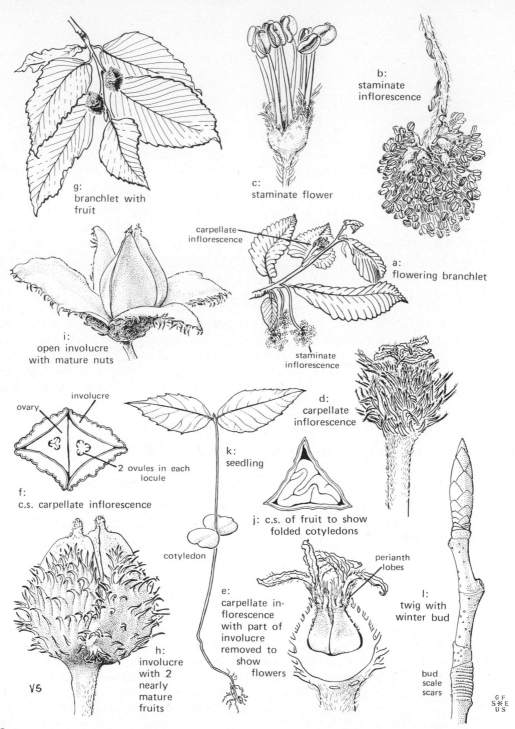

g: branchlet with fruit

b: staminate inflorescence

c: staminate flower

carpellate inflorescence

a: flowering branchlet

i: open involucre with mature nuts

staminate inflorescence

d: carpellate inflorescence

ovary

involucre

2 ovules in each locule

f: c.s. carpellate inflorescence

k: seedling

cotyledon

j: c.s. of fruit to show folded cotyledons

perianth lobes

e: carpellate inflorescence with part of involucre removed to show flowers

l: twig with winter bud

h: involucre with 2 nearly mature fruits

bud scale scars

√5

cupule or involucre that may be muricate, bristly, or spiny, some genera with 3 nuts enclosed in an involucre (*Castanea*). Seeds with large embryo; endosperm absent.

Aids for Field Identification: Trees or shrubs; monoecious; male flowers in catkins; fruit enclosed in cupule or bur.

Illustrations: 21, 25, 48. *References:* 8, 21, 25, 31, 36, 40, 48.

GERANIACEAE (Figure C.16) (Geranium family)

Magnoliopsida (Dicotyledoneae). Five to eleven genera and approximately 756 species, cosmopolitan. Selected Genera: *Geranium, Pelargonium, Erodium, Sarcocaulon*.

Primarily herbs, sometimes suffrutescent, shrubby, or arborescent. Leaves simple and lobed or dissected or compound, alternate or opposite, stipulate, venation mostly palmate. Inflorescence cymose or umbellate or flowers solitary. Flowers bisexual, actinomorphic or zygomorphic; perianth biseriate; calyx of 5 (4 or 8) sepals, distinct or connate to the middle, imbricate or valvate, the dorsal one sometimes spurred and the spur adnate to the pedicel (as seen in cross section); corolla of 5 (rarely 2, 4, 8, or 0) petals, distinct, hypogynous or subperigynous, often alternating with nectariferous glands; androecium of 5 to 15 stamens in 1 to 3 whorls, staminodes often present, distinct or basally connate, anthers 2-celled, longitudinally dehiscent; gynoecium of 1 pistil, ovary superior, syncarpous, 3- to 5-lobed, carpels and locules 3 to 5(8), placentation axile, styles 3 to 5, slender and usually beaklike, stigmas 3 to 5. Fruit a schizocarp, dehiscing into 1 to 2 or many-seeded mericarps, the style adhering to the ovarian beak, often twisted. Seeds 1 to 2 in each locule; embryo mostly curved; endosperm thin or absent.

Aids for Field Identification: Herbs (sometimes suffrutescent); leaves palmately lobed or dissected, venation palmate; flowers 5-merous; carpels coiled.

Illustrations: 21, 25, 48. *References:* 8, 21, 25, 29, 30, 31, 33, 36, 40.

HAMAMELIDACEAE (Figure C.17) [including Altingiaceae] (Witch hazel family)

Magnoliopsida (Dicotyledoneae). Approximately 26 genera and 100 species, chiefly of the subtropics. Selected Genera: *Fothergilla, Hamamelis, Liquidambar, Loropetalum*.

Trees or shrubs often with stellate indumentum. Leaves deciduous or evergreen, simple, alternate (rarely opposite), stipules mostly paired and persistent, margins commonly with glandular teeth, sometimes palmately lobed. Inflorescence capitate, spicate or axillary. Flowers bisexual or unisexual (the species monoecious), actinomorphic or zygomorphic; perianth uniseriate or biseriate; calyx of 4 to 5 basally connate sepals, the tube more or less adnate to the ovary; corolla of 4 to 5 distinct petals appearing to be inserted on the calyx or petals absent, epigynous, imbricate, or valvate; androecium of mostly 4 to 8 (2) stamens, uniseriate, distinct, perigynous, anthers 2-celled, dehiscence longitudinal or valvular; gynoecium of 1 bicarpellate pistil, ovary inferior, half-inferior, or rarely superior, syncarpous, locules 2, placentation axile, styles and stigmas 2. Fruit a capsule, often woody. Seeds 1 or more per locule, sometimes winged; embryo straight; endosperm thin and fleshy.

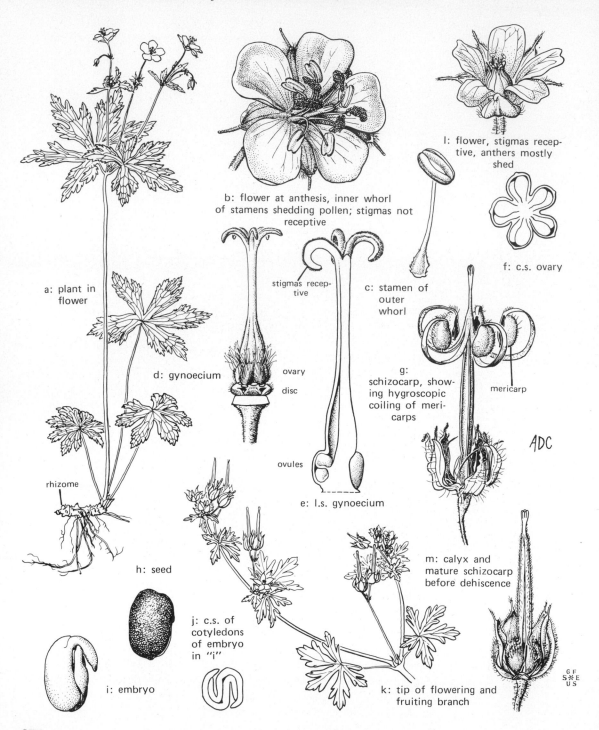

a: plant in flower

rhizome

b: flower at anthesis, inner whorl of stamens shedding pollen; stigmas not receptive

stigmas receptive

c: stamen of outer whorl

d: gynoecium

ovary

disc

ovules

e: l.s. gynoecium

l: flower, stigmas receptive, anthers mostly shed

f: c.s. ovary

g: schizocarp, showing hygroscopic coiling of mericarps

mericarp

ADC

h: seed

j: c.s. of cotyledons of embryo in "i"

i: embryo

k: tip of flowering and fruiting branch

m: calyx and mature schizocarp before dehiscence

GF
S*E
US

HAMAMELIDACEAE: Hamamelis. a-k, H. vernalis; l, H. virginiana

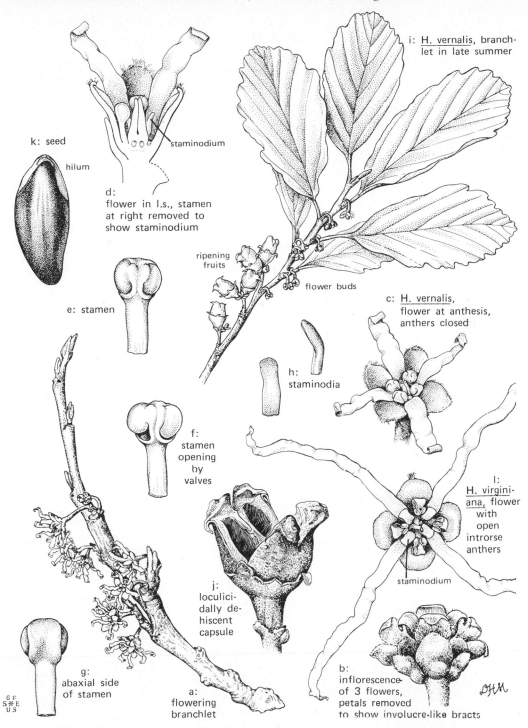

k: seed

hilum

d:
flower in l.s., stamen
at right removed to
show staminodium

staminodium

i: H. vernalis, branch-
let in late summer

ripening
fruits

flower buds

e: stamen

c: H. vernalis,
flower at anthesis,
anthers closed

h:
staminodia

f:
stamen
opening
by
valves

l:
H. virgini-
ana, flower
with
open
introrse
anthers

staminodium

j:
loculici-
dally de-
hiscent
capsule

g:
abaxial side
of stamen

a:
flowering
branchlet

b:
inflorescence
of 3 flowers,
petals removed
to show involucre-like bracts

DHM

G F
S✳E
U S

375

Aids for Field Identification: Trees or shrubs; leaves alternate, simple indumentum stellate; ovary mostly inferior or half-inferior; carpels 2; fruit a woody capsule.

Illustrations: 21, 25, 48.   *References:*   8, 12, 21, 25, 29, 30, 31, 36, 40.

HYDROCHARITACEAE   (Frogs bit family)

Liliopsida (Monocotyledoneae). Approximately 16 genera and 100 species, temperate and tropical fresh and/or saline waters. Selected Genera: *Elodea, Egeria, Hydrilla, Hydrocharis, Limnobium, Vallisneria, Thalassia, Halophila*.

Submerged or floating aquatic herbs, roots terrestrial or floating. Leaves basal or cauline, often ribbonlike, simple, alternate, opposite, or whorled, mostly sessile. Inflorescence umbellate or the flowers solitary, a bifid spathaceous bract or pair of opposite bracts usually present. Flowers bisexual or unisexual and the species dioecious, actinomorphic; perianth usually biseriate and with 3 distinct segments in each series (rarely 2); outer perianth segments (sepals) 3(2); often green, valvate; inner perianth segments (petals) mostly 3 and petaloid, imbricate; androecium of 1 to many stamens, inner sometimes staminodes, pistillode present in some staminate flowers, anthers 2-celled, dehiscence longitudinal; gynoecium of 1 pistil, ovary inferior, syncarpous, carpels 2 to many (commonly 3 to 6), unilocular, placentation parietal, the parietal zones sometimes poorly defined, style 1 but often with as many branches as placentae, stigmas as many as the style branches. Fruit berrylike, irregularly dehiscent or indehiscent. Seeds numerous; embryo straight; endosperm absent.

Aids for Field Identification: Aquatic herbs, free-floating or rooted; inflorescence umbellate or flowers solitary; perianth biseriate; ovary inferior; placentation parietal.

Illustrations: 21, 25.   *References:*   8, 21, 25, 29, 30, 31, 36, 40.

IRIDACEAE   (Iris family)

Liliopsida (Monocotyledoneae). Approximately 60 genera and 80 species, tropics and temperate regions with the chief centers of distribution in South Africa and tropical America. Selected Genera: *Belamcanda, Crocus, Freesia, Gladiolus, Iris, Ixia, Nemastylis, Romulea, Sisyrinchium, Tigridia*.

Perennial herbs or very rarely subshrubs. Stems solitary or several, rhizomes, bulbs, or corm often present. Leaves mostly basal and numerous, simple, generally equitant, mostly linear to ensiform. Inflorescence a raceme or panicle or flowers solitary. Flowers bisexual, actinomorphic or zygomorphic, usually very showy, subtended individually or in clusters by 2 spathelike bracts; perianth of 6 petaloid segments in 2 series (the outer whorl often distinguishable from the inner whorl by color, size, or texture), all generally connate basally into a tube (sometimes obsolete); androecium of 3 distinct stamens (perhaps representing the remaining outer whorl of an ancestral biseriate androecium), situated opposite the outer perianth segments and often inserted upon them, the anthers 2-celled, mostly basifixed, extrorse, dehiscing by vertical slits; gynoecium of 1 compound pistil, ovary inferior, tricarpellate, trilocular with axile placentation or rarely unilocular with 3 parietal placentae, the ovules few to many on each placenta (rarely solitary), anatropous, the style 1, often 3-branched, the stigmas 3, branches filiform to subulate or flabellate, entire or lobed or fimbriate, the style crests sometimes winged and petaloid, the stigmatic surface terminal or adaxial. Fruit a loculicidal capsule dehiscing by 3 valves. Seeds sometimes arillate; embryo small; endosperm copious.

Aids for Field Identification: Perennial herbs often from corms, bulbs, or rhizomes; leaves often equitant; perianth showy; floral parts in 3s; styles petaloid in some; ovary inferior.

Illustrations: 21, 25.   *References:* 8, 21, 25, 29, 30, 31, 36, 40.

JUGLANDACEAE (Figure C.18)   (Walnut family)

Magnoliopsida (Dicotyledoneae). Approximately 7 genera and 50 species, primarily in the north temperate zone but extending to the subtropics, specifically to India, Indochina, and South America. Genera: *Alfaroa, Carya, Engelhardtia, Juglans, Oreomunnea, Platycarya, Pterocarya.*

Trees (rarely shrubs). Leaves deciduous, pinnately compound, alternate (rarely opposite), exstipulate, usually resinously dotted beneath, aromatic. Inflorescence erect or pendulous catkins or spikes variously disposed in panicles and spikes or solitary. Flowers unisexual and species monoecious (dioecious in *Engelhardtia* spp.); perianth tepallate or absent; staminate flowers with 0 to 6 tepals, androecium of 3 to 100 stamens, filaments short, anthers basifixed, 2-celled, and longitudinally dehiscent, rudimentary gynoecium often present; pistillate flowers with 0 to 4 tepals (adnate to ovary), gynoecium of 1 pistil, ovary inferior, 2 carpels (occasionally 3), 1-loculed above and 2- to 4-loculed below, placentation apical, styles 2 or style 1 with 2 lobes, stigmas 2. Fruit a drupe, rarely a nut. Seed 1; embryo very large; endosperm absent.

Aids for Field Identification: Trees, monoecious; leaves pinnately compound and alternate; fruit with a thick husk.

Illustrations: 21, 25, 48.   *References:* 8, 21, 25, 29, 30, 31, 36, 40.

JUNCACEAE   (Rush family)

Liliopsida (Monocotyledoneae). Eight or nine genera and 400 species, worldwide but more numerous in temperate and cold montane areas in wet or damp habitats. Selected Genera: *Juncus, Luzula.*

Annual or perennial herbs (shrubby in *Prionium*). Stems erect, mostly short and leafy at the base, rhizomes erect or horizontal. Leaves mostly basal, flat or terete, often grasslike, mostly linear or filiform, lamellate or reduced to scales or sheaths, sheaths open or closed. Inflorescence in panicles, corymbs, cymes, heads or the flowers solitary. Flowers usually bisexual (if unisexual the species dioecious), actinomorphic; perianth mostly biseriate, segments 3 to each series (rarely only 3 segments in 1 series) and the segments undifferentiated, mostly greenish, reddish-brown, or yellowish, glumaceous, coriaceous (rarely scarious); androecium of 3 or 6 stamens, hypogynous, opposite the perianth segments, distinct, anthers 2-celled, basifixed, introrse, longitudinally dehiscent; gynoecium of 1 pistil, ovary superior, syncarpous, carpels 3, locules 1 or 3, placentation parietal or axile, styles 1 or 3, stigmas 3 or 1 with 3 lobes. Fruit a dry, many-seeded capsule, loculicidally dehiscent. Seeds small, often tailed; embryo straight; endosperm present.

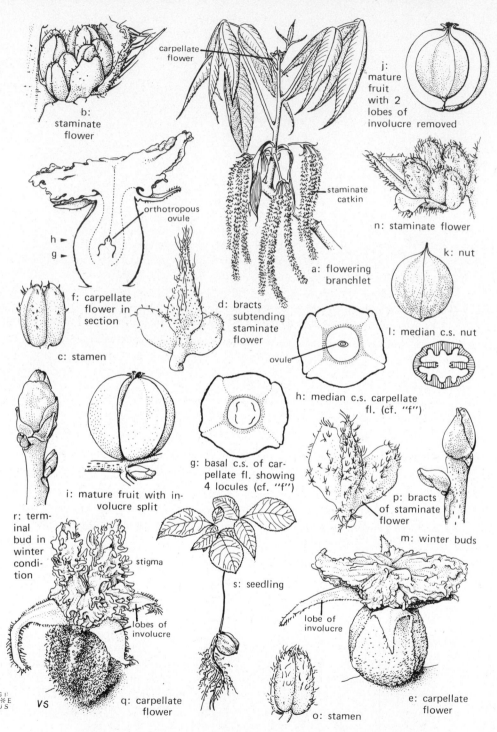

b: staminate flower

carpellate flower

j: mature fruit with 2 lobes of involucre removed

staminate catkin

n: staminate flower

orthotropous ovule

h ►
g ►

f: carpellate flower in section

d: bracts subtending staminate flower

a: flowering branchlet

k: nut

c: stamen

l: median c.s. nut

ovule

h: median c.s. carpellate fl. (cf. "f")

g: basal c.s. of carpellate fl. showing 4 locules (cf. "f")

i: mature fruit with involucre split

p: bracts of staminate flower

m: winter buds

r: terminal bud in winter condition

stigma

lobes of involucre

s: seedling

lobe of involucre

VS

q: carpellate flower

o: stamen

e: carpellate flower

378

Aids for Field Identification: Grasslike herbs of wet or damp sites; floral parts in 3s; fruit a many-seeded capsule.

Illustrations: 21, 25.   *References:* 8, 21, 25, 29, 30, 31, 36, 40.

LAMIACEAE (Figure C.19) [= Labiatae, including Tetrachondraceae]   (Mint family)

Magnoliopsida (Dicotyledoneae). Between 180 and 200 genera and 3,200 to 3,500 species, cosmopolitan, chief center in the Mediterranean region. Selected Genera: *Ajuga, Mentha, Thymus, Rosmarinus, Salvia, Scutellaria, Lamium, Monarda, Coleus, Ocimum, Perilla, Prunella, Stachys, Teucrium, Plectranthus, Origanum, Nepeta, Glechoma, Leonurus, Leonotis, Ballota.*

Annual or perennial herbs, sometimes shrubs, trees, or vines, often aromatic. Stems often quadrangular. Leaves mostly simple (sometimes pinnately dissected or compound), opposite or whorled, exstipulate. Inflorescence variable (flowers solitary and axillary or in paired cymes, cincinnal, congested, and headlike or rarely in simple racemes). Flowers bisexual, zygomorphic (rarely actinomorphic or nearly so); perianth biseriate; calyx of 5 persistent sepals, synsepalous, actinomorphic to zygomorphic, lobes typically 5 (sometimes 2 or obsolete), tube often ribbed; corolla sympetalous, often 4- to 5-lobed, often bilabiate (3 + 2 or with no clear indication of the number of individual petals), imbricate; androecium of 2 or 4 stamens, epipetalous, often didynamous, staminode sometimes present; anthers 1- or 2-locular, the loculi often divergent and the connective much developed; hypogynous nectariferous disk often present; gynoecium of 1 pistil, ovary superior, syncarpous, carpels 2 (each usually deeply lobed and appearing as 4), locules 2 but often appearing as 4 due to the intrusion or constriction of the ovary wall, style 1, mostly gynobasic (arising from inner base of the carpel lobes), bifid, stigmas 2. Fruit a schizocarp (1 to 4 nutlets, free or in pairs). Seeds 4 per ovary; embryo straight; endosperm scanty or absent.

Aids for Field Identification: Aromatic herbs or shrubs; stems square; leaves opposite, decussate; corolla zygomorphic; style gynobasic; fruit schizocarp (1 to 4 nutlets).

Illustrations: 21, 25, 48.   *References:* 8, 21, 25, 29, 30, 31, 36, 40.

LILIACEAE (Figure C.20) [includes Alliaceae, Asparagaceae, Trilliaceae; excludes Agavaceae, Amaryllidaceae, Smilacaceae]   (Lily family)

Liliopsida (Monocotyledoneae). Approximately 226 genera and 2,655 species, cosmopolitan. Selected Genera: *Allium, Aloe, Asparagus, Calochortus, Camassia, Clintonia, Colchicum, Convallaria, Erythronium, Fritillaria, Hemerocallis, Kniphofia, Lilium, Medeola, Muscari, Polygonatum, Scilla, Smilacina, Trillium, Tulipa, Uvularia, Veratrum.*

Predominantly perennial herbs. Stem erect or often modified into rhizomes, bulbs, corms, tubers, or into cladophylls. Leaves basal or cauline, simple, mostly lamellate but sometimes reduced to scales or sheaths (phyllodes), alternate or whorled (opposite in *Scolyopus*). Inflorescence various (often in racemes) or flowers solitary. Flowers bisexual (rarely unisexual and then the species usually dioecious), actinomorphic or infrequently slightly zygomorphic, often large and showy; perianth in 2 usually very similar series of 3 distinct segments each (rarely 4 or more than 6), generally petaloid and hence known as

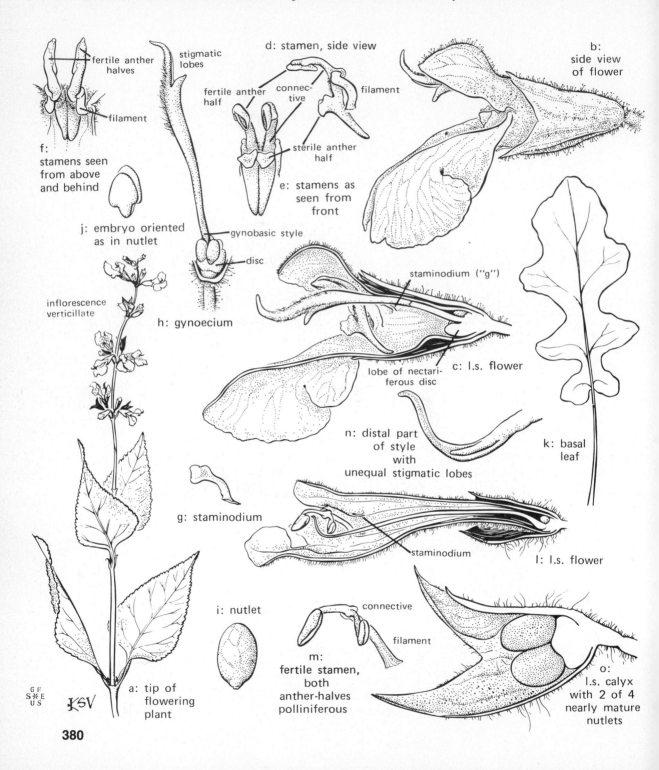

fertile anther halves

filament

f:
stamens seen
from above
and behind

j: embryo oriented
as in nutlet

stigmatic
lobes

d: stamen, side view

fertile anther
half

connec-
tive

filament

sterile anther
half

e: stamens as
seen from
front

b:
side view
of flower

inflorescence
verticillate

gynobasic style

disc

h: gynoecium

staminodium ("g")

lobe of nectari-
ferous disc

c: l.s. flower

n: distal part
of style
with
unequal stigmatic lobes

k: basal
leaf

g: staminodium

staminodium

l: l.s. flower

a: tip of
flowering
plant

i: nutlet

connective

filament

m:
fertile stamen,
both
anther-halves
polliniferous

o:
l.s. calyx
with 2 of 4
nearly mature
nutlets

G F
S ❋ E
U S

KSV

380

LILIACEAE: Smilacina and Maianthemum. a-g, S. racemosa; h-l, M. canadense

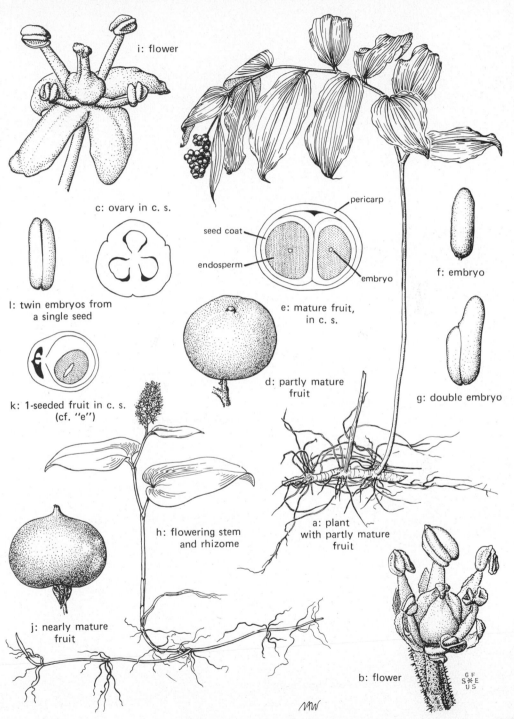

i: flower

c: ovary in c. s.

l: twin embryos from a single seed

k: 1-seeded fruit in c. s. (cf. "e")

pericarp

seed coat

endosperm

embryo

e: mature fruit, in c. s.

f: embryo

g: double embryo

d: partly mature fruit

h: flowering stem and rhizome

a: plant with partly mature fruit

j: nearly mature fruit

b: flower

tepals; androecium of 6 stamens (rarely 3 or to 12), opposite the perianth segments, the filaments distinct or connate, anthers 2-celled, basifixed or versatile, extrorse or introrse, dehiscing by vertical slits (rarely by a terminal pore); gynoecium of 1 compound pistil, ovary usually superior, carpels 3, locules usually 3 with axile placentation (rarely 1 with parietal placentation), the ovules mostly numerous and biseriate, the style usually 1 (rarely 3), divided or trifid, the stigmas 1 or 3. Fruit a septicidal or loculicidal capsule or a berry. Seeds with straight or curved embryo; endosperm copious.

Aids for Field Identification: Perennial herbs often from corms, bulbs, or rhizomes; perianth showy; floral parts in 3s, stamens 6; ovary superior.

Illustrations: 21, 25, 48.   *References:* 8, 21, 25, 29, 30, 31, 36, 40.

MAGNOLIACEAE (Figure C.21) [excluding Illiciaceae and Schisandraceae]
                                                                    (Magnolia family)

Magnoliopsida (Dicotyledoneae). Ten or 12 genera and circa 230 species, southeastern North and Central America, West Indies, Venezuela, Brazil, East Asia, Malay Archipelago, New Guinea.

Key to Tribes: (Selected Genera)

Fruit a follicetum (aggregation of follicles) or if otherwise, not samaroid; anther introrse
       or latrorse; seeds arillate . . . . Magnolieae
Fruit a samaracetum (aggregation of samaras); anthers extrorse; seeds not arillate
       . . . . Liriodendreae

   Tribe 1.   Magnolieae (11 genera):   *Magnolia, Michelia, Manglietia, Talauma*
   Tribe 2.   Liriodendreae (1 genus, 2spp.):   *Liriodendron*

Trees or shrubs. Leaves deciduous or evergreen, simple, alternate, stipulate and stipules often enclosing the young buds and leaving a circular scar, margins mostly entire, petioled, pinnately veined. Flowers solitary, axillary or terminal, mostly large and showy, bisexual (except in *Kmeria*); perianth cyclic or spiral, actinomorphic, sepals and petals not always differentiated; calyx often of 3 sepals, distinct; corolla of 6 to many petals, distinct; androecium of numerous, distinct stamens, mostly spirally disposed or inserted, anthers 2-celled (locular) and longitudinally dehiscent; gynoecium of numerous (rarely as few as 1 to 3), apocarpous, unicarpellate, unilocular pistils, often spirally arranged on an elongate axis, placentation marginal, style 1, stigma 1. Fruit a follicle or aggregation of follicles (follicetum), samara or aggregation of samaras (samaracetum) or rarely a berry. Seeds 1 to 2 per fruit, the funicle often elongate, testa mostly arillate; embryo minute; endosperm copious and oily.

Aids for Field Identification: Trees or shrubs; twigs with circular stipular scars; flowers conspicuous, floral parts numerous and mostly spirally arranged; gynoecium apocarpous; fruit a samaracetum or follicetum.

Illustrations: 21, 25, 48.   *References:* 8, 21, 25, 29, 30, 31, 36, 38, 40, 45.

MALVACEAE (Figure C.22)   (Mallow family)

Magnoliopsida (Dicotyledoneae). Between 75 and 85 genera and 1,000 to 1,500 species, widely distributed in tropics and temperate regions. Selected Genera: *Gossypium, Malva, Hibiscus, Sida, Malvastrum, Sidalcea, Kosteletzkya, Abutilon.*

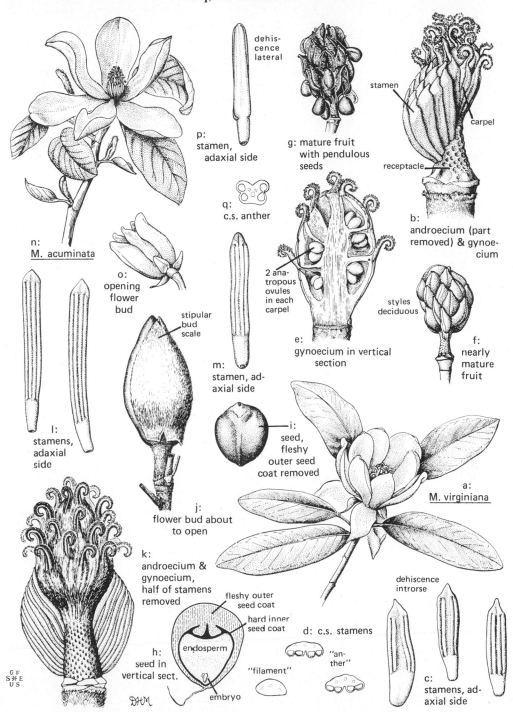

MAGNOLIACEAE: Magnolia. a-i, M. virginiana; j-l, M. grandiflora; m, M. tripetala; n-q, M. acuminata

dehiscence lateral

p: stamen, adaxial side

g: mature fruit with pendulous seeds

stamen

carpel

receptacle

b: androecium (part removed) & gynoecium

q: c.s. anther

2 anatropous ovules in each carpel

e: gynoecium in vertical section

styles deciduous

f: nearly mature fruit

n: M. acuminata

o: opening flower bud

m: stamen, adaxial side

stipular bud scale

i: seed, fleshy outer seed coat removed

l: stamens, adaxial side

j: flower bud about to open

a: M. virginiana

k: androecium & gynoecium, half of stamens removed

fleshy outer seed coat

hard inner seed coat

d: c.s. stamens

dehiscence introrse

endosperm

h: seed in vertical sect.

"filament"

"anther"

embryo

c: stamens, adaxial side

G F S✳E U S

DHM

383

MALVACEAE: Kosteletzkya. a-f, K. virginica var. altheifolia

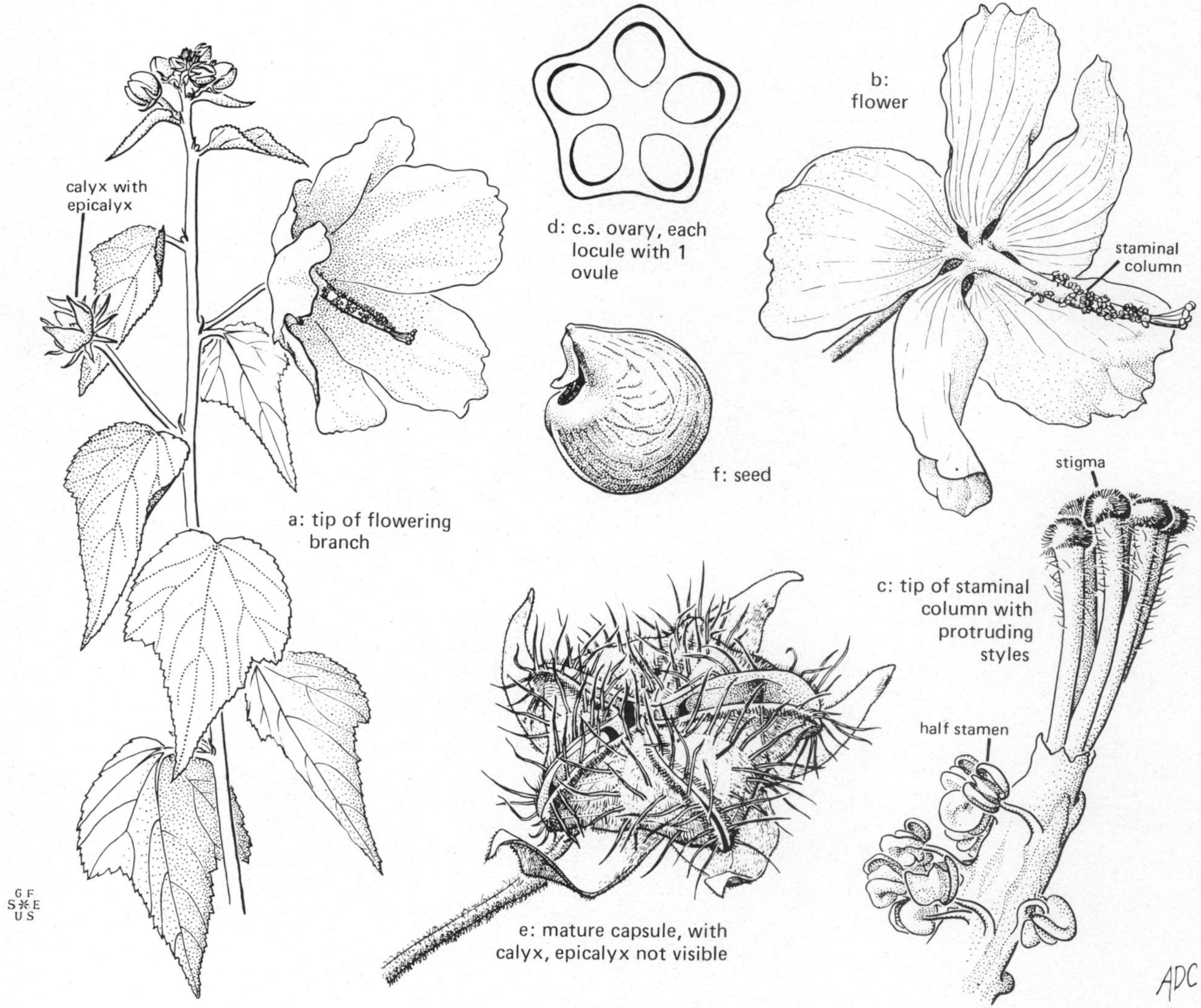

staminal
column

b:
flower

stigma

c: tip of staminal
column with
protruding
styles

half stamen

d: c.s. ovary, each
locule with 1
ovule

f: seed

e: mature capsule, with
calyx, epicalyx not visible

a: tip of flowering
branch

calyx with
epicalyx

ADC

G F
S ✳ E
U S

Trees, shrubs, or herbs often with stellate or lepidote indumentum and mucilaginous sap. Leaves simple, often palmately lobed and veined, alternate, stipulate. Inflorescence cymose or flowers solitary in the leaf axils. Flowers bisexual (rarely unisexual and the species dioecious), actinomorphic; perianth biseriate, epicalyx often present; calyx of 3 to 5 sepals, commonly 5, distinct or basally connate, valvate; corolla of 5 petals, distinct but basally adnate to the staminal column, imbricate or convolute; androecium of numerous stamens, monadelphous, anthers 1-celled, dehiscence longitudinal; gynoecium of 1 pistil (rarely more), ovary superior, syncarpous (rarely unicarpellate), carpels 1 to (but commonly) 5, locules 2 to many (rarely 1) and more commonly 5, placentation axile, style 1 and branched above or as many as the carpels, stigmas as many or twice as many as carpels. Fruit a capsule or schizocarp, or more rarely a berry or samara. Seeds often comose or pubescent; embryo straight or curved; endosperm scanty or absent.

Aids for Field Identification: Shrubs or herbs, frequently with stellate trichomes; perianth conspicuous, sepals persistent; stamens monadelphous; anthers 1-locular; carpels 5 or more.

Illustrations: 21, 25, 48.   *References:* 8, 21, 25, 29, 30, 31, 36, 40.

NYMPHAEACEAE [including Euryalaceae, Barclayaceae, and Cabombacae]
(Water lily family)

Magnoliopsida (Dicotyledoneae). Six genera and circa 83 species, cosmopolitan. Genera: *Barclaya, Cabomba, Euryale, Nuphar, Nymphaea, Ondinea, Victoria.*

Aquatic, annual, or perennial herbs. Leaves simple, alternate, mostly peltate or cordate. Flowers solitary, bisexual, actinomorphic; perianth biseriate; calyx of 3 to 6 sepals (larger than petals and yellow in *Nuphar*), distinct; corolla 3 to many petals (absent in *Ondinea*), distinct, inner petals often petaloid staminodia; androecium of 3 to many distinct stamens, filaments often extending as a sterile appendage beyond the anther sacs, anthers 2-celled with longitudinal dehiscence; gynoecium of one syncarpous pistil or 3 to many apocarpous pistils, ovary superior (inferior in *Victoria*), carpels 1 to many, locules 1 to many, placentation parietal or laminar, stigmas mostly 1 and discoid or 5 to 35 and radiate. Fruit a leathery, tardily dehiscent berry or follicle. Seeds 1 to many per locule; embryo straight; endosperm starchy.

Aids for Field Identification: Aquatics; long-petioled, peltate, or pseudopeltate leaves with latex; floral parts many, except sepals 3; placentation laminar or parietal.

Illustrations: 21, 25, 48.   *References:* 8, 21, 25, 29, 30, 31, 36, 40, 46.

OLEACEAE
(Olive family)

Magnoliopsida (Dicotyledoneae). Approximately 29 genera and 600 species, cosmopolitan, especially temperate and tropical Asia. Selected Genera: *Olea, Syringa, Ligustrum, Chionanthus, Fraxinus, Jasminum, Forsythia, Menodora, Schrebera, Phillyrea, Osmanthus, Linociera, Notelaea.*

Trees or shrubs, occasionally vines. Leaves simple or pinnately compound, typically opposite, exstipulate. Inflorescence axillary or terminal, racemose, paniculate, or thyrsiform. Flowers bisexual (unisexual in some; e.g., *Fraxinus* spp., and the species dioecious or polygamodioecious), actinomorphic; perianth biseriate (apetalous in some *Fraxinus* spp.); calyx typically of 4 (rarely more or absent) sepals, synsepalous; corolla typically of 4 (occasionally more or absent) petals, sympetalous (sometimes very deeply lobed or divided and seemingly apopetalous; distinct in *Fraxinus* spp.); androecium typically of 2 stamens, distinct (often petalostemonous), usually epipetalous, anthers 2-celled (locular), often apiculate by extension of the connective, dehiscing longitudinally; gynoecium of 1 compound pistil, ovary superior, the carpels and locules 2, placentation axile, the ovules usually 2 in each locule, anatropous, the style 1 or absent, stigmas 1 to 2. Fruit a berry (*Ligustrum*), drupe (*Olea, Jasminum*), loculicidal capsule (*Forsythia*), circumscissile capsule (*Menodora*), or samara (*Fraxinus*). Seeds with a straight embryo, the radicle sometimes hidden within the base of the cotyledons; endosperm firm to fleshy.

Aids for Field Identification: Trees or shrubs; leaves opposite; sepals and petals 4, each series connate; stamens 2; ovary 2-locular and superior; fruit a samara or drupe.

Illustrations: 21, 25, 48.   *References:* 8, 21, 25, 29, 30, 31, 36, 40, 44.

ONAGRACEAE (Figure C.23)   (Evening primrose family)

Magnoliopsida (Dicotyledoneae). Approximately 21 genera and 640 species, temperate regions and tropics. Selected Genera: *Circaea, Clarkia (Godetia), Epilobium, Fuchsia, Gaura, Gayophytum, Lopezia, Ludwigia, Oenothera, Zauschneria.*

Annual or perennial herbs, rarely shrubs or trees, sometimes aquatic. Leaves simple, alternate or opposite, stipules absent or if present caducous. Inflorescence spicate, racemose, or paniculate or flowers solitary and axillary. Flowers bisexual, actinomorphic, or rarely zygomorphic; perianth usually biseriate (corolla absent in some *Ludwigia* spp.); hypanthium present, adnate to the ovary, often forming an elongate tube above the ovary; calyx of 4 (2 to 6) sepals, distinct at tip of hypanthium, valvate, sometimes persistent; corolla of 4 (2 to 6 or rarely 0) petals, distinct at tip of hypanthium, usually clawed, convolute, or imbricate; androecium of as many or twice as many stamens as the petals (in *Lopezia*, 1 fertile stamen and 1 staminode), antisepalous or if in 2 whorls diplostemonous, distinct, inserted upon the hypanthium, the anthers typically 2-celled, sometimes each cell divided into 2 or several false cells by transverse plates, longitudinally dehiscent; gynoecium of 1 compound pistil, ovary inferior, carpels and locules 4 (2 or 5), placentation axile (septa sometimes incomplete), ovules 1 to many in each locule, anatropous, style 1, stigma 1 and capitate or with radiate branches. Fruit usually a loculicidal capsule, berry, achene, or nut. Seeds usually numerous (rarely solitary), often comose; embryo straight or nearly so; endosperm absent.

Aids for Field Identification: Perennial herbs; flowers 4-merous; hypanthium often elongated into tube above the inferior ovary; anthers longitudinally dehiscent.

Illustrations: 21, 25, 48.   *References:* 8, 21, 25, 27, 29, 30, 31, 36, 40.

ORCHIDACEAE (Figure C.24)   (Orchid family)

Liliopsida (Monocotyledoneae). Approximately 735 genera and 17,000 species, cosmopolitan but primarily tropical.

ONAGRACEAE: Oenothera. a-k, O. missouriensis

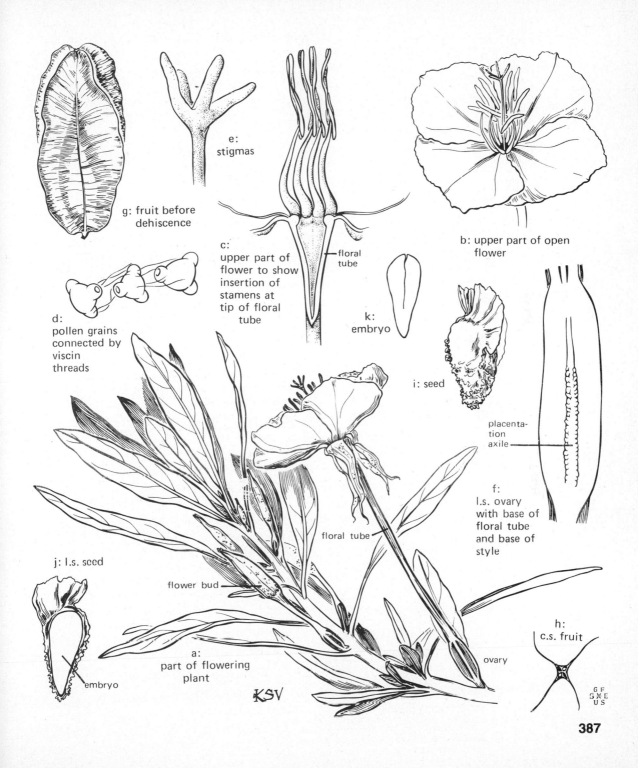

g: fruit before dehiscence

e: stigmas

c: upper part of flower to show insertion of stamens at tip of floral tube

floral tube

b: upper part of open flower

d: pollen grains connected by viscin threads

k: embryo

i: seed

placenta-tion axile

f: l.s. ovary with base of floral tube and base of style

j: l.s. seed

floral tube

flower bud

embryo

a: part of flowering plant

ovary

h: c.s. fruit

KSV

387

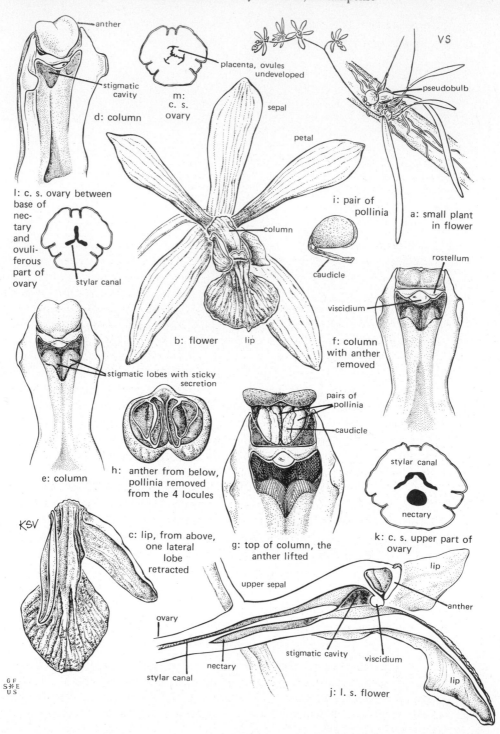

anther

stigmatic cavity

d: column

m: c. s. ovary

placenta, ovules undeveloped

sepal

petal

VS

pseudobulb

i: pair of pollinia

a: small plant in flower

l: c. s. ovary between base of nectary and ovuliferous part of ovary

stylar canal

column

caudicle

rostellum

viscidium

f: column with anther removed

b: flower

lip

stigmatic lobes with sticky secretion

e: column

h: anther from below, pollinia removed from the 4 locules

pairs of pollinia

caudicle

stylar canal

nectary

k: c. s. upper part of ovary

KSV

c: lip, from above, one lateral lobe retracted

g: top of column, the anther lifted

upper sepal

lip

anther

ovary

stigmatic cavity

viscidium

lip

stylar canal

nectary

j: l. s. flower

GF S✳E US

388

Selected Genera (see reference 28):

Subfamily I.   APOSTASIOIDEAE
   Tribe Apostasieae: *Apostasia*
Subfamily II.   CYPRIPEDIOIDEAE
   Tribe Cypripedieae: *Cypripedium*
Subfamily III.   NEOTTIOIDEAE
   Tribe Neottieae: *Epipactis, Goodyera, Listera, Spiranthes*
Subfamily IV.   ORCHIDOIDEAE
   Tribe Orchideae: *Habenaria, Orchis*
Subfamily V.   EPIDENDROIDEAE
   Tribe Gastrodieae: *Cleistes, Isotria, Pogonia, Vanilla*
   Tribe Epidendreae: *Calopogon, Cattleya*
   Tribe Malaxideae: *Liparis, Malaxis*
   Tribe Vandeae: *Aplectrum, Corallorhiza, Cymbidium, Tipularia*

Terrestrial, saprophytic, or epiphytic, perennial herbs from fibrous roots, rhizomes, or tuberous roots. Stems leafy or scapose, epiphytes with base of stem resembling a bulb and aerial roots with velamen (parchmentlike covering with apparent moisture-holding qualities). Leaves simple, mostly alternate (often distichous) and often with sheathing bases, saprophytes often with scalelike, achlorophyllous leaves. Inflorescence spicate, racemose, paniculate, or flowers solitary. Flowers usually perfect, zygomorphic, bracteate, mostly resupinate (twisted 180° or upside down); perianth biseriate, perianth fusion exhibits a range from partial connation to adnation; calyx of 3 green or petaloid segments (medial segment is often variously modified); corolla of 3 petals, the labellum (medial petal) usually larger and often with a basal spur or sac; androecium of 2 fused, lateral stamens or 1 terminal stamen (stigmas, style, and stamens are adnate to form a column or gynandrium), anthers 2-celled with introrse dehiscence, pollen granular or bound by viscin threads into masses (pollinia); gynoecium of 3 fused carpels, ovary inferior, usually 1-locular with parietal placentation (rarely 3-locular with axile placentation), style 1 (portion of column), stigmas usually a depression in column and usually divided into 3 lobes, one of which is usually modified into a rostellum that lies between the fertile stigmas and the anther. Fruit a capsule, dehiscence longitudinal, valves hygroscopic. Seeds numerous, very small, mycotrophic; embryo undifferentiated; endosperm absent.

Aids for Field Identification: Perianth conspicuous, zygomorphic; floral parts in 3s (stamens 1 or 2); gynandrium and pollinia present; ovary inferior.

Illustrations: 21, 25.   *References:* 8, 21, 25, 28, 29, 30, 31, 36, 40.

PLANTAGINACEAE   (Plantain family)

Magnoliopsida (Dicotyledoneae). Three genera and approximately 250 species, cosmopolitan. Genera: *Plantago, Bougueria, Littorella.*

Annual or perennial herbs. Leaves simple, mostly basal, alternate or opposite, often sheathing at the base, venation seemingly parallel, often not differentiated into petiole and blade, exstipulate. Inflorescence a bracteate head or spike, bracteoles absent. Flowers mostly bisexual, actinomorphic; perianth biseriate; calyx of 3 sepals, synsepalous, the tube sometimes deeply lobed or parted, valvate, herbaceous or membranous, hypogynous; corolla of 4 petals, sympetalous, the corolla tube 3 to 4 lobed, imbricate, scarious;

androecium of 4 (rarely 1 to 2) stamens, epipetalous and alternate with the corolla lobes, filaments often long, anthers 2-celled, longitudinally dehiscent, versatile; gynoecium of 1 pistil, ovary superior, syncarpous, carpels 2, locules 1 to 4 (commonly 2), placentation axile or basal (rarely free central), style 1, stigma 1, often 2-cleft. Fruit a circumscissile capsule or bony nut. Seeds 1 or more in each locule, often mucilaginous; embryo mostly straight, small; endosperm fleshy.

Aids for Field Identification: Herbs with simple, basal leaves; flowers 4-merous, corolla sympetalous, scarious; fruit commonly a circumscissile capsule.

Illustrations: 21, 25, 48.   *References:* 8, 21, 25, 29, 30, 31, 36, 40.

POACEAE (Figure C.25) [= Gramineae]   (Grass family)

Liliopsida (Monocotyledoneae). Approximately 600 genera and between 7,500 and 10,000 species, cosmopolitan and in almost every type of habitat.

Selected Genera (see reference 15):

Subfamily I.   FESTUCOIDEAE
Tribe 1.   Festuceae: *Bromus, Festuca, Poa, Dactylis, Lolium, Vulpia*
Tribe 2.   Aveneae: *Sphenopholis, Trisetum, Aira, Avena, Holcus, Calamagrostis, Agrostis, Polypogon, Cinna, Anthoxanthum, Phalaris, Alopecurus, Phleum*
Tribe 3.   Triticeae: *Elymus, Sitanion, Hystrix, Hordeum, Agropyron, Triticum, Secale*
Tribe 4.   Meliceae: *Melica, Glyceria, Catabrosa, Schizachne*
Tribe 5.   Stipeae: *Stipa, Oryzopsis, Piptochaetium*
Tribe 6.   Brachyelytreae: *Brachyelytrum*
Tribe 7.   Diarrheneae: *Diarrhena*
Tribe 8.   Nardeae: *Nardus*
Tribe 9.   Monermeae: *Monerma, Parapholis*

Subfamily II.   PANICOIDEAE
Tribe 10.   Paniceae: *Digitaria, Leptoloma, Anthaenantia, Stenotaphorum, Brachiaria, Axonopus, Eriochloa, Paspalum, Panicum, Oplismenus, Echinochloa, Sacciolepis, Rhynchelytrum, Setaria, Pennisetum, Cenchrus, Amphicarpum*
Tribe 11.   Andropogoneae: *Imperata, Miscanthus, Saccharum, Erianthus, Sorghum, Sorghastrum, Andropogon, Arthraxon, Dichanthium, Bothriochloa, Schizachyrium, Eremochloa, Heteropogon, Manisuris, Tripsacum, Zea, Coix*

Subfamily III.   ERAGROSTOIDEAE
Tribe 12.   Eragrosteae: *Eragrostis, Tridens, Triplasis, Erioneuron, Munroa, Redfieldia, Calamovilfa, Lycurus, Sporobolus, Muhlenbergia, Blepharoneuron*
Tribe 13.   Chlorideae: *Eleusine, Dactyloctenium, Leptochloa, Gymnopogon, Tripogon, Ctenium, Hilaria, Schedonnardus, Cynodon, Chloris, Trichloris, Bouteloua, Buchloe, Cathestecum, Aegopogon, Tragus, Spartina*

POACEAE

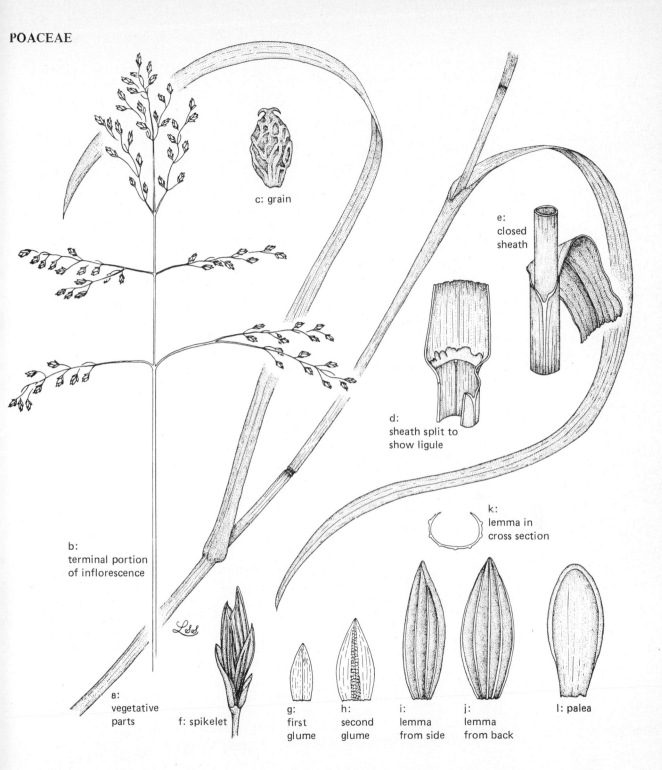

c: grain

e: closed sheath

d: sheath split to show ligule

k: lemma in cross section

b: terminal portion of inflorescence

a: vegetative parts

f: spikelet

g: first glume

h: second glume

i: lemma from side

j: lemma from back

l: palea

391

Tribe 14.   Zoysieae: *Zoysia*
Tribe 15.   Aeluropodeae: *Distichlis, Allolepis, Monanthochloe*
Tribe 16.   Unioleae: *Uniola*
Tribe 17.   Pappophoreae: *Pappophorum, Enneapogon, Cottea*
Tribe 18.   Orcuttieae: *Orcuttia, Neostapfia*
Tribe 19.   Aristideae: *Aristida*

Subfamily IV.   BAMBUSOIDEAE
Tribe 20.   Bambuseae: *Arundinaria, Bambusa, Schizostachyum*
Tribe 21.   Phareae: *Pharus*

Subfamily V.   ORYZOIDEAE
Tribe 22.   Oryzeae: *Oryza, Leersia, Zizania, Zizaniopsis, Luziola, Hydrochloa*

Subfamily VI.   ARUNDINOIDEAE
Tribe 23.   Arundineae: *Arundo, Phragmites, Cortaderia, Molinia*
Tribe 24.   Danthonieae: *Danthonia, Schismus*
Tribe 25.   Centotheceae: *Chasmanthium*

Annual or perennial herbs, rarely woody and arborescent. Roots fibrous, adventitious (the primary or seminal roots often short-lived). Stems (culms) erect, ascending, prostrate or creeping, commonly branched at the base, rhizomes and stolons common, internodes typically hollow, nodes closed and swollen, terete. Leaves simple, alternate, typically 2-ranked, composed of a blade, sheath, and ligule, the sheath encircling the culm and the margins free and overlapping or sometimes connate; blade usually flat, long and narrow, linear to lanceolate, rarely with a constricted petiolar base, parallel veined; ligule commonly present at the junction of the blade and sheath (adaxial surface), typically membranous or reduced to a ring of hairs or absent. Inflorescence—basic unit a spikelet, composed of 1 or more flowers (florets) and their subtending bracts on an axis (rachilla), spikelets sessile or pedicelled and arranged in spikes, panicles or racemes (compound inflorescences), two bracts commonly present at the base of each spikelet, the lower one called the first glume and the upper one the second glume, in some genera the first glume is reduced or absent or occasionally both glumes are absent. Flowers (florets) bisexual or unisexual (fertile floret is one with a functional pistil regardless of presence or absence of stamens), borne above the glumes and subtended by 2 bracts—the lemma, which often resembles the glumes and is greenish, keeled, nerved or awned, and the palea, which is partially enclosed by the lemma and is often thin and 2-nerved or 2-keeled. The flower (floret) proper occurs in the axil of the palea, perianth absent or reduced to 2 to 3 fleshy or hyaline scales (lodicules); androecium of 1 to 6 (rarely more), usually 3, hypogynous stamens, anthers 2-celled, basifixed, longitudinally dehiscent; gynoecium of 1 pistil, ovary superior, syncarpous, carpels 3, unilocular, styles 1 to 3, commonly 2, stigmas typically 2, often plumose. Fruit a caryopsis or rarely a nut, berry, or utricle. Seed 1 per fruit; embryo variable in size, placed on the abaxial side of fruit and lateral to the endosperm; endosperm copious and starchy.

Aids for Field Identification: Herbs, culm (stem) with hollow internodes, terete; leaves mostly flat and 2-ranked, leaf sheath usually open; fruit a grain.

Comparison of morphological differences between Poaceae and Cyperaceae

	Poaceae	*Cyperaceae*
Stems	Terete or flattened	Triangular (at least just below the inflorescence)
Leaves	2-ranked	3-ranked
Leaf sheath	Mostly open	Mostly closed
Bract immediately subtending the flower	Usually 2-nerved	With odd number of nerves
Fruit	Usually a caryopsis	Usually an achene, never a caryopsis
Embryo	Lateral to the endosperm	Embedded in the endosperm

Key to Poaceae, Cyperaceae, and Juncaceae:

Perianth (calyx and corolla) present and consisting of 6 stiff, greenish or brownish segments, flowers not associated with chaffy or scaly bracts (glumes or lemmas); stamens 3 or 6; fruit a many-seeded capsule...............................JUNCACEAE
(Rushes)

Perianth absent or consisting of 2 to 6 bristles or scales, flowers associated with 1 to 4 chaffy scalelike bracts (scales, glumes, lemmas); stamens typically 3(6); fruit 1-seeded, not a capsule.

    Leaves in 2 vertical rows (2-ranked), sheaths typically open with overlapping edges; stems usually round in cross section with hollow internodes; each flower (floret) of spikelet typically contained between two bracts (lemma and palea)....POACEAE
(Grasses)

    Leaves in 3 vertical rows (3-ranked), sheaths typically closed (tubular); stems often triangular in cross section, with solid or with pithy partitioned internodes; each flower of spikelet in axil of a single bract ....CYPERACEAE
(Sedges)

Illustrations: 1, 15, 17, 21, 25.   *References:* 6, 8, 15, 17, 21, 25, 29, 30, 31, 36, 40.

POLEMONIACEAE [including Cobaeaceae]   (Phlox family)

    Magnoliopsida (Dicotyledoneae). Approximately 15 genera and 300 species, chiefly North American, a few in Chile, Europe, and North Asia. Selected Genera: *Collomia, Gilia, Eriastrum, Gymnosteris, Ipomopsis, Langloisia, Leptodactylon, Linanthus, Loeselia, Navarretia, Phlox, Polemonium, Siphonella, Cobaea.*

    Annual to perennial herbs, rarely vines, shrubs, or small trees. Leaves mostly simple, alternate or opposite, exstipulate, margins entire, palmatifid, or pinnatifid. Inflorescence usually cymose, often corymbose or capitate, or flowers rarely solitary. Flowers bisexual, actinomorphic (rarely somewhat bilabiate); perianth biseriate; calyx of 5 sepals, synsepalous, often lobed; corolla of 5 petals, sympetalous, often lobed, salverform to rotate; androecium of 5 stamens, distinct, epipetalous, antisepalous, the anthers 2-celled, longitudinally dehiscent, introrse; intrastaminal disk usually present; gynoecium of 1 compound pistil, ovary superior, carpels 3(2 or 5), locules 3(2 or 5), placentation axile, style 1, filiform, terminal, often trifid at apex, stigmas (2) 3. Fruit a loculicidal capsule (rarely indehiscent). Seeds sometimes with a mucilaginous coat; embryo straight or slightly curved; endosperm abundant and fleshy to firm to fleshy.

Aids for Field Indentification: Flowers 5-merous, carpels and locules usually 3; corolla often salverform or rotate; corolla lobes contorted in bud; ovules and seeds usually numerous.

Illustrations: 21, 25. *References:* 8, 21, 25, 29, 30, 31, 36, 40, 42.

POLYGONACEAE   (Smartweed family)

Magnoliopsida (Dicotyledoneae). Approximately 40 genera and 800 species, chiefly north temperate, a few tropical, arctic, and southern. Selected Genera: *Chorizanthe, Eriogonum, Rumex, Rheum, Calligonum, Polygonum, Fagopyrum, Polygonella, Antenoron, Coccoloba, Antigonon, Brunnichia, Triplaris.*

Mostly herbs, some shrubs or climbers (rarely trees). Stems often with swollen nodes. Leaves simple, alternate (rarely whorled or opposite), entire base of petiole often with a stipular membranous or scarious sheath (ocrea) clasping the stem. Inflorescence basically cymes or cymules disposed in spikes, racemes, panicles, or heads. Flowers mostly bisexual (when unisexual the species monoecious or dioecious), actinomorphic; perianth mostly biseriate; sepals (tepals) 3 to 6, often petaloid, imbricate, often enlarged and becoming membranous in fruit; corolla absent; androecium of mostly 6 to 9 (rarely more or less) stamens in two series, filaments free or basally connate, anthers 2-locular, longitudinally dehiscent; gynoecium of 1 syncarpous pistil, ovary superior, carpels 2, 3, or 4, unilocular, placentation basal, style 1, stigmas 2 to 4. Fruit angled, round, flat or winged achene (nutlet). Seed 1 per fruit; embryo mostly curved; endosperm abundant, mealy.

Aids for Field Identification: Herbs with swollen nodes; ocreae mostly present; calyx (tepals) petaloid and biseriate; petals absent; fruit mostly lenticular or triangular, 1-seeded, unilocular, achene.

Illustrations: 21, 25, 48. *References:* 8, 16, 21, 25, 29, 30, 31, 40.

POTAMOGETONACEAE   (Pondweed family)

Liliopsida (Monocotyledoneae). One or two genera and approximately 100 species, cosmopolitan.

Genera (see reference 21):

Cauline leaves alternate; fruit drupaceous ....*Potamogeton*
Cauline leaves opposite; fruit achenial ....*Groenlandia*

Aquatic perennial herbs of fresh water, submerged or floating, often with creeping stems or rhizomes. Leaves alternate or opposite, immersed leaves often thin and the floating leathery, shapes quite variable, simple, stipules present and sheathing at the base, the sheaths open apically and free or adnate to the petiole at the base. Inflorescence peduncled, axillary spikes. Flowers bisexual or unisexual and the species monoecious or dioecious, actinomorphic; perianth of 4 segments, distinct, valvate, clawed; androecium of 4 stamens inserted on the claws of the perianth segments, anthers 2-celled, extrorse, sessile; gynoecium of 4, sessile, unilocular, unicarpellate pistils (carpels). Fruit indehiscent achenes or drupelets. Seeds 1 per ovary; endosperm absent.

Aids for Field Identification: Fresh-water aquatic perennial herbs; leaves 2-ranked; inflorescence axillary spikes (sometimes few flowered); tepals 4; fruits 1-seeded.

Illustrations: 21, 25. *References:* 8, 21, 25, 29, 30, 31, 36, 40.

PRIMULACEAE [including Coridaceae] (Primrose family)

Magnoliopsida (Dicotyledoneae). Twenty-one to 28 genera and between 800 and 1,000 species, cosmopolitan but chiefly northern hemisphere. Selected Genera: *Primula, Androsace, Hottonia, Dodecatheon, Cyclamen, Lysimachia, Trientalis, Glaux, Anagallis, Samolus, Centunculus.*

Annual or perennial herbs (rarely shrubby). Leaves mostly simple, variously lobed or dissected, alternate, opposite, or whorled, basal or cauline, exstipulate. Inflorescence paniculate, umbellate, racemose or the flowers solitary and terminal or axillary, bracteate. Flowers bisexual, actinomorphic (rarely zygomorphic), often heterostylous; perianth biseriate; calyx mostly of 5 sepals (4 to 9), connate at least at the base, persistent; corolla mostly 5-merous (absent in *Glaux*), petals mostly united, the tube sometime split nearly to the base, imbricate; androecium of mostly 5 stamens, antipetalous, an outer whorl sometimes represented by staminodes, often epipetalous, anthers 2-celled, longitudinally dehiscent, introrse; gynoecium of 1 pistil, ovary superior (rarely half-inferior), syncarpous, carpels 5, locule 1, placentation free-central, style 1 (often heterostylous), stigma capitate. Fruit a variously dehiscent capsule or pyxis. Seeds often numerous, small; embryo small and straight; endosperm hard or fleshy.

Aids for Field Identification: Herbs, flowers 5-merous, sympetalous; stamens opposite the petals; placentation free-central.

Illustrations: 21, 25, 48. *References:* 5, 8, 21, 25, 29, 30, 31, 36, 40.

RANUNCULACEAE (Figure C.26) (Buttercup family)

Magnoliopsida (Dicotyledoneae). Approximately 50 genera and 1,900 species, chiefly north temperate regions.

Key to Subfamilies and Tribes: (Selected Genera)

  A. Carpels with more than 1 ovule; fruit a follicle or berry . . HELLEBOROIDEAE
      B. Flowers zygomorphic; petals conspicuous; seeds often transversely ridged
           with scales or winged . . . . Delphineae
      BB. Flowers actinomorphic; petals small and narrow or absent; seeds neither
           scaly nor winged . . . . Helleboreae
  AA. Carpels with 1 ovule; fruit an achenecetum, rarely a berry
. . . . RANUNCULOIDEAE
    C. Leaves alternate; sepals imbricate; corolla absent or present.
      D. Flowers subtended by an involucre or leaves remote from calyx; sepals
          persistent at anthesis and petaloid; corolla absent . . . . . . Anemoneae
      DD. Flowers not subtended by an involucre; calyx mostly caducous; corolla
          mostly present . . . . Ranunculeae

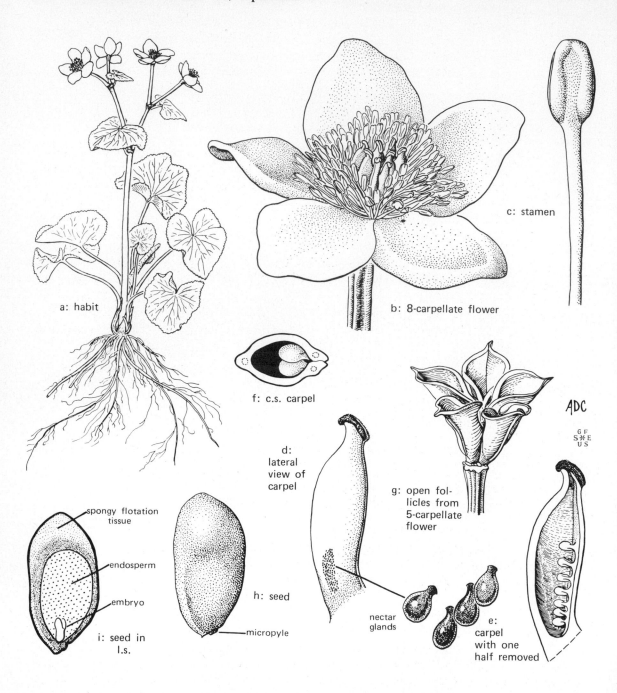

a: habit

b: 8-carpellate flower

c: stamen

f: c.s. carpel

d: lateral view of carpel

g: open follicles from 5-carpellate flower

ADC

spongy flotation tissue

endosperm

embryo

i: seed in l.s.

h: seed

micropyle

nectar glands

e: carpel with one half removed

CC.   Leaves opposite; sepals induplicate-valvate or rarely partly imbricate; corolla absent or represented only by outer staminodes . . . . Clematideae

Tribe 1.   Delphineae: *Aconitum, Delphinium*
Tribe 2.   Helleboreae: *Actaea, Aquilegia, Caltha, Coptis, Helleborus, Isopyrum*
Tribe 3.   Anemoneae: *Anemone, Barneoudia, Hepatica*
Tribe 4.   Ranunculeae: *Adonis, Myosurus, Ranunculus, Thalictrum*
Tribe 5.   Clematideae: *Clematis, Clematopsis*

Annual or perennial herbs, occasionally shrubs or vines (*Clematis*). Leaves usually basal and cauline, with the latter reduced upward, often palmately compound (rarely pinnate or simple), usually alternate (opposite in *Clematis, Ranunculus* spp.), exstipulate. Inflorescence a cyme, raceme, or panicle or flowers solitary. Flowers bisexual (rarely unisexual and the species then monoecious or dioecious), actinomorphic or zygomorphic (in Delphineae); perianth biseriate and dichlamydeous or uniseriate with corolla absent or perianth not differentiated; calyx of few to many sepals, which are often petaloid, distinct; corolla of few to many petals, or lacking, distinct, nectariferous glands usually present; androecium of many, spirally arranged, distinct stamens, anthers 2-celled and longitudinally dehiscent; gynoecium of 3 to many (rarely 1), spirally arranged, distinct, simple pistils, ovary superior, carpel 1, locule 1, placentation marginal, style and stigma 1. Fruit a follicle, an achene, a berry, or rarely a capsule. Seed with minute embryo; endosperm copious.

Aids for Field Identification: Herbs; leaves often palmately compound and basal; petals often absent, reduced, or modified; stamens numerous, spirally arranged; pistils numerous, distinct, spirally arranged, unilocular, and unicarpellate.

Illustrations: 21, 25, 48.   *References:*   8, 20, 21, 25, 29, 30, 31, 36, 40.

ROSACEAE (Figure C.27) [excluding Chrysobalanaceae and Neuradaceae](Rose family)

Magnoliopsida (Dicotyledoneae). Approximately 100 genera and 2,000 species, cosmopolitan.

Selected Genera:

Subfamilies (see reference 40):

Subfamily I.   SPIRAEOIDEAE: *Spiraea, Quillaja, Holodiscus*
Subfamily II.   PYROIDEAE (POMOIDEAE): *Pyrus, Malus, Sorbus*
Subfamily III.   ROSOIDEAE: *Adenostoma, Agrimonia, Alchemilla, Dryas, Fragaria, Geum, Kerria, Rhodotypos, Rosa, Rubus, Potentilla*
Subfamily IV.   PRUNOIDEAE: *Prunus*

Mostly trees, shrubs, and perennial herbs. Leaves mostly alternate, simple or compound, usually stipulate, however these may be caducous or adnate to the petiole. Inflorescence extremely variable, primarily racemose or cymose. Flowers usually perfect (if imperfect, the species dioecious), usually actinomorphic, epicalyx often present and alternate with calyx lobes; perianth usually biseriate (bases adnate to form a hypanthium, which is often lined with a nectariferous disk), perigynous to epigynous (seldom hypogynous); calyx of 5 connate sepals (free or adnate to ovary); corolla of 5 petals (rarely

ROSACEAE subfamily ROSOIDEAE: Potentilla. a-h, P. canadensis; i, P. simplex; j, P. recta; k-m, P. tridentata

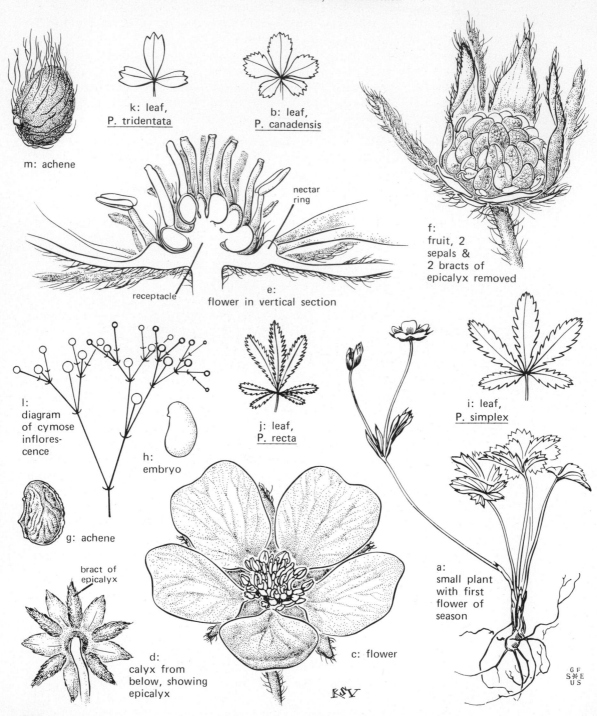

m: achene

k: leaf, P. tridentata

b: leaf, P. canadensis

nectar ring

f: fruit, 2 sepals & 2 bracts of epicalyx removed

receptacle

e: flower in vertical section

l: diagram of cymose inflorescence

h: embryo

j: leaf, P. recta

i: leaf, P. simplex

g: achene

a: small plant with first flower of season

bract of epicalyx

d: calyx from below, showing epicalyx

c: flower

absent) inserted on rim of the hypanthium; androecium usually of numerous, distinct stamens in whorls of 5, anthers bilocular and usually longitudinally dehiscent; gynoecium of 1 to many pistils, apocarpous to syncarpous, these at the base of, on the sides of, or enclosed by the floral cup (pistils free from or adnate to the floral cup), ovary superior to inferior (with intermediates), carpels 1 to many, compound pistils with 2 to 5 locules, placentation axile, styles (or style branches) and stigmas as many as carpel number. Fruits diverse, dry, or fleshy, aggregates, etc. Seeds few to several per locule; embryo small; endosperm usually absent.

Aids for Field Identification: Trees, shrubs, or perennial herbs; leaves stipulate; perianth 5-merous, hypanthium present; stamens in 5s; ovary superior to inferior; endosperm absent.

Illustrations: 21, 25, 48. *References:* 8, 21, 25, 29, 30, 31, 34, 36, 40.

RUBIACEAE (Madder family)

Magnoliopsida (Dicotyledoneae). Approximately 500 genera and 6,000 species, most are tropical but a number are temperate, and *Galium* has a few Arctic species. Selected Genera: *Galium, Cephalanthus, Diodia, Hedyotis, (Oldenlandia, Houstonia), Coffea, Cinchona, Cephaelis, Gardenia, Rubia, Mitchella, Asperula, Pinckneya.*

Trees or shrubs, sometimes vines, infrequently herbs (those in north temperate regions are mostly herbs). Leaves simple, opposite or whorled, stipules present and interpetiolar or intrapetiolar (sometimes foliaceous and indistinguishable from the leaves or reduced to glandular setae), margins usually entire. Inflorescence basically a dichasial cyme (only the central flower present in some, as *Gardenia*), the dichasia sometimes aggregated into globose heads. Flowers bisexual, usually actinomorphic (rarely zygomorphic and somewhat bilabiate); perianth biseriate; calyx of 4 to 5 sepals, connate; corolla of 4 to 5 petals, connate, usually salverform, rotate, or infundibular; androecium of as many stamens as petals, usually distinct (sometimes syngenesious), epipetalous on the corolla tube, antisepalous, the anthers 2-celled (locular), dehiscing longitudinally, introrse; gynoecium of 1 compound pistil, ovary typically inferior, carpels 2 or more, locules usually 2, placentation typically axile or seemingly basal (rarely unilocular with parietal placentation), the ovules usually numerous in each locule, the style 1 and slender, often bifurcate, the stigmas usually 1 on each style branch and linear or solitary and 2-lobed. Fruit a berry, drupe, loculicidal or septicidal capsule, or schizocarp (separating into 1-seeded segments in *Galium*). Seeds sometimes winged; embryo straight or curved; endosperm usually copious and fleshy or rarely cartilaginous (absent in some).

Aids for Field Identification: Herbs or shrubs; leaves opposite or whorled, stipulate; flowers 4- or 5-merous, carpels usually 2, ovary inferior. The family closely resembles Caprifoliaceae, and no single feature distinguishes the two. However, the Rubiaceae typically have stipules, whereas the Caprifoliaceae usually have none.

Illustrations: 21, 25, 48. *References:* 8, 21, 25, 29, 30, 31, 36, 40.

SALICACEAE (Willow family)

Magnoliopsida (Dicotyledoneae). Two or perhaps three genera and circa 530 species, mostly in north temperate regions. Genera: *Salix, Populus, Chosenia.*

Trees, shrubs, or subshrubs. Leaves deciduous, simple, alternate, generally petioled, stipulate, and the stipules free but often caducous. Inflorescence erect or pendulous, the dense catkins often produced before the leaves. Flowers unisexual (species mostly dioecious), borne in the axil of a bract and often with a small cupular disk or 1 to 2 nectariferous scales (perhaps representing the calyx); perianth absent; staminate flowers of 2 or more stamens, filaments distinct or basally connate, anthers 2-celled and vertically dehiscent; pistillate flowers with 1 sessile or short stipitate compound pistil, ovary superior, carpels 2 to 4, locule 1, placentation parietal or basal, style 1 or with 2 to 4 branches, stigmas 2 to 4. Fruit a 2- to 4-valved capsule. Seeds mostly small, numerous and comose; embryo straight; endosperm little or none.

Aids for Field Identification: Trees or shrubs; dioecious; flowers in catkins; fruit a capsule; seeds comose.

Illustrations: 21, 25, 48.   *References:* 8, 21, 25, 29, 30, 31, 36, 40.

SCROPHULARIACEAE (Figure C.28)   (Snapdragon family)

Magnoliopsida (Dicotyledoneae). Approximately 220 genera and 2,700 species, cosmopolitan.

Selected Genera (see reference 40):

Subfamily I.   VERBASCOIDEAE
   Tribe Verbasceae: *Celsia, Verbascum*
   Tribe Aptosimeae: *Aptosimum*

Subfamily II.   SCROPHULARIOIDEAE (ANTIRRHINOIDEAE)
   Tribe Calceolarieae: *Calceolaria*
   Tribe Hemimerideae: *Alonsoa*
   Tribe Antirrhineae: *Antirrhinum, Linaria*
   Tribe Scrophularieae (Cheloneae): *Chelone, Collinsia, Paulownia, Penstemon*
   Tribe Manuleae: *Sutera, Zaluzianskia*
   Tribe Gratioleae: *Gratiola, Mimulus, Torenia*
   Tribe Selagineae: *Hebenstretia, Selago*

Subfamily III.   RHINANTHOIDEAE
   Tribe Digitalideae: *Digitalis, Veronica*
   Tribe Gerardieae: *Agalinis (Gerardia)*
   Tribe Rhinantheae: *Castilleja, Melampyrum, Pedicularis*

Mostly herbs or shrubs (seldom vines or trees), some are saprophytic (chlorophyllous or achlorophyllous) or parasitic. Leaves simple to pinnately lobed, alternate to opposite (infrequently whorled), exstipulate. Inflorescence variable, bracts and bracteoles usually present. Flowers bisexual, usually zygomorphic; perianth biseriate; calyx of 4 to 5 connate sepals; corolla of 4 to 5 connate petals, the tube sometimes very short and often bilabiate, the anterior petals frequently spurred or with a gibbous to saccate base; androecium commonly of 4 (seldom 2 or 5) didynamous, epipetalous stamens that are alternate with the lobes, a staminode is frequently present, filaments distinct, anthers bilocular (occasionally connivent), mostly with longitudinal dehiscence (rarely poricidal); nectariferous disk usually present; gynoecium of 1 compound pistil, ovary superior, car-

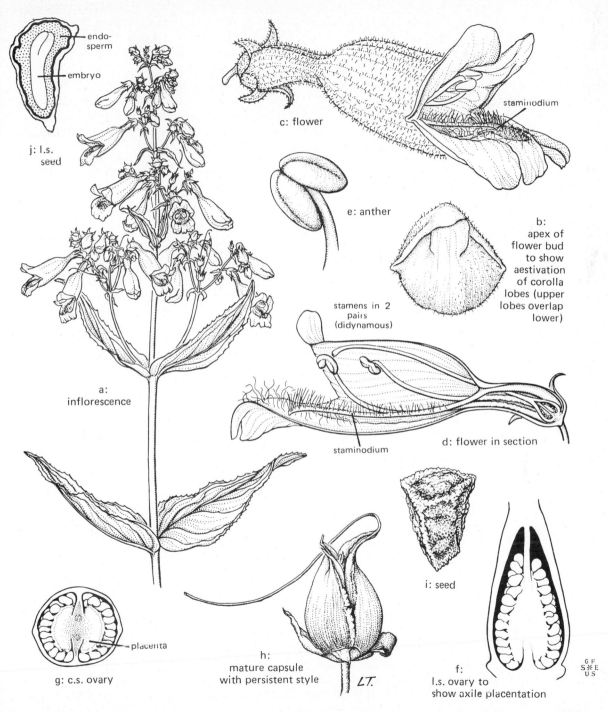

endo-
sperm

embryo

j: l.s.
seed

c: flower

staminodium

e: anther

b:
apex of
flower bud
to show
aestivation
of corolla
lobes (upper
lobes overlap
lower)

stamens in 2
pairs
(didynamous)

a:
inflorescence

staminodium

d: flower in section

i: seed

placenta

g: c.s. ovary

h:
mature capsule
with persistent style

LT.

f:
l.s. ovary to
show axile placentation

GF
S¾E
US

401

pels 2, locules usually 2 or infrequently unilocular, placentation axile, style 1, stigmas bilobed or 2. Fruit usually a capsule with septicidal, loculicidal, or poricidal dehiscence (rarely an indehiscent capsule or a berry). Seeds numerous; embryo straight or curved; endosperm fleshy.

Aids for Field Identification: Herbs or shrubs; leaves alternate or opposite; perianth 5-merous with zygomorphic corolla; stamens 4(2 or 5) and epipetalous; ovary superior; fruit capsular.

Illustrations: 21, 25, 48.   *References:* 8, 21, 25, 29, 30, 31, 36, 40.

SOLANACEAE   (Nightshade family)

Magnoliopsida (Dicotyledoneae). Approximately 90 genera and 2,000 species, tropics and temperate regions, the chief center of distribution being Central and South America. Selected Genera: *Atropa, Browallia, Capsicum, Chamaesaracha, Datura, Hyoscyamus, Lycium, Lycopersicon, Nicandra, Nicotiana, Petunia, Physalis, Salpichroa, Schizanthus, Solanum.*

Annual, biennial, or perennial herbs, shrubs, trees, vines, or lianas. Leaves simple, typically alternate, exstipulate. Inflorescence generally cymose. Flowers bisexual, usually actinomorphic; perianth biseriate; calyx of 5(4 or 6) sepals, synsepalous, often lobed, persistent, sometimes enlarging in fruit; corolla of 5 petals, sympetalous, often lobed, usually plicate or convolute; androecium usually of 5 unequal stamens, sometimes 4 and didynamous and with a staminode or rarely 2, distinct (the anthers sometimes connivent), epipetalous, antisepalous, the anthers 2-celled (1 cell sometimes undeveloped), longitudinally or poricidally dehiscent; hypogynous disk usually present; gynoecium of 1 compound pistil, ovary superior, carpels 2, locules 2 (sometimes apparently 3 to 5 by false septa), placentation axile, the ovules numerous, style 1 and terminal, stigma 2-lobed. Fruit a berry or a septicidal capsule. Seeds smooth or foveolate; embryo curved or annular and embedded in the endosperm; endosperm copious, fleshy, semitransparent.

Aids for Field Identification: Herbs or shrubs; flowers 5-merous, ovary 2-locular or falsely 3- to 5-locular; anthers frequently connivent and poricidal; fruit a berry or septicidal capsule. The family is closely affiliated with the Scrophulariaceae and can best be distinguished from it by the actinomorphic, often plicate corolla and usually 5 stamens.

Illustrations: 21, 25, 48.   *References:* 8, 21, 25, 29, 30, 31, 36, 40.

TYPHACEAE   (Cattail family)

Liliopsida (Monocotyledoneae). One genus with circa 10 to 15 species, mostly temperate and tropical. Genus: *Typha.*

Marsh or lake herbs, perennial from creeping rhizomes. Leaves simple, mostly basal, alternate, epetiolate, parallel-veined and long-linear in shape. Inflorescence an androgynous spike, each floret subtended by a caducous bractlike spathe. Florets unisexual; perianth of bristles; staminate flowers with 2 to 5 distinct or monadelphous stamens, connate filaments with long silky trichomes, anthers linear; pistillate flowers with 1 pistil, ovary superior and long-stipitate, carpel 1, locule 1, placentation pendulous, style 1 (filiform), stigma 1 (linear to spatulate or rhomboidal, 1-sided). Fruit an achene with a persistent style. Seed 1; embryo narrow; endosperm mealy.

Aids for Field Identification: Flowers densely crowded in a terminal spike; perianth of bristles.

Illustrations: 21, 25.   *References:* 8, 21, 25, 29, 30, 31, 36, 40.

ULMACEAE (Elm family)

Magnoliopsida (Dicotyledoneae). Approximately 15 genera and 200 species, mostly tropical and temperate zones. Selected Genera: *Celtis, Planera, Trema, Ulmus, Zelkova.*

Trees or shrubs with sympodial stems and watery sap. Leaves simple, usually alternate, 2-ranked, bases oblique, margins entire to serrate, caducous stipules paired. Inflorescence cymose, fasciculate or flowers solitary; flowers borne on wood of previous season in Ulmoideae and on twigs of current season in Celtoideae. Flowers usually unisexual (species then monoecious) but bisexual in *Ulmus,* actinomorphic to zygomorphic; perianth spiraled, uniseriate, with hypanthium; calyx of 4 to 8 connate sepals; corolla absent; androecium of 4 to 8 distinct stamens arising from the hypanthium, anthers 2-locular with longitudinal dehiscence (introrse or extrorse); gynoecium of 1 sessile or stipitate pistil, syncarpous, ovary superior, carpels 2, usually unilocular, placentation pendulous, styles and stigmas 2. Fruit a samara or drupe. Seed 1; embryo curved or straight; endosperm absent.

Aids for Field Identification: Trees or shrubs with watery sap; leaves 2-ranked with oblique bases; fruit a samara or drupe.

Illustrations: 21, 25, 48.   *References:* 8, 9, 21, 25, 29, 30, 31, 36, 40.

VERBENACEAE [including Avicenniaceae, Chloanthaceae, Stilbeaceae, Symphoremaceae] (Verbena family)

Magnoliopsida (Dicotyledoneae). Approximately 99 genera and 3,151 species, almost all tropical and subtropical. Selected Genera: *Avicennia, Callicarpa, Chloanthes, Citharexylum, Clerodendrum, Duranta, Lantana, Phyla, Priva, Stachytarpheta, Stilbe, Stylodon, Tetraclea, Verbena, Vitex.*

Herbs, shrubs, trees, or vines, the stems typically quadrangular. Leaves usually simple, sometimes palmately or pinnately compound, opposite or whorled, exstipulate. Inflorescence variable, usually determinate and cymose, typically bracteate, the bracts sometimes colored. Flowers usually bisexual, zygomorphic (rarely actinomorphic); perianth biseriate; calyx of 5(4, 6, or 8) sepals, synsepalous, usually lobed or toothed, persistent; corolla with as many petals as sepals, sympetalous, typically lobed with the lobes generally unequal, salverform (rarely bilabiate); androecium of 4(2 or 5) stamens, distinct, didynamous, epipetalous, antisepalous, staminodes sometimes present, the anthers 2-celled, longitudinally dehiscent, introrse; gynoecium of 1 compound pistil, ovary superior, carpels mostly 2(4 or 5), locules as many as carpels or twice as many by false septation, placentation axile, the ovules usually 2 in each carpel, style 1 and terminal (rarely depressed between the lobes of the ovary), stigma lobes usually as many as carpels. Fruit generally a drupe, less commonly a capsule or schizocarp (nutlets). Seeds with a straight embryo; endosperm usually absent, if present fleshy.

Aids for Field Identification: Herbs, shrubs, or small trees; stems often 4-angled; leaves opposite; flowers zygomorphic, 5-merous, ovary 2- or 4-locular; ovary not lobed and style terminal; fruit a drupe or schizocarp (4 nutlets). The Verbenaceae are closely related and similar to the Boraginaceae and the Lamiaceae. The borages have alternate leaves, a 4-lobed ovary, and a gynobasic style. The mints have a 4-lobed ovary, a gynobasic style, and inflorescence which is often a bracteate thyrse or verticillaster.

Illustrations: 21, 25, 48.　　*References:* 8, 21, 25, 29, 30, 31, 36, 40.

VIOLACEAE　　　　　　　　　　　　　　　　　　　　　　　　(Violet family)

Magnoliopsida (Dicotyledoneae). About 22 genera and 900 species, cosmopolitan. Selected Genera: *Anchietea, Hybanthus, Rinorea, Viola.*

Herbs or shrubs (rarely vines). Leaves usually simple, mostly alternate, stipules leafy or small. Inflorescence racemose or flowers solitary. Flowers bisexual, zygomorphic to actinomorphic, sometimes cleistogamous, bracteolate; perianth biseriate; calyx of 5 distinct, imbricate sepals; corolla of 5 distinct, imbricate petals, the lower often larger and spurred or gibbous; androecium of 5 distinct, connate, or connivent stamens, anthers 2-celled with longitudinal dehiscence; gynoecium of 1 syncarpous pistil, ovary superior, carpels 3 to 5, unilocular, 3 to 5 parietal placentae, style 1, stigmas variable but usually truncate. Fruit a loculicidal capsule or berry. Seeds smooth, winged, or with tomentum; embryo straight; endosperm fleshy.

Aids for Field Identification: Shrubs or herbs; corolla spurred; placentation parietal; fruit a capsule.

Illustrations: 21, 25, 48.　　*References:* 2, 8, 21, 25, 29, 30, 31, 36, 40.

FAMILY STUDY WORKSHEET

Family: _____

A. Selected Example.

PLANT:
Habit _____
Duration _____
Other _____

LEAVES:
Type _____
Duration _____
Arrangement _____
Venation _____
Other _____

INFLORESCENCE:
Type _____

FLOWER: Type _____

CALYX:
Sepal number _____
Sepal fusion _____
Symmetry _____
Other _____

COROLLA:
Petal number _____
Petal fusion _____
Symmetry _____
Other _____

ANDROPERIANTH:
Position _____

ANDROECIUM:
Stamen number _____
Stamen fusion _____
Stamen arrangement _____
Stamen position _____
Other _____

GYNOECIUM:
Pistil number _____
Style number _____
Stigma number _____
Ovary number _____
Locule number _____
Placenta position _____
Carpel number _____
Ovary position _____
Other _____

FRUIT: Type _____

FLORAL FORMULA:

SCIENTIFIC NAME:

REFERENCE (Manual): _____

B. Other Representatives.

_____ Ca Co S P   _____ Ca Co S P
_____ Ca Co S P   _____ Ca Co S P
_____ Ca Co S P   _____ Ca Co S P

C. Family Synthesis. Diagnostic features, special structures, comparison, notes, etc.

Figure C.29

The Structure of Vascular Plants

Phytography is a branch of botany that deals with the descriptive terminology of plants and their component parts to provide an accurate and complete vocabulary for description, identification, and classification.* This classification of terms represents a compromise between the traditional approaches to phytography and effective teaching-learning organization. The major organs and their component parts (a) are basic morphology. The use of types (b) and structural types (c) conforms closely to accepted practice.

This morphological glossary is based primarily upon those found in Lawrence (1951), Stearn (1966), Johnson (1931), Featherley (1954), Correll and Johnston (1971), Swartz (1971), Porter et al. (1973), and Radford et al. (1974). Some classifications and definitions of terms are based on special papers that are cited in the literature.

OUTLINE OF SYSTEM OF CLASSIFICATION FOR TERMS

A. ANGIOSPERM STRUCTURAL PARTS AND TYPES

I. The plant
   a. Plant parts
   b. Plant types
II. Roots
   a. Root parts
   b. Root types
   c. Root structural types

*See Chapter 5 for general descriptive terms related to color, size, number, fusion, shape, disposition, surface, texture, symmetry, and temporal phenomena.

   III.   Stems
- a. Stem parts
- b. Stem types
- c. Stem structural types

   IV.   Buds
- a. Bud parts
- b. Bud types
- c. Bud structural types

   V.   Leaves
- a. Leaf parts
- b. Leaf types
- c. Leaf structural types
- d. Petiole and petiolule structural types
- e. Stipule and stipel types

   VI.   Inflorescences
- a. Inflorescence parts
- b. Inflorescence types

   VII.   Flowers
- a. Flower parts
- b. Flower types

   VIII.   Perianth and hypanthium
- a. Perianth parts
- b. Hypanthium parts
- c. Perianth fusion types
- d. Perianth and androecium position types
- e. Perianth and hypanthium structural types

   IX.   Androecium
- a. Androecial parts
- b. Androecial fusion types
- c. Stamen parts
- d. Stamen arrangement types
- e. Stamen position types
- f. Stamen structural types
- g. Anther parts
- h. Anther types
- i. Anther attachment
- j. Pollen

   X.   Gynoecium
- a. Gynoecial parts
- b. Gynoecial fusion types
- c. Carpel parts
- d. Carpel types
- e. Stigma types
- f. Style types
- g. Style position types
- h. Ovule parts
- i. Ovule types
- j. Ovule position types
- k. Placenta position types
- l. Ovary position types

XI. Fruits
    a. Fruit parts
    b. Fruit structural types
XII. Seeds and seedlings
    a. Seed parts
    b. Seed types
    c. Embryo parts
    d. Embryo types
    e. Aril structural types and special seed surface features
    f. Seedling parts
    g. Seedling types

B. SEXUAL PHENOMENA

I. Sex
    a. Flower sex
    b. Inflorescence sex
    c. Plant sex
    d. General sexual terms

C. DEVELOPMENT AND PATTERNS

I. Growth and development
    a. Growth regions
    b. Growth periodicity
    c. General development
    d. Developmental shoot types
II. Patterns
    a. Symmetry
    b. Arrangement systems
    c. Branching patterns

D. GYMNOSPERM GLOSSARY

I. Vegetative and reproductive structures
    a. Vegetative structures
    b. Reproductive structures

E. LOWER VASCULAR PLANT GLOSSARY

I. Sporophyte
    a. Stems
    b. Leaves
    c. Reproductive structures
II. Gametophyte
III. Asexual reproduction

A. ANGIOSPERM STRUCTURAL PARTS AND TYPES

I. The Plant

a. Plant Parts

bud   Immature vegetative or floral shoot or both, often covered by scales.

flower   Reproductive structure of flowering plants with or without protective envelopes, the calyx and/or corolla; short shoot with sporophylls and with or without sterile protective leaves, the calyx and corolla.

fruit   Matured ovary of flowering plants, with or without accessory parts.

leaf   A photosynthetic and transpiring organ, usually developed from leaf primordium in the bud; an expanded, usually green, organ borne on the stem of a plant.

root   An absorbing and anchoring organ, usually initially developed from the radicle and growing downward.

seed   Matured ovule of seed plants.

stem   A supporting and conducting organ usually developed initially from the epicotyl and growing upward.

b. Plant Types

(Classification based on habit)

herb   A usually low, soft, or coarse plant with annual aboveground stems.

shrub   A much-branched woody perennial plant, usually without a single trunk.

tree   A tall, woody perennial plant usually with a single trunk.

vine or liana   An elongate, weak-stemmed, often climbing annual or perennial plant, with herbaceous or woody texture.

II. Roots

a. Root Parts

root cap   Parenchymatous, protective apex of root.

root hair   Lateral, absorbing outgrowth of the epidermal cell.

secondary root   Lateral root with root cap and hairs, derived from the pericycle.

b. Root Types

(Classification based on position and origin)

adventitious   Arising from organ other than root; usually lateral.

primary   From radicle of embryo; tip of main axis.

secondary   From pericycle within the primary or secondary root; lateral.

c. Root Structural Types   **(Figure D.1)**

aerating or knee   Vertical or horizontal aboveground roots.

aerial   Fibrous, adventitious roots, frequently with an adhesive disk; a crampon.

ROOT STRUCTURAL TYPES and STEM PARTS

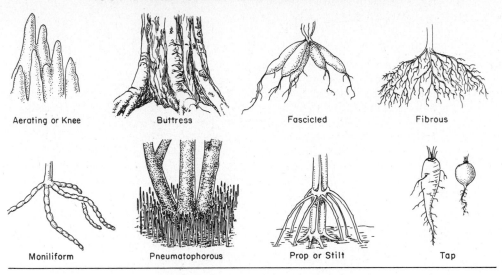

Aerating or Knee   Buttress   Fascicled   Fibrous

Moniliform   Pneumatophorous   Prop or Stilt   Tap

TWIG SURFACE PARTS

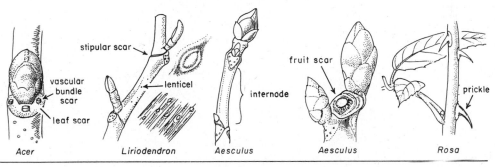

stipular scar   fruit scar   prickle

vascular bundle scar   lenticel   internode

leaf scar

Acer   *Liriodendron*   *Aesculus*   *Aesculus*   *Rosa*

THORNS   PITH

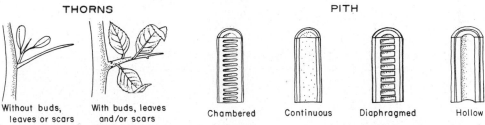

Without buds, leaves or scars   With buds, leaves and/or scars   Chambered   Continuous   Diaphragmed   Hollow

WOOD

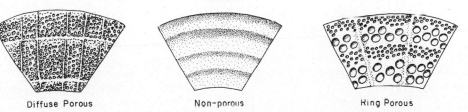

Diffuse Porous   Non-porous   Ring Porous

Figure D.1

buttress   Roots with boardlike or planklike growth on upper side, presumably a supporting structure.

contractile   Roots capable of shortening, usually drawing the plant or plant part deeper into the soil, usually with a wrinkled surface.

fascicled   Fleshy or tuberous roots in a cluster.

fibrous   With fine, threadlike, or slender roots.

fleshy   Succulent roots.

haustorial   Absorbing roots, within host of some parasitic species.

moniliform   Elongate roots with regularly arranged swollen areas.

pneumatophorous   With spongy, aerating roots, usually found in marsh plants.

prop or stilt   Adventitious, supporting roots usually arising at lower nodes.

tap   Persistent, well-developed primary root.

tuberous   Fleshy roots resembling stem tubers.

III. Stems

a. Stem Parts   (Figure D.1)

1. Twig Surface Parts

bud   An immature shoot.

internode   A section or region of stem between nodes.

leaf scar   A mark indicating former place of attachment of petiole or leaf base.

lenticel   A pore in the bark.

node   Region of stem from which a leaf, leaves, or branches arise.

prickle   A sharp-pointed outgrowth from the epidermis or cortex of any organ.

stipular scar   A mark indicating former place of attachment of stipule.

terminal bud scale scar rings   Several marks in a ring indicating former places of attachment of bud scales.

vascular bundle or trace scar   A mark indicating former place of attachment within the leaf scar of the vascular bundle or trace.

2. Major stem parts

Bark   Tissues of plant outside wood or xylem.

    exfoliating bark   Bark cracking and splitting off in large sheets.

    fissured bark   Split or cracked bark.

    plated bark   Split or cracked bark with flat plates between fissures.

    ringed bark   Split or cracked bark with circular fissures.

    shreddy bark   Soft but coarse fibrous bark, usually shallowly fissured.

    smooth bark   Bark without fissures.

    winged bark   Bark with one or more thin, flat longitudinal expansions or elongate plates.

pith   Centermost tissue of stem, usually soft.

    chambered pith   Solid core of pith cells absent, only distinct partitions present.

    continuous pith   With solid core of parenchyma or pith cells.

    diaphragmed pith   With solid core of pith cells and distinct partitions.

    hollow pith   Disintegrated pith with a large central cavity.

    spongy pith   Porous, easily compressible pith.

wood   Xylem consisting of vessels and/or tracheids, fibers, and parenchyma cells.

    annual ring   Usually one year's growth of wood; spring and summer wood.

> **diffuse porous wood**   Annual rings with vessels or pores more or less evenly distributed.
> **nonporous wood**   Annual rings with tracheids only, no vessels produced in spring or summer wood.
> **ring porous wood**   Annual rings with vessels or pores usually in the spring wood, in a well-defined circular band.

b.   Stem Types   (Figure D.2)

(Classification based on habit, direction of growth, or position)

arborescent   Treelike in appearance and size.
ascending   Inclined upward.
cespitose   Short, much-branched plant forming a cushion.
clambering   Sprawling across objects, with climbing structures.
climbing   Growing upward by means of tendrils, petioles, or adventitious roots.
columnar   Erect with a stout main stem or trunk.
decumbent   Reclining or lying on the ground with tips ascending.
dichotomous   With equally forked branches or stems.
eramous   With unbranched stems.
erect   Upright.
fastigiate   Strictly erect and parallel.
fruticose   Shrubby.
geniculate   Abruptly bent at a node, zigzag.
procumbent, prostrate, or reclining   Trailing or lying flat, not rooting at the nodes; *humistrate*.
ramose   Branched.
repent   Creeping or lying flat and rooting at the nodes.
soboliferous   With loosely clumped shoots arising some distance apart from rhizomes or underground suckers.
stoloniferous   With runners or propagative shoots rooting at the tip producing new plants; bearing stolons; sarmentose.
strict   Stiff and rigid.
supine   Prostrate, with parts oriented upward.
trailing   Sprawling on ground, usually with adventitious roots.
twining   Coiling around an object.
virgate   Wandlike; long, slender, and straight.

c.   Stem Structural Types   (Figure D.2)

bulb   A short, erect, underground stem surrounded by fleshy leaves.
bulbel   A small bulb produced from the base of a larger bulb.
bulbil   A small bulb or bulblike body produced on aboveground parts.
bulblet   A small bulb, irrespective of origin.
caudex   A short, thick, vertical or branched perennial stem, usually subterranean or at ground level.
cladode (phylloclad)   A flattened main stem resembling a leaf.
corm   The enlarged, solid, fleshy base of a stem with scales; an upright underground storage stem.
cormel   Small corm produced at base of parent corm.
culm   Flowering and fruiting stems of grasses and sedges.
pachycauly   Short, thick, frequently succulent stems, as in some cacti.

STEM TYPES, STEM STRUCTURAL TYPES and BUD TYPES

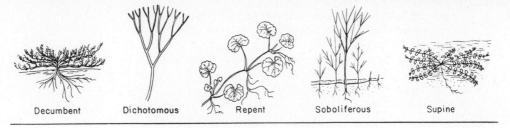

Decumbent   Dichotomous   Repent   Soboliferous   Supine

STEM STRUCTURAL TYPES

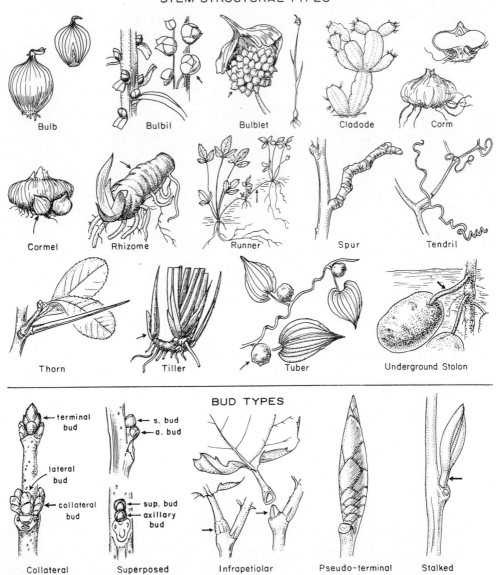

Bulb   Bulbil   Bulblet   Cladode   Corm

Cormel   Rhizome   Runner   Spur   Tendril

Thorn   Tiller   Tuber   Underground Stolon

BUD TYPES

terminal bud
lateral bud
collateral bud

Collateral

s. bud
a. bud
sup. bud
axillary bud

Superposed   Infrapetiolar   Pseudo-terminal   Stalked

Figure D.2

414

primocane   The first-year, nonflowering stem, as in most blackberries; a turion.

pterocauly   Winged stems.

rhizome   A horizontal underground stem.

rootstock   A term applied to miscellaneous types of underground stems or parts.

runner or stolon   An indeterminate, elongate, aboveground propagative stem, with long internodes, rooting at the tip forming new plants.

sarcocauly   Fleshy stems.

scape   A naked flowering stem with or without a few scale leaves, arising from an underground stem.

sclerocauly   Hard, dryish stems.

spur   A short shoot on which flowers and fruits or leaves are borne.

sucker   A shoot arising below ground or from an old stem, usually fast-growing and adventitious; surculose.

tendril   Long, slender, coiling branch, adapted for climbing (most tendrils are leaf structures).

thorn   A sharp-pointed branch.

tiller   A grass shoot produced from the base of the stem.

tuber   A thick storage stem, usually not upright.

turion   An overwintering bud, as in *Lemna*

underground stolon   A determinate elongate underground propagative stem with long internodes forming a bulb or tuber at the tip.

IV. Buds

a. Bud Parts

bud primordium   Meristematic tissue that gives rise to a lateral bud.

flower primordium   Meristematic tissue that gives rise to a flower.

leaf primordium   Meristematic tissue that gives rise to a leaf.

promeristem   Apical growing or meristematic tissue that gives rise to other bud parts.

scale   Protective leaf on outside of bud.

b. Bud Types   **(Figure D.2)**

(Classification based on position and arrangement)

accessory   Buds lateral to or above axillary buds.
   collateral   Bud(s) lateral to axillary bud.
   superposed   Bud(s) above axillary bud.
axillary or lateral   In axils of leaves or leaf scars.
infrapetiolar or subpetiolar   Axillary bud surrounded by base of petiole.
pseudoterminal   Bud appearing apical but is lateral near apex, developing with death or nondevelopment of terminal bud.
terminal   At apex or end of stem.

c. Bud Structural Types

(Classification based on composition and cover)

flower   Contains flower primordia; will give rise to one or more flowers.

leaf   Contains leaf and stem primordia; will give rise to branch with leaves.

mixed   Contains flower, leaf, and stem primordia; will give rise to branch with leaves and flower(s).

naked   Shoot and/or flower primordia not surrounded by scales.

protected   Shoot and/or flower primordia surrounded by scales.

V. Leaves

a. Leaf Parts

blade   Flat, expanded portion of leaf.

leaflet   A distinct and separate segment of a leaf.

ligule   An outgrowth or projection from the top of the sheath, as in the Poaceae.

midrib   The central conducting and supporting structure of the blade of a simple leaf.

midvein   The central conducting and supporting structure of the blade of a leaflet.

petiole   Leaf stalk.

petiolule   Leaflet stalk.

pulvinus   The swollen base of a petiole or petiolule.

rachilla   Secondary axis of compound leaf.

rachis   The main axis of a pinnately compound leaf.

sheath   Any more or less tubular portion of the leaf surrounding the stem or culm, as in the Poaceae.

stipels   Paired scales, spines, or glands at the base of petiolule.

stipules   Paired scales, spines, glands, or bladelike structures at the base of a petiole.

b. Leaf Types   (Figure D.3)

(Classification based primarily on arrangement of leaflets)

Note: This classification is based on discrete segments of leaves or leaflets, but the terms with ''compound'' are equally applicable to segments, divisions, etc., of any structure with a blade; for example, palmately divided, pinnately cleft, etc.

bifoliolate, geminate, or jugate   With two leaflets from a common point.

bigeminate, bijugate   With two orders of leaflets, each bifoliolate; doubly paired.

bipalmately compound   With two orders of leaflets, each palmately compound.

bipinnately compound   With two orders of leaflets, each pinnately compound.

biternate   With two orders of leaflets, each ternately compound.

compound   With leaf divided into two or more leaflets.

decompound   A general term for leaflets in two or more orders—bi-, tri-, and so on—pinnately, palmately, or ternately compound.

imparipinnately compound   *Odd-pinnately* compound, with a terminal leaflet.

interruptedly pinnately compound   With smaller and larger leaflets alternating along the rachis.

palmately compound   With leaflets from one point at end of petiole.

palmate-pinnate   With first-order leaflets palmately arranged, second-order pinnately arranged.

paripinnately compound   *Even-pinnately* compound, without a terminal leaflet.

pinnately compound   With leaflets arranged oppositely or alternately along a common axis, the rachis.

simple   With leaf not divided into leaflets.

tergeminate   With three orders of leaflets, each bifoliolate, or with geminate leaflets ternately compound.

LEAF TYPES and LEAF STRUCTURAL TYPES

COMPOUND PALMATE and COMPOUND PINNATE

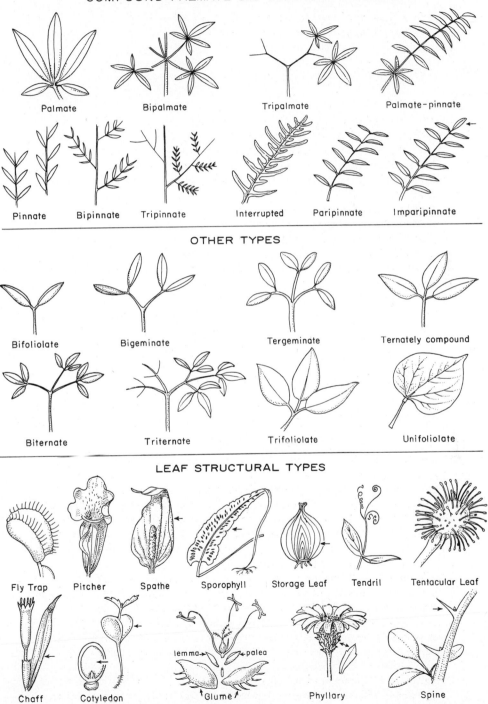

Palmate   Bipalmate   Tripalmate   Palmate-pinnate

Pinnate   Bipinnate   Tripinnate   Interrupted   Paripinnate   Imparipinnate

OTHER TYPES

Bifoliolate   Bigeminate   Tergeminate   Ternately compound

Biternate   Triternate   Trifoliolate   Unifoliolate

LEAF STRUCTURAL TYPES

Fly Trap   Pitcher   Spathe   Sporophyll   Storage Leaf   Tendril   Tentacular Leaf

Chaff   Cotyledon   lemma   palea   Glume   Phyllary   Spine

Figure D.3

ternately compound   With leaflets in groups of three.

trifoliolate   With three leaflets, pinnately compound with terminal petiolule longer than lateral; or palmately compound with petiolules equal in length.

tripalmately compound   With three orders of leaflets, each palmately compound.

tripinnately compound   With three orders of leaflets, each pinnately compound.

triternate   With three orders of leaflets, each ternately compound.

unifoliolate   With single leaflet with a petiolule distinct from the petiole of the whole leaf, as in *Cercis*.

c.   Leaf Structural Types   (Figure D.3)

bract   Modified leaf found in the inflorescence.

bracteole or prophyllum   Small leaf, usually on a pedicel.

chaff or pale   Scale or bract at base of tubular flower in composites.

complete   Leaf with blade, petiole, and stipules.

cotyledon   Embryonic leaf.

elaminate   Without blade.

epetiolate   Without petiole, leaf sessile.

epetiolulate   Without petiolule, leaflet sessile.

epicalyx   Group of leaves resembling sepals below the true calyx.

exstipellate   Without stipels.

exstipulate   Without stipules.

fly trap   Hinged, insectivorous leaf, as in *Dionaea*.

glume   Bract, usually occuring in pairs, at the base of the grass spikelet.

incomplete   Leaf without one or more parts: blade, petiole, stipules.

lemma   Outer scale subtending grass floret.

palea   Inner scale subtending grass floret.

phyllary   One of the involucral leaves subtending a capitulum, as in composites.

phyllode   Flattened bladelike petiole or midrib.

pitcher   Ventricose to tubular insectivorous leaf, as in *Sarracenia*.

scale   Small, nongreen leaf on bud or modified stem.

spathe   An enlarged bract enclosing an inflorescence.

spine   Sharp-pointed petiole, midrib, vein, or stipule.

sporophyll   A spore-bearing leaf.

storage leaf   Succulent, fleshy leaf.

tendril   Usually a coiled rachis or twining leaflet modification.

tentacular   Glandular-haired or tentacle-bearing insectivorous leaf, as in *Drosera*.

d.   Petiole and Petiolule Structural Types

channeled or canaliculate   With a longitudinal groove.

inflated   Swollen or thickened, as in *Eichhornia*.

pericladial   With a sheathing base, as in the Apiaceae.

petiolate   With a petiole.

petiolulate   With a petiolule.

phyllodial   Flattened and bladelike.

pulvinal   With a swollen base, as in the Fabaceae.

sessile or absent   Without petiole or petiolule.
winged   With flattened bladelike margins

e.   Stipule and Stipel Types

(Classification based primarily on attachment or function)

adnate   With stipule attached to petiole.
basal   With stipules attached near base of petiole.
interpetiolar   With connate stipules from two opposite leaves.
lateral   With stipules adnate to petiole and free part of stipules located along the petiole.
median   With stipules adnate to petiole with free part of stipules near middle of petiole.
photosynthetic   Bladelike and green.
sheathing or protective   Enclosing a bud or flower.
stipellate   With stipels.
stipulate   With stipules.
vestigial   Minute; a remnant.

VI.   Inflorescences

a.   Inflorescence Parts   (Figure D.4)

bract   Modified, usually reduced, leaf in the inflorescence.
bracteole or bractlet   A secondary or smaller bract.
cupule   Fused involucral bracts subtending flower, as in *Quercus*.
epicalyx   A whorl of bracts below but resembling a true calyx.
flower   Modified reproductive shoot of angiosperms.
involucel   Small involucre; secondary involucre.
involucre   A group or cluster of bracts subtending an inflorescence.
pedicel   Individual flower stalk.
peduncle   Main stalk for entire inflorescence.
perigynium   Saclike bract subtending the pistillate flower, as in *Carex*.
phyllary   Individual bract within involucre.
rachilla   Central axis of a grass or sedge spikelet.
rachis   Major axis within an inflorescence.
ray   Secondary axis in a compound inflorescence.
scape   Naked peduncle.
spathe   A sheathing leaf subtending or enclosing an inflorescence.

b.   Inflorescence Types   (Figure D.4)

(Classification based primarily on arrangement and development)

Note: Inflorescence types are essentially secondary arrangements. The primary arrangement (alternate, opposite, or whorled) of the individual flowers should be indicated; and the tertiary arrangement of sessile-flowered inflorescences should be noted, for example, spikelets racemose or heads umbellate. *Determinate inflorescences* have the central flower maturing first with the arrest of the elongation of the central axis; *indeterminate inflorescences* have the lateral or lower flowers matur-

INFLORESCENSE TYPES

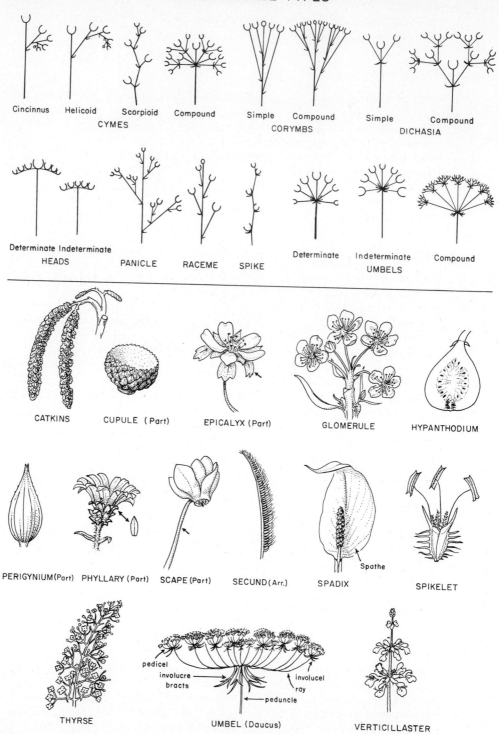

Cincinnus Helicoid Scorpioid Compound
CYMES

Simple Compound
CORYMBS

Simple Compound
DICHASIA

Determinate Indeterminate
HEADS

PANICLE RACEME SPIKE

Determinate Indeterminate
UMBELS

Compound

CATKINS CUPULE (Part) EPICALYX (Part) GLOMERULE HYPANTHODIUM

PERIGYNIUM(Part) PHYLLARY (Part) SCAPE (Part) SECUND(Arr.) SPADIX

Spathe

SPIKELET

THYRSE

pedicel
involucre
bracts

involucel
ray

peduncle

UMBEL (Daucus)

VERTICILLASTER

Figure D.4

420

ing first without the arrest of the elongation of the central axis. A satisfactory classification of inflorescences has not been developed.

1. *Inflorescences with sessile flowers*

ament or catkin   A unisexual spike or elongate axis with simple dichasia that falls as a unit after flowering or fruiting.

capitulum or head   A determinate or indeterminate crowded group of sessile or subsessile flowers on a compound receptacle or torus.

glomerule   An indeterminate dense cluster of sessile or subsessile flowers.

hypanthodium   An inflorescence with flowers on a wall of a concave capitulum, as in *Ficus*.

spadix   Unbranched, indeterminate inflorescence with flowers embedded in the rachis.

spike   Unbranched, indeterminate, elongate inflorescence with sessile flowers.

spikelet   A small spike; the basic inflorescence unit in grasses and sedges.

2. *Unbranched inflorescences with pedicellate flowers*

cincinnus   A tight, modified helicoid cyme in which pedicels are short on the developed side.

corymb   A flat-topped or convex indeterminate cluster of flowers.

cymule   A simple, small dichasium.

helicoid cyme   A determinate inflorescence in which the branches develop on one side only, appearing simple.

raceme   Unbranched, indeterminate inflorescence with pedicelled flowers.

scorpioid cyme   A zigzag determinate inflorescence with branches developed on opposite sides of the rachis alternately.

simple cyme or dichasium   A determinate, dichotomous inflorescence with the pedicels of equal length.

umbel   A determinate or indeterminate flat-topped or convex inflorescence with the pedicels arising at a common point.

3. *Branched inflorescences with pedicellate flowers*

compound corymb   A branched corymb.

compound cyme   A branched cyme.

compound umbel   An umbel with primary rays or peduncles arising at a common point with a secondary umbel arising from the tip of the primary rays; a branched umbel.

panicle   Branched inflorescence with pedicelled flowers.

thyrse   A many-flowered inflorescence with an indeterminate central axis and with many opposite lateral dichasia.

verticillaster   Whorled dichasia at the nodes of an elongate rachis.

4. *General inflorescence terms and types*

cyathium   A pseudanthium subtended by an involucre, frequently with petaloid glands, as in *Euphorbia*.

monochasium   A cymose inflorescence with one main axis.

pleiochasium   Compound dichasium in which each cymule has three lateral branches.

pseudanthium   Several flowers simulating a simple flower but composed of more than a single axis with subsidiary flowers.

scapose   With a solitary flower on a leafless peduncle or scape, usually arising from a basal rosette.

secund   One-sided arrangement.

solitary   One-flowered, not an inflorescence.

umbellet   The secondary umbel in a compound umbel.

VII.   Flowers

a.   Flower Parts   (Figure D.5)

accessory organs   The calyx and corolla.

androecium   One or more whorls or groups of stamens; all stamens in flower.

androgynophore   The stipe or column on which stamens and carpels are borne.

calyx   The lowermost whorl of modified leaves; sepals.

carpel   The female sporophyll within flower; floral organ that bears ovules in angiosperms; unit of compound pistil.

clinanthium   The compound receptacle of the composite head.

corolla   The whorl of petals located above the sepals.

disk   A discoid structure developed from receptacle at base of ovary or from stamens around the ovary.

essential organs   The androecium and gynoecium.

gynoecium or pistil   The whorl or group of carpels in the center or at the top of the flower; all carpels in a flower.

gynophore   The stipe of a pistil or carpel.

hypanthium   The fused or coalesced basal portion of floral parts (sepals, petals, stamens) around the ovary.

pedicel   The flower stalk.

perianth   An aggregation of tepals or combined calyx and corolla.

petal   A corolla member or segment; a unit of the corolla.

polyphore   A receptacle or torus bearing many distinct carpels, as in *Rosa*.

receptacle or torus   The region at end of pedicel or on axis to which flower is attached.

sepal   A calyx member or segment; a unit of the calyx.

stamen   The male sporophyll within the flower; the floral organ that bears pollen in angiosperms.

tepal   A perianth member or segment; term used for perianth parts undifferentiated into distinctive sepals and petals.

whorl   A cyclic or acyclic group of sepals, or petals, or stamens, or carpels.

b.   Flower Types   (Figure D.6)

(Classification based on evolutionary flower-pollinator relationships. From Leppik [1957])

actinomorphic   Flowers with radial symmetry and parts arranged at one level; with definite numbers of parts and size; for example, *Anemone, Caltha*.

amorphic or paleomorphic   Flowers without symmetry; usually with an indefinite number of stamens and carpels, and usually subtended by bracts or discolored upper leaves; for example, *Salix discolor, Echinops ritro* (mostly fossil forms).

haplomorphic   Flowers with parts spirally arranged at a simple level in a semispheric or hemispheric form; petals or tepals colored; parts numerous; for example, *Nymphaea, Magnolia*.

pleomorphic   Actinomorphic with numbers of parts reduced; for example, *Tripogandra*.

FLOWER PARTS

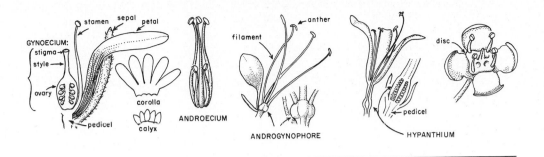

GYNOECIUM:
stigma
style
ovary
pedicel

stamen sepal petal

corolla

calyx

ANDROECIUM

filament anther

ANDROGYNOPHORE

pedicel

HYPANTHIUM

disc

PERIANTH and HYPANTHIUM PARTS

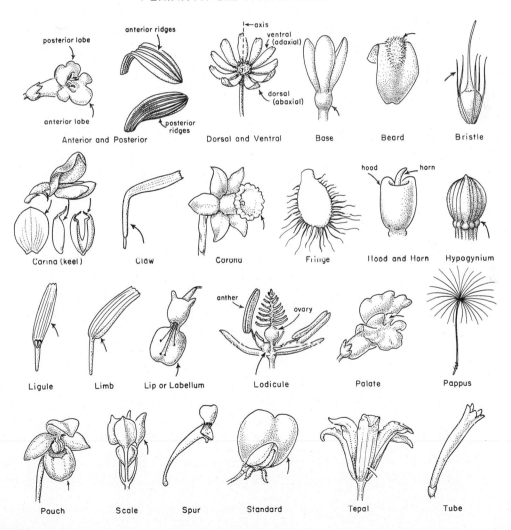

posterior lobe

anterior lobe

Anterior and Posterior

anterior ridges

posterior ridges

axis
ventral (adaxial)

dorsal (abaxial)

Dorsal and Ventral

Base

Beard

Bristle

Carina (keel)

Claw

Corona

Fringe

hood horn

Hood and Horn

Hypogynium

Ligule

Limb

Lip or Labellum

anther ovary

Lodicule

Palate

Pappus

Pouch

Scale

Spur

Standard

Tepal

Tube

Figure D.5

423

FLOWER TYPES

FLORAL EVOLUTION of ANGIOSPERMS (TYPE CLASSES)

(Adapted from Leppik)

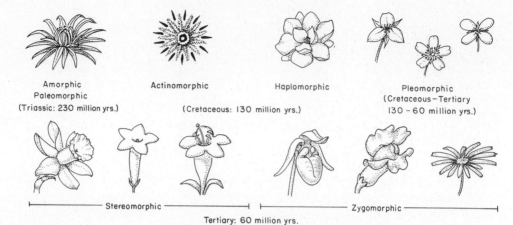

Amorphic
Paleomorphic
(Triassic: 230 million yrs.)

Actinomorphic

(Cretaceous: 130 million yrs.)

Haplomorphic

Pleomorphic
(Cretaceous–Tertiary
130 – 60 million yrs.)

|——————— Stereomorphic ———————| |——————— Zygomorphic ———————|

Tertiary: 60 million yrs.

PERIANTH STRUCTURAL TYPES

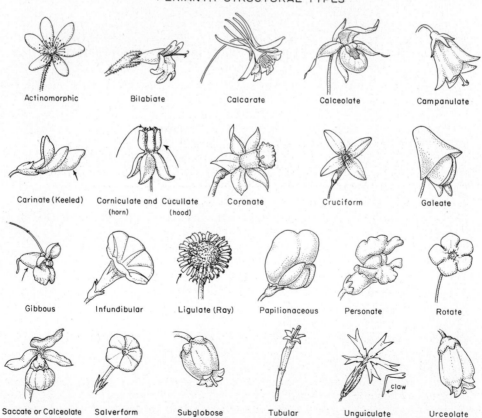

Actinomorphic   Bilabiate   Calcarate   Calceolate   Campanulate

Carinate (Keeled)   Corniculate and Cucullate   Coronate   Cruciform   Galeate
                    (horn)   (hood)

Gibbous   Infundibular   Ligulate (Ray)   Papilionaceous   Personate   Rotate

Saccate or Calceolate   Salverform   Subglobose   Tubular   Unguiculate   Urceolate
                                                              claw

Figure D.6

424

stereomorphic Flowers three-dimensional with basically radial symmetry; parts many or reduced and usually regular; for example, *Narcissus, Aquilegia*.

zygomorphic Flowers with bilateral symmetry; parts usually reduced in number and irregular; for example, *Cypripedium, Salvia*.

VIII. Perianth and Hypanthium

a. Perianth Parts

(Figure D.5)

anterior lobes The lobes away from axis, toward the subtending bract; abaxial lobes.

anterior ridges, lines, or grooves The lines, grooves, or ridges in or on the dorsal side, abaxial, within the perianth.

base Bottom or lower portion.

beard A tuft, line, or zone of trichomes.

bristle A stiff, strong trichome, as in the perianth of some members of the Cyperaceae.

callosity A thickened, raised area, which is usually hard; a *callus*.

carina Keel.

claw The long, narrow petiolelike base of a sepal or petal.

corona A crown; any outgrowth between the stamens and corolla that may be petaline or staminal in origin.

dorsal side Back or abaxial side, or the lower side of a perianth part.

faucal area The throat area.

fringe The modified margin of a petal, sepal, tepal, or lip.

hood A cover-shaped perianth part, usually with a turned-down margin.

horn A curved, pointed, and hollow protuberance from the perianth.

hypogynium Perianthlike structure of bony scales subtending the ovary, as in *Scleria* and other members of the Cyperaceae.

keel The two united petals of a papilionaccous flower; any structure ridged like the bottom of a boat.

ligule The strap-shaped portion or a ray or ligulate corolla.

limb Expanded portion of corolla or calyx above the tube, throat, or claw.

lip or labellum Either of two variously shaped parts into which a corolla or calyx is divided usually into an upper and lower lip, as in the Lamiaceae and Orchidaceae.

lobe Any usually rounded segment or part of the perianth.

lodicule Scalelike perianth part in the Poaceae; hyaline scales at base of ovary in the Poaceae.

palate The raised area in the throat of a sympetalous corolla.

pappus Bristly or scaly calyx in the Asteraceae.

petal A corolla member or segment; a unit of the corolla.

posterior lobe The lobe next to axis, away from the subtending bract; adaxial lobe.

posterior ridges, lines, or grooves The lines, grooves, or ridges in or on the ventral side, adaxial, within the perianth.

pouch or sac A bag-shaped structure.

scale Small, scarious to coriaceous, flattened bodies within the perianth, as in the Cyperaceae and Asteraceae.

sepal A calyx member or segment; a unit of the calyx.

spur A tubular or pointed projection from the perianth.

standard, banner, or vexillum The upper, usually wide petal in a papilionaceous corolla.

tepal A member or segment of perianth in which the parts are not differentiated into distinct sepals and petals.

throat An open, expanded tube in the perianth.

tube   The cylindrical part of the perianth.

ventral side   Top side or upper side of a perianth part.

wing   Lateral petals, as in the Fabaceae; a flattened extension, appendage, or projection from a perianth part.

b.   Hypanthium Parts

base   Bottom or lower portion of the hypanthium.

limb   Free, flared portion of the hypanthium.

neck   Narrowed portion of hypanthium, between the base and a flared limb.

tube   cylindrical part of the hypanthium.

c.   Perianth Fusion Types   (Figure D.12)

(Classification based primarily on fusion of parts)

achlamydeous   Without perianth.

apetalous   No petals or corolla.

apopetalous or choripetalous   With separate petals.

aposepalous or chorisepalous   With separate sepals.

asepalous   No sepals or calyx.

chlamydeous   With perianth.

dichlamydeous   With perianth composed of distinct calyx and corolla.

homochlamydeous   With perianth composed of similar parts, each part a tepal.

sympetalous   With fused petals.

synsepalous   With fused sepals.

d.   Perianth and Androecium Position Types   (Figure D.12)

(Classification based on insertion of floral parts—corolla, calyx, and androecium—the androperianth)

epigyny   The condition in which the sepals, petals, and stamens are attached to the floral tube above the ovary with the ovary adnate to the tube or hypanthium.

epihyperigyny   The condition in which the sepals, petals, and stamens are attached to the floral tube or hypanthium surrounding the ovary; a combination perigyny and partly inferior ovary.

epihypogyny   The condition in which the sepals, petals, and stamens are attached about halfway from the base of the ovary to the partly adnate hypanthium tube; half-inferior insertion of parts.

epiperigyny   The condition in which the sepals, petals, and stamens are attached to the floral or hypanthium cup above the ovary, with the lower part of the hypanthium completely adnate to the ovary.

hypanepigyny   The condition in which the sepals, petals, and stamens are attached to the elongate floral tube or hypanthium above the inferior ovary, as in *Oenothera*.

hypogyny   The condition in which the sepals, petals, and stamens are attached below the ovary.

perigyny   The condition in which the sepals, petals, and stamens are attached to the floral tube or hypanthium surrounding the ovary with the tube or hypanthium free from the ovary.

e. Perianth and Hypanthium Structural Types (Figure D.6)

actinomorphic   With radially arranged perianth parts; raylike figure.
bilabiate   Two-lipped; with two unequal divisions.
calcarate   Spurred.
calceolate   Slipper-shaped, as in the corolla of *Cypripedium*.
campanulate   Bell-shaped; with flaring tube about as broad as long and a flaring limb.
carinate   Keeled.
corniculate   Horned.
coronate   Tubular or flaring perianth or staminal outgrowth; petaloid appendage.
cruciate   Four separate petals in cross form.
cucullate   Hooded.
galeate   Helmet-shaped, as one sepal in *Aconitum*.
gibbous   Inflated on one side near the base.
globose   Round.
infundibular   Funnel-shaped.
lingulate or ray   Strap-shaped.
papilionaceous   With large posterior petal (banner or standard), two lateral petals (wings), and usually two connate lower petals (keel); as in the Fabaceae.
personate   Two-lipped, with the upper arched and the lower protruding into corolla throat.
rotate   Wheel-shaped, with short tube and wide limb at right angles to tube.
saccate   Pouchlike.
salverform   Trumpet-shaped, with slender tube and limb nearly at right angles to tube.
subglobose   Almost round or spherical.
tubular   Cylindrical.
unguiculate   Clawed.
urceolate   Urn-shaped.
ventricose   Inflated on one side near the middle.

IX. Androecium

a. Androecial Parts

stamen   Male sporophyll within the flower; floral organ that bears pollen in angiosperms.
staminodium   Sterile stamen, may be modified as a nectary or petaloid structure.
staminal disk   A fleshy, elevated cushion formed from coalesced staminodia or nectaries.

b. Androecial Fusion Types (Figure D.7)

(Classification based primarily on fusion of parts)

apostemonous   With separate stamens.
diadelphous   With two groups of stamens connate by their filaments.
gynandrial or gynostemial   With fused stamens and carpels (stigma and style) as in the Orchidaceae.
monadelphous   With one group of stamens connate by their filaments.
petalostemonous   With filaments fused to corolla, anthers free.
polydelphous   With several groups of stamens connate by their filaments.
syngenesious   With fused anthers.

ANDROECIUM

STAMEN POSITION

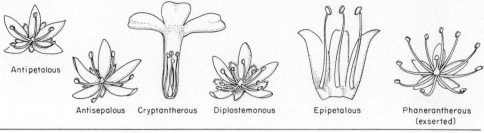

Antipetalous   Antisepalous   Cryptantherous   Diplostemonous   Epipetalous   Phanerantherous (exserted)

STAMEN ARRANGEMENT   ANDROECIAL TYPES

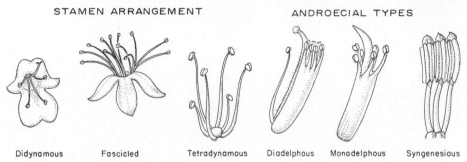

Didynamous   Fascicled   Tetradynamous   Diadelphous   Monadelphous   Syngenesious

STAMEN STRUCTURAL TYPES   THECAL ARRANGEMENT

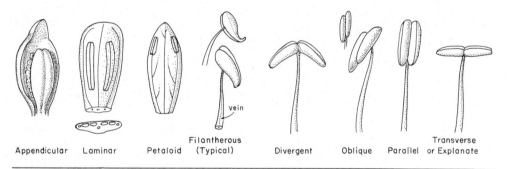

Appendicular   Laminar   Petaloid   Filantherous (Typical)   Divergent   Oblique   Parallel   Transverse or Explanate

vein

ANTHER TYPES and ATTACHMENT

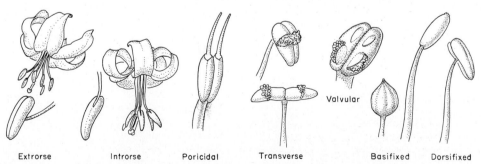

Extrorse   Introrse   Poricidal   Transverse   Valvular   Basifixed   Dorsifixed

Figure D.7

428

c. Stamen Parts

> **anther**   Pollen-bearing portion of stamen.
> **filament**   Stamen stalk.

d. Stamen Arrangement Types   (Figure D.7)

> **didymous**   With stamens in two equal pairs.
> **didynamous**   With stamens in two unequal pairs.
> **tetradynamous**   With stamens in two groups, usually four long and two short.
> **tridynamous**   With stamens in two equal groups of three.

e. Stamen Position Types   (Figure D.7)

> **allagostemonous**   Having stamens attached to petal and torus alternately.
> **antipetalous**   Opposite the petals.
> **antisepalous**   Opposite the sepals.
> **diplostemonous**   With stamens in two whorls, outer opposite the sepals, inner opposite petals.
> **epipetalous**   With stamens attached to or inserted on petals or corolla.
> **episepalous**   With stamens attached to or inserted on sepals or calyx.
> **obdiplostemonous**   With stamens in two whorls, outer opposite petals, inner opposite the sepals.
> **phaenantherous**   With stamens exserted.

f. Stamen Structural Types   (Figure D.7)

> (Classification based primarily on structure of filament and anther)

> Note: In this classification, intermediate types of stamens do occur; shapes, apices, and bases of anthers should be described separately and independently of stamen type.

> **appendicular**   Typical stamen with a variously shaped or modified, protruding connective, as in *Viola*.
> **filantherous**   Stamen with distinct anther and filament with or without thecal appendages, as in *Rhexia* or *Vaccinium*.
> **laminar**   Leaflike stamen without a distinct anther and filament but with embedded or superficial microsporangia, as in *Degeneria*.
> **petalantherous**   With a terminal anther and distinctly petaloid filament, as in *Saxifraga*.
> **petaloid**   Petal-like stamen without distinct anther and filament but with marginal microsporangia, as in *Magnolia nitida*.

g. Anther Parts   (Figure D.7)

> **connective**   Filament extension between thecae.
> **locule**   Compartment of an anther
> **pollen grain**   Young male gametophyte.
> **pollen sac**   Male sporangium.
> **theca**   One-half of anther containing two pollen sacs or male sporangia.

h.   Anther Types   **(Figure D.7)**

(Classification based on dehiscence)

longitudinal   Dehiscing along long axis of theca.
extrorse   Dehiscing longitudinally outward.
introrse   Dehiscing longitudinally inward.
latrorse   Dehiscing longitudinally and laterally.
poricidal   Dehiscing through a pore at apex of theca.
transverse   Dehiscing at right angles to long axis of theca.
valvular   Dehiscing through a pore covered by a flap of tissue.

i.   Anther Attachment   **(Figure D.7)**

basifixed   Anther attached at its base to apex of filament.
dorsifixed   Anther attached dorsally and medially to apex of filament.
subbasifixed   Anther attached near its base to apex of filament.
versatile   Dorsifixed but anther seemingly swinging free on the filament.

j.   Pollen

dyads   Grains occurring in clusters of two.
filiform   Threadlike.
monad   Grains occurring singly.
pollinia   Grains occurring in uniform coherent masses.
polyad   Grains occurring in groups of more than four.
tetrads   Grains occurring in groups of four.

X.   Gynoecium

a.   Gynoecial Parts   **(Figure D.8)**

carpel   Female sporophyll within flower; floral organ that bears ovules.
carpopodium   Short, thick, pistillate stalk.
locule   Ovary cavity.
ovary   Ovule-bearing part of pistil.
placenta   Ovule-bearing region of ovary wall.
stigma   Pollen-receptive portion of pistil.
stipe   Pistillate stalk.
style   Attenuated, nonovule-bearing portion of pistil between stigma and ovary.

b.   Gynoecial Fusion Types   **(Figure D.8)**

(Classification based on fusion)

apocarpous   With carpels separate.
semicarpous   With ovaries of adjacent carpels partly fused, stigmas and styles separate.
syncarpous   With stigmas, styles, and ovaries completely fused.

GYNOECIUM

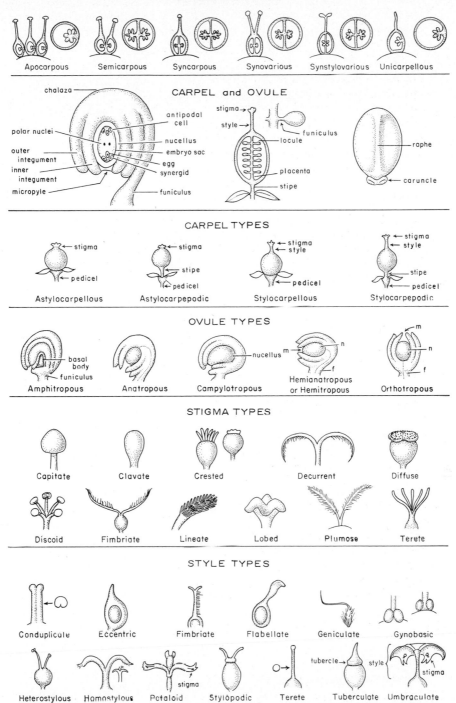

GYNOECIAL TYPES

Apocarpous Semicarpous Syncarpous Synovarious Synstylovarious Unicarpellous

CARPEL and OVULE

chalaza

antipodal cell

polar nuclei

nucellus

outer integument

embryo sac

inner integument

egg

synergid

micropyle

funiculus

stigma

style

funiculus

locule

placenta

stipe

raphe

caruncle

CARPEL TYPES

stigma

pedicel

Astylocarpellous

stigma

stipe

pedicel

Astylocarpepodic

stigma
style

pedicel

Stylocarpellous

stigma
style

stipe

pedicel

Stylocarpepodic

OVULE TYPES

basal body
funiculus

Amphitropous

Anatropous

nucellus

Campylotropous

m

n

f

Hemianatropous
or Hemitropous

m

n

f

Orthotropous

STIGMA TYPES

Capitate Clavate Crested Decurrent Diffuse

Discoid Fimbriate Lineate Lobed Plumose Terete

STYLE TYPES

Conduplicate Eccentric Fimbriate Flabellate Geniculate Gynobasic

Heterostylous Homostylous Petaloid stigma Stylopodic Terete tubercle style Tuberculate stigma Umbraculate

Figure D.8

431

synovarious   With ovaries of adjacent carpels completely fused, styles and stigmas separate.
synstylovarious   With ovaries and styles of adjacent carpels completely fused, stigmas separate.
unicarpellous or stylodious   With solitary, free carpel in gynoecium.

c.   Carpel Parts   (Figure D.8)

funiculus   Stalk that attaches the ovule to the placenta.
locule   Ovary cavity.
ovary   Ovule-bearing part of carpel in simple ovary.
ovule   Embryonic seed consisting of integument(s) and nucellus.
placenta   Ovule-bearing region of ovary wall.
stigma   Pollen-receptive portion of carpel.
stipe, podogyne, or carpopodium   Basal stalk.
style   Attenuated portion of carpel between stigma and ovary.

d.   Carpel Types   (Figure D.8)

(Classification based on presence or absence of style and stipe)

astylocarpellous   Without a style and a stipe.
astylocarpepodic   Without a style, with a stipe.
stylocarpellous   With a style and without a stipe, the normal carpel.
stylocarpepodic   With a style and a stipe.

e.   Stigma Types   (Figure D.8)

(Classification based on shape)

capitate   Headlike.
clavate   Clubshaped.
crested or cristate   With a terminal ridge or tuft.
decurrent   Elongate, extending downward.
diffuse   Spread over a wide surface.
discoid   Disklike.
fimbriate   Fringed.
lineate   In lines, stigmatic surface linear.
lobed   Divided into lobes.
plumose   Featherlike.
terete   Cylindrical and elongate.

f.   Style Types   (Figure D.8)

(Classification based primarily on shape)

astylous   Style absent.
conduplicate   Folded with a longitudinal groove.
cristate   Crested.
eccentric   Off-center style.

fimbriate   Fringed.

flabellate   Fan-shaped.

geniculate   Bent abruptly.

gynobasic   Attached at base of ovary in central depression.

heterostylous   With styles of different sizes or lengths or shapes within a species.

homostylous   With styles of same sizes or lengths and shapes.

involute   With margins infolded longitudinally, with groove present.

petaloid   Petal-like.

stylopodic   With a stylopodium or discoid base, as in the Apiaceae.

terete   Cylindrical and elongate.

tuberculate   With hard, swollen, persistent base or tubercle.

umbraculate   Umbrella-shaped, as in *Sarracenia*.

g.   Style Position Types

gynobasic   At the base of an invaginated ovary.

lateral   At the side of an ovary.

subapical   At one side near apex of ovary.

terminal or apical   At the apex of the ovary.

h.   Ovule Parts   (Figure D.8)

chalaza   End of ovule opposite micropyle.

embryo sac   Female gametophyte.

integuments   Outer covering of ovule; embryonic seed coat.

micropyle   Hole through integument(s).

nucellus   Female sporangium within ovule; megasporangium in seed plants.

raphe   Longitudinal ridge on outer integument.

i.   Ovule Types   (Figure D.8)

(Classification based on orientation of ovule body in relation to the funiculus and micropyle)

amphitropous   With body bent or curved on both sides so that the micropyle is near the medially attached funiculus.

anatropous   With body completely inverted so that funiculus is attached basally near adjoining micropyle area.

campylotropous   With body bent or curved on one side so that micropyle is near medially attached funiculus.

hemianatropous or hemitropous   With body half-inverted so that funiculus is attached near middle with micropyle terminal and at right angles.

orthotropous or atropous   With straight body so that funicular attachment is at one end and micropyle at other.

j.   Ovule Position Types   (Figure D.12)

(Based on position of ovule in locule and orientation of the micropyle and raphe—adapted from Bjornstad [1970])

epitropous, dorsal   Ovule pendulous or hanging, micropyle above, raphe dorsal (away from ventral bundle).

epitropous, ventral   Ovule pendulous or hanging, micropyle above, raphe ventral (toward ventral bundle).

heterotropous   Ovule position not fixed in ovary.

hypotropous, dorsal   Ovule erect, micropyle below, raphe dorsal (away from ventral bundle).

hypotropous, ventral   Ovule erect, micropyle below, raphe ventral (toward ventral bundle).

pleurotropous, dorsal   Ovule horizontal, micropyle toward ventral bundle, raphe above.

pleurotropous, ventral   Ovule horizontal, micropyle toward ventral bundle, raphe below.

k.   Placenta Position Types (Placentation)   **(Figure D.12)**

axile   With the placentae along the central axis in a compound ovary with septa.

basal   With the placenta at the base of the ovary.

free-central   With the placenta along the central axis in a compound ovary without septa.

laminate   With the placenta over the inner surface of the ovary wall.

marginal or ventral   With the placenta along the margin of the simple ovary.

parietal   With the placentae on the wall or intruding partitions of a unilocular compound ovary.

pendulous, apical, or suspended   With the placenta at the top of the ovary.

l.   Ovary Position Types

inferior   Other floral organs attached above ovary with hypanthium adnate to ovary.

half-inferior   Other floral organs attached around ovary with hypanthium adnate to lower half of ovary.

superior   Other floral organs attached below ovary.

XI.   Fruits

a.   Fruit Parts

carpophore   Floral axis extension between adjacent carpels, as in the Apiaceae.

ectocarp or exocarp   Outermost layer of pericarp.

endocarp   Innermost differentiated layer of pericarp.

funiculus   Seed stalk.

mericarp   A portion of fruit that seemingly matured as a separate fruit.

mesocarp   Middle layer of pericarp.

pericarp   Fruit wall.

placenta   Region of attachment of seeds on inner fruit wall.

replum   Persistent septum after dehiscence of fruits, as in the Brassicaceae.

retinaculum, jaculator, or echma   A persistent indurated, hooklike funiculus in the fruits of Acanthaceae.

rostellum or beak   Persistent stylar base on fruit.

seed   A matured ovule.

septum or dissepiment   Partition.

b.   Fruit Structural Types

(Classification based primarily on origin, texture, and dehiscence; types grouped as simple, aggregate, multiple, accessory. No satisfactory classification of fruits has been developed.)

1. Simple fruits

(Fruit derived from the ovary of a solitary pistil in a single flower)

aa. Dry indehiscent fruit types   **(Figure D.9)**

(Fruits that do not split open at maturity)

achene   A one-seeded, dry, indehiscent fruit with seed attached to fruit wall at one point only, derived from a one-loculed superior ovary.

balausta   Many-seeded, many-loculed indehiscent fruit with a tough, leathery pericarp, as in *Punica*.

calybium   A hard, one-loculed dry fruit derived from an inferior ovary, as in *Quercus*.

capsule, indehiscent   Dry fruit derived from a two-or-more-loculed ovary, as in *Peplis*.

caryopsis or grain   A one-seeded dry, indehiscent fruit with the seed coat adnate to the fruit wall, derived from a one-loculed superior ovary.

cypsela   An achene derived from a one-loculed, inferior ovary.

nut   A one-seeded, dry, indehiscent fruit with a hard pericarp, usually derived from a one-loculed ovary.

nutlet   A small nut.

samara   A winged, dry fruit.

utricle   A small, bladdery or inflated, one-seeded, dry fruit.

bb. Dry dehiscent fruit types   **(Figure D.9)**

(Fruits that split open at maturity)

capsule   Dry, dehiscent fruit derived from a compound ovary of two or more carpels.

diplotegium   A pyxis derived from an inferior ovary.

follicle   A dry, dehiscent fruit derived from one carpel that splits along one suture.

legume   A usually dry, dehiscent fruit derived from one carpel that splits along two sutures.

loment   A legume that separates transversely between seed sections.

silicle   A dry, dehiscent fruit derived from two or more carpels that dehisce along two sutures and that has a persistent partition after dehiscence and is as broad as, or broader, than long.

silique   A silicle-type fruit that is longer than broad.

cc. Capsule types   **(Figure D.9)**

(Classification based on type of dehiscence)

acrocidal capsule   One that dehisces through terminal slits or fissures, as in *Staphylea*.

anomalicidal or rupturing capsule   One that dehisces irregularly, as in *Ammannia*.

basicidal capsule   One that dehisces through basal slits or fissures, as in some species of *Aristolochia*.

circumscissle capsule or pyxis   One that dehisces circumferentially, as in *Plantago*.

denticidal capsule   One that dehisces apically, leaving a ring of teeth, as in *Cerastium*.

indehiscent capsule   One that does not dehisce at maturity, as in *Peplis*.

loculicidal capsule   One that dehisces longitudinally into the cavity of the locule, as in *Epilobium*.

operculate capsule   One that dehisces through pores, each of which is covered by a flap, cap, or lid, as in *Papaver*.

FRUIT TYPES

DRY INDEHISCENT FRUIT TYPES

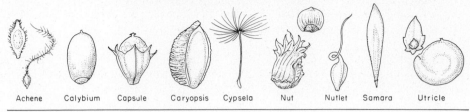

Achene Calybium Capsule Caryopsis Cypsela Nut Nutlet Samara Utricle

DRY DEHISCENT FRUIT TYPES

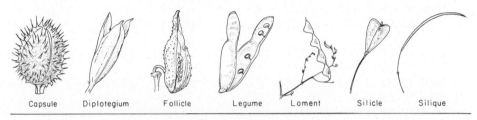

Capsule Diplotegium Follicle Legume Loment Silicle Silique

CAPSULE TYPES

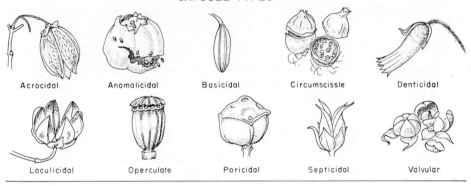

Acrocidal Anomalicidal Basicidal Circumscissle Denticidal

Loculicidal Operculate Poricidal Septicidal Valvular

SCHIZOCARPIC FRUIT TYPES

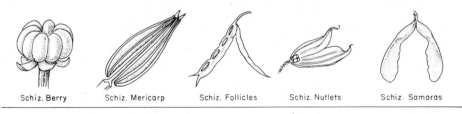

Schiz. Berry Schiz. Mericarp Schiz. Follicles Schiz. Nutlets Schiz. Samaras

FLESHY FRUIT TYPES

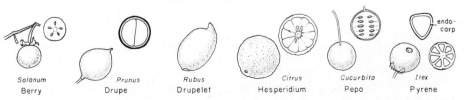

endo-carp

Solanum
Berry

Prunus
Drupe

Rubus
Drupelet

Citrus
Hesperidium

Cucurbita
Pepo

Ilex
Pyrene

Figure D.9

436

436

  poricidal capsule One that dehisces through pores, as in *Triodanis*.
  septicidal capsule One that dehisces longitudinally through the septa, as in *Penstemon*.
  valvular or septifragal capsule One with valves breaking away from the septa, as in *Ipomoea*.

dd. Schizocarpic fruit types   **(Figure D.9)**

  (Fruits derived from a simple, two or more-locular compound ovary in which the locules separate at
  fruit maturity, simulating fruits derived from the ovaries of simple pistils)

  schizocarpic achenes Separating achenes that are one-seeded, dry, indehiscent fruits with seed
    attached to fruit wall at one point only, derived from a superior ovary, as in *Sidalcea*.
  schizocarpic berries Separating berries that have a fleshy pericarp, as in *Phytolacca*.
  schizocarpic carcerules Separating carcerules that are dry, few-seeded, indehiscent locules, as in
    Althaea.
  schizocarpic follicles Separating follicles that are dry, dehiscent fruits derived from one carpel,
    splitting along one suture, as in Apocynaceae.
  schizocarpic mericarps Separating mericarps that are dry, seedlike fruits derived from an inferior
    ovary, as in the Apiaceae.
  schizocarpic nutlets Separating nutlets that are dry, indehiscent four-parted fruits with a hard
    pericarp around a gynobasic style, as in the Boraginaceae and Lamiaceae.
  schizocarpic samaras Separating samaras that are winged, dry fruits, as in *Acer*.

ee. Fleshy fruit types   **(Figure D.9)**

  amphisarca A berrylike succulent fruit with a crustaceous or woody rind, as in *Lagenaria*.
  berry Fleshy fruit with succulent pericarp, as in *Vitis*.
  drupe A fleshy fruit with a stony endocarp, as in *Prunus*.
  drupelet A small drupe, as in *Rubus*.
  hesperidium A thick-skinned septate berry with the bulk of the fruit derived from glandular hairs,
    as in *Citrus*.
  pepo A berry with a leathery nonseptate rind derived from an inferior ovary, as in *Cucurbita*.
  pyrene Fleshy fruit with each seed surrounded by a bony endocarp, as in *Ilex*.

2. Aggregate fruit types   **(Figure D.10)**

  (A group of separate fruits developed from carpels of one flower)

  achenecetum An aggregation of achenes, as in *Ranunculus*.
  baccacetum An aggregation of berries, as in *Actaea*.
  drupecetum An aggregation of drupelets, as in *Rubus*.
  follicetum An aggregation of follicles, as in *Caltha*.
  samaracetum An aggregation of samaras, as in *Liriodendron*.

3. Multiple fruit types

  (Fruits on a common axis that are usually coalesced and derived from the ovaries of several flowers)

  bibacca A fused double berry, as in *Lonicera*.
  sorosis Fruits on a common axis that are usually coalesced and derived from the ovaries of several
    flowers, as in *Morus*.

FRUIT TYPES, Cont'd., SEED PARTS and EMBRYO TYPES

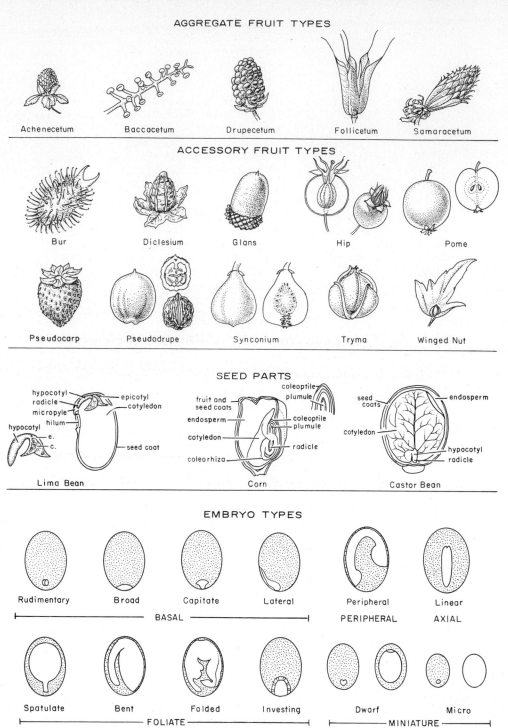

AGGREGATE FRUIT TYPES

Achenecetum Baccacetum Drupecetum Follicetum Samaracetum

ACCESSORY FRUIT TYPES

Bur Diclesium Glans Hip Pome

Pseudocarp Pseudodrupe Synconium Tryma Winged Nut

SEED PARTS

hypocotyl
radicle
micropyle
epicotyl
cotyledon
hilum
hypocotyl
e.
c.
seed coat

Lima Bean

fruit and
seed coats
endosperm
cotyledon
coleorhiza
coleoptile
plumule
coleoptile
plumule
radicle

Corn

seed
coats
cotyledon
endosperm
hypocotyl
radicle

Castor Bean

EMBRYO TYPES

Rudimentary Broad Capitate Lateral Peripheral Linear

BASAL PERIPHERAL AXIAL

Spatulate Bent Folded Investing Dwarf Micro

FOLIATE MINIATURE

Figure D.10

438

syconium   A syncarp with the achenes borne on the inside of a hollowed-out receptacle or peduncle, as in *Ficus*.

4.   Accessory fruit types   **(Figure D.10)**

(Fruits derived from simple or compound ovaries and some nonovarian tissues, as the hypanthium; classification arranged alphabetically; types of accessory structures given in parentheses below.)

bur (involucre)   Cypsela enclosed in dry involucre, as in *Xanthium*.

coenocarpium (various structures)   Multiple fruit derived from ovaries, floral parts, and receptacles of many coalesced flowers, as in *Ananas*.

diclesium (calyx)   Achene or nut surrounded by a persistent calyx, as in *Mirabalis*.

glans (involucre)   Nut subtended by a cupulate, dry involucre, as in *Quercus*.

hip (receptacle and hypanthium)   An aggregation of achenes surrounded by an urceolate receptacle and hypanthium, as in *Rosa*.

pome (receptacle and hypanthium)   A berrylike fruit, adnate to a fleshy receptacle, with cartilaginous endocarp, as in *Malus*.

pseudocarp (receptacle)   An aggregation of achenes embedded in a fleshy receptacle, as in *Fragaria*.

pseudodrupe (involucre)   Two- to four-loculed nut surrounded by a fleshy involucre, as in *Juglans*.

synconium (receptacle, possibly peduncle)   Multiple fruit surrounded by a hollow, compound, fleshy receptacle, as in *Ficus*.

tryma (involucre)   Two- to four-loculed nut surrounded by a dehiscent involucre at maturity, as in most species of *Carya*.

winged nut (bract)   Nut enclosed in a winglike bract, as in *Carpinus*.

XII.   Seeds and Seedlings

a.   Seed Parts   **(Figure D.10)**

aril   Outgrowth of funiculus, raphe, or integuments or fleshy integuments or seed coat, a *sarcotesta*.

chalaza   Funicular end of seed body.

embryo   Young sporophyte consisting of epicotyl, hypocotyl, radicle, and one or more cotyledons.

endosperm   Food reserve tissue in seed derived from fertilized polar nuclei, or food reserve derived from megametophyte in gymnosperms.

hilum   Funicular scar on seed coat.

micropyle   Hole through seed coat.

perisperm   Food reserve in seed derived from diploid nucellus or integuments.

raphe   Ridge on seed coat formed from adnate funiculus.

seed coat   Outer protective covering of seed.

b.   Seed Types

(Classification based on type of nourishing tissue)

cotylespermous   With food reserve in cotyledon, derived from zygote.

endospermous or albuminous   With food reserve in endosperm or albumen, derived from fertilized polar nuclei.

> **hypocotylespermous or macropodial** With food reserve stored in hypocotyl, derived from zygote.
>
> **perispermous** With food reserve in perisperm, derived from diploid nucellus or integuments.

c. Embryo Parts (Figure D.10)

> **coleoptile** Protective sheath around epicotyl in grasses.
> **coleorhiza** Protective sheath around radicle in grasses.
> **cotyledon** Embryonic leaf or leaves in seed.
> **epicotyl** Apical end of embryo axis that gives rise to shoot system.
> **hypocotyl** Embryonic stem in seed, located below cotyledons.
> **plumule** Embryonic leaves in seed derived from epicotyl.
> **radicle** Basal end of embryo axis that gives rise to root system.

d. Embryo Types (Figure D.10)

(Classification based primarily on embryo shape, size, and position—adapted from Martin [1946])

> **bent** Foliate embryo with expanded and usually thick cotyledons in an axile position bent on the hypocotyl in a jacknife position.
> **broad** Basal, globular, or lenticular embryo in copious endosperm.
> **capitate** More or less basal, headlike, or turbinate embryo in copious endosperm.
> **dwarf** Axial embryo variable in size relative to seed, small to nearly total size of seed; seeds 0.2–2 mm long.
> **folded** Foliate embryo with cotyledons usually thin and extensively expanded and folded in various ways.
> **investing** Axial embryo usually erect with thick cotyledons overlapping and encasing the somewhat dwarfed hypocotyl; endosperm wanting or limited.
> **lateral** Basal or basolateral embryo, discoid, or lenticular, usually surrounded by copious endosperm.
> **linear** Axial embryo several times longer than broad, straight, curved or coiled; cotyledons not expanded; endosperm present or absent.
> **micro** Axial embryo in minute seeds, less than 0.2 mm long; minute and undifferentiated to almost total size of seed.
> **peripheral** Peripheral embryo, large and elongate, arcuate, annular, spirolobal, or straight; cotyledons narrow or expanded; perisperm central or lateral.
> **rudimentary** Basal, small, nonperipheral embryo in small to large seed; relatively undifferentiated; endosperm copious.
> **spatulate** Foliate, erect embryo with variable cotyledons, thin to thick and slightly expanded to broad.

e. Aril Structural Types and Special Seed Surface Features

1. *Aril Structural Types*

> **arillate** General term for an outgrowth from the funiculus, seed coat, or chalaza, or a fleshy seed coat.
> **carunculate** With an excrescent outgrowth from integuments near the hilum, as in *Euphorbia*.
> **fibrous** With stringy or cordlike seed coat, as mace in *Myristica*.
> **funicular** With a persistent elongate funiculus attached to seed coat, as in *Magnolia*.

       sarcous With a fleshy seed coat.
       strophiolate With elongate aril or strophiole in the hilum region.

2. Special seed surface features

       (See C. VII. Surface in Chapter 5)

       alate Winged.
       circumalate Winged circumferentially.
       comose With a tuft of trichomes.
       coronate With a crown.
       crested With elevated ridge or ridges; raphal.
       umbonate With a distinct projection usually from the side.
       verrucose Warty.

f. Seedling Parts

       (Specialized parts only—adapted from Duke [1969])

       cataphyll Rudimentary scale leaf produced by seedling, usually in cryptocotylar species.
       collet External demarcation between hypocotyl and root.
       eophyll Term applied to first few leaves with green, expanded lamina developed by seedlings; transitional type leaves developed before formation of adult leaves.
       metaphyll Adult leaf.

g. Seedling Types

       (Classification based on position of cotyledons in germination)

       cryptocotylar or hypogeous With the cotyledons remaining inside the seed; seed usually remaining below ground.
       phanerocotylar or epigeous With the cotyledons emergent from seed, usually appearing above ground.

B. SEXUAL PHENOMENA

I. Sex

a. Flower Sex   **(Figure D.11)**

       (Classification based on sexual condition of the individual flower)

       imperfect or unisexual With stamens or carpels absent in the flower.
       neuter or agamous Without stamens and carpels in flower, or sex organs abortive.
       perfect or bisexual With both stamens and carpels or pistils in the flower.
       pistillate, carpellate, or female With pistils or carpels only in the flower.
       staminate or male With stamens only in the flower.

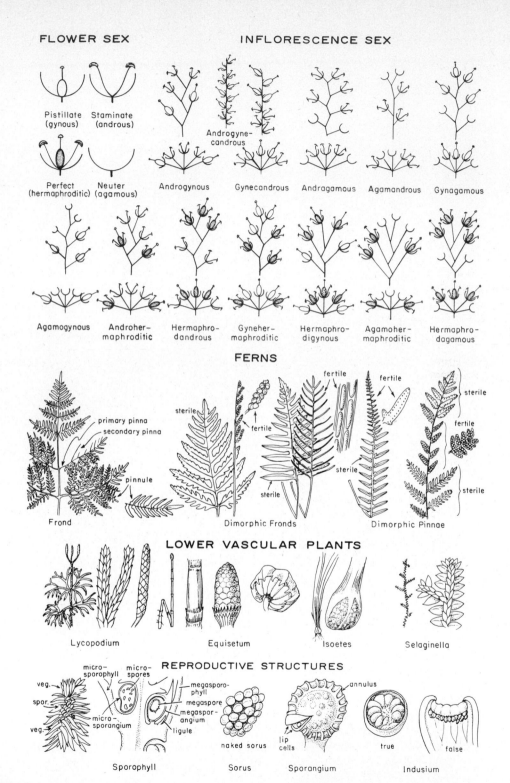

FLOWER SEX

Pistillate (gynous) Staminate (androus)

Perfect (hermaphroditic) Neuter (agamous)

INFLORESCENCE SEX

Androgynecandrous

Androgynous Gynecandrous Andragamous Agamandrous Gynagamous

Agamogynous Androhermaphroditic Hermaphrodandrous Gynehermaphroditic Hermaphrodigynous Agamohermaphroditic Hermaphrodagamous

FERNS

primary pinna
secondary pinna

pinnule

Frond

sterile fertile

Dimorphic Fronds

fertile fertile

sterile

Dimorphic Pinnae

sterile

fertile

sterile

LOWER VASCULAR PLANTS

Lycopodium Equisetum Isoetes Selaginella

REPRODUCTIVE STRUCTURES

veg.
spor.
veg.

microsporophyll
microspores
microsporangium

megasporophyll
megaspore
megasporangium
ligule

annulus

lip cells

Sporophyll naked sorus
Sorus Sporangium true false
Indusium

Figure D.11

442

b. Inflorescence Sex (Figure D.11)

(Classification based on sexual condition of individual flowers in different parts of inflorescence; arrangement of flowers by sexual condition in an inflorescence)

Note: If plant sex is staminate or pistillate or hermaphroditic or neuter, then inflorescence sex would be respectively the same. The terms in parentheses below are examples of the sexual condition of one inflorescence type, the head, as in composites.

agamandrous Inflorescence with neuter flowers inside or above and staminate outside or below (agamandrocephalous).

agamogynous Inflorescence with neuter flowers inside or above and pistillate outside or below (agamogynecephalous).

agamohermaphroditic Inflorescence with neuter flowers inside or above and hermaphroditic outside or below (agamohermaphrodicephalous).

andragamous Inflorescence with staminate flowers inside or above and neuter flowers outside or below (andragamocephalous).

androgynecandrous Inflorescence with staminate flowers above and below pistillate, as in the spikes of some species of *Carex*.

androgynous Inflorescence with staminate flowers inside or above and pistillate outside or below (androgynecephalous).

androhermaphroditic Inflorescence with staminate flowers inside or above and hermaphroditic outside or below (androhermaphrodicephalous).

gynagamous Inflorescence with pistillate flowers inside or above and neuter flowers outside or below (gynagamocephalous).

gynecandrous Inflorescence with pistillate flowers inside or above and staminate outside or below, as in spikes of some species of *Carex*.

gynehermaphroditic Inflorescence with pistillate flowers inside or above and hermaphroditic outside or below (gynehermaphrodicephalous).

hermaphrodagamous Inflorescence with hermaphroditic flowers inside or above and neuter outside or below (hermaphrodagamocephalous).

hermaphrodandrous Inflorescence with hermaphroditic flowers inside or above and staminate outside or below (hermaphrodandrocephalous).

hermaphrodigynous Inflorescence with hermaphroditic flowers inside or above and pistillate outside or below (hermaphrodigynecephalous).

c. Plant Sex

(Classification based on sexual condition of all flowers on the plant(s) within the species)

androdioecious Some plants with staminate flowers and some with perfect flowers.

andromonoecious Plant with staminate and perfect flowers.

diclinous Plant with imperfect flowers; stamens and carpels in separate flowers.

dioecious Plant with all flowers imperfect, but staminate and pistillate on separate plants.

female, carpellate, or pistillate Plant with pistillate flowers only.

gynodioecious Some plants with perfect flowers and some with pistillate.

gynomonoecious Plant with pistillate and perfect flowers.

hermaphroditic or monoclinous Plant with all flowers perfect.

male or staminate Plant with staminate flowers only.

monoecioius Plant with all flowers imperfect, but staminate and pistillate flowers on same plant.

polygamous   Plant with perfect and imperfect flowers.

polygamodioecious   Plants dioecioius, but with some perfect flowers on staminate or pistillate plants or both.

polygamomonoecious   Plant monoecious, but with some perfect flowers.

trioecious   Plants staminate, pistillate or perfect.

d.   General Sexual Terms

agamous or neuter   Without sex; sexual organs abortive.

heterocephalous, heterocymous, or heterospicous   Heads, cymes, spikes with flowers of different sexual conditions. Note: other inflorescence word stems could be used for appropriate inflorescence type.

heterosexual   Inflorescences or flowers within the plant with different sexual conditions.

homocephalous, homocymous, or homospicous   Heads, cymes, spikes with flowers sexually uniform.

homosexual   Inflorescences or flowers sexually uniform.

C.   DEVELOPMENT AND PATTERNS

I.   Growth and Development

a.   Growth Regions

adaxial or abaxial   Growth region localized on top or bottom of a leaf part or any dorsiventral structure.

apical or terminal   Growth region at the apex of the structure.

basal   Growth region at the base of a blade, as in grasses.

intercalary   Growth region near the base of an internode or base of blade.

interstitial   Growth all over in an organ, no localized meristems, as in some fruits.

marginal   Growth region near the edge of a leaf.

b.   Growth Periodicity

accrescent   Growing after flowering or bud development has occurred, as the sepals in *Hypericum* and bud scales in *Carya*.

determinate   Growth of plant parts, the size of which is limited by cessation of meristematic activity during the year.

indeterminate or evergrowing   Continual growth of plant parts, not limited by a cessation of meristematic activity.

intermittent   A renewal and cessation of meristematic activity that produces clusters of stems and/or leaves along an axis.

c.   General Development

acropetal   Developing upward, toward apex.

basipetal   Developing downward, toward base.

centripetal   Developing from the outside inward, or from bottom upward.

centrifugal   Developing from the inside outward, or from top downward.

determinate   Central flower develops first, arresting development of primary axis.

indeterminate   Lateral flowers develop first with primary axis continuing to elongate or develop.

d.   Developmental Shoot Types

(Classification based on origin and development of shoots)

dwarf shoots or spurs   Shoots that develop from preformed buds that have very short internodal lengths or intervals.

epicormic shoots or water sprouts   Shoots that develop from dormant lateral buds on the trunk that have very long and frequently variable internodal lengths or intervals.

flushing shoots   Shoots that develop from mature terminal buds several times during a season. Terminal bud will develop shoot with new terminal bud, which will develop more shoots and a terminal bud, which will develop more shoots and a terminal bud—several times in a season with several flushes of growth.

heterophyllous shoots   Shoots that develop from winter buds that do not contain the primordia of all the leaves to develop during the year.

lammas shoots   Abnormal late season shoots that develop from the terminal bud, not a recurring phenomenon as in flushing shoots.

long bud shoots   Abnormal buds or shoots that elongate, then have arrested growth without the development of leaves and lateral branches.

long shoots   Normal shoots that develop from terminal or axillary buds that have normal internodal lengths or intervals.

preformed shoots   Normal shoots that develop from winter buds that contain primordia of all leaves that will expand during the season.

proleptic shoots   Abnormal late-season shoots that develop from the lateral buds immediately beneath the terminal.

sucker shoots   Shoots that develop from adventitious buds on old stumps or roots, usually after cutting or injury, that have elongate internodal lengths and intervals.

syleptic shoots   Abnormal shoots that develop from lateral buds before they have reached maturity.

II.   Patterns

a.   Symmetry

actinomorphic or radial   With floral parts that radiate from center like spokes on wheel.

asymmetric   Without regularity in any dimension.

dorsiventral   Planate and having distinct dorsal and ventral surfaces, the two usually different.

equilateral   With halves or sides equal in shape and size.

inequilateral   With halves or sides unequal in shape and size.

irregular   With floral parts within a whorl dissimilar in shape and/or size.

regular   With floral parts within a whorl similar in shape and size.

spherical   With multidimensional radial symmetry.

zygomorphic or bilateral   With floral parts in two symmetrical halves.

b. Arrangement Systems

(Based on Leppik [1961])

anthotaxis Arrangement of sporophylls, primarily reproductive in function.
carpotaxis Arrangement of fruits, reproductive in function.
phyllotaxis Arrangement of leaves, primarily photosynthetic in function.
rhizotaxis Arrangement of roots.
semataxis Arrangement of semaphylls (petals, sepals, tepals), primarily advertising (pollinator attracting) in function.
Fibonaci phyllotaxis A fundamental type of leaf arrangement expressed as a fraction in which each succeeding fraction is the sum of the two previous numerators and the sum of the two previous denominators, that is, $\frac{1}{2}, \frac{1}{3}, \frac{2}{5}, \frac{3}{8}, \frac{5}{13}, \frac{8}{21}$, etc. The numerator represents the number of turns or spirals around a stem before one leaf is directly above another and the denominator represents the number of leaves in the turns or spirals before one is directly above the other; $\frac{2}{5}$ phyllotaxy would mean two twists and five leaves before one leaf is directly above the other or an angle of divergence of 144° between succeeding leaves in the stem ($\frac{2}{5}$ of 360°). According to Leppik, anthotaxis and semataxis do not necessarily follow the same pattern, with anthotaxis in *Michelia* cited as being in $\frac{2}{7}, \frac{3}{7}, \frac{3}{8}$, and $\frac{4}{10}$ systems of arrangement.

c. Branching Patterns   **(Figure D.12)**

(Adapted from Halle [1971])

dichotomous Branching into two equal parts.
flabellate Fan-shaped branching.
monopodial Branching with a main axis and reduced or missing laterals; *excurrent*.
>   a. Without lateral branches. Main axis ending in a cluster of large leaves, for example, *Carica papaya*.
>   b. With continuous lateral branches. Main axis beset with laterals, for example, *Trema lamarckiana*.
>   c. With continuous but self-pruning lateral branches *(excurrent)*. Main axis with upper laterals present and lower have fallen off, for example, *Pinus*.
>   d. With short-lived laterals. Main axis beset with phyllomorphic branches, appearing as compound leaves that may be deciduous, for example, some tropical Rubiaceae, *Pycnanthus*.
>   e. With spiral or whorled main laterals on main axis and distichous branches, for example, *Pycnanthus*.
>   f. With spiral or whorled main laterals and dichotomous branches, for example, *Phyllanthus*.
>   g. With tiered or clustered lateral branches. Main axis with periodic lateral branch production, for example, *Bombax*.

sympodial Branching without a main axis but with many more- or less-equal laterals.
>   a. Leader displacement. Successive laterals take over leader growth, producing umbrella-shaped trees, for example, *Acacia Melia;* deliquescent.
>   b. Leader displacement by two equal laterals (dichotomous substitution). Production of a forked branching system, for example, *Manihot*.
>   c. Leader displacement as a short root *(apposition)*. Usually a single lateral becomes dominant, for example, *Terminalia*.

Note: Other types of stem growth and branching patterns occur, particularly in tropical woody plants, but no adequate classification has been devised.

OVULE POSITION

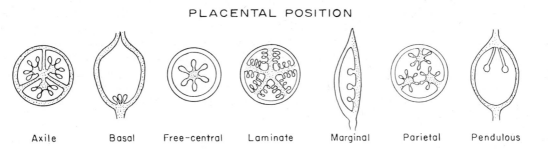

Epitropus dorsal   Epitropus ventral   Hypotropous dorsal   Hypotropous ventral   Pleurotropous dorsal   Pleurotropous ventral

PLACENTAL POSITION

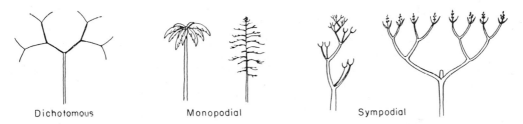

Axile   Basal   Free-central   Laminate   Marginal   Parietal   Pendulous

BRANCHING PATTERNS

Dichotomous   Monopodial   Sympodial

Figure D.12

D. GYMNOSPERM GLOSSARY*

I. Vegetative and Reproductive Structures

a. Vegetative Structures

acicular leaf   Needlelike, long and slender, for example, *Pinus*.

awl-shaped leaf   Subulate; narrow, flat, stiff, sharp-pointed, usually less than $\frac{1}{2}$ inch long, for example, *Juniperus*.

fascicle   Cluster of needles borne on a minute, determinate short shoot in the axil of a primary leaf (bract), for example, *Pinus*.

fascicle sheath   Closely imbricated bud scales at the base of the fascicle of needles, for example, *Pinus*.

*Adapted from Hardin, J. W., in Radford, A. E. et al., 1974, *Vascular Plant Systematics,* Chapter 6, pp. 151–152.

linear leaf   Narrow, flattened, triangular, or quadrangular leaf usually $\frac{1}{2}$ to 2 inches long, for example, *Taxus, Picea*.

long shoot   Elongated internodes, rapid annual growth.

multinodal shoot   Spring shoot developing from the terminal winter bud and producing two or more whorls of branches; the cones are partly lateral in the middle of the shoot, for example, *Pinus echinata*.

needle   Acicular; slender, elongated leaf, usually over 2 inches long, for example, *Pinus*.

peg (sterigmata)   Lateral stem projection to which leaf is attached and persistent after leaf dehiscence, that is, there is an abscission layer between peg and leaf. Leaf may be sessile, for example, *Picea*, or petiolate, for example, *Tsuga*, on the peg.

scale leaf   Small, usually appressed and imbricate, for example, *Juniperus, Thuja*.

short shoot   Very short or inconspicuous internodes with growth very slow if at all.

> **determinate**   No continued growth; shoot with leaves drops as a unit, for example, *Pinus fascicle*.

> **indeterminate**   Continued growth by an apical meristem; spur branch, for example, *Larix, Cedrus, Ginkgo*.

uninodal shoot   Spring shoot developing from the terminal winter bud and producing only one internode with one whorl of branches at the end; the cones are subterminal at the end of the shoot, for example, *Pinus resinosa*.

b.   Reproductive Structures

apophysis   Exposed outer surface of either an ovuliferous scale or megasporophyll as seen when the cone is closed.

aril   An outgrowth from the stem forming a fleshy covering of the seed, for example, *Taxus, Torreya;* or rudimentary at base of the fleshy seed, for example, *Cephalotaxus*.

bract   Modified leaf subtending the ovuliferous scale; may be distinct or fused to the scale.

cone (strobilus)   Aggregation of sporangia-bearing structures at tip of the stem (either sporophylls or scales in the gymnosperms).

> **female cone (megasporangiate strobilus)**   Bearing ovules or seeds.

> **male cone (microsporangiate strobilus)**   Bearing pollen sacs (microsporangia).

epimatium   Fleshy covering of the seed and more or less fused with the integument; arising from the chalazal end of the ovule like an additional integument, for example, *Podocarpus*.

megasporophyll   Modified leaf-bearing ovules, for example, *Zamia*.

microsporophyll   Modified leaf-bearing microsporangia or pollen sacs.

ovuliferous scale   Highly modified lateral branch in the axil of a leaf (bract), and bearing ovules. May be flat or peltate, woody or fleshy, for example, Pinaceae.

receptaculum   A fleshy structure below the seed formed from the bases of bracts and the swollen receptacle or cone axis, for example, *Acmopyle*, and *Podocarpus* spp.

umbo   Projection, with or without spine or prickle, on the apophysis of the cone scale.

E.   LOWER VASCULAR PLANT GLOSSARY*   (Figure D.11)

I.   Sporophyte

a.   Stems

1.   *Branching*

axil   With branches arising from buds in leaf axil.

*Adapted from Mickel, J. T., in Radford, A. E. et al., 1974, *Vascular Plant Systematics*, Chapter 6, pp. 153–156.

dichotomous   With branches forking into two more or less equal parts.
epipetiolar   With branches arising from buds on the petiole.
monopodial   Having one main axis of growth.
random   With branches arising from buds without relation to leaves.
sympodial   With branches more or less equal without a main axis.

2. *Phyllotaxis*

distichous   With leaves in two rows.
polystichous   With leaves in several rows.

3. *External features*

aerial stem   An erect stem arising from a horizontal rhizome.
articulate   Generally meaning having a joint as in leaves, leaflets, or stems, as in herterophyllous
    species of *Selaginella,* or having a swollen area, often discolored, at the point of branching of
    the stem.
jointed   With stems that can be pulled apart easily at the nodes, as in *Equisetum.*
node   Point on the stem where leaves are attached; or the point of branching of the stem.
rameal sheath   Leaf sheath on the stem joints, as in *Equisetum.*
tubercules   Silica deposits on the stem ridges, as in *Equisetum.*

4. *Internal features*

canals   As in *Equisetum.*
    central canal   The large centrally located air space in the stem.
    carinal canal   A canal beneath a stem ridge associated with a vascular bundle.
    vallecular canal   A canal beneath a stem groove.
stele   The central primary vascular system of the stem and associated tissues, delimited from the
    cortex by endodermis and pericycle.
    actinostele   A protostele having a xylem core in the form of radiating ribs, as viewed in
        transverse section.
    dictyostele   A dissected solenostele with each individual bundle a meristele.
    eustele   A dissected siphonostele with phloem only to the outside of the xylem.
    plectostele   A protostele dissected into anastomosing, platelike units.
    protostele   Stele having a solid column of vascular tissue with xylem centrally located.
    siphonostele   A stele having vascular tissue in the form of a hollow cylinder, with a central
        pith.
    solenostele   A siphonostele having phloem both internal and external to the xylem.

b.   Leaves

1. *Duration*

evergreen   Bearing green leaves through the winter.
marcescent   The leaves of short duration, dying at the end of the growing season.

2. *Megaphylls*

(Leaves with a branching vein system that are associated with nodal leaf gap)

aa. General

crozier The coiled, developing leaf of a fern.
frond The leaf of a fern.

bb. Parts of a leaf

blade The expanded portion of a leaf.
costa The midvein of a minor division of a fern leaf.
lamina The leaf tissue other than the veins or axes.
pinna A primary division of a fern leaf.
pinnule A secondary division of a fern leaf.
rachis The axis of a compound fern blade.
segment The ultimate division or unit of a dissected fern leaf.
stipe The petiole of a fern leaf.
stipe bundles The vascular bundles of the fern petiole.

cc. Venation patterns

areoles The spaces formed by a vein network.
false veins Small veinlike areas of thick-walled cells in the leaves of some lower vascular plants.
free No veins uniting to form a network.
included veinlets Veins ending inside areoles.
net Veins uniting to form a network.

dd. Leaf division

bipinnate Twice pinnate.
pectinate Pinnatifid with closely set segments; comblike.
pinnate Compound, with the leaflets arranged on both sides of a common axis.
pinnate-pinnatifid Pinnate with pinnatifid pinnae.
pinnatifid Pinnately cut, more than halfway to the midvein.
simple Undivided.

ee. Blade architecture

anadromous Having the first lobe or segment of a pinna arising basiscopically in compound leaves.
catadromous Having the first lobe or segment of a pinna arising acroscopically in compound leaves.
imparipinnate Pinnate with a conform terminal leaflet.
palmate Radiately lobed or divided.
pedate Palmately cut or divided, with the lower pair basiscopically exaggerated.
pinnate Compound, with the leaflets arranged on both sides of a common axis.

ff. Dimorphism

(With leaves of two types in ferns, the fertile being of different size, shape, or dissection than the vegetative)

complete dimorphism Leaves completely fertile or vegetative.
partial dimorphism Leaves with only a part modified as fertile.

slight dimorphism   Fertile and vegetative leaves only slightly different appearance.
vegetative frond   Frond lacking sporangia.

3. *Microphylls*

(Leaves with only one vein and no leaf gap)

axillary leaves   Leaves borne in the axils of branches, as in heterosporous species of *Selaginella*.
bast bundles (peripheral strands)   Bundles of thick-walled cells parallel to the midrib, as in *Isoetes*.
lacuna   Chamber or internal air space.
lateral leaf   Leaf on the side of the stem, as in heterophyllous species of *Selaginella*.
ligule   A small, membranous outgrowth or projection at the base of the leaf, appearing above the sporangium in fertile leaves, as in *Selaginella* and *Isoetes*.
median leaf   Leaf on top of stem, as in heterophyllous species of *Selaginella*.
seta   A hairlike extension of the leaf, as in homophyllous species of *Selaginella*.

c. Reproductive Structures

annulus   Thick-walled ring of cells on the sporangium.
elater   One of four elongate appendages on the spores, as in *Equisetum*.
eusporangiate   In which the sporangium develops from a great amount of leaf tissue as opposed to only one or a few cells.
exospore or exine   Outer spore wall layer.
fovea   Pit or depression containing the sporangium in the leaf base of *Isoetes*.
heterosporous   Having two kinds of spores, usually differing in size.
homosporous   Having spores of only one kind.
indusium   A flap of tissue covering a sorus.
    false indusium   A folded leaf margin protecting the sorus.
    inferior indusium   An indusium attached beneath the sorus with the sporangia appearing above it.
    true Indusium   An epidermal outgrowth protecting the sorus.
leptosporangiate   Having the entire sporangium develop from a periclinal division of a superficial cell or small group of cells.
lip cells   The line of cells between which the sporangium dehisces.
massula   A clump of microspores, as in *Azolla*.
megasporangium   The sporangium in which megaspores are produced.
megaspore   A spore that gives rise to a female gametophyte.
megasporophyll   A leaf bearing one or more megasporangia.
microsporangium   The sporangium in which microspores are produced.
microspore   A spore that gives rise to a male gametophyte.
microsporophyll   A leaf bearing one or more microsporangia.
paraphyses   Hairs or hairlike structures in the sorus.
perispore or perine   An outer covering of some fern spores, with a different configuration than that of the exospore.
receptacle   Point on a leaf where sporangia are attached.
sorus   A cluster of sporangia.
sporangiophore   The umbrella-shaped, sporangium-bearing unit of the strobilus, as in *Equisetum*.
sporangium   A spore case.

spore shapes
    monolete   Bean-shaped, with a single scar line.
    trilete   Basically tetrahedral, but often appearing round or triangular, with three scar lines forming a Y.
sporocarp   A hard, nutlike structure containing the sporangia in heterosporous ferns.
sporophyll   A leaf-bearing sporangia.
stomium   Lip cell region of a fern sporangium.
strobilus   Stem with short internodes and spore-bearing appendages; a cone.
velum   The membranous flap covering the sporangium, as in *Isoetes*.

II. Gametophyte

antheridium   The male sex organ producing the sperm.
archegonium   The female sex organ producing the egg.
prothallus   Gametophyte of lower vascular plants.
rhizoid   A hairlike absorptive organ on gametophytes and rarely on sporophytes.

III. Asexual Reproduction

apogamy   Producing sporophytes from a gametophyte without fertilization.
apospory   Producing gametophytes directly from a sporophyte without producing spores.
bulblet   A small, budlike vegetative propagule produced on the leaves of some ferns.
gemma   A vegetative reproductive bud borne on the stem, as in *Lycopodium;* a multicellular reproductive propagule on gametophytes, as in ferns.

SUGGESTED READING

Bocquet, G. 1959. The campylotropous ovule. Phytomorphology 9:222–227.

Bjornstad, I. N. 1970. Comparative embryology of Asparagoideae-Polygonateae, Liliaceae. Nytt Magasin for Botanikk 17(3-4):169–207.

Correll, D. S., and M. C. Johnston. 1971. Manual of the Vascular Plants of Texas. Texas Research Foundation, Renner.

Cronquist. A. 1968. The Evolution and Classification of Flowering Plants. Houghton-Mifflin Company, Boston.

Davis, P. H. and V. H. Heywood. 1963. Principles of Angiosperm Taxonomy. D. Van Nostrand, Princeton.

Duke, J. A. 1969. On tropical tree seedlings 1. Seeds, seedlings, systems, and systematics. Annals Missouri Botanical Garden 56(2):125–161.

Faegri, K. and L. van der Pijl. 1966. The Principles of Pollination Ecology. Pergamon Press, New York.

Featherly, H. I. 1954. Taxonomic Terminology of Higher Plants. Iowa State College Press, Ames.

Fernald, M. L. 1950. Gray's Manual of Botany, eighth edition. American Book Company, New York.

Gray, P. 1967. The Dictionary of Biological Sciences. Reinhold Publishing Corporation, New York.

Halle, F. 1971. Architecture and growth of tropical trees exemplified by the Euphorbiaceae. Biotropica 3(1):56–62.

Hickey, L. J. 1973. Classification of the architecture of dicotyledonous leaves. American Journal of Botany 60(1):17–33.

Jackson, B. D. 1928. A Glossary of Botanic Terms, fourth edition, J. B. Lippincott Company, Philadelphia.

Johnson, A. M. 1931. Taxonomy of the Flowering Plants. The Century Company, New York.

Lawrence, G. H. M. 1951. Taxonomy of Vascular Plants. The Macmillan Company, New York.

Leppik, E. E. 1957. A new system for classification of flower types. Taxon 6:64–67.

———. 1961. Phyllotaxis, anthotaxis, and semataxis. Acta Biotheoretica 14:1–28.

Lindley, J. 1848. An Introduction to Botany, fourth edition. London.

Martin, A. C. 1946. The comparative internal morphology of seeds. American Midland Naturalist 36:513–660.

Porter, D. M., R. W. Kiger, and J. E. Monahan. 1973. A guide for contributors to Flora North America, Part II: An outline and glossary of terms for morphological and habitat description (provisional edition). FNA Report 66. Department of Botany, Smithsonian Institution, Washington, D.C.

Radford, A. E. et al. 1974. Vascular Plant Systematics. Harper & Row, New York.

Rickett, H. W. 1944. The classification of inflorescences. Botanical Review 10:187–231.

Stearn, W. T. 1966. Botanical Latin. Thomas Nelson and Sons Ltd., London.

Swartz, D. 1971. Collegiate Dictionary of Botany. The Ronald Press Company, New York.

Systematics Association Committee for Descriptive Terminology. 1960. I. Preliminary list of works relevant to Descriptive Biological Terminology. Taxon IX(8):245–257.

appendix *E*

Collection and Field Preparation of Plant Specimens*

Herbarium specimens are permanent records of a species (or population) as it occurred at a given time and place. The future value and use of any specimen largely depends on the care with which the collector selects, collects, and prepares the specimens. The following directions and suggestions on specimen preparation, field equipment, and field records are given to assist collectors in preparing high-quality herbarium specimens accompanied by adequate field notes.

A. COLLECTING OBJECTIVES AND PLANNING

The specimens to be collected depend on the objectives of the collector. The types of material to be collected will determine the techniques used, amount of material collected, and type of field data recorded. The following suggestions can greatly aid the collector in getting maximum benefit from valuable field time.

1. Outline objectives for a specific expedition.
2. Prepare a list of all equipment to be utilized on field expedition.
3. Obtain localities from herbarium specimens if specific materials are needed. It is best, when possible, to have several localities in the event some areas have been disturbed. Check flowering and fruiting dates on the specimens in the herbarium.
4. Check local weather conditions and seasonal progress in the areas to be visited.
5. Collect duplicate specimens (replicates) except in the case of very rare or protected plants (check state laws and conservation lists). Other plants to be avoided

*Adapted from Massey, J. R., in Radford, A. E., et al., 1974, *Vascular Plant Systematics*, Harper & Row, New York, Chapter 18, pp. 387–398.

include those noxious weeds or parasites and their host plants under state or federal quarantine. Duplicates provide sufficient material for dissection or, if needed, verification by a specialist.

6. Prepare voucher specimens (typical herbarium specimens) when collecting materials for cytological, anatomical, and other studies.

7. Obtain collecting permits and other necessary permission in advance. If a trip is planned to a specific area such as a national park, foreign country, or other area, it is wise to obtain permits prior to the trip and it is also advisable to obtain permission for collecting on private lands whenever possible.

8. Collect and document as many different species as possible in floristic studies. Do not overlook the inconspicuous (so-called belly plants) or those plants that are difficult to collect or identify. Collect thoroughly in each habitat.

B. SUPPLIES AND EQUIPMENT

1. Field press. A press typically consists of two hardwood frames, with each frame made as follows:

   4 wood strips $\frac{3}{4} \times \frac{1}{4} \times 18$ inches
   5 wood strips $\frac{3}{4} \times \frac{1}{4} \times 12$ inches

   The five short strips should be equally spaced on the four larger strips, which are also equally spaced and nailed, or riveted securely at the intersection of the strips. The completed frame should be 12×18 inches. Repeat for the second frame.

   Some collectors prefer the cheaper and often stronger 12×18-inch plywood press boards cut from $\frac{3}{4}$-inch fir plywood. A number of types of presses may be purchased from biological supply houses.

2. Driers (blotters). Excellent driers may be made by cutting sheets 11×16 inches from lightweight builders deadening felt (unsaturated) or from heavy blotting paper. Driers are also available from biological supply houses.

3. Newsprint. Cut paper 22×16 inches and fold to 11×16 inches. Many use newspapers as found on the newsstand but unused newsprint may be purchased in rolls from local printers. Biological supply houses usually offer precut papers.

4. Press straps (webbing straps). A pair of strong web straps (parachute or cinch type) with claw buckles are excellent for field purposes. Sash cord or rope is often also used. The minimum length for press straps is 4 feet.

5. Field notebook. A pocket-size book that will not disintegrate when wet and pencil or pen with waterproof ink are necessary items. Some prefer to keep two books—one taken to the field and another that remains in a safe place and into which field notes are copied. If field labels are used, one is placed in the paper with the specimen and a copy kept by the collector. Field notebooks should be permanent.

6. String tags. Waterproof string tags are useful for labeling plants that are not pressed immediately following collection.

7. Diggers and clippers. Both pruning shears or garden clippers and digging tools are necessary. A trowel (preferably with a steel shank), geologist pick, dandelion digger, or heavy sheath knife are excellent for field use. A small trench shovel and pocket knife are also useful.

8. Hand lens. For field observations and identifications a small $5\times$ or $10\times$ lens is desirable. These are generally available from bookstores and biological supply houses.

9. Collection bottles. Glass or plastic bottles with leakproof screw caps are often desirable for collecting some materials. The size used depends on the materials to be collected. Small vial-type bottles are ideal for collecting floral buds, flowers for clearing, and other materials to be preserved in liquid preservatives.

10. Liquid preservative. The type of solution used will, of course, depend on future use and type of material. For general anatomical purposes and materials such as wood, leaves, flowers, and the like, a mixture of formalin-acetic acid-alcohol (FAA) is widely used and may be prepared in the following manner:

ethyl alcohol (70 percent)	90 cc
formalin (commercial strength)	5 cc
glacial acetic acid	5 cc

    Cytological materials are often fixed in a 6:3:1 mixture of chloroform, 95 percent ethyl alcohol, and glacial acetic acid or in Carnoy's fluid (3:1 absolute ethyl alcohol and glacial acetic acid). Carnoy's containing 95 percent ethyl alcohol is useful for clearing leaves. When using the 6:3:1 mixture, the glacial acetic acid should not be added until materials are ready to be fixed.

    After materials have been in fixative for 24 to 48 hours, the fixative should be discarded and materials stored in 70 percent alcohol. At Kew, a solution of 50 percent alcohol, 5 percent formalin, 5 percent glycerol, and 40 percent water is used for storing plant materials. Each vial should contain a field label. Use pencil and slips of bond paper; the numbers should correspond to those in the field notebooks.

11. Vascula and collecting bags. Plant materials not pressed in the field immediately may be stored in a metal container (vasculum) in folds of wet paper. Due to the cost and bulkiness of this container, many now prefer to use plastic bags (turkey or fertilizer) or rubber-lined canvas bags (military laundry bags). Local conditions will determine in part the type of bag to be used, but avoid exposure to sun, particularly if clear plastic bags are used. All materials should be carefully labeled to avoid confusion when materials are pressed.

12. Waxed paper. A roll of waxed paper, available in supermarkets, is useful in pressing many plants, particularly those that deliquesce or are viscid. A single sheet placed over the material can greatly facilitate the removal of materials from the pressing paper.

13. Manila coin envelopes. Small coin envelopes are excellent for collecting seeds, individual flowers, leaves, pollen, pollinators, and other materials that cannot or should not be pressed. These should always be carefully labeled with the collection number.

14. Trays, cans, jars for living materials. It is often desirable to collect living materials for experimental work in the garden or greenhouse. A variety of containers may be used and will depend on the type of material to be collected and the length of time materials must be stored.

15. Cardboard storage boxes. Merrill cases or other cardboard containers that can be purchased flat make excellent storage for dry materials removed from the press in the field. It is advisable to use an insecticide or repellent if the materials are to be stored for an extended period of time.

16. Insecticides and repellents. Moth crystals or flakes (naphthalene or paradichlorobenzene) may be used as repellents. Considerable quantities of paradichloro-

benzene in an airtight container may be used as an insecticide. An excellent treatment is to place PDB in the folds of the specimen paper, tie specimens into bundles, and seal in a plastic bag.

17. **Maps.** Highway, topographic, and geologic maps are often very useful for obtaining localities for particular species. Detailed county road maps can be purchased, usually at modest cost, from state highway departments.

18. **Camera.** Photographs are of considerable value when collecting woody or other materials where entire plants are not pressed.

19. **Color charts.** Several color charts are available for determination of flower colors in the field, for example, Horticulture Society Color Chart, Nickerson Color Fan, and Horticultural Color Chart.

C. SELECTION OF MATERIAL

Vigorous, typical specimens are to be selected. Avoid insect-damaged plants. Specimens should be representative of the population but should include the range of variation of the plants, not those that best fit the press. Roots, bulbs, and other underground parts should be carefully excavated and the dirt removed with care. In most cases, flowering and/or fruiting materials are necessary for identification purposes. Many collectors prefer to add extra flowers and fruits to their collections when possible to avoid dissection of the specimen proper. Plants too large for a single sheet may be divided and pressed as a series of sheets (see discussion below). In collecting large herbs, shrubs, and trees, different types of foliage, flowers, and fruits should be collected from the same plant. Collect sufficient material to fill an herbarium sheet and still leave enough room for the label. Bark and wood samples are often desirable additions when collecting woody plants. Proper identification of many plants depends on several different characteristics—some roots, others seed or mature fruits, some flower color (which should be noted in the fieldbook).

D. PRESSING PLANT SPECIMENS

I. Arranging and Preparing Specimens

After the specimens have been dug or cut, they should be pressed as soon as possible (see E. Field Storage of Specimens for information on field storage of fresh specimens). The care given a specimen in pressing will largely determine its future value. Specimens should be placed in a single fold of newsprint or other suitable absorbent lightweight paper. Plants too large to fit the 11 × 16-inch fold of paper may be bent into a V, N, or M figure. Bruise the stem before bending and it will be less apt to break. Specimens should not protrude from the fold of paper. Protruding parts will likely have to be removed when specimens are mounted. A specimen may be trimmed to reduce bulk and expose certain characters advantageously if sufficient material (e.g., leaf petioles, branch bases, etc.) is left so that the pattern of branching, leaf arrangement, and other features are readily discernible. When pressing large plants, make several sheets rather than a single sheet with a crumpled mass of material. Arrange specimens in such a way that some upper and some lower surfaces of the leaves are exposed. Spread flowers or inflorescences to show as many surfaces or views as possible. Section some flowers longitudinally and press flat to exhibit the inner parts and thereby reduce the need for dissection of the finished specimens.

Excessively bulky and fleshy parts such as stems and fruits may be split and both parts included. Fruits should be sectioned when possible in such a way that both transverse and longitudinal sections are included. Succulent plants (e.g., cacti) may be split and the fleshy inner parts removed. Some collectors prefer dipping these plants in boiling water, xylol, or benzene before pressing. Salt may be applied to cut surfaces to hasten drying. Each sheet should bear a collector's number that refers to notes in the collector's field notebook. When preparing a series of sheets for parts of one specimen, number each sheet with the collection number and the number of parts (e.g., #429, sheet 1 of 3); duplicates should simply bear the collection number. Do not include more than one species in a single sheet. Collect sufficient material to make a full sheet but avoid crowding or overlapping of specimens in the specimen paper. See Smith (1971) for an excellent illustrated account of specimen preparation.

II. Arranging the Press

Specimens in the specimen paper are placed between two driers in the press. A ventilator is often inserted before the next specimen paper and driers are added. For each specimen pressed, a drier unit (two driers with ventilator between them) is added to the press. The usual sequence is ventilator, drier, specimen in specimen paper, drier, ventilator, drier, and so on. As the successive specimens are added and the press built, every effort should be made to keep the press level for even distribution of pressure when the press straps or weights are applied. To do so, use alternate corners of the sheet for bulky roots or other parts. Sheets or pads of polyurethane sponges are very useful when placed over or around bulky specimens.

E. FIELD STORAGE OF SPECIMENS

Although it is desirable to press collections immediately, it may not always be practical. Delicate materials should be pressed as soon as possible and other specimens properly stored. Vascula, plastic bags, or rubberized bags can be used for storage if specimens are first wrapped in moistened paper. Specimens may be kept in good shape without spoilage in the containers if they are kept moist and are not packed too tightly. Supersaturation with water or drying out will spoil specimens. The specimen bags should be kept as cool as possible and a conscious effort should be made to park a specimen-loaded vehicle in the shade on field trips.

On returning from the field, specimens in bags should be put into a cold room until the plants are pressed. Although immediate pressing is again desirable, plants can be kept in this manner for several days.

Plastic gallon or half-gallon milk cartons or jugs make excellent water containers for field use. It is a good practice to wrap specimens collected at one locality in an uncut, numbered (with a wax pencil before paper is soaked), double sheet of newspaper so that one end is open and the other closed; bind with a rubber band; tag with a locality number that is entered in the field notebook; moisten and place upright in a bag; and then place the bag upright in the field vehicle. Moisten by sprinkling water over the open top of specimen bags as needed. If it is necessary to keep wet materials for long periods of time, as is necessary for collectors in tropical regions, one of several preservatives may be used:

1. 2 parts commercial formaldehyde (40 percent):3 parts water.
2. 1 part formaldehyde:2 parts 70 percent alcohol
3. 40 to 50 percent alcohol
4. 1 to 2 percent aqueous solution of oxyquinoline sulfate

Specimens are dipped, sprayed, or brushed with one of these solutions and enclosed in airtight packages. For additional details consult Lawrence (1951), and Smith (1971).

F.   DRYING FIELD COLLECTIONS

Plants should be placed in the press and the press closed and tightened. The faster the drying process, the less difficulty with mold, mildew, and loss of color. Plants should be sweated in the field press for 12 to 24 hours and the press opened. Any last arrangement of the specimens must be made at this time and the wet driers changed. For exceptional results, driers should be changed at least three times during the first 48 hours. In many areas, blotters may be dried in the sun (usually 1 hour is sufficient). If specimens are to be dried without artificial heat, blotters should be changed daily until specimens are dry. Automobile luggage carriers are excellent means of drying specimens provided ventilators with open ducts are used between the blotters.

If artificial heat is used, there should be maximum airflow through the press. Use double-faced corrugated cardboard or aluminum ventilators. Never attempt to dry specimens in an oven. A metal or wooden box with an open top that will accommodate a press sideways (corrugations pointing up) and equipped with an electric heater with a fan makes an excellent drying chamber. A collapsible drying frame (either a wooden box or metal frame with canvas shirt) may be used in the field and a camp stove or lantern used as a heat source. Electric heating coils or light bulbs may also be used as heat sources, but a fan should be installed either in or above the chamber. Special drying cabinets are sold, but most lack sufficient ventilation for proper drying of specimens.

G.   DATA TO ACCOMPANY SPECIMENS

I. Field Notes

As mentioned earlier, every collector should keep a field book. This book is not simply a road log. Each species collected at a given place and time should be given a collection number. The best system to use is a chronological one beginning with number one and continuing from there. Avoid elaborate numbering systems with prefixes and cryptic notations or abbreviations. Do not use the same number for any other collection. All duplicates or replicates should bear the same collection or collector's number. Although some abbreviations may be useful and efficient in the field, these should be fully written out when permanent labels are made from the field notes. A specimen without field data (at least locality and date) is of little scientific value.

Data to be recorded in the field notebook should include collector's number (for reference), exact locality, approximate altitude, nature of the habitat (type of soil, mois-

Herbarium of the University of North Carolina

TEXAS
Bee Co.

Pyrrhopappus multicaulis DC.

Fine hard sandy loam, grassy roadside, eight
miles northeast of Beeville at Medio Creek.

Flowers pale sulfur yellow, tinged rose-violet
outside; anthers black (11 AM); common.

C. A. Lawson   928   28 March 1967

HERBARIUM OF THE UNIVERSITY OF NORTH CAROLINA

Det.   Date

HERBARIUM OF THE UNIVERSITY OF CALIFORNIA

See pollen slide collection for permanent
mount prepared from this specimen.

Prepared by: Date

PLANTAE EXSICCATAE GRAYANAE

1427. Sibara virginica (L.) ROLLINS

In Rhodora, xliii. 481 (1941) et in Contrib. Gray Herb.
clxv. 135 (1947); Fernald, Gray's Man. (ed. 8) 723 (1950).

Syn. *Cardamine virginica* L. Sp. Pl. 656 (1753). *Arabis
virginica* (L.) Poir. Encycl. Suppl. i. 413 (1810); Hopkins
in Rhodora, xxxix. 80 (1937). *Cardamine ludoviciana* Hook.
in Journ. Bot. i. 191 (1834). *Arabis ludoviciana* (Hook.)
C. A. Mey. in Fisch. & Mey. Ind. Sem. Hort. Petrop. ix.
60 (1843). *Planodes virginicum* (L.) Greene, Leafl. Bot.
Obs. ii. 221 (1912).

Waste places; Tobler Road, off Lowe's Ferry Pike,
Knox County, TENNESSEE.

Coll. A. J. SHARP (no. 2444)   May 6, 1947

ANNOTATION LABEL

Det. Ronald L. McGregor

Pressed   by P. C. Hutchison;

field number   ; cultivation number

Grown from imported plants of the above collection at
University of California Botanic Garden, Berkeley.

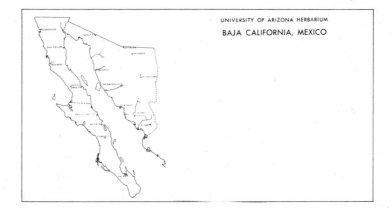

UNIVERSITY OF ARIZONA HERBARIUM
BAJA CALIFORNIA, MEXICO

FLORA of QUEBEC, CANADA
The Gaspe' Peninsula
Herbarium of Keene State College

County

No.
July 1972
Collector, D. E. Boufford

PLANTS OF THE ROCKY MOUNTAIN REGION
Colorado

Alt.   ft.

Reed C. Rollins, No.

Figure E.1

ture conditions, slope exposure, light conditions), associated species, and other pertinent information. With reference to the plant proper, record those features that will not be evident from the pressed specimen (height, branching, depth of root system, odor, etc.) and those features that may be lost in drying, for example, flower color. Flower color may best be determined by using a color chart. The more complete the field notes, the more complete the permanent label can be and the greater the information content of the specimen.

II. Permanent Label   **(Figure E.1)**

The permanent label is the label affixed to the mounting sheet with a specimen. Information included in addition to the name of the plant and authority (e.g., *Claytonia caroliniana* Michaux) must come from the collector's field notebook. Do not abbreviate or use symbols. Specimens are now used throughout the world and symbols and abbreviations are often difficult to translate. Be specific in giving localities. If local names are used, give some reference to a city, major highway, or easily located reference point. Minimal data for labels should include date, locality, county, state, collector, and collection number. Labels, unless done by offset printing, should be typewritten using a carbon ribbon or written in longhand using india or other permanent ink. Paper should be of high-rag content—preferably 100 percent.

H. IDENTIFICATION OF COLLECTIONS

Materials should be identified using the appropriate manuals, floras, and monographs. If it is necessary to have materials identified or verified by a specialist, one of the duplicates is sent with a label to the specialist. Generally, the specialist will keep the specimen unless he has agreed to do otherwise. In the event that a duplicate specimen is identified by someone else, the collector should enter the plant name on the label followed by the name of the specialist and the date that the duplicate was determined, for example,

> *Ipomopsis rūbra* (L.) Wherry
>   duplicate determined by Verne Grant, 1974

Once materials have been labeled and identified, they are ready to be mounted.

SUGGESTED READING

Davis, P. H. and V. H. Heywood. 1963. Principles of Angiosperm Taxonomy. Van Nostrand Company, New York.

Lawrence, G. H. M. 1951. Taxonomy of Vascular Plants. The Macmillan Company, New York.

Smith, C. E., Jr. 1971. Preparing herbarium specimens of vascular plants. Agriculture Information Bulletin No. 348. U. S. Government Printing Office, Washington, D. C.

Epilogue: Principles and Conclusions

The systematist as a scientist and philosopher rationally and critically investigates the concepts, methods, truths, and assumptions related to his or her field of study to develop a system of principles for the guidance and enlightenment of students, researchers, and scholars in systematics.*

The principles of general systematics (A. General Systematic Principles) form a broad conceptual base for the study of taxonomy. The working principles in the constituent element system (B. Constituent Element System) furnish a detailed model for the study of any topic in systematics. Since expert knowledge of any subject comes with a thorough understanding of its principles, the system of working principles, when applied, should provide a sound foundation for the comprehension of the major concepts in systematics. Application of the working principles to major types of taxonomic research and some summaries are presented as examples of the constituent element approach in B. I. Summary for Floristic Study, II. Summary for Revisionary Studies of Plant Taxa, and III. Summary for a Paleoethnobotanical Study. The conclusions (C. Conclusions) represent a synoptic summary of the concepts, processes, principles, and question-exercise approach of this text. D. Plant Taxonomy: Sequential Saviors by Duane Isely provides a persuasive argument for studying fundamentals in a basic course in systematics rather than detailed treatments of the current trends in the discipline. This section supplies insight into the ebb and flow of modern systematic research and aptly illustrates the bandwagon syndrome. A

---

*The operational credo used in the organizing and writing of this book is expressed well by Isely in the last three paragraphs of D. A Professional Taxonomic Perspective in the Epilogue. The basic philosophy and fundamental principles underlying the entire text are summarized in VI. A Hierarchical Resumé for Systematists in the Prologue.

final comment (E. The Major Goal of the *Fundamentals of Plant Systematics*) is made on the primary goal of this book.

A. GENERAL SYSTEMATIC PRINCIPLES

1. Systematics is based on the precise description and exact definition of character states with evolution as the logical foundation for systematic relationships.
2. Systematics uses many characters from all types of evidence and all fields of botany for the description, identification, and classification of plants.
3. Systematics involves characterization and delimitation of taxa that make identification and classification schemes possible.
4. Systematics involves hierarchical arranging or ordering of data and taxa that make classification systems feasible.
5. Systematics is involved in the determination of relationships of taxa that are basic to presumed phylogenetic classification.
6. Systematics provides the system for naming and names for all taxonomic groups.
7. Systematics uses the species as the base and a hierarchical arrangement of named taxa as the framework to which all taxonomic information is attached.
8. Systematics involves all phases and levels of learning: recall, recognition, application, interpretation, generalization, and evaluation for the development, communication, and use of taxonomic information.
9. Systematics is a holistic science involving field, laboratory, garden, herbarium, and library studies in the accumulation and use of quantitative and qualitative data for taxonomic purposes.
10. Systematics uses the intellectual approaches of observation and description, experiment and analysis, and synthesis and theory for the procurement and interpretation of taxonomic data.
11. Systematics is an intellectual pursuit that provides meaningful explanations of the diversity of nature.
12. Systematics is a fundamental discipline to botanical endeavor as well as an integrative field, using all aspects of botany, in its own development.
13. Systematics is a science that contributes to the advancement of related disciplines.
14. Systematics is a historically based but dynamic and ever-changing field. A steady stream of data incorporated into the taxonomy of plants requires reordering of descriptions, revision of indentification schemes, reevaluation of classification systems, and perception of new relationships.

B. CONSTITUENT ELEMENT SYSTEM

The principles related to each major topic are summarized in each chapter (see C. Conclusions, item 5). The summaries represent a set of working principles for comprehending each subject. The questions and exercises following each summary provide applications of principles for understanding each topic treated.

In the summaries, the definitions (what) are usually based on structure or composition; the purpose (why) provides the functional reasons for the study of the concept; and the operations (how) include the fundamental developmental procedures involved in the subject as a process. Structure, function, and development constitute the logical basis for understanding the concept, subject, or topic.

The basic premise supplies the proposition for the study of the taxonomic subject or is the inclusive fundamental principle for the study of the subject. The fundamental principle(s) refer(s) to the basis from which any action proceeds or is changed and is the foundation for comprehending the concept, subject, or topic.

The guiding principles or guidelines are the rules controlling the activities related to the subject. The rules are the considered recommendations for accomplishing the functional goals linked with the topic. The guidelines provide the practical approach to understanding the concept, subject, or topic.

The basic assumptions are the suppositions or premises related to the subject of study that are accepted as true. Suppositions are verbalized for the sake of logical reasoning in the study of the concept. An understanding of a topic requires a knowledge of the basic assumptions pertinent to the study. Articulated assumptions are an integral part of the context for study.

This set or system of working principles provides a practical methodology for an understanding of the fundamentals of any taxonomic study and also furnishes a foundation for the development and testing of hypotheses for systematic research. Summaries for floristic and revisionary studies along with general and limited study hypotheses are presented in I. Summary for Floristic Study and II. Summary for Revisionary Studies of Plant Taxa. A summary for a study peripherally related to plant systematics, as in III. Summary for a Paleoethnobotanical Study, shows the application of the constituent element system to a general problem.

I. Summary for Floristic Study

Definitions for the Study of Floristics *Floristics* is a study of plant species diversity in relation to habitat diversity within an area. *Vegetation pattern* study is the study of community diversity in relation to habitat diversity within an area.

Purpose of Floristics To provide an inventory of the plant species, communities, and habitat diversity within an area; to facilitate identification of the plant species in such an area; to provide estimates of diversity of plant species within an area; to observe relationships between species-community distribution and habitat; to provide predictive models for site-species relationships; to add basic information about species biology to systematics; and to provide a foundation for taxonomic and ecological research.

Operations in Floristics To list species within an area; to study their ecological and biological characteristics; to describe site-species relationships in terms of ecological characteristics; to interpret flora and communities as to origin, migration, and evolution; and to prepare predictive models for species-community/habitat relationships.

Basic Premises for the Study of Floristics That species diversity is related to habitat diversity within an area under natural competitive conditions and that predictive site-species relationships can be determined.

Fundamental Principles for the Study of Flora and Vegetation

1. Plant species diversity is related to total habitat diversity within an area.
2. Habitat diversity is related to the diversity of species, climate, geology, hydrology, soils, and topography within an area.
3. Communities are recurring combinations under similar habitat conditions within an area at a given time.
4. Communities are the results of the interaction of species diversity and habitat diversity in an area at a given time.
5. Flora and vegetation of a region is the total species and community diversity, respectively, resulting from biological evolution and habitat interaction.

Guidelines for Floristic Work

1. Species should be determined and listed in study area by habitat.
2. Voucher specimens of each species should be collected to serve as permanent records of a species or population as it occurred at a given time and place.
3. Voucher specimens should properly represent the variants in a given local population.
4. Flora should be analyzed for life form, diaspore type, sociability, duration, and phenology.
5. Indicator species should be determined for specific environmental factors and species that are exclusive to restricted communities or habitats should be indicated.
6. Flora should be analyzed for indigenous and introduced species, disjuncts, relicts, range extensions, clinal variation, and new variations.
7. Unique features within a flora should be indicated, for example, unusually large specimens of a species, largest fruit, smallest flower, rarest taxa, endemics, most variable population, largest tree, and oldest plant.
8. Flora of study area should be compared with that of larger immediate area and disjunct, climatically similar areas as to similarities and/or differences in species and larger taxa.
9. Flora should be interpreted as to origin, migration, and evolution.

Guidelines for Vegetation and Flora Pattern Study

1. Voucher specimens for each species found in each community should be collected.
2. Patterns (communities) in which flora occurs should be determined.
3. Pattern (communities) distribution should be described.
4. Communities should be mapped in relation to slope, exposure, elevation, and so on.
5. Unique pattern compositions and distributions should be indicated.
6. Community composition and distribution of communities of area should be compared with those in similar climatic regions.
7. Communities should be interpreted as to origin, evolution, and migration.

Basic Assumptions for the Study of Floristics

1. Each species is adapted selectively and uniquely to its habitat.
2. References exist for the identification of taxa and the determination of abiotic features.

3. The herbarium is an institutional center available for reference work and a documentation service for floristic research and study.

A General Hypothesis for Floristic Studies   Floristic diversity is related to habitat diversity within an area.

Hypotheses for Limited Floristic Studies (Examples)

1. Floristic diversity within a county is related to habitat diversity in the county.
2. Floristic diversity in savannahs is related to habitat diversity within savannahs.
3. Floristic diversity in the chestnut oak forest is related to habitat diversity within the chestnut oak forest.
4. Floristic diversity over diabase is related to habitat diversity over diabase.
5. Floristic diversity in an estuary is related to habitat diversity in the estuary.
6. Floristic diversity on a mountain range is related to the habitat diversity in the mountain range.

II. Summary for Revisionary Studies of Plant Taxa*

Definition   A revision is a treatment of selected taxa throughout at least a major portion of their range, including a study of nomenclature and classification along with descriptions based on several types of evidence.

Purpose of a Revisionary Study   To improve the classification, to determine relationships, to characterize more comprehensively and delimit the taxa effectively, to ascertain the correct name(s) of the taxa named, and to produce a means of identifying the taxa.

Operations in a Revisionary Study

1. Select taxa to be studied as problems are recognized in a taxon or group.
2. Select characters to be evaluated.
3. Observe and record the character states of the selected characters.
4. Note the character correlations and discontinuities of variation of the character states.
5. Circumscribe the taxa on the basis of the observed variation and character correlations.
6. Determine the relative position and rank of the circumscribed taxa.
7. Revise, if necessary, the nomenclature of the taxa based on their circumscription.
8. Produce a key or similar device for the identification of the taxa.

Basic Premise for Revisionary Studies   Problems recognized in previous identification schemes, descriptions, or classification; the availability of new evidence; and the development of new techniques provide the bases for reexamining the character variation and correlations within the taxa to improve the taxonomy of the group.

*Contributed by Deborah Qualls, Department of Biology, University of North Carolina at Chapel Hill. (This constituent element summary for revisionary studies of taxa was an exercise in a class in advanced systematics in which the students selected a subject of their choice for summary.)

Fundamental Principles for a Revisionary Study

1. Discontinuities in variation of characters provide clues to species relationships that form the basis of classification.
2. As more information is obtained about variation within and among the taxa, a better classification can be derived.
3. Variation of characters in many fields of evidence may be used; traditionally morphological characters have been used most frequently because they are easily observed and readily show variation.

Guiding Principles for a Revisionary Study

1. The work previously done forms a basic foundation for further study; therefore, the literature pertinent to the treatment should be carefully studied.
2. Characters must be studied throughout the entire range of the taxon as well as among the different taxa to determine variation within each taxon and to determine any intergradation between two or more taxa.
3. Characters should be selected from all parts of the organism and all stages of the life cycle; characters should be selected from several fields of evidence.
4. As changes are made in the circumscription, delimitation, and classification of taxa, the nomenclature should be reevaluated for accuracy and compliance with the rules of botanical nomenclature.

Basic Assumptions for a Revisionary Study

1. As knowledge increases about the taxa, the previous classification of the taxa needs to be reevaluated.
2. Variation can be observed and studied in individuals, populations, and taxa; within this variation discontinuities exist.
3. From a study of this variation, logical relationships among individuals, populations, and taxa may be inferred.

Hypothesis for a Revisionary Study Recognition of problems and new evidence within a taxon shows the need for improvement of the taxonomy of the group.

III. Summary for a Paleoethnobotanical Study*

Definitions for Paleoethnobotany *Archaeobotany* is the study of plant remains derived from archaeological contexts. *Paleoethnobotany* is the analysis and interpretation of the direct interrelationships between humans and plants as manifested in the archaeological record.

Purpose of the Study of Paleoethnobotany To elucidate cultural adaptation to the plant world; to determine the impact of plants on a prehistoric human population; and to determine the impact of a human population on prehistoric environments.

*Contributed by Gloria Caddell, Department of Biology, University of North Carolina at Chapel Hill. (This constituent element summary for the study of paleoethnobotany was an exercise in a class in advanced systematics in which the students selected a subject of their choice for summary.)

Operations in Paleoethnobotany   To recover plant remains systematically from an archaeological site; to analyze those that can be securely dated; to sort, identify, and quantify the botanical remains; to record and compare attributes and measures of morphological variation with a reference collection that reflects the range of variation in local plant populations; to determine the uses of plants; to interpret the plant remains as to their relative importance to prehistoric population; and to describe prehistoric environments.

Basic Premise for the Study of Paleoethnobotany   Botanical evidence is fundamental to an analysis of the dynamic functioning of a prehistoric culture in relation to its environment.

Fundamental Principles for the Study of Paleoethnobotany

1. The kinds of plants used and how they are used is determined by cultural definitions and biological constraints. A culture defines appropriate plant resources.
2. The use of such plant resources modifies to some degree the structure and composition of the local plant communities.
3. Plant remains from a site are a by-product of specific human cognitive patterns of behavior and accidental inclusions from the local plant environment.
4. Plant evidence expresses some aspects of past societies.
5. The material correlates of a social organization should be detectable with plant evidence.
6. The composition of botanical samples may reflect general trends in plant use through time.
7. The most important plant resources should be commonly recovered in contexts indicative of food preparation, consumption, or storage.
8. The floral samples from an archaeological site directly reflect neither the importance of each food item in the diet nor the entire range of plants utilized.
9. The botanical remains from an archaeological site may give some indication of the vegetation in the immediate vicinity and also the degree of disturbance of the area.
10. Ethnographic analogy may aid in the interpretation of plant uses in the past.

Guiding Principles or Guidelines for Paleoethnobotanical Studies

1. The kinds and quantities of plant remains present on an archaeological site are affected by the nature of the plant parts, the methods of plant gathering and processing and the means and rate of utilization by a prehistoric population, postdepositional activities on a site, and the recovery, laboratory processing, and identification procedures used by the archaeologist or archaeological botanist.
2. Field collecting and laboratory processing procedures must be kept consistent and attempts should be made to analyze analogous features to limit interpretative discrepancies.
3. An attempt should be made to recover all size classes of botanical remains.
4. Methods should be used to ensure that the contents of botanical samples as nearly as possible represent the composition of plant remains present on a site.

5. One or even several features (pits, hearths, structures) from a cultural period cannot be expected to reflect all the economic activities of that cultural period.

6. All identifications should be verified by comparison with a documented reference collection.

7. Because archaeobotanical remains are initially sorted into homologous parts and not into taxonomic species, a voucher collection must reflect this difference.

8. Once the genera or species are identified, their population characteristics should be described. Attributes and measures of morphological variation should be recorded.

9. Presence of an item, ubiquity of an item, density of an item in samples, and the proportion of the recovered botanical remains formed by an item may be used to determine the importance of an item to subsistence.

10. Utilization of a plant cannot always be inferred from the presence of a part in an archaeological sample.

11. The proportions of different plant components in one assemblage may not indicate their relative importance, but a change in the proportions from one assemblage to another may indicate changes in their frequency or utilization.

12. If a particular plant could not grow in habitats on or near a site, its presence there could indicate utilization.

13. Caution should be used when interpretations of seasonal occupation are based on plant remains. The season in which a plant part matures may not necessarily be the season in which it was utilized.

14. Only general statements about subsistence and settlement patterns may be confidently made on the basis of floral remains from archaeological sites.

15. Documentation of plant samples must be complete and accurate.

16. Data should be stored in a manner in which it can be efficiently retrieved.

17. Results should be disseminated to the archaeological community.

Basic Assumptions for the Study of Paleoethnobotany

1. Recognizable plant taxa exist, and they may be identified from carbonized plant remains.

2. Patterns of human subsistence behavior can be inferred from botanical remains from an archaeological site.

3. The geological processes that operate in the present were also operative in the past.

4. The plants represented in the archaeological record had the same habitat requirements in the past as in the present.

5. The dietary significance of plants in the past is the same as in the present.

A General Hypothesis for a Paleoethnobotanical Study   A paleoethnobotanical study will provide the facts for the elucidation of the cultural adaptation to the plant world; for the determination of the impact of plants on a prehistoric human population; and for the determination of the impact of a human population on prehistoric environments.

C. CONCLUSIONS

Organizational and Operational Resume

1. The basic premise for the study of plant systematics is that in the tremendous variation in the plant world, with active past and present speciation, conceptu-

ally discontinuous units exist that are (and can be) identified, classified, described, named, and logically related.

2. The fundamental principle for plant systematics (and many other disciplines)—the principle that provides the basis for active study of all taxonomic concepts, subjects, and topics as well as changes in the subject system—is the question approach: what is it, how do you do it, why do you do it, or what are the definition(s), process(es), and reason(s) involved in the study?

3. The guidelines or guiding principles establish the conditions (when-how-where) for the development of the product(s) or result(s) associated with the system related to each taxonomic concept, subject, or topic.

4. The basic assumptions supply the fundamental context for the study of each taxonomic concept, subject, or topic.

5. The constituent element system for the study of each taxonomic concept, subject, or topic—definitions, parts, kinds, reasons, processes, and pertinent principles—leads to a basic understanding of each.

6. The question-exercise approach to the study of plant systematics produces usable knowledge and practical application for each taxonomic concept, subject, or topic considered.

7. The references provide the resources for perspective on the taxonomic subject of study and supply the basic observational, analytical, experimental, and synthetical procedures for testing the hypotheses evolving from the background study of the taxonomic topic of concern.

Informational and Institutional Comments

8. New and old evidence from all fields—morphology, anatomy, embryology, palynology, cytology, ultrastructure, reproductive biology, genetics, physiology, ecology, geography, chemistry, and geology—is used for the selection of characters and determination of character states for group(s) of organisms for a more comprehensive characterization of new and old taxa, for a more effective diagnostic delimitation of new and old taxa, for the elucidation of pertinent intra- and intertaxon relationships, and for a probable better understanding of the evolutionary trends, patterns, and pathways within taxa.

9. The botanical institutions—the library, herbarium, garden, and laboratory—are the focal points for systematic research, training, reference work, and documentation service.

10. The fundamentals of vascular plant systematics—taxonomic concepts, processes, and principles—provide skills and values relevant to making the education of the student a challenging and ever-expanding process that leads to the realization of academic potential.

Basic Relationships and Fundamentals

11. The fundamental components of taxonomy and evolution are conceptually independent but intricately related.

12. All types of taxonomic systems for any subject have a hierarchy with two or more ranks and one or more positions at each rank.

The Scientific Method for Systematic Study and Research

13. The constituent element paradigm for the fundamental components of systematics is an effective model for the scientific study of most taxonomic problems.

The application of the model aids in the identification and definition of the problem, the procurement of data relevant to the problem, the formulation of a hypothesis for the solution of the problem, and the empirical testing of the hypothesis.

Summary for Taxonomic and Systematic Activities

14. Taxonomists construct taxa that are reliably delimited and circumscribed at their respective positions and ranks; make taxa that are easily and positively identifiable; form taxa that are recognizable in nature; establish taxa that are predictable in character correlations under natural conditions and occurrences; and provide taxa with a particular circumscription, position, and rank with one correct name.

15. Systematists ascertain types, patterns, and trends in variation within populations, species, and higher taxa; classify, identify, describe, and name; and observe, analyze, synthesize, and evaluate to determine relationships of taxa based on similarity and differentiating characteristics from all fields of evidence.

16. Systematists study speciation to understand the biological mechanisms by which species arise, determine the nature (origin) of existing species, understand the species biology of present populations, and determine the environmental factors that seemingly control the stability of existing species populations.

17. Systematists as phylogenists select organisms, analyze the characteristics of the group of organisms by selecting and defining characters and character states, assess homology, define morphoclines, hypothesize primitive versus derived states, determine the phylogenetic branching pattern of the taxa by grouping monophyletic taxa on the basis of shared derived features; evaluate and reevaluate character state changes and groupings of the cladogram; use the cladogram to devise a classification scheme and to deduce biogeographical history and also to use the indicated character state changes to hypothesize past evolutionary events.

The Future of Systematics

18. Much research in systematics remains to be done. New species await discovery, description, and naming in many parts of the world. Monographs of many groups of plants have to be made. Revisions of many groups treated more than a century ago are sorely needed. Relatively unknown floristic provinces, particularly in the tropics, still exist that need exploration and study. Floras based on soil conditions are barely underway in most of the world. Detailed analyses of diversity as aids in the determination of evolutionary relationships and as bases for the analyses of distribution patterns have just begun. An understanding of speciation in many groups is yet to be made.

19. The basic fields of evidence used in systematic studies need redefinition, new conceptual bases, and the use of modern techniques for increased precision, greater accuracy, and easier application. Many types of evidence, other than structural, remain to be applied to taxonomic problems of all sorts. The principles of taxonomy need much thought and reevaluation. Studies in the biology

of species have not been made for the vast majority of the species of the earth. Mathematically based schemes for the determination of relationship have been developed but are not yet widely used. Classification shows some maturity but new relationships are, and will continue to be, discovered with the incorporation of new ideas and data into systems synthesis and theory. Many anomalous groups remain to be accurately placed in schemes of classification.

20. Systematics will continue to develop as its researchers and scholars assimilate the pertinent and the useful from the many expanding fields of evidence on which it is based. The science of taxonomy will incorporate more mathematical logic, improved methods, and more refined intuition into its activities. Creative and dynamic systematics will provide a sound foundation as an integrating and problem-solving discipline in the future progress of the life sciences and society.

D. A PROFESSIONAL TAXONOMIC PERSPECTIVE*

Nonwarranty*

These remarks, placed in written form at the request of Albert E. Radford, approximate a portion of an ad lib lecture presented to a graduate plant taxonomy class. They epitomize the difference between the spoken word and its appearance in cold print. For, when spoken, it is the manner of delivery (as is the case with President Reagan) rather than the substance that mediates the effectiveness. What presumably sounded well enunciated may reproduce primarily as trivial or even superficial. As this is written, there has been no time for maturation of enunciation and no research has been done to hone the precision and accuracy of statement. So be it.

Plant Taxonomy: Sequential Saviors*

The understanding of a graduate (and undergraduate) student about the state of the art is commonly imperfect. No doubt mine, 45 years ago, was deficient. But, or so it seems, the science was yet primarily one of the herbarium and field. There was much interest in plant distribution as it related to phyletic interpretation—and especially as it concerned glaciers and continental drift. Indeed, we were emerging from a taxonomic Pleistocene when the doings of the glaciers could, with appropriate ingenuity, be used to explain what needed to be explained (look at Rhodora, a botanical journal, in the thirties and early forties). Or if not, continental drift could do it. (Though subsequently discredited, continental drift was never entirely abandoned by biologists, and now with the cooperation of the earth scientists, it has seen a revival in explanations of the history of floras and their species.) Beyond glaciers and drift, taxonomists who made slides and used the compound microscope (commonly called comparative morphologists), as Arthur Eames, were contributing major new insights in family relationships. The taxonomic world, with the wide dissemination of Besseyan classification through the Poole textbook, was becoming increasingly uncomfortable with the traditional Englerian system.

Many said taxonomy was archaic, stagnate, or even now unnecessary. It needed saviors. They came.

*This section was contributed by Duane Isely, Iowa State University, Ames, Iowa.

The first group possibly was composed of the chromosome counters. Then it was found that one could identify parts of chromosomes and conduct genome analyses that, in some instances, threw new light on plant relationships. There emerged *cytotaxonomy*. The cytotaxonomists proclaimed that, working with the basic inherited matter of life, they could provide data and insights unavailable to those who studied corpses in a plant morgue. Taxonomy at last had a definitive bulwark for postulating relationships and no dissertation was complete without the appropriate chromosome ritual tally, prefaced of course, by a detailed accounting of what was fixed in what for how many hours.

But this was not enough.

In a classic paper "The Structure and Origin of Species" Camp and Gilly in 1943 coined the word biosystematics. *Biosystematics* has taken many forms. The only definition on which all practitioners might agree is "My taxonomy (i.e., biosystematics) is better than yours." In general, biosystematics included studies of population dynamics supplemented by cytotaxonomic and genetic data. The biosystematists presumably could provide substantive evidence for their postulates rather than seat-of-the-pants elusive interpretations of the classical taxonomists based primarily on interpretation of observations. Regardless, however, whatever the methodology, theses came to be identified not as "a revision of *Radfordiodendron*" but rather as "a biosystematic study of the genus *Radfordiodendron*."

But this was not enough.

Science is experimental. This is proclaimed in the introduction of every biology textbook. An investigator has a hypothesis. The investigator conducts experiments to prove or disprove the hypothesis. This only is science. Those failing are but pseudoscientists or perhaps even fakers. And where did that leave taxonomy? Perhaps an occult art? Taxonomists, second-rate citizens, have panted for 100 years for experiments to better their status in the eyes of the Lord. The *experiment* among other things, came with the distinguished work of Clausen, Keck, and Hiesey (CKH). They conducted transplant *experiments* to prove a hypothesis that different races of a species growing in contrasting environments had a genetic basis. Although they did not bugle the splendor of their light, as others bringing new groceries have done, their contemporaries did it for them. Seizing the CKH program as a drowning man grasps the proverbial straw, *experimental taxonomy* was born and set out to conquer the world. Dozens of papers and theses poured out in the next decade, cluttering up journals with their lengthy tables and verbosity. Lacking the input of genius of CKH (particularly of Clausen) and, in many cases, the amenable physical surroundings (the contrasting California environments) for their research, they were mostly about as exciting as watching concrete set. And CKH brought the ecotype concept to maturity: a model for the mechanism of speciation, a conceptual basis for the pregnancy of taxa analogous to that, in the broader amphitheatre of biology, which Darwin provided a hundred years earlier. Ecotypes, in the 1960s blossomed everywhere and became a password for respectability.

But this was not enough; in fact it was all for naught; suzerainty passed elsewhere.

For CKH and all of their predecessors worked primarily with plants and their visible parts. But it is the chemicals within these plants that, providing them with their various ways, are the fundamental markers of like and unlike. *Chemotaxonomy* (otherwise known as spot taxonomy) was not a new idea; it burst on the scene with violence in the latter fifties because a reasonably simple technique, two-dimensional paper chromatography, became available, allowing one easily to detect differences between the substances (usually secondary metabolites) of organisms. Its boil-like eruption in the United States was in large part a consequence of the pioneering work of Turner and Alston, with the former as unexcelled flag waver. And chemotaxonomy continues as a holy church moving now to the expensive luxury of amino acid sequencing and the yet more extravagant efforts of determining nucleotide sequence in DNA and RNA. Here at last we are studying the ultimate units that will relate, in unequivocal terms, the phyletic structure of the living world.

But this was not enough.

Numerical taxonomy (and by other names) was proclaimed by chief caliphs Sokal and Sneath in many papers but particularly from the podium of the first edition of their text. All before were idiots. One counts, measures, or otherwise quantifies, without discrimination, at least 40 plant (or bacterial or animal) attributes, pours this through the appropriate computer programs and the answers come clear and evident in the form of clusters and nonclusters—a rigorous method (*rigorous* means my work as contrasted with yours) free of the subjectivity of classical taxonomy. And as the prophets of yore proclaimed, *computer taxonomy* begat *phenetic taxonomy*. Phylogeny is verbal spinach, nonsense; classification must be based on the totality of shared and nonshared attributes. Renouncing the sins of their forefathers, more and more taxonomists now spend much of their lives in front of a blinking screen, with the consequence that they run red lights on their way home from work.

But this was not enough. For after Jesus came Allah and after him, Joseph Smith. For now it is *cladistics*. It is the subscience of branchology; phylogeny, not phenology, is the goal. If you are sufficiently parsimonious, if you can learn a new terminology somewhat akin to Sanskrit, and if you can find dendrons you have it made. Sokal and Sneath were but sub-Neanderthal blunderers.

But this will not be enough.

As of 1985, all of these except the perhaps recently dissident cladists have been reasonably absorbed into the metabolism of the discipline, and all have contributed to the growth and strength of that organism. But none has lived up to his or her original adolescently proclaimed virtues; none has any exclusive patrimony for what has followed.

Where and when, one asks, will the lightning strike next. I cannot answer because the ways of the gods are mysterious and hidden; their pleasure must be awaited.

In the meantime, there are still those who look at plants and attempt to consider them on a holistic basis. These plants are the products of the interaction of their genes and pseudogenes and of the reticulate interactions of several generations of metabolic progeny that follow diverse and tortuous developmental pathways to produce an elm tree—or a frog. And perhaps this ultimate genotype, the elm in toto, is a better marker of its phyletic nature than samples of any of the thousands of assembly lines that have participated in its production. This does not assert that knowledge of basic metabolic sequences is not frequently helpful in interpreting why an elm is an elm and not an oak—or a frog—but the elm itself is not displaced from its central podium about which all other information must be *realistically* related.

E. THE MAJOR GOAL OF THE FUNDAMENTALS OF PLANT SYSTEMATICS

The major goal for students of the fundamentals of plant systematics is the acquisition of basic knowledge, skills, and values usually associated with the study of the subject.

Competent treatment of the material in text and class aids in the procurement of information on systematic concepts, processes, and principles. Experience in answering questions, applying principles, working out exercises, and solving problems helps develop skills in problem analysis, thoroughness in investigation, accuracy in documentation, and proficiency in postulation and testing of hypotheses. Disciplined, analytic, holistic, and creative thinking about systematic topics by informed and trained students promotes comprehension of those subjects, skillful solutions to perplexing problems, and the development of scholarly values.

It is hoped that the organization and presentation of the systematic material in this text and the approach to its study have helped the student attain the major goal as well as develop a greater awareness of the significance of the subject and a deeper appreciation of the applicability of the study of plant systematics to everyday living.

QUESTIONS ON FUNDAMENTALS OF SYSTEMATICS

1. What are the principles for the study of principles?

2. What is a logical classification for the types of principles? Give the basis for your classification.

3. How can the term principle be defined from structural, functional, practical, and theoretical standpoints?

4. Under what conditions does a principle become a theory, guideline, premise, or supposition? Give examples.

5. What are the bases for classifying principles as general or working?

6. What are the principles for the study of questioning?

7. What are the principles pertinent to the study of exercises as a method of learning?

8. What criteria are used for establishing positions, ranks, identities, circumscriptions, and names of taxa?

9. What criteria are used for determining kind, number, admissibility and weighting of characters used in a taxonomic study?

10. Why should the reasons for studying any subject, topic, or concept have a fundamental basis? Why should the definition of a subject have a structural basis? Why should operations for the production of any product have guidelines?

11. What is the basic premise for the study of systematics? Why is it considered basic?

12. Why is identification, nomenclature, and circumscription considered part of the classification process? Taxonomic process?

13. Why do systematists consider all people taxonomists?

14. Why is a knowledge of variation in populations considered basic to the study of speciation and phylogeny?

15. How does systematics provide a focal point for the training of students in biology and the study of organismic diversity?

EXERCISES IN GENERAL SYSTEMATICS

1. Use an unabridged dictionary and Chapter 1 of the text to determine the meanings of the following terms, and then indicate how the terms in each group can be hierarchical. (a) Systematics, taxonomy, classification; (b) description, characterization, circumscription; (c) nomenclature, name, scientific name, binomial, epithet; (d) recognize, identify, name; (e) data, information, knowledge, wisdom; (f) evidence, character set, character, character state; (g) descriptive, distinctive, diagnostic; (h) order, rank, position; (i) documentation, reference, citation; (j) flora, manual, monograph; (k) hierarchy, rank; (1) chapter, section, paragraph.

2. Study the taxonomic fundamentals diagram in the Prologue and then give the bases for the relationships and interrelationships. Consider the following questions in determining relationships (see prospectus and conspectus in the diagram):

   a. Does classification precede identification? Does description precede identification? Does naming precede classification?

   b. How should the fundamental procedures be ranked from a standpoint of priority?

   c. How are the linkage triangles appropriate?

   d. How do the processes depend on one another?

   e. Why should each of these processes be considered systems? Do all form a system?

   f. Which of the fundamentals should be studied first in a course? Why?

3. Study the taxonomic evidence diagram in the Prologue and then indicate the relationships of the various fields of study. Consider the following questions in determining relationships (see prospectus and conspectus):

   a. How are morphology, genetics, and ecology used as taxonomic evidence?

   b. How is morphology related to evolution? To genetics? To history?

   c. How is taxonomy the pedestal for study of all fields of evidence? The acme?

   d. Why should each of these fields of evidence be considered a system? Do all make a system? How?

4. Study the taxonomic aspects diagram in the Prologue and then indicate the taxonomic reasons for the design (see prospectus and conspectus).

5. What is the significance of the title of the chart? In what respects could the diagrams be called a taxosystem? The taxonomic aspects, the aspectus; taxonomic evidence, the taxogram; taxonomic fundamentals, the trirelator?

6. The following sequence represents an analysis-synthesis review of the text as applied to (a) classification, (b) identification, (c) nomenclature, (d) description, (e) information, (f) documentation, (g) variation, (h) speciation, and (i) phylogeny. List, describe, or determine the basic definition of each as a system, basic types of each as a system, basic processes in the development of each as a system, basic processes in the use of each as a system, basic product(s) derived from the development of each as a system, basic products derived from the use of each as a system, basic hierarchies in each as a system, basic terms for each as a system, basic reasons for change in system, basic reasons for evaluating each as a system, basic premise(s) for the study of each, and basic guiding principles for the development and use of each as a system.

7. Prepare a constituent element summary for a study of your choice.

8. Systematics is the study of relationships among taxa. Prepare a constituent element summary for a study of the phylogenetic relationships among selected taxa of your choice.

Index

— N.Y. Bot Garden

— Montreal Bot. Garden

R. C. Vrijenhoek — Rutgers

932-2804